Coloniality in the Cliff Swallow

Cliff swallows (*Hirundo pyrrhonota*).

Coloniality in the Cliff Swallow
The Effect of Group Size on Social Behavior

*Charles R. Brown and
Mary Bomberger Brown*

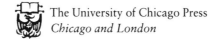
The University of Chicago Press
Chicago and London

Charles R. Brown is associate professor of biology at the University of Tulsa. Mary Bomberger Brown is research associate in biology at the University of Tulsa.

The University of Chicago Press, Chicago 60637
The University of Chicago Press, Ltd., London
© 1996 by The University of Chicago
All rights reserved. Published 1996
Printed in the United States of America

05 04 03 02 01 00 99 98 97 96   1 2 3 4 5

ISBN: 0-226-07625-3 (cloth)
  0-226-07626-1 (paper)

Library of Congress Cataloging-in-Publication Data

Brown, Charles Robert, 1958 Sept. 22–
    Coloniality in the cliff swallow : the effect of group size on social behavior / Charles R. Brown and Mary Bomberger Brown.
        p.   cm.
    Includes bibliographical references and index.
    1. Hirundo pyrrhonota—Behavior.   2. Social behavior in animals.
I. Brown, Mary Bomberger.   II. Title.
QL696.P247B76   1996
598.8'13—dc20   95-44561
                                                                            CIP

∞ The paper used in this publication meets the minimum requirements of the American National Standard for Information Sciences—Permanence of Paper for Printed Library Materials, ANSI Z39.48-1984.

To the cliff swallows of Whitetail Creek

Contents

Preface xi

1 **Introduction** 1
   1.1 Overview 1
   1.2 Background 2
   1.3 Major Questions 4
   1.4 The Cliff Swallow 7
   1.5 Organization of This Book 10
   1.6 Summary 12

2 **Field Methods and Data Analysis** 13
   2.1 Measuring Colony Size and Use 13
   2.2 Checking Contents and Status of Nests 17
   2.3 Capturing and Banding Birds 19
   2.4 Sexing and Measuring Birds 22
   2.5 Color-Marking and Observing Birds at Nests 24
   2.6 Sampling Ectoparasites 26
   2.7 Fumigation 31
   2.8 Measuring Densities, Positions, and Ages of Nests 34
   2.9 Measuring Within-Colony Breeding Synchrony 35
   2.10 Data Analysis 36
   2.11 Summary 38

3 **Study Site and Study Population** 40
   3.1 The Platte River Valley 40
   3.2 Climate 42
   3.3 Historical Occurrence and Current Distribution of Cliff Swallows 46
   3.4 Types of Colonies 49

viii • Contents

    3.5 Overview of the Study Population   57
    3.6 Summary   67

4   Ectoparasitism   68

    4.1 Background   68
    4.2 Ectoparasites of Cliff Swallows   69
    4.3 Effects of Colony Size and Density   73
    4.4 Effects of Date   77
    4.5 Effects of Nest Age, Size, and Spatial Position   79
    4.6 Variance in Ectoparasite Load   90
    4.7 Effects of Ectoparasites on Nestlings   93
    4.8 Ectoparasitism of Adults   102
    4.9 Transmission of Ectoparasites   106
    4.10 Behavioral Responses to Ectoparasitism   112
    4.11 Summary   116

5   Competition for Nest Sites   118

    5.1 Background   118
    5.2 Fighting for Nest Sites   120
    5.3 Nest Building and Colony Size   126
    5.4 Disadvantages of High Nest Density: Being Buried Alive   130
    5.5 Trespassing among Neighbors   132
    5.6 Egg Destruction by Conspecifics   141
    5.7 Summary   147

6   Misdirected Parental Care
Extrapair Copulation, Brood Parasitism, and Mixing of Offspring   149

    6.1 Background   149
    6.2 Extrapair Copulation   151
    6.3 Brood Parasitism   167
    6.4 Mixing of Mobile Offspring   175
    6.5 Summary   185

7   Shortage of Suitable Nesting Sites   187

    7.1 Background   187
    7.2 Nest Spacing within Colonies   189
    7.3 Total Substrate Size of Colony Sites   192
    7.4 Annual Use of Colony Sites   196
    7.5 Nest Spacing Comparisons with Barn Swallows   202
    7.6 Are Cliff Swallow Nesting Sites Limited?   204
    7.7 Summary   208

8 Avoidance of Predators   210

- 8.1 Background   210
- 8.2 Natural History of Predation on Cliff Swallows   212
- 8.3 Detection of Predators in relation to Colony Size   218
- 8.4 Individual Alertness and Group Vigilance   220
- 8.5 Mobbing of Predators   225
- 8.6 Breeding Synchrony and the Swamping of Predators   226
- 8.7 Predation in relation to Nest Position   233
- 8.8 Nest Packing and Vulnerability to Predators   234
- 8.9 Attraction of Predators to Colonies   236
- 8.10 Do Cliff Swallows Avoid Predators by Nesting Colonially?   239
- 8.11 Summary   241

9 Social Foraging 1
Natural History, Food Distribution, and Mechanisms of Information Transfer   243

- 9.1 Background   243
- 9.2 Natural History of Foraging Behavior   245
- 9.3 Cliff Swallow Food Sources   247
- 9.4 Spatiotemporal Variability in Foraging Locations   251
- 9.5 Information Transfer at the Colony Site   260
- 9.6 Food Calls   266
- 9.7 Social Foraging to Avoid Predators?   268
- 9.8 Summary   272

10 Social Foraging 2
Effects of Colony Size   275

- 10.1 Departure Frequencies and Waiting Intervals   276
- 10.2 Food Delivery Rates   280
- 10.3 Amount of Food Delivered   283
- 10.4 Nestling Body Mass   286
- 10.5 Adult Body Mass   290
- 10.6 Starvation during Cold Weather   300
- 10.7 Differences in Local Food Resources among Colony Sites   304
- 10.8 Foraging Costs: Increased Travel Distances and Search Areas   306
- 10.9 Foraging Costs: Prey Capture Rates and Nestling Starvation   311
- 10.10 Benefits of Information Transfer   315
- 10.11 The Geometrical Model as an Explanation for Coloniality   317
- 10.12 Summary   319

11 Reproductive Success 321
   11.1 Background 321
   11.2 Variation in Clutch Size 323
   11.3 Effects of Year and Colony Size 333
   11.4 Effects of Date 339
   11.5 Effects of Nest Spatial Position 345
   11.6 Effects of Nest Age and Size 350
   11.7 Variance in Reproductive Success 353
   11.8 Age-Specific Differences in Reproductive Success 355
   11.9 Summary 358

12 Survivorship 360
   12.1 Background 360
   12.2 Methods for Estimating Survivorship 362
   12.3 Effects of Age, Sex, and Year 368
   12.4 Effects of Colony Size 373
   12.5 Effects of Body Mass 380
   12.6 Effects of Ectoparasites 386
   12.7 Other Correlates of First-Year Survivorship 391
   12.8 Other Correlates of Adult Survivorship 395
   12.9 Summary 397

13 Colony Choice 399
   13.1 Background 399
   13.2 Behavior during Colony Selection 400
   13.3 Sorting of Birds among Colonies 409
   13.4 Histories of Breeding Colony Choices by Individuals 419
   13.5 Natal Dispersal and Colony Choice by Yearlings 432
   13.6 Switching between Colonies within a Season 437
   13.7 Predictability of Colony Size 441
   13.8 Conclusions about Colony Choice 445
   13.9 Summary 447

14 The Evolution of Coloniality 449
   14.1 The Historical Causes of Coloniality 449
   14.2 Lifetime Reproductive Success 456
   14.3 Variation in Colony Size 463
   14.4 Within-Group Asymmetries: Spatial and Temporal Effects 469
   14.5 Relatedness and Reciprocity 474
   14.6 Conclusions and Unanswered Questions 478
   14.7 Summary 481

   Appendix 483
   References 485
   Index 529

Preface

When we arrived at the Cedar Point Biological Station on a wintry April day in 1982 to begin our study of cliff swallows, we could scarcely have imagined we would still be there twelve years later. Most of the credit of course goes to the birds themselves. This study would never have lasted for over a decade had the cliff swallows' natural history not been so fascinating; had their social behavior not been so relevant to a variety of general issues in behavioral ecology; or had the birds not been so utterly tolerant of people catching them in nets, painting their heads, spraying their nests with insecticide, scaring them with rubber snakes, drawing their blood, rooting around in their nests, handling their eggs and young, and generally pestering them to no end!

Although cliff swallows qualify as one of the world's greatest study animals, work of this sort could not have been sustained for long without the assistance of numerous humans. Special recognition must go to our research assistants, who cheerfully endured long hours in the field helping to collect the data used in this book: Cathy Boersma Anderson, Carol Brashears, Karen Brown, Rachel Budelsky, Barbara Calnan, Sasha Carlisle, Beth Chasnoff, Miyoko Chu, Zina Deretsky, Laurie Doss, Kristen Edelmann, Jessica Thomson Fiorillo, Ellen Flescher, Jennifer Grant, Winnie Hahn, Leila Hatch, Audrey Hing, Jerri Hoskyn, Laura Jackson, Deborah Johnson, Veronica Johnson, Mike Kostal, Bara MacNeill, Kathi Miller, Christine Mirzayan, Laura Molles, Laura Monti, Cecily Natunewicz, Charlene Patenaude, Barbara Raulston, Craig Richman, Sarah Rosenberg, Todd Scarlett, Martin Shaffer, Lora Sherman, Karen Van Blarcum, and Zoë Williams. Other field assistance was provided by Josef Kren, Bruce Rannala, William Scharf, Linda Seitz, and Peter Walsh in the course of their own research in the study area. We were assisted in the analysis of data by Carol Brashears (chapter 4), Kristen Edelmann (chapter 9), and Jerri Hoskyn (chapter 7). We acknowledge especially the efforts of Winnie

Hahn, who single-handedly updated all the analyses in chapter 12 so that the mark-recapture data from the 1992 season could be included.

We owe much to the University of Nebraska–Lincoln for providing us with living and laboratory space each year at the Cedar Point Biological Station. This research would not have occurred had Cedar Point not happened to be at probably the best spot in North America for cliff swallows and had John Janovy Jr. not so warmly encouraged us to come there in the first place and supported our work in so many ways in the early years. We thank him and the other Cedar Point directors—Joan Darling, Anthony Joern, Kathleen Keeler, William Scharf, and Linda Vescio—for their cooperation. The staff at Cedar Point was always helpful, especially Ron Randall, who went beyond his capacity as maintenance director to keep our living quarters functional, our equipment repaired, and our trucks running.

We might never have begun a study of cliff swallows without John Hoogland's suggestion in the spring of 1981 that we do so, or without his and Paul Sherman's classic paper on bank swallows as a model in the early years. Our thoughts on coloniality have been shaped by discussions with many people over the past decade, especially John Hoogland, Henry Horn, Tony Ives, Harry Power, the Princeton population biology group from 1981 to 1985, Bruce Rannala, Scott Robinson, Sievert Rohwer and his 1988–89 lab group, Dan Rubenstein, Bridget Stutchbury, Peter Walsh, and David Wilcove.

We are grateful to others who have helped us in various ways: the McGinleys, Oren Clary, and Vern Kaulson for access to their land; the Nebraska Public Power District and the Central Nebraska Public Power and Irrigation District for access to their structures containing swallow colonies; the U.S. Fish and Wildlife Service and the Nebraska Game and Parks Commission for permits; the Chevron Chemical Company for providing fumigant; Mathew Werner and the University of Nebraska–Lincoln's High Plains Climate Center for providing climatological data; Art Gingert for allowing use of his photographs; Roger Pradel for providing the SURGE software; Cluff Hopla and Jenella Loye for advice on cliff swallow ectoparasites; Bob Hargesheimer for building our bird blinds; Rich Boardman and Larry Gall for computer aid; John Maisano and Sally Pallatto for illustrations; Bill Sacco for making photographic prints; Nancy Carrignan for word processing; Steve Zack for help with the "big net"; Anthony Joern for identifying grasshoppers; Kyle Hoagland and Ron Randall for piloting the annual boat tours of the Lake McConaughy colonies; and William Gergits and Therion Corporation for conducting the preliminary DNA profile analyses. Much of this work was done while we held appointments at Yale University, and we thank the Department of Biology and the Peabody Museum of Natural History for the support and freedom necessary to conduct this research.

We acknowledge with gratitude the funding agencies that had faith in us and kept us afloat financially: the National Science Foundation (BSR-8407329, BSR-8600608, BSR-9001294, BSR-9015734, DEB-9224949, and a Graduate Fellowship to CRB), the Erna and Victor Hasselblad Foundation, the National Geographic Society (3244-85, 3996-89, 4545-91), the American Philosophical Society, Yale and Princeton Universities, the Cedar Point Biological Station, the Chapman Fund of the American Museum of Natural History, the National Academy of Sciences, Sigma Xi, Alpha Chi, and the University of Tulsa.

Several colleagues kindly read earlier drafts of various chapters and suggested improvements: Dale Clayton, Mike Deacon, James Gibbs, Cluff Hopla, Paul Johnsgard, Robert Kaul, Walter Koenig, Jenella Loye, Ron Mumme, Harry Power, Bruce Rannala, Mark Reynolds, Scott Robinson, William Scharf, Tom Seeley, Linda Seitz, Bridget Stutchbury, Peter Walsh, Harrington Wells, Mathew Werner, Gerald Wilkinson, David Winkler, and the members of the 1994 Tulsa behavioral ecology seminar. We thank especially Paul Harvey, John Hoogland, Judith May, Robert May, Douglas Siegel-Causey, and an anonymous reviewer for reading the entire manuscript and offering numerous constructive criticisms. We are grateful to Susan Abrams and Alice Bennett of the University of Chicago Press for their assistance, and special thanks go to our editor, Christie Henry, for guiding the manuscript through production and providing many useful suggestions.

Finally, our deepest debt is to our parents, Raymond and Kathryn Brown and Donald and Ruth Bomberger, who helped in ways far too numerous to mention. Their patience and understanding in nurturing our childhood dreams and ambitions are what really led us to the cliff swallows along the Platte River and have kept us there.

1 Introduction

> Legends and stories abound in San Juan Capistrano, some written, some told from generation to generation, tales filled with laughter and adventure, with whimsey and with deep religious faith. And the legend of the [cliff] swallows is the favorite of them all.
> Lydian Bruton (1975)

1.1 OVERVIEW

Three virtually identical concrete bridges 0.8 km apart span an irrigation canal in southwestern Nebraska. Four thousand cliff swallows move onto one, constructing their mud nests side by side underneath the bridge's protective overhang, covering every inch of available surface. Not a single cliff swallow breeds at the second bridge, which is within sight of the first. At the third bridge, only 150 cliff swallows settle. Many of the same birds return the next year but redistribute themselves so that the first bridge has none, the second bridge has 3,000 birds, and the third bridge has only 30, while some of the swallows move to more distant colonies of other sizes.

This actual sequence of events illustrates in a nutshell the major issue on which our research of the past twelve years has focused and the principal theme of this book: Why do cliff swallows live in colonies? And why do these colonies vary so much in size?

These deceptively simple questions have not been easy to answer. They have required extensive data on cliff swallow social behavior, plus long-term information on the birds' natural history, ecology, and demography. We do not pretend to understand fully, even after twelve years, why cliff swallows choose to live in colonies as they do or all the socioecological costs and benefits of coloniality. As with any long-term study, more questions have emerged as the earlier ones were answered. However, we have learned enough to increase our understanding of how group size may affect social behavior and demography and to provide insights into the evolution of coloniality in cliff swallows and other animals. We hope these insights will prove useful to readers interested in other group-living animals.

1.2 BACKGROUND

Vertebrate social systems during the breeding season can be broadly classified into three major types: territorial, cooperative or communal, and colonial. This classification is based primarily on an increasing degree of conspecific association in use of space, and some species may exhibit more than one type within or between populations. The evolution of territoriality is most easily explained in terms of economic defensibility: when the costs of defending critical resources are less than the benefits of excluding others, animals should maintain exclusive use of certain areas (Brown 1964; Wittenberger 1981). Similarly, the evolution of cooperative or communal systems—in which related individuals typically remain together as an extended family and only certain ones may actually breed—can be understood in terms of inclusive fitness theory (Hamilton 1964) and the notion of habitat limitation. When suitable habitat is limited or other circumstances portend poor reproductive prospects for some individuals, these animals may raise their inclusive fitness by deferring breeding and remaining as part of an extended family on familiar terrain and presumably "helping" to raise siblings or half-siblings (Emlen 1982a,b, 1994; Woolfenden and Fitzpatrick 1984; Brown 1987; Koenig and Mumme 1987, reviewed in Stacey and Koenig 1990a).

The evolution of coloniality, as represented by cliff swallows living in groups of largely unrelated individuals and spatially restricted to particular breeding sites, is less well understood. The two principal questions are why these animals aggregate in the first place and why the resulting groups vary in size. For a variety of reasons, no general patterns have emerged, despite numerous field studies over the last half century (reviewed in Burger 1981; Birkhead 1985; Wittenberger and Hunt 1985; Brown, Stutchbury, and Walsh 1990; Siegel-Causey and Kharitonov 1990, 1996). Speculation as to the causes of animal coloniality began as early as the fourth century B.C. with Aristotle's *Historia animalium* (Thompson 1910) and continued with Allee (1931, 1938, 1951) and Darling (1938). More recently research has been directed primarily by Alexander's (1971, 1974) suggestion that the three principal selective forces leading to social living are the ability of individuals in a group to avoid predators better, their ability to find food more easily, and the need for individuals to aggregate where limited, critical resources (e.g., nesting sites) are found. Alexander (1971, 1974) also pointed out that social animals probably experience two automatic costs of group living: increased transmission of parasites and disease and increased competition for limited resources such as food or nesting sites. Alexander's ideas were anticipated in part by Allee (1931, 1938, 1951) and Lubin (1974) and were brought to widespread attention when Hoogland and Sherman (1976) published a pioneering study of the effect of group size on behavior and reproductive success in colonially nesting bank swallows. Hoogland and Sherman's approach illustrated the power of using natural variation in group size to evaluate many of the potential socioecological costs and benefits of group living. Another advantage of this approach is that the data gathered are consistent with the way the effects of group size have generally been modeled by theoreticians (e.g., Pulliam 1973; Sibly 1983; Pulliam and Caraco 1984; Giraldeau and Gillis 1985; Giraldeau 1988; Giraldeau and Caraco 1993; Higashi and Yamamura 1993; Rannala and Brown 1994). Other studies have also used natural variation in group size to evaluate potential costs and benefits of coloniality (e.g., Veen 1977; Hoogland 1979a,b, 1981, 1995; Rypstra 1979; Robinson 1985; Brown and Brown 1986, 1987, 1988a; Van Vessem and Draulans 1986; Moller 1987a; Shields and Crook 1987; Brown 1988a; Burger and Gochfeld 1990, 1991; Spiller 1992; Wiklund and Andersson 1994; Tyler 1995). However, a limitation of this paradigm is that it does not address why colonies vary in size in the first place (Brown, Stutchbury, and Walsh 1990), and this important question has largely been neglected.

We do not attempt a comprehensive review of the empirical work done on avian coloniality; such reviews are available from Burger (1981), Birkhead (1985), Wittenberger and Hunt (1985), Kharitonov and Siegel-Causey (1988), Brown, Stutchbury, and Walsh (1990), and Siegel-Causey and Kharitonov (1990, 1996), with the last two authors also providing summaries of the extensive work on coloniality done by Russian ecologists. These reviews indicate that the presumed benefits of coloniality (predator avoidance, increased foraging efficiency, aggregation at limited resources) apply in various species to differing degrees, and predator avoidance at least generally increases with group size. Less attention has been paid to the costs of coloniality, although some species have been shown to experience various disadvantages of grouping, especially those associated with parasite transmission (section 4.1). The major conclusions that emerge from this literature are that few (if any) species have been studied intensively enough or long enough to evaluate all (or even most) of the potentially important costs and benefits of coloniality as a function of group size; that we still do not understand the basis for the variation in breeding colony size so commonly observed; and that therefore we do not as yet have a general understanding of the evolution of avian coloniality, a point also emphasized by Birkhead (1985) and Wittenberger and Hunt (1985). Once we better understand the evolutionary basis for coloniality, we will likely gain insights into the evolution of sociality more generally.

It is important to note that in discussing animal coloniality in this chapter and throughout this book we are referring to assemblages of largely unrelated individuals that generally exhibit no "helping" or cooperative breeding among adults. This distinction is an important one, because it means that indirect (kin) selection, the theoretical framework used in studies of cooperative breeders (Stacey and Koenig 1990b), does not directly apply in analyses of the origin and maintenance of coloniality among nonrelatives. However, any theoretical framework developed for coloniality, presumably based on the advantages and disadvantages of living in groups, should also be directly applicable to cooperative breeders, most of which live in small to moderate-sized groups of related individuals. This was illustrated well in Hoogland's (1995) study of black-tailed prairie dogs, colonial rodents that live in extended family groups. The costs and benefits of coloniality have variously affected many aspects of prairie dog social behavior, including infanticide, communal nursing, nepotistic alarm calling, and avoidance of inbreeding. Cooperatively breeding species have attracted considerable interest (e.g., Brown 1987; Stacey and Koenig 1990a), yet in birds coloniality among noncooperative breeders is more common than true cooperative or communal breeding. Lack (1968) estimated that at least 14% of the world's 8,600 to 9,000 species of birds breed colonially, whereas only about 2% are known to breed cooperatively (Stacey and Koenig 1990b). The rest presumably are territorial to some degree or have not been studied.

1.3 MAJOR QUESTIONS

Studying the evolution of coloniality is complicated by the problem that most adaptations to coloniality represent simultaneously dependent and independent variables: "Each adaptive trait or selective pressure must be analyzed separately, although in nature all interact simultaneously in a complex web of interwoven selective pressures and adaptive responses. Our task then becomes one of trying to tease apart the pieces of the large picture into coherent parts and then to resynthesize these parts into a meaningful whole" (Wittenberger and Hunt 1985, 2). The approach we have used in this study is to (1) describe the socioecological costs and benefits of coloniality in cliff swallows as a function of group size; (2) evaluate the alternative reproductive options that become available to the animals once they have formed groups; (3) investigate the demographic consequences of living in colonies and especially the effect of colony size on life history parameters such as survivorship; and (4) examine the observed patterns of colony choice by individuals and consider the hypotheses about why group size varies.

1.3.1 The Costs and Benefits of Coloniality

We ask which socioecological costs and benefits of coloniality may apply to cliff swallows. Once identified, we ask how these vary with colony size as a measure of their importance. Major costs we address are increased ectoparasite transmission; increased competition for nesting sites, mating opportunities, nesting material, and food; increased interference among conspecifics resulting in loss of eggs; increased attraction of predators; and increased probability of misdirecting parental care toward unrelated individuals. Major benefits we address are increased opportunities to gain information on the location of food sources; increased avoidance of predators; increased opportunities to parasitize the parental care provided by other individuals; and aggregation in a limited nesting habitat.

Where possible, we evaluate both the *average* effect of these costs and benefits in groups of various sizes and the *variance* in the effect within and among colonies. Not all individuals in a single group may experience a given cost or benefit equally, and not all groups of the same size may experience similar average effects.

1.3.2 Alternative Reproductive Options Available in Colonies

Group living affords numerous opportunities for individuals to engage in various alternative reproductive tactics. The presence of many simultaneously breeding conspecifics spatially restricted to the relatively small area of a colony site presumably increases opportunities for both sexes to engage in extrapair copulation (Gladstone 1979; Westneat, Sherman, and Morton 1990; Birkhead and Moller 1992; Moller and Birkhead 1993a) and for females to commit conspecific brood parasitism (Hamilton and Orians 1965; Brown 1984; Rohwer and Freeman 1989). These reproductive options may either supplement an individual's reproductive success, assuming that the perpetrator also maintains a nest and provisions its own young, or enable an individual to achieve some, perhaps limited, reproductive success when it could not establish a nest or attract a mate of its own.

For studies of coloniality, the critical issue is whether these tactics increase in either frequency or success with colony size. There is increasing evidence that many territorial species that often occur at relatively low population densities also experience high rates of extrapair copulation (reviewed in Birkhead and Moller 1992) and brood parasitism (reviewed in Yom-Tov 1980; Rohwer and Freeman 1989). A high incidence of these behaviors may not be an automatic consequence of colonial living, although it seems likely that their frequency should be greatest among colonial species.

Regardless of the relation between group size and the average incidence of extrapair copulation or conspecific brood parasitism, these behaviors may have an important effect on the evolution of coloniality if certain individuals in a colony are more or less likely to perpetrate them or suffer their costs. For example, Morton, Forman, and Braun (1990) showed that young male purple martins breeding in colonies were more likely than older males to be cuckolded, probably by those same older males. They also found a trend for young birds to be more likely to suffer conspecific brood parasitism. In this case young individuals were disproportionately victimized, suggesting that there must be powerful compensatory benefits of coloniality for them. The opportunities to victimize younger individuals can be viewed as a benefit of living in colonies for older birds. Morton, Forman, and Braun (1990) suggested that martins form colonies largely *because* of the attendant opportunities for older males to engage in extrapair copulation (and see Wagner 1993). Although this hypothesis alone is inadequate to explain the evolution of coloniality, because it does not explain why the younger birds would join a colony in the first place, it underscores the importance of evaluating which individuals experience the costs and benefits of coloniality and their consequences. The martin example also illustrates nicely that alternative reproductive tactics may be an important incidental benefit of living in a colony for certain individuals, and this added benefit may affect the payoffs to be expected in groups of various sizes.

Cliff swallows commonly brood parasitize the nests of neighbors (Brown 1984; Brown and Brown 1988a, 1989) and engage in extrapair copulation (Butler 1982; Brown 1985a). We ask to what degree the incidence and success of extrapair copulation and conspecific brood parasitism vary with cliff swallow colony size. We also evaluate whether certain classes of individuals are more or less likely to either perpetrate these behaviors or be victimized by them.

1.3.3 Demographic Consequences of Coloniality

The socioecological costs and benefits, and the consequences of any alternative reproductive strategies made possible by coloniality, manifest themselves in an individual's reproductive success and survival. To date, what we know about the demographic consequences of coloniality are confined to several species in which annual reproductive success has been measured for individuals in different-sized colonies (Wittenberger and Hunt 1985; Brown, Stutchbury, and Walsh 1990; section 11.1). No clear trends are apparent, some species showing an increase in annual reproductive success with colony size, others a decrease, still others no relationship, and a few suggesting that intermediate-sized colonies are best (Brown, Stutchbury, and Walsh 1990). However, not a single study, to our knowledge, has followed colonial individuals over their lifetimes and measured or estimated lifetime reproductive success and survivorship as a function of either natal or breeding colony size. There have been a number of excellent long-term studies on colonial species per se (e.g., Bryant 1988b, 1989; Cooke et al. 1983; Cooke and Rockwell 1988; Ollason and Dunnet 1988; Coulson 1968, 1988; Thomas and Coulson 1988), but these have largely focused on individuals at a single colony and hence cannot address the long-term demographic effects of different group sizes.

We use mark-recapture methods to follow the histories of banded cliff swallows over an eleven-year period and calculate survivorship for birds occupying natal and breeding colonies of different sizes. Using these calculations and data on annual reproductive success, we estimate lifetime reproductive success as a function of group size (see Vehrencamp, Koford, and Bowen 1988). We also measure correlates of annual reproductive success, to evaluate the fitness effects of the presumed costs, benefits, and alternative reproductive tactics of coloniality.

1.3.4 Variation in Group Size

Virtually all colonial bird species exhibit variation in colony size within and between populations (Brown, Stutchbury, and Walsh 1990). The reasons for this variation are unknown, and few if any empirical or theoretical studies have explicitly addressed this issue. At least three major hypotheses are possible (Brown, Stutchbury, and Walsh 1990): size variation reflects a distribution of individuals such that each animal receives an equal fraction of the locally available resources (an "ideal free" distribution, sensu Fretwell and Lucas 1970); differences in individuals' phenotypic qualities influence what constitutes an "optimal" colony size for them; and limitations on the ability to sample, assess, and predict future colony size prevent individuals from always settling in the group that is best for them, causing size variation. These hypotheses are discussed more fully in Brown, Stutchbury, and Walsh (1990), although few data exist for any species to let us test them.

The range in colony sizes exhibited by cliff swallows is among the greatest seen in land birds. In our study area, colonies range from 2 nests to 3,700 nests, with some birds nesting solitarily. We address the possible cause(s) of this variation by examining the colony choices birds make over their lifetimes. Do individuals always settle in colonies of similar sizes, and if not, do age or other phenotypic characters affect the observed settlement patterns? Are there environmental correlates of colony size that might suggest something like an ideal free matching of colony size to ecological resources? Is there evidence that certain individuals "do better" in colonies of particular sizes? How much information on potential colony choices is available to a cliff swallow at the time it selects a colony site in the spring? Answers to these questions are necessary before we can understand the observed size variation and its role in the evolution of coloniality in cliff swallows and other species.

1.4 THE CLIFF SWALLOW

The cliff swallow, *Hirundo pyrrhonota*, is a 20–28 g passerine identified in the field by its characteristic square tail, orange rump patch, and in most populations, triangular white forehead patch (see frontispiece). During the breeding season, the species is widely distributed and common throughout most of western North America, from southern Alaska to central Mexico (AOU 1983; Brown and Brown 1995). The birds are less common east of the Great Plains, occurring there in a local, patchy distribution, although they have been increasing across the eastern United States in recent years. Cliff swallows are migratory, arriving in the southern and coastal parts of their breeding range in March and in most other areas by early May. They depart for their wintering range during late July, August, and early September. In our Nebraska study area, the first birds appear in mid-April, and the last ones usually leave by mid-August.

Cliff swallows migrate through Central America and northern South America, staying east of the Andes. They seldom occur as far east as the Bahamas, Cuba, or the Virgin Islands (AOU 1983; Brown and Brown 1995). One bird banded in our study area during the breeding season was recovered in El Salvador on 6 October, while it was presumably still migrating south. The species' wintering range is in south-central South America from southern Brazil (São Paulo province) and possibly southeastern Paraguay south to central and south-central Argentina (AOU 1983; Ridgely and Tudor 1989; Fjeldså and Krabbe 1990; Brown and Brown 1995). Most of the birds apparently winter in lowlands along the Rio Paraná and Rio Uruguay north and northwest of Buenos Aires, with some individuals occasionally recorded as far west as Tucumán province of Argentina and as far south as Tierra del Fuego (Hudson 1951; AOU 1983; Ridgely and Tudor 1989; Brown and Brown 1995). They avoid the high Andes (Fjeldså and Krabbe 1990). One cliff swallow from our study area was recovered in São Paulo province, Brazil, on 2 January, apparently on its wintering range. Little is known of the cliff swallow's natural history during the winter; it seems to occur mostly in marshes, open agricultural areas, and grassy savannas below 1,000 m, and it travels and roosts in large flocks (Hudson 1951; Ridgely and Tudor 1989; Brown and Brown 1995). The species is not known to breed while in South America, although this possibility should not be ruled out, given that the congeneric North American barn swallow has recently been discovered nesting on its Argentine wintering range (Martinez 1983; Ridgely and Tudor 1989).

Cliff swallows historically nested primarily underneath rocky outcroppings on the sides of steep cliffs and canyons in mountainous regions of western North America from the Lower Sonoran Zone through the Transition Zone to about 3,000 m, but rarely at higher altitudes (Grinnell and Miller 1944; Brown and Brown 1995). They have occasionally been reported to build their nests underneath limbs on the sides of large trees in California (Grinnell and Willett in Dawson 1923). These birds construct an enclosed gourd- or retort-shaped mud nest about the size of a large cantaloupe. Each mud pellet used to construct the nest is collected by the birds at a mud source along a riverbank or pond shore, sometimes up to

1 km away, transported back in the bird's beak, and molded into the nest. The birds apparently assess mud quality (Kilgore and Knudsen 1977; Robidoux and Cyr 1989), selecting mud with a relatively high clay and silt content that maximizes the nest's strength and reduces the likelihood of its crumbling when dry. In sites relatively protected from the elements and depending on the quality of the mud used to construct the nests, some nests can survive intact for many years. Some nests in our study area are at least thirteen years old, requiring only slight additions of mud to the entrance neck in any given year. Cliff swallows often reuse nests that are intact from the previous summer.

Relatively recently, cliff swallows in many areas have begun using artificial nesting sites such as bridges, dams, highway culverts, buildings, and other structures that provide a protective overhang and vertical substrate for nest attachment. This switch to alternative nesting sites has enabled them to expand their range and colonize areas where they presumably formerly did not occur, such as the southeastern United States. However, unlike the ecologically similar barn swallow, which now uses artificial nesting sites almost exclusively (Speich, Jones, and Benedict 1986), the cliff swallow still commonly uses natural sites in most areas of its historical range. In our study area, for instance, colonies are found both on natural cliff faces and on bridges and in highway culverts. The birds seem to require colony sites near relatively open areas for foraging and a water source to provide mud (Emlen 1941, 1952).

Cliff swallows are extremely social throughout the breeding season. Although solitarily nesting pairs are occasionally found, the birds usually breed in dense colonies. Nests are typically placed side by side and often share walls with adjacent nests. Colonies commonly range into the hundreds of nests, and occasionally groups of several thousand nests are recorded. Smaller colonies tend to be the rule in more easterly locations where the birds have not been established as long, but relatively small ones are common even in the center of the cliff swallow's range in the western United States. Cliff swallows forage in groups that may at times consist of 1,000 birds or more (Emlen 1952; Brown 1988b), gather mud for their nests in groups (Brown and Brown 1987), preen and sunbathe in groups, and migrate and winter in large flocks (Brown and Brown 1995). Breeding within a colony is highly synchronous, with almost every individual in the colony engaged in the same general activity (e.g., nest building) at any given time (Myres 1957; Emlen 1952; Brown and Brown 1987). These birds feed entirely on aerial insects caught in flight. Cliff swallows have a limited vocal repertoire; only four types of vocalizations are commonly used (Samuel 1971a; Brown 1985b). The species' natural history has been well described by Emlen (1952, 1954), Mayhew (1958), Samuel (1971b), Grant and Quay (1977), and Brown and Brown (1995).

Cliff swallows of North America exhibit substantial morphological variation. At least six subspecies have been described at various times, although probably only four at most are valid and marked intergradation occurs among them (AOU 1957; Behle 1976; Browning 1992; Brown and Brown 1995). The variation tends to be clinal, with a pattern of larger birds (in terms of wing length) in the north (Alaska and western Canada) to smaller birds in the south (southern Arizona and northern Mexico). Northern birds tend to have a larger and whiter forehead patch, with more southern birds showing smaller patches of light cinnamon color (southwestern Utah, Arizona, and parts of New Mexico and Texas) to deep cinnamon or chestnut (northern Mexico) on the forehead. Northern birds have lighter rump patches than Arizona or Mexican birds (Behle 1976). *Hirundo p. pyrrhonota*, the most widespread subspecies and the one occurring in our study area, is intermediate in size and exhibits a relatively large white to cream forehead patch and a rusty orange rump patch.

The North American cliff swallow was formerly placed in the genus *Petrochelidon*, a distinct grouping of red-rumped, square-tailed swallows that all nest colonially and build similar retort-shaped mud nests. The *Petrochelidon* group is widespread, with species breeding on all continents except Europe and Antarctica: *pyrrhonota* in North America, *fulva* in Central and South America, *preussi, rufigula,* and *spilodera* in Africa, *fluvicola* in Asia, and *ariel* in Australia (Sharpe and Wyatt 1885–94; Turner and Rose 1989). These species are markedly similar in behavior and ecology. Although the American Ornithologists' Union (1983) merged *Petrochelidon* into the genus *Hirundo* because of hybridization between the cave swallow, *P. fulva*, and the barn swallow, *H. rustica,* in a limited area of overlap in central Texas, *Petrochelidon* should probably be retained for the seven species of true cliff swallows. A molecular phylogeny of the swallow family (Sheldon and Winkler 1993; Winkler and Sheldon 1993) suggests that *Petrochelidon* is more derived than *Hirundo,* and *Petrochelidon* may be at least as distinct as the *Delichon* group (the house martins) of Europe and Asia.

1.5 ORGANIZATION OF THIS BOOK

Analysis of the cliff swallow's social behavior and the effects of group size has proved to be complex, with many parts of the story interconnected. Although this complexity has helped sustain our fascination with this bird and this problem for over a decade, the interrelatedness of the components presents challenges in presenting this material in a coherent way. The four major questions described earlier in this chapter constitute the framework for this study, but because of their scope it is not possible to address each one exclusively in a separate, self-contained chapter. Data and analyses relevant to more than one of these questions must often be given in the same chapter. We have grouped related material as much as possible, although decisions on the order in which material is presented and in chapter content were necessarily arbitrary in some instances. We hope readers will not be too distracted by our liberal cross-referencing of material between chapters.

As with any ongoing long-term study, we have been faced with the issue of when to terminate data collection. All analyses use minimally all relevant data collected during a ten-year period from the beginning of the study in 1982 through 1991. The material on survivorship (chapter 12) and colony choice (chapter 13), requiring that we follow specific individuals for as many years as possible, include data collected through 1992. We used data from 1993 only in chapter 3 in characterizing climatological patterns and the study population, in section 4.8 on the number of swallow bugs found on adult birds, in section 5.3.3 on distances the birds traveled to gather mud, in section 13.2 on radio telemetry, in section 13.7 on daily population changes at colony sites, and in section 14.3.1 on the percentage of birds occupying each colony size. Most of our data analysis was done during 1991 and 1992, and the text was written during the 1992–93 and 1993–94 academic years. The large number of analyses, and their size and complexity, prevented us in most cases from continuing to add data after 1992.

This book is not intended to be a review of all of our past cliff swallow research. Material in our previous works is presented here only when relevant to the effects of colony size and the four major questions posed at the outset. In most cases where previously published material is presented, we have updated the earlier analyses with additional data, in some cases reanalyzing the data in different ways and perhaps reaching different conclusions. Sample sizes, analyses, and conclusions presented in this book should be considered more definitive than those in earlier works.

In chapter 2 we begin with a description and discussion of the major field methods used during this study and the general techniques we employ for data analysis. In chapter 3 we describe the study site in some detail, presenting information on the ecology of the area, the types of cliff swallow colonies found there, and characteristics of the study population. In chapter 4 we address ectoparasitism, a major cost of coloniality and a subject that has been studied the most thoroughly. In chapter 5 we turn to competition for nesting sites and interference among neighbors. Misdirected parental care and parasitism of care provided by other individuals are analyzed in chapter 6, in which the associated costs and benefits are evaluated. In chapter 7 we explore shortage of suitable nesting habitat as a potential cause of coloniality in cliff swallows. The major benefits of coloniality are addressed in chapter 8, on the avoidance of predators, and in chapters 9 and 10, where we take up social foraging and information centers. The demographic consequences of colonial living are explored in chapter 11 on annual reproductive success and chapter 12 on survivorship. Patterns of colony choice by individual birds are presented and discussed in chapter 13. In chapter 14 we examine the evolution of coloniality, based on the material presented in earlier chapters, summarize our conclusions, and highlight the important unanswered questions about coloniality in cliff swallows and other animals.

1.6   SUMMARY

The central theme of this book is the effect of group size on social behavior and demography in the colonially breeding cliff swallow. Why do these birds live in colonies, and why do those colonies vary in size? Considerable empirical work on avian coloniality over the past twenty-five years reveals few general patterns, and we lack long-term studies focusing on animals in different-sized groups. In our study, which has lasted twelve years and is continuing, we use natural variation in cliff swallow colony size to measure the socioecological costs and benefits of coloniality as a function of group size. Other major issues we address are what sort of alternative reproductive options (such as parasitizing the parental care provided by conspecifics) become available to cliff swallows living in a group, how colony size affects demographic parameters such as survivorship and reproductive success, and how patterns of colony choice by individuals may lead to the observed variation in colony size. Cliff swallows are migratory passerines that breed in western North America and winter in poorly known locations in south-central South America. The birds build gourd-shaped mud nests, occasionally nesting solitarily but usually in colonies of variable size. They breed underneath overhanging rock ledges on the sides of cliffs and on artificial structures such as bridges. Cliff swallows are extremely social at all times and breed synchronously within a colony. There are four valid subspecies in North America, with north-to-south clinal variation in morphology. The cliff swallow is part of the *Petrochelidon* group of swallows, which share similar morphology, behavior, and ecology and are distributed worldwide. This study is based on data collected beginning in 1982 and updates the analyses and in some cases changes the conclusions in our previously published work on cliff swallows.

2 *Field Methods and Data Analysis*

> I always look up at the swallow nests when I'm under a bridge. . . . I find myself wondering how long the nest numbers will remain visible, whether fifteen or twenty years from now some student will ask some teacher what they mean. . . . The teacher will say those numbers mean that a scientist came here one time looking for the costs and benefits of living in large colonies.
> John Janovy Jr. (1994)

In this chapter we describe some of the field methods used in this study and the general methods and philosophy of data analysis we have adopted. We describe procedures used in collecting basic data relevant to analyses in more than one chapter, with methods for specific analyses included in the individual chapters that follow.

2.1 MEASURING COLONY SIZE AND USE

Colony size throughout this book refers to the maximum number of active nests at a site during a breeding season. An active nest is any nest in which at least one egg was laid. We determined colony size at some sites by checking nest contents directly (section 2.2), providing an exact count of active nests. At other sites where nests were inaccessible for nest checks, we estimated the number of active nests in the colony by counting all the nests in sections of colonies where birds were observed to be settled and multiplying those counts by the number of sections with similar nest densities that were obviously active. We also counted or estimated the number of birds present at colony sites during prolonged alarm responses (to us) when presumably most of the birds living there appeared. These counts or estimates of birds were used to refine colony size estimates based on the number of visible nests at colony sites where active nests could not be counted directly. To test the accuracy of colony size estimates based on the number of birds observed at a site, for twenty-three colonies we compared estimated sizes based on the number of birds with the actual counts of active nests based on nest checks (table 2.1). Agreement between estimates and exact counts was close, averaging only a 3.4% difference. Thus we consider our estimated colony sizes to be accurate and comparable to those obtained by checking nest contents.

We collected colony size information each time we visited a colony site

Table 2.1 Cliff Swallow Colony Size at Sites Where We Counted Active Nests by Examining Nest Contents and Independently Estimated Size by Number of Birds Present

Colony Site	Colony Size from Nest Checks	Colony Size from Number of Birds Present	Percentage Difference
9125	1	1	0
9128	1	1	0
9132	5	3	−40.0
9160	6	5	−16.7
9260	11	10	−9.0
9232	30	30	0
9032	42	45	+7.1
8841	68	65	−4.4
9124	68	60	−11.8
9041	86	85	−1.2
8941	125	150	+20.0
9130	137	140	+2.2
9241	165	160	−3.0
9030	190	200	+5.3
9176	250	220	−12.0
8930	260	250	−3.8
9170	296	300	+1.3
9223	380	400	+5.2
8705	1,028	1,000	−2.7
9205	1,478	1,500	+1.5
9105	1,747	1,700	−2.7
8905	2,200	2,000	−10.0
9005	2,350	2,250	−4.2

for any reason. No size was considered definitive until the number of arrivals early in the year had stabilized and the colony had stopped increasing (section 13.7.2). Most analyses use a single colony size for each site each year. Size did vary slightly during the season, mostly owing to birds' losing their nests or nest contents and then leaving the colony. It was not logistically feasible to visit and estimate size at all colonies at similar intervals throughout the season, but departures owing to nest failure seemed to be infrequent at most sites. Therefore we consider use of a single colony size for each site each year to be a fair relative means of comparing colonies.

In nine colonies, however, there was mass desertion by a substantial fraction of the residents after egg laying (perhaps in response to ectoparasitism; section 4.9.1), leading to a drastic reduction in colony size at each site. In these instances we used a smaller colony size in any analyses pertaining to the birds that remained, such as the analysis of the effect of colony size on adult body mass for birds feeding nestlings (section 10.5.2). As a result, there may occasionally be slight discrepancies in colony sizes between different analyses. We also used smaller colony sizes (relative to earlier in the year) for late-season analyses involving colonies where we experimentally reduced colony size (Brown 1988a).

We visited most colony sites repeatedly throughout the breeding season; those in the primary study area that extended from Lewellen to Paxton, Nebraska (section 3.3), were generally visited at least once a week from mid-April or early May through late July or early August, and we visited some sites almost daily. For the colony sites in this primary study area, we were usually able to determine the date they became active (first birds arrived) to within one to six days. We repeatedly verified the inactive status of those sites that were not occupied in a given year. Colony sites in the secondary study area, a region extending 30–50 km east and west of the primary study area (section 3.3), were visited on a more erratic schedule, depending on need. However, we visited each colony in the secondary study area at least once each year during the second half of June specifically to determine colony size. Where colony size had to be estimated, we visited only on warm, sunny days when all residents of a colony were likely to be present. Late June was the best time to score both colony size and whether a site was in fact used that year, because by then most birds were already feeding nestlings and no longer moving between sites, and most of the relatively few late-starting colonies had at least colonized sites and begun nest building. In rare instances, however, late colonies commenced in early to mid-July, and some of these might have been overlooked in our secondary study area if we did not happen to visit a site again after late June.

Potential colony sites (e.g., bridges, cliff faces) were obvious (e.g., fig. 3.5), and many if not most of the potential colony sites in the primary and secondary study areas (including ones not used previously) were checked for use each year. However, each year we discovered one or more active colony sites we had been unaware of, and thus a relatively few colonies in the study area may have been undetected each year. Any undetected colonies in all likelihood were small ones, since the large colonies were more conspicuous and hence less likely to be overlooked. Thorough searching of the study area for colonies did not begin until 1984, and the best data on colony use are for 1987–93.

In some colonial species it can be hard to define a "colony," especially when individuals are spread out in varying densities over a relatively large area (e.g., Nettleship 1972; Newton and Campbell 1975; Davis and Dunn 1976; Parsons 1976; Coulson and Dixon 1979; P. Buckley and F. Buckley 1980a; Schmutz, Robertson, and Cooke 1983; Emms and Verbeek 1989; Clarke and Fitz-Gerald 1994). This was not a problem in cliff swallows, because the birds usually occupied a clearly defined site (e.g., a bridge) with at least 1 km between it and the next suitable site. For artificial sites, we defined a cliff swallow colony as any single bridge, highway culvert, building, or other structure containing nests, subject to the following modification. If birds occupying bridges or culverts within sight of each other interacted socially at all, they were considered part of the same colony. Conversely, if different birds occupying the same bridge or culvert did not appear to ever interact socially, they were considered to represent separate colonies. Social interaction among birds was judged in a qualitative sense by observing, for example, whether individuals from different parts of the bridge or culvert tended to forage together or engaged in the same alarm response to a predator.

We found that cliff swallows living on the same bridge or culvert usually behaved as a single group, whereas birds on different bridges or culverts generally tended to forage separately and responded separately to predators. There were exceptions: one particularly long culvert underneath a four-lane interstate highway in some years had birds nesting in clusters at both ends but not in the middle, and the birds at each end acted independently of each other (perhaps because of the frequent and noisy automobile traffic on the interstate, which seemed to deter them from flying above the colony and may have prevented their hearing alarm calls given at the opposite end). In this case the birds at each end of the culvert were considered to be separate colonies. Another exception involved birds occupying parallel bridges on the same interstate, which often interacted substantially with each other, even though the bridges were physically separate. In this case the birds on each bridge were clearly part of the same colony in a social sense.

We used the same criterion of social interaction for birds at cliff nesting sites. If birds occupying a group of nests interacted socially with those from another group, they were considered to be part of the same colony, regardless of the distance between them along the cliff. If they did not interact socially, they were judged to be separate colonies. That cliff swallow colony sites usually were clearly isolated from the next nearest, plus the criterion of social interaction we applied in less obvious instances, enabled us to define colonies in what we regard as the most biologically appropriate way. The definition of colony used in this book does not differ from the one in our earlier cliff swallow papers (e.g., Brown and Brown 1986, 1987; Brown 1988a), although it is perhaps more clearly stated here.

What criteria did we use to select colony sites for study? Selection was based only on accessibility and proximity to the Cedar Point Biological Station, our base of operations. All colonies in our primary study area that permitted the safe checking of nests were studied, regardless of size. Similarly, all colonies in the primary study area where it was possible to catch birds were included in our mark-recapture sampling program (chapters 12, 13). Exclusion of sites in the primary study area was based strictly on their being inaccessible to us. Outside the primary study area, we sometimes selected particular sites for study, usually based on colony size, in most cases to "fill in" data for a colony size class that may not have been well represented or for other analyses that may have required a given colony size.

2.2 CHECKING CONTENTS AND STATUS OF NESTS

We observed nest contents with a dental mirror and penlight inserted through each nest's mud neck (fig. 2.1). Occasionally we had to chip small bits of dried mud off the neck to insert the mirror, but the birds quickly repaired any damage with fresh mud. Slight enlargements of nest entrances for research purposes had no effect on the birds' reproductive success in a study specifically designed to examine investigator-induced disturbance in cliff swallows (Hamilton and Martin 1985). We did not detect any adverse effects of nest entrance enlargements either.

We began checking the status of nests as soon as a colony site was first occupied by cliff swallows. Where new nests were being built, we charted the progress of a nest using categories denoting how much of the nest had been built (fig. 2.2; see Emlen 1954): categories included "partial bottom," meaning only a partial mud ledge had been constructed; "bottom,"

Fig. 2.1 The contents of cliff swallow nests were observed by positioning a dental mirror inside the nest's tubular entrance and shining a flashlight onto the mirror.

Fig. 2.2 Cliff swallow nests in various stages of completion, illustrating some of the categories we used to chart nest building progress: *a*, partial bottom; *b*, bottom; *c*, one-half present; *d*, no neck; *e*, neck incomplete; *f*, largely complete nest.

meaning a full nest bottom large enough for birds to sit in; "1/4, 1/3, 1/2, or 3/4 present," denoting the approximate amount of nest built; "no neck" or "neck incomplete," the stages immediately preceding nest completion; and finally "complete," denoting a nest with an apparently full neck (although birds would add fresh mud to necks of finished nests throughout the breeding season, even while feeding nestlings). Eggs were sometimes laid in nests that were only half complete, so we always checked for eggs in any nest that was half finished or more. Presence of fresh mud on a nest was noted, since this was an excellent sign that the nest was active and often the only indication before egg laying for nests that were intact and complete when first occupied in the spring.

Nest checks were done every other day at most colonies. At some colonies we checked nests daily during the peak egg laying periods to maximize the probability of detecting conspecific brood parasitism (section 6.3). In some instances weather or other logistical problems caused us to miss a check at a colony, leading to a two- or three-day interval between successive checks. We always checked nests after 08:00 mountain daylight time (MDT) to avoid disrupting normal egg laying patterns; cliff swallows appear to follow the usual hirundinid pattern of laying (at least nonparasitically) at dawn or in the early daylight hours (Brown 1984). To avoid undue disturbance and stress on the birds, we also did not check nests in cold or rainy weather. All nest checks at most sites could be accomplished in an hour or less, and we spent no longer than 1.5 hours per daily check at even the largest colonies. At colonies larger than 1,000 nests, we used two teams of observers to expedite nest checking and minimize disturbance, although in the large colonies birds habituated to us quickly and often continued normal activity a few nests away from where we were working.

Except for some large colonies primarily in the early years of the study, we usually checked all the nests in a colony. Where we did not check all the nests, we attempted to select sections at random and checked all nests within those sections. Nest or section selection was hardly ever completely random, however, because logistical problems such as high water in places could prevent access to certain parts of a colony.

Nest checks were continued throughout the birds' incubation period in order to determine, among other things, the hatching date for each nest. Knowing hatching dates allowed us to specify age of nestlings. We did not use hatching date to "backdate" to a laying date in nests with unknown clutch initiation times, because incubation periods in cliff swallows are variable, ranging from 12 to over 20 days (see Brown and Brown 1988a). If a nest had nestlings that were still wet and bright red or if eggshells and unhatched eggs remained, that day was considered the hatching date for the nest, and the nestlings that day were classified as 1 day old. If no eggshells or unhatched eggs remained, the nest was assumed to have hatched the preceding day (between the nest checks), and the nestlings were classified as 2 days old. Upon hatching, checking at a nest was terminated until the nestlings were 10 days old (section 10.4), at which time we weighed and banded them, examined them for ectoparasites (section 2.6), and scored how many nestlings remained. To prevent premature fledging, no further checking of a nest was done after the nestlings were 10 days old.

We reached nests on bridges and cliff faces with aluminum ladders and usually waded into culverts where nests were low enough to reach without ladders. Nests were marked by writing numbers with chalk on the nearby concrete substrate (in the cases of bridges and culverts) or by driving nails with numbered heads into the cliff face.

2.3 CAPTURING AND BANDING BIRDS

From 1982 to 1992 we captured 54,373 cliff swallows in our study area and banded them with U.S. Fish and Wildlife Service bands. The band number for each individual was unique, permitting us to follow that bird's history from year to year.

2.3.1 Nestlings

The vast majority of nestlings were banded at 10 days of age when they were weighed and checked for ectoparasites (section 2.6). A relatively few were banded at older ages—for example, when we collected blood samples for parentage studies (Brown and Brown 1988b). In 1990–92 other older nestlings were banded at colonies where we had not checked nests earlier and thus did not know exact hatching times. The rationale for banding these nestlings was simply to mark birds that would be of known age (in years) if recaptured in subsequent seasons. For each nestling banded at 10 days of age we recorded brood size, measured as the number of surviving nestlings in the nest at that time; number of ectoparasites present on its body (section 2.6); body mass measured to the nearest 0.5 g with a Pesola scale; and whether it had any morphological abnormality or wound on its body. Because data collected on the nestlings within a given nest were not independent, we calculated an average body mass or ectoparasite count per nest based on all the nestlings sampled in that nest. These means were used in all analyses of nestling body mass (e.g., section 10.4) and ectoparasitism (e.g., section 4.3.1). Other variables were also known for each nest where we banded nestlings: the date the nest became active, the dates egg laying and hatching began, the size and age of the nest (section 2.8), a measure of the nest's synchrony within the colony (section 2.9), the original clutch size, whether brood parasitism or egg loss attributed to conspecifics had occurred there, the nest's spatial position in the colony (section 2.8), and in some cases the identity (band numbers) of the adult nest owners.

To reach nestlings for banding we sometimes had to remove more of the nest's neck than was necessary when we simply checked contents (section 2.2). In the early years of the study, before we became adept at removing nestlings without breaking the nests' necks, we used fresh mud to repair any damage we did. The birds usually added to our repair jobs with mud of their own, and we could document no adverse effects of one-time neck breakage and temporary removal of nestlings. By 1987 we had become more skilled at reaching into nests and could usually extract nestlings with virtually no nest damage.

2.3.2 Adults

Adult cliff swallows were captured primarily in mist nets at the colony sites. We used two basic netting techniques: (1) stationary nets were set between poles in front of the entrances of culverts (as in Bullard 1963) or along an open side of a bridge; and (2) two persons would stretch a net between them (affixed to poles), carry the net to the top of a bridge or cliff, in unison drop the net over the side of the bridge or cliff, and then, after birds became entangled, raise the net and carry it a short distance away for processing (fig. 2.3; see Lueshen 1962). This latter "drop netting" technique was used at sites where the bridge's or cliff's height above the ground (or water depth) prevented us from setting a stationary net. In our study area more colonies had to be drop netted than could be netted with stationary nets. Birds were removed from a net and placed in holding bags

Fig. 2.3 At colony sites such as this culvert where deep water prevented our setting a mist net on one end, cliff swallows were caught by "drop netting." The mist net was carried onto the top of the culvert and quickly dropped over the side. The birds flushing from the culvert became entangled, after which the net was carried away for processing the birds.

until each was processed. Holding and handling time for each bird was generally short (5 to 15 minutes), although occasionally when large numbers of birds were caught at once, the last individuals to be processed had, unavoidably, been in the holding bag up to 60 minutes. Cliff swallows are quite tolerant of handling, and of over 100,000 birds processed during this study (counting recaptures of previously banded birds), fewer than 0.04% died as a result of capture and handling.

Some adult cliff swallows were also captured by plugging nest entrances at night with cotton, trapping the owners (which sleep there) inside (Brown and Brown 1988b). This technique required that we quietly enter a colony several hours after dark, plug the nests, and return at dawn to extract the nest occupants. Nest plugging was quite effective because both nest owners usually sleep in the nest, even as the nestlings approach fledging. The primary advantage of this method was that nest ownership of adults, and parentage (Brown and Brown 1988b), could be assigned, resulting in exact information on reproductive success for the adults caught that way. The disadvantages, however, were several: entering colonies at night was risky, because making an inadvertent noise could cause a major colony disturbance at night, when the birds seemed disoriented if flushed out of their nests; extracting adults at dawn was time consuming, and therefore only about thirty nests could be plugged at a time—if more were done, the last birds to be extracted the next morning would have waited inside their nest for 3 to 4 hours, too long a time; and owners sometimes deserted if nests were plugged before hatching. In contrast, we did not document any desertions by color-marked birds at known nests when these birds were caught in mist nets, regardless of nesting stage. Capture in a mist net is apparently less traumatic to cliff swallows than capture in a nest at night, and therefore we concentrated on mist netting to capture the bulk of the adults in this study.

All adults were banded and weighed to the nearest 0.5 g with a Pesola scale, and beginning in 1986 all were sexed (section 2.4). Some adults were also sampled for ectoparasites (section 2.6). These same data were recorded for each previously banded bird each time it was recaptured either in the year of banding or in subsequent years. The amount of information known for each adult varied, therefore, depending on how often it had been recaptured and whether it had been color-marked (section 2.5) and observed (or captured) at a nest.

Adult cliff swallows were caught at colonies at regular intervals throughout the season. Each time a bird was caught it was assigned to one of three periods that represented approximately its colony's stage: "early," the period when most of the birds in the colony were either building nests or laying eggs; "mid," when most were incubating; and "late," when most were feeding nestlings. Because of the high degree of synchrony within a colony (section 8.6.1), most colony residents were at the same nesting stage at any given time. Therefore these classifications based on the stage of the colony as a whole were probably accurate for most of the individuals (for which we did not know nest location within the colony and thus exact nest status). More detailed information on adult capture effort and sampling schedules at colonies is presented in sections 10.5.1 and 12.2.1.

2.3.3 Juveniles

Juvenile cliff swallows (birds that had fledged that season) often were caught during mist netting in the latter part of each summer. Juveniles were banded and weighed and in a few cases examined for ectoparasites using parasite-sampling jars (section 2.6). Most juveniles were caught at the same colony where they had hatched (especially the younger juveniles that clearly had been flying only a short time), although there was some evidence that older juveniles moved between colonies after fledging (section 6.4.2). Banded juveniles represented birds of known age (in years) when recaptured in subsequent seasons.

2.4 SEXING AND MEASURING BIRDS

Cliff swallows are sexually monomorphic in plumage, so it is a challenge to determine an individual's sex in the field. Adults captured through 1985 were not sexed at the time of banding, although the sex of some was determined from observations of copulation behavior at nests. Any bird originally banded in 1982–85 was sexed if it was recaptured anytime after 1985. All birds originally banded in 1986–92 were sexed at the time of banding.

We used a combination of brood patch and cloacal protuberance to sex cliff swallows. Breeding females show well-developed brood patches (Mayhew 1958). Development of the brood patch coincides with egg laying in most birds (Bailey 1952). We found that females continued to show brood patches throughout the breeding season in our study area; near the end of the season some began to grow feathers over the patch, but even in these cases the brood patch was obvious. A rule of thumb was that after 15 May any bird showing an obvious brood patch was sexed as a female.

Males exhibited a cloacal protuberance, an enlargement of the cloacal wall caused by the seminal sacs that swell during the breeding season (Drost 1938; Wolfson 1952; Lake 1981). Many males already seemed to show cloacal protuberances on their arrival in the spring and continued to exhibit them throughout the season, although they became noticeably smaller in July and August. Any bird with a cloacal protuberance at any time during the breeding season was sexed as a male. Enlarged brood patches and cloacal protuberances were not found together on the same bird, supporting the assumption that these were sex-specific characters. The major uncertainty in sexing occurred early in the season before about 15 May: birds without either a brood patch or an obvious cloacal protuberance sometimes were caught. In most cases we sexed these individuals as females.

Errors in sexing did occur, usually in the scoring of the cloacal protuberance. In some individuals the protuberance was not obvious immediately upon feeling the bird's cloaca, perhaps obscured by the lips of the cloaca, which themselves sometimes felt hard in females and may have led to occasional errors in sexing. Some males also exhibited bare areas on the lower belly that, though clearly not brood patches, occasionally contributed to sexing error when we and our assistants were processing large numbers of birds quickly. Errors were infrequent, however. We dissected 39 birds (21 males and 18 females) that died accidentally and had earlier been sexed based on presence of a brood patch or cloacal protuberance: only 4 of these 39 birds had been incorrectly sexed (10.2%). The four errors were females that we had incorrectly sexed as males. There was no significant difference among the sexes in accuracy ($\chi^2 = 0.85$, df $= 1$, $p = .36$). Therefore we conclude that the accuracy of our sex determinations of adult cliff swallows in the field was about 90%.

The accuracy of the sex determinations in our analyses was probably higher than 90%, however. Many individuals were caught more than once during a season or over their lifetimes. If an apparent error had been made on one out of a total of three or more sexing occasions, we used the sex reported most often. If the bird had been caught only twice with different sexings (or, for example, if it had been caught four times with sexings twice as a male and twice as a female), we did not include that individual in any analysis using sex. Thus we corrected many of the sexing errors through multiple captures, and if a sexing discrepancy was not clearly resolvable, we disregarded that individual.

We discovered that cliff swallows exhibit one slight sexual dimorphism in plumage. Birds of both sexes show dark blue feathering at the base of the throat. In some birds (of both sexes) this blue patch is very small and virtually the entire throat is dark chestnut. At the other extreme, some birds have an extensive blue patch that covers more than half the throat. Males consistently have more blue on the throat than females. This character alone was not sufficient to sex birds because there is substantial individual variation, but if we had a pair of birds suspected to be male and female (for example, having been caught sleeping together in a nest), the one with more blue invariably proved to be the male. We used this criterion to supplement our other sexing methods where possible. The variation in extent of blue is perhaps age related, although we did not systematically record extent of blue on birds of known age.

In 1989 we measured the unflattened wing chord of adult cliff swallows to determine whether body size differences existed among individuals in our population. Wing length is often used as an index of body size (e.g., James 1970; Behle 1976; Payne 1984; but see Rising and Somers 1989 and Freeman and Jackson 1990). There was little variation, and no significant difference, between males and females or between yearling birds and birds older than one year (ages known for birds banded as nestlings or juveniles). For one-year-old males ($N = 130$), mean wing length was 110.06 mm ($\pm$ 0.19); for one-year-old females ($N = 96$), 109.83 mm ($\pm$ 0.26); for males two years old and over ($N = 238$), 110.33 mm ($\pm$ 0.16); for females two years old and over ($N = 176$), 110.10 mm ($\pm$ 0.18). Although there was a suggestion that slightly larger birds might have been more likely to settle in smaller colonies (section 13.3.4), we found no other relationships involving wing length. Because the variation in our population seemed so low, we did not further analyze wing length.

2.5 COLOR-MARKING AND OBSERVING BIRDS AT NESTS

We color-marked adult cliff swallows by painting their white forehead patches in individually unique one-, two-, and three-color combinations (fig. 2.4). When cliff swallows are at their nests, often only the head is visible, making colored leg bands impractical. We used paint-dispensing

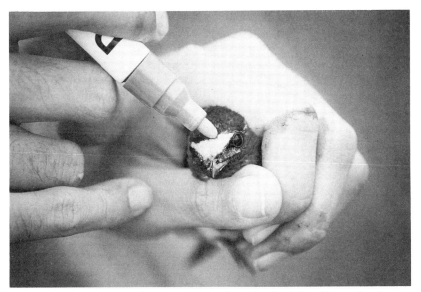

Fig. 2.4 Paint-dispensing marking pens were used to color the cliff swallows' white forehead patches in distinctive combinations. This bird's forehead patch was painted in vertical thirds with light orange, light blue, and light green. The shape of the forehead patch, and its contrast with the otherwise dark head and throat, remained unchanged by the color-marking.

marking pens, primarily the UniPaint and Decocolor brands. Colors we used were light blue, light green, silver, pink, orange, red, yellow, white (unpainted feathers), black, and occasionally brown and gold. These colors permitted enough unique combinations to color-mark as many birds (>500) as we desired in each colony. Except for black, the colors were light enough that the shape of a bird's forehead patch remained unchanged even when painted, and without binoculars it was difficult to notice that a bird had even been marked. We saw no evidence that other cliff swallows reacted differently to marked birds, or that marking changed a cliff swallow's behavior in any way, based on qualitative comparisons of the activities of marked and unmarked birds.

Most color-marking was done at several selected colonies near the beginning of the nesting season. Individuals were netted, marked, and subsequently observed to determine where in the colony they lived. As with other procedures done to them, color-marking seemed not to cause cliff swallows that already owned nests to abandon them, although marking possibly may have caused some swallows that did not own nests at the time to leave and settle elsewhere. Paint marks stayed fresh for about ten days and were identifiable on most birds for at least two weeks. In some cases we initiated a second round of capturing two to three weeks after the initial marking session to refurbish the markings.

Sections of colonies were designated for observation, and netting and capture of birds in those areas were emphasized in order to mark as many residents as possible. Sections chosen tended to be toward the edges of colonies, primarily because colonies in culverts were the most feasible for observation and it was difficult to see birds in the relatively dark interiors of culverts. Observers were positioned in chairs near culvert entrances, about 5–10 m from the nests. Cliff swallows quickly habituated to our presence and seemed to ignore us whenever we were sitting still, so blinds were not necessary. Each observer focused on twenty to forty-five contiguous nests and determined nest ownership by noting which color-marked birds were consistently associated with which nests. In most cases about 75% of the nest owners in the sections of colonies being observed were color-marked. We concentrated our observation during the nest building, egg laying, and early incubation periods, principally to document trespassing behavior (section 5.5) and conspecific brood parasitism (section 6.3; Brown 1984; Brown and Brown 1988a, 1989). During this time we observed the focal nests intensively, being present for approximately 75% of daylight hours. All interactions among color-marked nest owners and unmarked birds were recorded. Additional, less intensive observation was conducted at some colonies during nestling feeding periods to quantify the number of food deliveries by parents (section 10.2).

2.6 SAMPLING ECTOPARASITES

2.6.1 Swallow Bugs

Cimicid swallow bugs were sampled in two major ways: by collecting recently vacated cliff swallow nests and counting all bugs present in them, and by counting the number of bugs on the bodies of adult and nestling swallows. The latter method provided a relative index of bug parasitism among nests and colonies, while the former let us estimate the absolute level of parasitism per nest.

Counts of bugs in nests were taken two to seven days after nestling swallows had fledged. Only nests that had earlier contained nestlings were sampled. We removed these nests in their entirety from the substrate and placed them in plastic bags (fig. 2.5). If large portions of a nest were lost during collection, the nest was not used. At many nests, removal exposed dense aggregations of bugs that had wedged themselves between the nest and the substrate (fig. 2.5). Many of these bugs were not clinging to any nesting material and hence were not bagged; thus, at the time of nest collection we also estimated the number of bugs left in each nest's "scar"

Fig. 2.5 Populations of swallow bugs in cliff swallow colonies were sampled by collecting entire nests, bagging them, and later counting all the bugs present. The darker region on the substrate above the researcher's hand is a dense concentration of swallow bugs—probably several hundred—that was wedged between the back of the nest and the substrate.

upon the substrate. These estimates were included in each nest's total bug count. In most cases nests were left bagged for one to four days before bug counting commenced. Bugs were counted by placing each nest in a pan and sifting through the nest materials by hand, breaking up chunks of dried mud to expose bugs. Each nest took from fifteen to sixty minutes to count. There was no obvious bug mortality from bagging if nests were processed within four days of collection; bugs left bagged for longer periods began to die. Since cliff swallow nests persist from year to year and are often reused (section 4.10.3), we considered our removal of nests for bug sampling to be destructive and of potential influence on the birds' behavior in subsequent years. Therefore nest collection was necessarily limited in scope. The long time required to process a heavily infested nest also forced us to limit nest collection.

Bug counts on nestling cliff swallows were done when each nestling was 10 days old, at which time nestlings were removed from their nests, weighed, banded, and replaced (section 2.3.1). Nestlings were sparsely feathered at that age, and they could be visually examined for ectoparasites in less than one minute (fig. 2.6). Any parasite present anywhere on a bird's body was obvious. We did not remove parasites from the nestlings. The bugs present on the nestlings' bodies represented only a fraction of

28 • *Field Methods and Data Analysis*

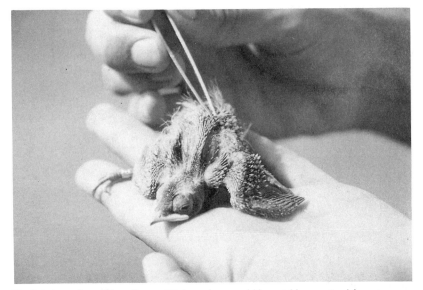

Fig. 2.6 A nestling cliff swallow at 10 days of age could be quickly examined for ectoparasites owing to its sparse feathering. All swallow bugs and fleas anywhere on the bird's body were easily seen and counted.

the total bugs in the nest, and we assumed that this fraction was constant among nests and colonies and therefore provided a relative measure of bug parasitism. This assumption was justified based on a sample of sixty-five nests collected (from eleven colonies) for which we had earlier measured parasitism on nestlings. The mean number of bugs counted per nestling in a nest correlated significantly with the nest's total parasite load (fig. 2.7). Using counts from nestlings enabled us to measure relative degree of bug parasitism for thousands of nests and nestlings over the course of the study without destroying nests.

Ectoparasitism of adult cliff swallows was assessed using ectoparasite-sampling jars, as described by Wheeler and Threlfall (1986). A bird was placed in a wide-mouthed jar with its head protruding from a hole in the lid; the bird's head rested on a rubber collar attached to the jar lid, and the rest of its body was suspended inside the jar (fig. 2.8). Several drops of ether were placed on filter paper at the bottom of the jar. The ether killed the parasites on the bird's body (except the head) or in its feathers, causing them to fall off the bird into the jar, whereupon they were counted and removed. This technique was quite effective in sampling swallow bugs, fleas (section 2.6.2), and lice (section 4.2.3). When we knew one or more swallow bugs were present on an adult—for example, when we saw them attached to a bird's brood patch—the bugs fell off within two minutes of

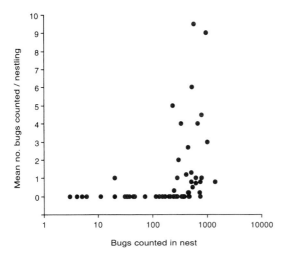

Fig. 2.7 Mean number of swallow bugs counted per nestling at 10 days of age for each cliff swallow nest as a function of the total number of bugs in the nest as determined from nest collection. There was a significant correlation between the number of bugs counted on nestlings and a nest's total bug load ($r_s = .62$, $p < .001$, $N = 65$ nests).

Fig. 2.8 Ectoparasites on the bodies of adult cliff swallows were sampled by suspending the birds in jars containing ether, which killed the parasites hidden in the feathers. The parasites fell off the birds and were easily counted on the filter paper at the bottom of the jar. Cliff swallows seemed unaffected by the procedure, which required each bird to remain in the jar for twenty minutes.

the bird's being placed in the jar. However, each bird was kept in the jar for a full twenty minutes (Wheeler and Threlfall 1986). No birds were injured or killed by this procedure, and it seemed to have no effect on their behavior. Some individuals were recaptured within an hour of their release from a jar. We did not sacrifice birds to see how many (if any) parasites remained on them after they were in the jar, but Fowler and Cohen (1983) reported that this method removed 80–90% of the total ectoparasites known to be present on other species.

More recently we have experimented with ways of counting swallow bugs on the outsides of nests without removing the nests from the substrate. Bugs clustered at the nest entrance or along the bottom of the nest are highly visible and can be counted at certain times of the year, depending on whether a nest is occupied by cliff swallows. We have used data collected in this fashion for a few analyses (e.g., section 4.9.1), although we are still evaluating these sampling methods. The advantage of counting bugs on the outsides of nests is that the method is not destructive, and the seasonal pattern of bug parasitism can be charted for the same nests. The disadvantage is that the method, like the counts on nestlings, is only a relative index of parasitism, because not all bugs present in a nest are on the outside of it at any given instant.

2.6.2 Fleas

Ceratophyllid bird fleas are much more mobile than swallow bugs, and consequently they were more difficult to sample. Total fleas in a nest could not be determined by collecting nests, because some jumped out of the nest during collection and many jumped out of the pan during counting. We thus resorted to relative measures to estimate flea parasitism among nests and colonies.

Fleas on adult cliff swallows were sampled using the ectoparasite jars described in section 2.6.1. Most fleas seemed to fall off adults during the first five minutes of sampling. We could not, of course, be sure that all fleas on a bird fell off in the jar, although we suspect that some, though killed by the ether, may have become wedged in the bird's feathers. Flea counts were taken from nestlings at 10 days of age at the same time we counted swallow bugs. Once on a nestling, fleas seemed reluctant to jump off it even when we handled the bird; when disturbed they tended to try to hide in the few feathers a 10-day-old nestling possessed.

Fleas were more likely to jump from nests when disturbed, and they often jumped on us when we were near nest entrances. We used their tendency to jump from nests to sample the number of fleas clustered at entrances of unoccupied nests. Our method was that adapted by Hopla and Loye (1983) from Humphries (1969). A black card coated with honey and mounted on a stick was placed about 5 cm from a nest's entrance (fig. 2.9). The card mimicked a bird blocking the entryway and stimulated the nega-

Fig. 2.9 Fleas clustering at the entrances of cliff swallow nests could be sampled by placing a card coated with honey in front of the entrance. The card simulated a bird's presence at the entrance and caused the fleas to jump from the nest onto the card.

tively phototaxic fleas to jump onto it, where they were trapped by the honey and could be counted. A card was held in front of each nest for twenty seconds, time enough for large numbers of fleas to leap onto it. Visual inspection of nests before and after sampling indicated that virtually all the fleas at the entrance leaped onto the card, but we could not know if there were other fleas inside the nest. Thus the card technique was considered a relative means of sampling. Card sampling could be used only early in the season before birds had arrived at colonies or at sites unused in a given year. Once a site was occupied by cliff swallows, the fleas present at nest entrances disappeared, presumably having jumped onto birds that either occupied the nests or passed nearby while investigating the nests.

2.7   FUMIGATION

Swallow bugs were experimentally removed from cliff swallow nests by application of an insecticide. We used a short-lived 1,2-dibromo-2,2-

Fig. 2.10 A short-lived acaricide (naled) was used to fumigate cliff swallow colonies. Using a hand sprayer, we lightly but thoroughly applied the fumigant as a fine mist to the outsides of the nests and the surrounding unoccupied substrate (concrete wall) up to about 30 cm from the active nests. We tried to avoid getting any spray into the interior of the nests.

dichloroethyl dimethyl phosphate acaricide, called naled (also known as Dibrom, manufactured by the Chevron Chemical Company). This insecticide successfully removed ectoparasites without harming avian hosts in a study of purple martins (Moss and Camin 1970) and in an earlier study of cliff swallows (Chapman and George 1991). We never detected any adverse effects on cliff swallows. Naled was diluted 1 part to 170 parts water (by volume) (Chapman and George 1991). We used hand sprayers to apply it as a fine mist to the outside of swallow nests and the surrounding substrate (fig. 2.10). Naled works primarily as a contact pesticide and has only a short residual fumigant action (Berg 1981). However, for convenience we use the term fumigation throughout this book in referring to the application of naled to nests.

For studies of the effect of swallow bugs on nestling swallow growth and survivorship (section 4.7; also Brown and Brown 1986), we fumigated nests daily or every other day, beginning as soon as birds established ownership of nests and continuing until nestlings fledged or approached fledging. In other studies (e.g., nest and colony use patterns in section 4.10, the colony reduction experiment in section 10.7), we fumigated colonies once or twice a week throughout the season.

To assess the fumigant's effectiveness, in 1984 in one of the colonies most heavily infested by swallow bugs, we collected 17 fumigated nests after nestlings fledged and compared their swallow bug counts to those from 50 nonfumigated nests in the same colony collected at the same time. Fumigated nests contained 22.76 (± 1.22) bugs per nest, whereas nonfumigated nests contained 707.80 (± 67.30) bugs per nest. Naled thus was effective against swallow bugs, and we suspect that the only bugs surviving in fumigated nests were wedged in crevices between nests where the insecticide may not have penetrated.

In cases where we compared fumigated and nonfumigated nests within the same colony (e.g., section 4.7), a colony was arbitrarily divided in half, and the "control" half of each colony was not fumigated. A sticky insect barrier known as Tree Tanglefoot was applied to the substrate between the two halves to prevent bugs from leaving the fumigated nests and entering the nonfumigated ones (fig. 2.11). In addition, all fumigated nests were surrounded by Tanglefoot applied to the substrate to prevent new bugs from immigrating (along the substrate) into the fumigated nests over the course of the season.

The fumigant's effect on fleas was unclear but probably minor. Initially

Fig. 2.11 Fumigated cliff swallow nests were surrounded by a sticky insect barrier applied to the concrete wall of the culvert, preventing any swallow bugs from entering or exiting along the substrate. Some bugs, however, may have entered or left the fumigated nests on the bodies of the adult swallows, and for this reason we were forced to fumigate nests repeatedly throughout the breeding season. Where possible, the insect barrier was applied at least 30 cm from the nests, to reduce the chances that cliff swallows who were prospecting for nesting sites near existing nests would alight on the sticky barrier and soil their feathers.

we detected no consistently significant differences in flea counts on nestlings from fumigated versus nonfumigated nests (table 6 in Brown 1985a). Subsequent analyses indicated a slight but consistent reduction in flea counts from fumigated nests, although the only direct within-colony comparisons were the ones done at six colonies in 1984 and reported in Brown (1985a). It seems possible that fumigation reduces flea numbers slightly. Thus we decided in general to present analyses using flea counts on nestlings mostly from nonfumigated nests, and we clearly note the instances where we used flea counts from fumigated nests.

Throughout this book, whenever we refer to a fumigated or a nonfumigated colony our reference is to sites where *nests* were fumigated or nonfumigated. Although in some cases ectoparasites were sampled from adult birds (sections 2.6, 4.8) and removed from them in the process, relatively few birds and colonies were involved in the adult sampling. Sites where ectoparasites were removed from adult birds were not considered to be fumigated colony sites (unless nests were fumigated) because the large populations of nest-based ectoparasites (e.g., section 4.3) were unaffected by the sampling of ectoparasites on adult cliff swallows.

2.8 MEASURING DENSITIES, POSITIONS, AND AGES OF NESTS

2.8.1 Nest Density

We calculated the nest density of each cliff swallow colony (also referred to as the colony density) by first measuring the total available substrate that appeared suitable for nest attachment. We considered the active nests closest to the colony edges to mark the outer limits of suitable substrate. Within that span we measured the total horizontal and vertical substrate that appeared suitable for nest attachment. This was most unambiguously done for bridges and culverts, where total concrete area could be calculated based on the number of beams and the vertical depth. These measurements allowed us to assign each colony site a total substrate size.

The number of cliff swallow nests on this expanse of substrate was determined, and we expressed density as nests per meter. Because most of our colonies were roughly linear, that is, single rows of nests with little vertical stacking (e.g., fig. 3.6a,d), nests per meter seemed to be the most realistic expression of nest density.

2.8.2 Nest Position

The positions of all active nests in each colony were mapped at the end of the nesting season. Relative nest locations were drawn on paper, and overlapping series of photographs at some colonies provided further documentation of nest positions. We measured in the field distances (in cm) beween all active nests within a colony. All nest-to-nest measurements were made between the centers of the nests' entrances. Because most colonies were linear, it was easy to designate a centermost nest (or two of them) with an equal number of neighbors on either side. For the few colonies that were less linear and more "honeycombed" (e.g., fig. 3.6b), the active nest(s) with an equal number of nests on all sides (regardless of position with respect to the geometric center of the colony's substrate) was considered the centermost. We directly measured or calculated each active nest's linear distance from the centermost nest, and we refer to this as a nest's distance from center. The measurements and maps allowed us to determine each nest's nearest neighbor distance—the shortest linear distance to a neighboring active nest. We also measured each nest's diameter as an index of total nest size. Nest diameter was measured by holding a meter stick in front of the nest at its widest point and taking the linear distance across the nest at that point.

In comparisons of center versus edge nests, edge nests were considered to be the ten sequentially placed active nests, beginning with the edgemost nest and moving inward. Center nests were the ten sequentially placed active nests, beginning with the centermost nest and moving outward. Center-edge comparisons were done only in colonies where nest array and colony size would yield roughly an equal number of edge and center nests on all sides of the colony's geometric center. This meant that most center-edge analyses were restricted to colonies that were roughly linear. For culvert colonies that were split in half for fumigation, the center nests were considered to be the innermost fumigated and nonfumigated nests, that is, the two nests that were separated by the Tanglefoot dividing line between the halves of the colony.

2.8.3 Nest Age

Each cliff swallow nest was classified as new or old. New nests were ones built in their entirety (that is, no previous nest remnant had been present at that location on the substrate) in a given year, or any nest built from an existing remnant that upon becoming active (known by deposition of fresh mud, section 2.2) was approximately half or less complete. Old nests were ones existing at the start of the nesting season that were more than half complete when becoming active. A nest was not included in any analysis of nest age if we did not know its exact age at the time it became active.

2.9 MEASURING WITHIN-COLONY BREEDING SYNCHRONY

The temporal position of each nest with respect to others in a colony is a measure of a nest's synchrony. To measure relative temporal position, we first ranked the clutch initiation dates (that is, date of first laying in a nest) of all nests in the colony. We determined the standard deviation of clutch initiation dates for the colony and the modal clutch initiation date. Each nest was then assigned, based on its clutch initiation date, to the appropriate number of standard deviations on either side of the modal date. For example, if the modal clutch initiation date in a colony was 1 June and the standard deviation of clutch initiation date was two days, a nest begun on 1 June would be classified as 0 (modal); one begun on 28 May would be classified as -2 (within two standard deviations before the mode); and one begun on 7 June would be classified as $+3$ (within three standard deviations after the mode).

This method allowed us to compare relative within-colony laying synchrony among nests in all colonies irrespective of how synchronous a site actually was. This let us pool data from different colonies, since all were measured on the same scale. For some analyses, we also computed each nest's within-colony hatching synchrony, measured the same way but with respect to the date a clutch hatched. Laying synchrony and hatching synchrony were the same for many nests. However, in colonies where substantial numbers of nests failed after egg laying but before hatching—especially if these failures were mostly among early or late nesters—a nest's relative temporal position during the nestling period could be very different from its position earlier during laying.

2.10 DATA ANALYSIS

For clarity we have kept the statistical analyses in the following chapters relatively simple and straightforward. We mostly use univariate tests, in some instances where others might apply various multivariate analyses. However, multivariate tests in ecology often are used incorrectly, seldom allow accurate inferences about causation, and when used correctly should be primarily of a preliminary, exploratory nature (James and McCulloch 1990). For these reasons we prefer a more conservative statistical approach emphasizing univariate analyses.

In keeping with this philosophy, we also have used mostly nonparametric statistical tests. Parametric tests require several assumptions about the data set's distribution and variance(s), and seldom are all of these assumptions valid (Potvin and Roff 1993) or directly tested before parametric tests are applied (Siegel 1956). By using nonparametric tests, we have avoided the potential problems of unmet assumptions. The commonly cited disadvantage of nonparametric tests—reduced power relative to parametric methods—can be overcome with larger sample sizes (Siegel 1956), and sample sizes for most analyses in this book are relatively large. Nonparametric methods often are more robust than their parametric counterparts, especially when the data contain outliers or are not normally distributed (Potvin and Roff 1993).

Readers may occasionally notice apparent discrepancies in sample sizes among different but related analyses. These instances resulted because complete information was not always available for each observation. Some analyses did not require complete information, permitting the use of additional data, whereas other analyses were more restrictive and required information that was not available for all observations. Statistical significance was set at $p = .050$, but in most cases we present exact probability values so that readers may judge the "significance" of any differences.

Various analyses require classifying data into categories. The most prominent example is where we establish colony size classes (e.g., 1–10 nests, 11–50 nests, 51–99 nests, etc.) in which data from several colonies of similar size are pooled whenever too few data are available, or there are too many total colonies, to consider each colony separately. These categories are arbitrary, but whenever used they are based on a biological intuition of what is reasonable gained from our twelve years of experience with cliff swallows. We justify these categories as far as possible, and in each case we experimented with slightly different groupings during preliminary analyses. In most instances, slight changes in category limits had little or no effect on the conclusions. Where such changes did affect the conclusions, either we so state or we elected not to use categories at all.

In some analyses we state which colony sites were used in collecting the relevant data. Our convention for naming sites is to use a four-digit number, the first two digits indicating the year and the last two digits the site. For example, colonies 8705 and 8805 were situated at the same physical site (05) in 1987 and 1988. When used, colony name thus also indicates the year the data were collected. We consider colonies active at the same site but in different years to be statistically independent.

Finally, we have made particular effort to avoid the sin of "pseudoreplication" (Hurlbert 1984). Pseudoreplication occurs in part whenever the actual physical space over which samples are taken is smaller or more restricted than the inference space implicit in the hypothesis being tested, and it is a common problem in ecological field studies (Hurlbert 1984). One way to avoid pseudoreplication is by clearly establishing the experimental unit in question. For example, if the issue is how adult body mass varies among colony sites of different sizes, the experimental unit is the colony site, not individuals within each colony, and the inference space is all the colony sites. The statistical test should be based only on the number of colony sites, with the individuals in each colony contributing to a single measure of body mass for their own colony site. Thus, in instances like this we averaged the data for all individuals within the colony and report a single mean value for each colony. Statistical tests are computed from the mean values for each colony, and therefore the effective sample size for statistical purposes is the number of colonies. (This weights large and small colonies equally in the analysis, although in most cases the mean values for large colonies are based on larger sample sizes than those for small colonies. For this reason, in most cases we also present standard errors of the means, to allow evaluation of the robustness of the mean values used in the statistical tests.) At other times, however, one might be interested in within-colony effects, in which case the experimental unit is each bird or nest and the inference space only the birds or nests within that colony. In these cases the effective sample size for statistical purposes would be the total number of birds or nests within the colony.

We believe it is important to specify exactly the sample size(s) each analysis is based on, so that readers may judge for themselves the strength of our conclusions. We indicate sample size in all tables and text presentations and in most figures. On a relatively few scatterplots, however, space did not permit sample size labeling. In scatterplots involving colony size as the independent variable, we generally used a logarithmic scale for colony size. This was done primarily for clarity, because linear plots compressed the representations of the smaller colonies (<100 nests) so severely that they could not be distinguished from each other. There were also biological reasons for our using a logarithmic scale and thus emphasizing any differences among the smaller colonies. For example, colonies of 2 nests were far more different, socially, from colonies of 75 nests than were colonies of 400 nests from colonies of 600 nests. Values reported in the text, tables, and figures are means ($\pm$ 1 SE) unless otherwise noted; where annual survival probabilities are presented (e.g., chapter 12), $\pm$ 1 SE is also given. Scientific names of organisms mentioned in the text are listed in the appendix.

2.11 SUMMARY

Colony size refers to the number of active cliff swallow nests and was determined by examining nest contents and directly counting active nests or by estimating the number of active nests based on portions of colonies examined or number of birds present. Our periodic visits to colony sites throughout the study area provided information on site use and size in a given year. A cliff swallow colony is defined as a group of nests, generally isolated on a single structure, in which the nest owners at least occasionally interacted while foraging or mobbing predators. Nests were checked regularly to determine when they became active and later to collect information on breeding phenology and reproductive success.

Adult, juvenile, and nestling cliff swallows were captured and banded, permitting us to follow the histories of individuals from year to year. Sex of cliff swallows was determined by whether individuals possessed a cloacal protuberance (males) or brood patch (females), a method that was about 90% accurate in the field. Cliff swallows were color-marked by painting their white forehead patches in distinctive color combinations, allowing us to observe individuals intensively and assign nest ownership by a bird's repeated presence at a nest.

Ectoparasitism by swallow bugs and fleas was assessed by counting the number of parasites on the bodies of 10-day-old nestlings and by the number falling off adults when the birds were placed in specially designed sampling jars. Bugs were also sampled by collecting entire cliff swallow nests and counting all bugs present, and fleas were sampled by holding a black card coated with honey near nest entrances to trap fleas that leaped onto it. Swallow bugs were experimentally removed from cliff swallow nests by regularly spraying nests with a dilute solution of the insecticide naled. Nest positions within colonies were mapped, and cliff swallow nests were classified as either new or old depending on how much of the nest was extant at the beginning of the breeding season. Within-colony breeding synchrony was measured in terms of a nest's clutch initiation date relative to the modal clutch initiation date for that colony. Data in this book tend to be analyzed with univariate, nonparametric statistical methods.

3 Study Site and Study Population

> In the summer-time colonies and colonies of [cliff] swallows come to build their nests under the jutting shelves of the bluffs. In some places we observed hundreds of their little mud edifices clustered together, and as we went past them the birds issued forth in clouds.
> J. Henry Carleton, 8 June 1845
> near Ash Hollow, Nebraska
> (Carleton 1943)

Our study site was centered in Keith County, Nebraska, on the southern edge of the Nebraska Sand Hills and within the Platte River Valley near the University of Nebraska's Cedar Point Biological Station. In this chapter we describe the major physiographic characteristics and climatic patterns of the region; we assess the cliff swallow's historical occurrence in the study area and its current distribution; we describe the types and locations of colonies; and we present a descriptive overview of the cliff swallow population found in the study area.

3.1 THE PLATTE RIVER VALLEY

The study site (elevation approximately 1,000 m above sea level) comprised an area approximately 150 × 50 km in southwestern Nebraska lying mostly in the floodplain of the North and South Platte Rivers (fig. 3.1). Most of the cliff swallow colonies were associated with one of these rivers in some way (section 3.3). Both rivers originate in Colorado as snow runoff from the Rocky Mountains. The drainage systems of the North and South Platte were probably established by the end of the Pliocene and their present courses during the Illinoian glaciation (about 550,000 years ago) (Johnsgard 1984). Each cuts a narrow valley through the area, the South Platte now about 70 m below the surrounding prairie tableland and the North Platte about 60 m below (as measured at Ogallala and calculated from Swinehart and Diffendal 1989). The slope of the South Platte Valley is relatively gentle, and within the study area it has no abrupt outcroppings or cliffs that at present provide any natural nesting sites for cliff swallows. The North Platte Valley has steep bluffs and rocky outcroppings along its southern side (fig. 3.2), presumably caused by erosion as the river established its present course (Layton and Buckhannan 1926), which cliff swallows use as nesting sites. These sandstone outcrop-

The Platte River Valley • 41

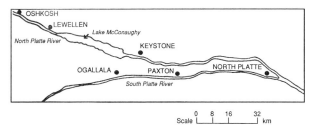

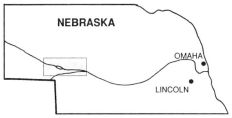

Fig. 3.1 Location of study site (*shaded box*) in southwestern Nebraska, U.S.A. Inset illustrates entire study area.

Fig. 3.2 Southern side of the North Platte River Valley in southwestern Nebraska, showing the bluffs and rocky outcroppings that historically served as nesting sites for cliff swallows (and still do for fewer birds). This view is northward along a canyon near the Ash Hollow area.

pings extend in a patchy distribution from near Keystone westward beyond the western boundary of the study area to at least the Wildcat Hills region southwest of Scottsbluff, Nebraska, a distance of approximately 250 km. The outcroppings generally face north and often support dense stands of cedars. The north side of the North Platte Valley contains fewer and smaller outcroppings that cliff swallows are not known to use as nesting sites.

In the 1930s Kingsley Dam was constructed (completed in 1941) on the North Platte River north of Ogallala, creating the 14,280 ha Lake McConaughy. The reservoir has hastened erosion along the south side of the North Platte Valley, probably increasing the number of outcroppings and cliffs along its southern shore. These outcroppings are often short lived, with large portions of them periodically crumbling, especially the overhangs cliff swallows nest under.

The flora of the Platte River Valley (and adjacent Sand Hills) contains elements characteristic of tallgrass prairies and deciduous woodlands to the east and shortgrass prairies and the Rocky Mountains to the west (Kaul, Challaiah, and Keeler 1983; Bleed 1989). The floodplain and valley walls support most of the trees that can be found in the region. Some of the more common species are eastern cottonwood, green ash, boxelder, hackberry, eastern red cedar, chokecherry, peachleaf willow, silver maple, Russian olive, and American elm (Kaul 1989; Sutherland and Rolfsmeier 1989). Land adjacent to the rivers was originally mostly shortgrass prairie, shading to mixed prairie in some areas, and much of it now tends to be used for farming and ranching. Representative flora found in open areas over which cliff swallows forage include *Amaranthus, Aster, Solidago, Xanthium, Lithospermum, Chorispora, Opuntia, Euphorbia, Astragalus, Dalea, Plantago, Penstemon, Bromus, Andropogon, Bouteloua, Buchloe, Stipa, Calamovilfa, Panicum,* and *Yucca* (Kaul, Challaiah, and Keeler 1983; Sutherland and Rolfsmeier 1989). The Platte River system in western Nebraska represents a hybrid zone for a variety of east-west species pairs of birds (e.g., Moore 1977; Rising 1983), and hybrid orioles, flickers, and buntings are sometimes seen in the study area.

3.2 CLIMATE

3.2.1 General Patterns

Climate is one physiographic factor likely to have a great impact on insectivorous birds such as cliff swallows. Short-term climatic changes can substantially affect the distribution and abundance of the aerial insects these birds feed on (Johnson 1969), and therefore climate may have a major influence on the birds' reproductive success, survival, and behavior. Swallows in general, including cliff swallows, are vulnerable to cold and rainy periods, which can cause widespread mortality (Kimball 1889; Bent 1942; Benton and Tucker 1968; Skead and Skead 1970; P. Stewart 1972; R. Stewart 1972; Brown 1976; DuBowy and Moore 1985; Krapu 1986; Elkins 1988; Littrell 1992).

The climatic patterns of our study area are those characteristic of the central Great Plains (Layton and Buckhannan 1926; Wilhite and Hubbard 1989). Precipitation in the study area averages 43–51 cm annually, with approximately 75% occurring during April–September and 50% during May–July (the cliff swallow's breeding season) (Wilhite and Hubbard 1989). The rainfall in spring and early summer is mainly the result of relatively frequent low-pressure disturbances that track across the study area, usually shifting farther to the north as the summer progresses. In July and August much of the rainfall comes from localized thunderstorms, and consequently rainfall amounts at that time may differ substantially over short distances (Wilhite and Hubbard 1989). Winter precipitation is usually snow. Annual and monthly precipitation in the study area, and in the entire state of Nebraska, is highly variable. At one Sand Hills site near Valentine, the annual range is from 73.43 cm in 1929 to 26.85 cm in 1974 (Wilhite and Hubbard 1989).

Average annual temperature in our study area averages about 9°C, with mean summer temperatures generally about 19–23°C. Daily maximum temperatures occasionally exceed 37.5°C (100°F) during July and early August. The average number of days per year with temperatures exceeding 26.7°C (80°F) in our study area is between 90 and 100 (Wilhite and Hubbard 1989). Wind patterns vary considerably because of the effect of local terrain features, but a few generalizations are possible. Winds are controlled to a large extent by the passage of frontal systems. Winds are generally southerly before the fronts pass and from the northwest afterward (Wilhite and Hubbard 1989). Data from North Platte, situated within the study area, show that over the year winds from the north and northwest are most common and tend to be consistently the strongest; westerly, southerly, and southeasterly winds are the next most common; and easterly, northeasterly, and southwesterly winds are least common (Wilhite and Hubbard 1989). Prevailing winds during the summer tend to have southerly components.

3.2.2 Temperature and Rainfall, 1982–93

In this section we present data on regional mean and maximum temperatures and rainfall collected each year of our study. These data let us evaluate to what degree yearly differences (if any) in cliff swallow biology reported in subsequent chapters may be attributed to climatic changes between years.

Meteorological data were collected on a statewide computer network (the Automated Weather Data Network) operated by the High Plains Climate Center at the University of Nebraska–Lincoln. Each reporting site within the state measured hourly and daily precipitation, air temperature, humidity, soil temperature, solar radiation, and wind speed and direction.

Table 3.1 Semimonthly Mean and High Temperatures (°C) Taken at Arthur, Nebraska, during the Years of the Study

	1982	1983	1984	1985	1986
1–15 April					
Mean	5.4	0.2	4.4	9.1	5.3
High	13.3	5.1	8.7	17.7	11.4
16–30 April					
Mean	5.9	8.7	5.3	12.1	10.0
High	11.9	14.7	11.1	18.2	17.0
1–15 May					
Mean	11.7	9.2	11.6	15.2	11.4
High	17.6	15.3	18.0	22.0	20.0
16–31 May					
Mean	13.5	12.3	14.9	16.6	10.6
High	19.3	19.2	22.2	24.4	18.2
1–15 June					
Mean	13.8	15.4	16.4	17.2	17.3
High	19.7	21.3	23.1	24.5	25.9
16–30 June					
Mean	19.3	19.8	20.8	19.3	22.2
High	25.1	25.3	29.0	28.2	30.7
1–15 July					
Mean	22.1	22.1	21.8	24.9	21.4
High	28.9	30.7	28.8	33.9	30.1
16–31 July					
Mean	24.4	23.0	22.9	19.7	21.2
High	32.3	29.3	30.4	27.9	29.4
1–15 August					
Mean	22.4	24.5	23.8	18.6	21.9
High	29.0	31.7	30.9	26.6	29.2
16–31 August					
Mean	21.8	24.7	23.1	19.6	20.6
High	28.1	31.8	30.7	28.1	27.9
Average					
Mean	16.0	16.0	16.5	17.2	16.2
High	22.5	22.4	23.3	25.2	24.0

Note: Daily means and highs were averaged for each date interval.

The reporting station we used was at Arthur, in Arthur County, about 48 km directly north of the center of the study area. Although another reporting station was available at North Platte near the eastern edge of the study area (fig. 3.1), we chose to use the one at Arthur because north-south gradients in climate are less pronounced than east-west gradients in the general region of the Sand Hills (Wilhite and Hubbard 1989). Data collected at Arthur should be representative of climatic conditions in our study area, especially for yearly comparisons. Even hourly correlations were strong between weather reported at Arthur and how cliff swallows

1987	1988	1989	1990	1991	1992	1993
4.9	8.7	5.3	5.4	9.1	8.9	3.5
11.9	16.3	13.2	11.4	16.4	16.3	7.7
15.9	8.3	12.7	10.2	6.8	10.1	9.3
25.1	15.5	20.5	16.8	11.6	17.9	16.7
16.5	13.1	12.1	9.7	12.7	16.3	13.2
22.8	20.9	19.7	16.2	19.5	25.2	20.1
15.1	16.4	15.8	14.7	16.3	12.3	14.4
21.2	22.3	23.4	20.7	20.9	19.3	21.0
19.9	20.6	15.8	18.7	18.9	15.9	15.4
27.4	27.7	22.7	25.5	24.9	21.9	21.0
20.1	24.7	19.7	22.3	21.2	19.2	18.4
27.6	32.7	27.4	30.3	27.1	25.1	25.3
20.6	22.8	25.1	21.7	21.3	19.2	19.7
28.1	29.8	33.2	28.9	28.9	25.7	25.3
25.2	22.2	21.6	20.5	23.2	18.7	20.9
27.4	29.8	28.9	27.6	31.1	25.1	26.8
22.5	23.2	21.4	20.3	20.6	20.5	20.2
29.6	31.0	27.9	27.5	27.2	27.8	26.1
17.9	21.0	21.1	23.9	24.0	17.4	19.9
24.6	29.0	27.2	32.3	32.6	24.2	25.5
17.8	18.1	17.1	16.7	17.4	15.9	15.5
24.6	25.5	24.4	23.7	24.1	22.8	21.5

might be expected to forage within the study area near Keystone (Brown, Brown, and Shaffer 1991).

Semimonthly temperature profiles during April to August (table 3.1) reveal that 1988 was the warmest year of the study. Daily maximum temperatures that year during the second half of June (the time when most cliff swallows were feeding nestlings) were over 5°C warmer than the overall average for the other years. June 1988 for the state of Nebraska as a whole was the fourth warmest June on record, dating to 1876 (when data collection began; M. Werner, pers. comm.). The coolest years were 1982,

Table 3.2 Semimonthly Total Rainfall (cm) Measured at Arthur, Nebraska, during the Years of the Study

	1982	1983	1984	1985	1986
1–15 April	0.00	0.00	1.02	0.00	0.00
16–30 April	0.00	2.21	5.89	0.00	0.00
1–15 May	8.43	4.95	2.72	0.00	3.12
16–31 May	2.92	3.45	3.73	0.51	4.47
1–15 June	3.33	2.44	2.74	1.12	2.64
16–30 June	0.00	0.00	4.52	1.02	0.41
1–15 July	1.12	2.62	3.20	0.00	0.41
16–31 July	4.80	5.00	2.34	2.13	0.20
1–15 August	1.52	0.41	0.61	0.71	0.41
16–31 August	2.01	0.61	0.00	2.21	0.10
Total	24.13	21.69	26.77	7.70	11.76

1983, 1992, and 1993. In particular, 1992 was perhaps the most anomalous year, climatically, during the study (table 3.1). Early May 1992 was extremely warm, being about 6°C warmer than the overall average of the other years, but the rest of the 1992 season was much cooler than normal, especially the month of July (table 3.1). For the state as a whole, July 1992 was the coldest on record (M. Werner, pers. comm.). Not surprisingly, the most widespread weather-related cliff swallow mortality we observed occurred in 1992 (section 10.6). April and early May high temperatures were the most variable from year to year (table 3.1), meaning that upon their arrival in the study area many cliff swallows faced several weeks of unpredictable weather.

Semimonthly rainfall totals for April–August (table 3.2) illustrate the extreme annual variability characteristic of the Nebraska Sand Hills and the Platte River Valley (also see Layton and Buckhannan 1926). The wettest year during the study was 1993, and the driest years were 1985 and 1986, averaging only about 40% of the rainfall of the other years. Again 1992 was an extreme year, being the wettest during the critical month of June (table 3.2). Rainfall presumably affects cliff swallows in a variety of ways: directly by preventing nest building and foraging when rain is heavy and exacerbating the physiological effects of cold temperatures when food is scarce, and indirectly by affecting reproduction and abundance of the aerial insects the birds feed on.

3.3 HISTORICAL OCCURRENCE AND CURRENT DISTRIBUTION OF CLIFF SWALLOWS

The first published reports of cliff swallows in the study area date from the mid-nineteenth century. Two naturalists associated with the United States Army, J. Henry Carleton in 1845 and George Suckley in 1859, reported cliff swallows at or near Ash Hollow, near present-day Lewellen. Carleton

1987	1988	1989	1990	1991	1992	1993
0.00	0.00	0.00	1.83	0.30	0.00	2.64
0.51	1.83	0.20	3.63	1.93	0.10	2.54
0.00	0.00	0.81	6.20	4.04	1.09	3.15
0.00	0.00	4.37	7.95	4.24	1.62	1.42
5.05	3.00	1.62	3.15	1.52	6.83	2.21
2.51	3.12	1.42	1.42	4.24	7.06	7.19
6.15	4.62	0.71	1.02	0.71	3.40	3.91
2.11	3.35	5.51	3.22	0.51	2.21	9.55
2.51	5.13	5.21	0.81	3.15	0.61	3.12
1.62	0.20	3.20	0.10	0.20	4.27	1.09
20.46	21.25	23.05	29.33	20.84	27.19	36.82

was a naturalist and writer who published a detailed account of an army expedition beginning at Fort Leavenworth, Kansas, that passed through western Nebraska en route to South Pass, Wyoming, to scout the territory and protect travelers on the Oregon Trail. Suckley was an army surgeon who made observations on natural history while on various expeditions throughout the American West in the 1850s. He often collected specimens for the Smithsonian Institution (Beidleman 1956). Also departing from Fort Leavenworth, Suckley's expedition was headed for Utah Territory.

Both expeditions crossed into the North Platte Valley at Ash Hollow (fig. 3.2), a popular crossing for Oregon Trail travelers. On 8 June 1845 Carleton noted cliff swallows on the bluffs along the south side of the North Platte River; his journal suggests he may have been several kilometers west of Ash Hollow at that point (Carleton 1943). On 14 July 1859 Suckley observed cliff swallows at Ash Hollow and along the North Platte River to the west (Beidleman 1956). Both men accurately described the birds' mud nests, so there is no doubt of their identifications. Each commented that many swallows were present; the birds must have been numerous enough to draw the attention of both men independently. Suckley also reported barn swallows nesting on cliffs in the Ash Hollow area, and Carleton described what appear to be rough-winged swallows there. From their observations alone, we can presume that cliff swallows were native to the study area and bred in relatively large numbers on the cliffs above the North Platte River well before the appearance of European settlement and modern structures such as bridges and highway culverts. Cliff swallows also were reported breeding in Lincoln County, Nebraska, in the 1890s (Ducey 1988) and as present in North Platte sometime probably before 1896 (Bruner, Wolcott, and Swenk 1904). Although these reports are not detailed, they suggest that cliff swallows also occurred in the eastern portion of the study area in the late nineteenth century.

Today cliff swallows are probably more abundant in the study area

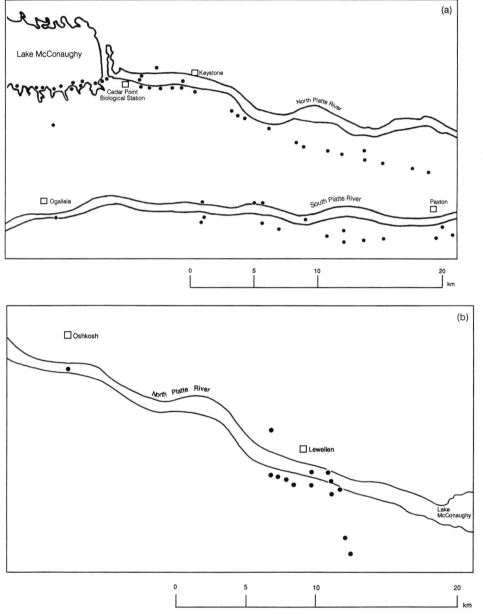

Fig. 3.3 Two portions of study area showing locations of cliff swallow colonies (*solid circles*) in Keith (*a*) and Garden (*b*) Counties. The areas illustrated are representative of the colony distribution and density found in the study area.

than during Carleton's and Suckley's era, principally because so many artificial nesting sites such as bridges are now available. However, the birds' distribution still roughly follows the North and South Platte Valleys (fig. 3.3). With the exception of one small colony along Blue Creek north of Oshkosh, all colonies known to us were within 6 km of one of the rivers, leading to a somewhat linear east-to-west arrangement throughout the study area (fig. 3.3). We systematically searched for cliff swallow colonies up to 45 km north and south of the North and South Platte Valleys (in areas that formerly were unbroken prairie and Sand Hills) but found none even though bridges apparently suitable as nesting sites exist in some areas. Thus the distribution of colonies today probably remains similar to that before European settlement of the area.

Our primary study area encompassed all colonies between Lewellen in Garden County on the west and Paxton in Keith County on the east (fig. 3.1). Within this area, one hundred colony sites were known (through 1992). Not all were active in a given year, and some were extant only in certain years, having fallen (in the case of cliff sites) or having been otherwise made unusable (section 4.10.2). Two sections of the primary study area showing extant colony sites as of 1992 are depicted in figure 3.3; these sections contain most of the colonies within the primary area and all the ones we studied intensively.

The secondary study area extended from Oshkosh in Garden County on the west to Maxwell (just east of North Platte) in Lincoln County on the east (fig. 3.1). This expanded study area contained an additional forty-four colony sites. With the exception of two west of Lewellen, these additional sites were all east of Paxton. Colonies in the secondary study area were visited occasionally during the breeding season, primarily to score use and size (section 2.1), although at a few we mist netted birds for the mark-recapture studies (chapter 12). In addition, in 1991 only we mist netted birds at a more distant site near Broadwater in Morrill County, Nebraska, about 45 km west-northwest of Oshkosh.

3.4 TYPES OF COLONIES

Cliff swallow colonies in our study area were either on natural cliff faces or on artificial structures. Among colonies of these two types, there were both similarities and differences, which were often great depending on the particular sites being compared.

3.4.1 Natural Sites

The natural cliff nesting sites in the study area are still largely where Carleton and Suckley observed them in the mid-nineteenth century (see fig. 3.2). The ones we studied were on outcroppings just west of Ash Hollow State Park along the North Platte River and on cliffs along the south side of Lake McConaughy, clustered mostly toward the southeastern end of the lake. We searched all outcroppings along the south side of the North Platte River from just west of Lewellen eastward to Lake McConaughy, and we thoroughly searched the south shore of the lake each year by boat. We have not found cliff swallows nesting on any of the cliffs east of Kingsley Dam (the eastern edge of Lake McConaughy), although there appear to be suitable cliffs east to about Keystone. We have not looked for the birds on natural nesting sites west of Lewellen, although they undoubtedly occur westward along the North Platte River, having been observed breeding, for example, on cliffs at Scottsbluff (Olson 1956).

Nests in the cliff colonies were underneath flat overhangs, ranging from 3 to 10 m or more above the ground or lake surface (fig. 3.4a). As was typical for the cliff swallow population in our study area as a whole, the size of colonies on cliff sites varied considerably, from 1 nest (solitary nesters) to 735 nests. Mean colony size for cliff sites was 85.8 ($\pm$ 11.3) nests ($N = 93$). These colonies exhibited several types of geometries. Some consisted of single, largely linear rows of nests (fig. 3.4b), whereas others were more honeycombed, with nests stacked, a given nest often sharing walls with several neighbors (fig. 3.4c). Most cliff colonies showed nest spacing intermediate between these two types, with some nests stacked two or three deep in tiers and others more isolated (fig. 3.4d).

3.4.2 Artificial Sites

Cliff swallows probably moved onto man-made structures as soon as these edifices appeared on the landscape. They were nesting on the sides of buildings in Kentucky as early as 1815 (Audubon 1831). The earliest we can find a record of cliff swallows' using artificial nesting sites in the vicinity of our study area was 1911 along the Platte River in Kearney County, Nebraska, about 150 km east of the study area. That year one pair of cliff swallows began nesting on the side of a wooden barn; the colony increased each year until it reached over 400 nests in 1925 (Jones 1933). The birds had moved onto artificial nesting sites in the study area itself at least by 1942, when a colony was using the concrete side of the spillway at Kingsley Dam (Mohler 1952), a colony site still active today. In 1942 cliff swallows were also nesting on a bridge over the North Platte River near Broadwater (Mohler 1946), where they still nest. The birds were probably using artificial structures in the study area as early as 1926, when Layton and Buckhannan (1926) reported that concrete and steel bridges and culverts were "common" on the main roads of Keith County. One bridge near Lewellen that at present supports a colony was installed about 1920 and is the oldest artificial colony site we are aware of.

Artificial nesting sites now used by cliff swallows in the study area include various types of bridges and culverts (fig. 3.5), buildings, and miscellaneous irrigation structures. Bridge sites tend to be along either the North or South Platte River or along the Sutherland Canal, an irrigation and power-generation supply canal that parallels first the North Platte River from Lake McConaughy to Paxton and then the South Platte River from Paxton to Sutherland. Bridges used by the birds are usually concrete. Although numerous wooden bridges (e.g., fig. 3.5a) are situated in the study area, we found only eleven colonies on wooden bridges through 1993. The birds commonly use relatively small concrete and metal bridges that span the Sutherland Canal (fig. 3.5b) and relatively large and tall highway bridges spanning the rivers (fig. 3.5c). Most of these sites are over water, but the birds will use bridges over land, such as one for the canal spanning a dry canyon (fig. 3.5d) or overpasses spanning highways. Bridge colonies showed the greatest diversity in size, ranging from 1 nest (solitary nesters) to 3,700 nests. Mean colony size on bridges was 604.6 ($\pm$ 47.9) nests ($N = 290$).

Highway culverts (fig. 3.5e) are box-shaped concrete structures built underneath roads for drainage. Most are essentially ditches, although a few have permanent streams or creeks flowing through them. Cliff swallows use only culverts that are relatively open and unobstructed by trees or vegetation on each end. Culverts vary considerably in height above the ground, and the minimum height for cliff swallow use seems to be about 2 m. Active culvert colonies are distributed somewhat randomly throughout the study area, although many are associated with Interstate 80 and U.S. Highway 30, which parallel the South Platte River throughout the study area, and U.S. Highway 26, which crosses and parallels the North Platte River in the western portion of the study area. Surprisingly, however, a number of culverts that looked to us like suitable nesting sites were never used by cliff swallows at any time during the study. Colony size for culverts (including the miscellaneous irrigation structures described below) ranged from 1 nest (solitary nesters) to 3,000 nests. Mean colony size for culverts was 309.3 ($\pm$ 29.4) nests ($N = 323$).

Although cliff swallows are well known for their propensity to nest on the sides of buildings in some parts of their range, we found only twenty colonies on buildings in our study area through 1993. These were on various structures: wooden barns in farmyards, the overhanging metal awnings of abandoned gas stations, cinderblock cabins with wooden rafters, and the sides of houses (until the human owners intervened and removed the birds' nests). Why cliff swallows chose these particular sites was not clear. Countless similar structures were present throughout the study area, most presumably suitable as nesting sites. Colony size on buildings ranged from 1 nest (solitary nesters) to 550 nests. Mean colony size was 105.1 ($\pm$ 34.8) nests ($N = 20$).

a

b

Fig. 3.4 Cliff swallow colonies on natural cliff sites along the south shore of Lake McConaughy, Keith County, Nebraska. Part (*a*) courtesy of Art Gingert, Wildlands Photography.

c

d

Fig. 3.5 Examples of types of artificial nesting sites used by cliff swallows in southwestern Nebraska.

d

e

f

Cliff swallows also used various irrigation structures as nesting sites, such as a concrete catwalk along a river diversion dam (fig. 3.5f) and the concrete walls above the water-release gates of the Sutherland Canal and Kingsley Dam. These sites were similar to culverts in structure and degree of nest exposure and therefore were combined with culverts for most analyses.

Nests on artificial sites were usually placed on a vertical wall where it joined a horizontal ceiling. The first nests built at a site were invariably placed against the ceiling, and subsequently the birds built nests in successive tiers below the first ones. Colony geometry of artificial sites generally was similar to that of natural sites. Some colonies contained nests in linear, largely single rows (fig. 3.6a), others were stacked in a honeycomb pattern (fig. 3.6b), and others showed an intermediate degree of stacking (fig. 3.6c). Nests in culverts were often spread out in a long single row along the top of each wall (fig. 3.6d). Bridges with multiple inner beams contained nests in uniform densities along each beam's surface (fig. 3.6e), or the birds sometimes grouped their nests in small clusters or pods at intervals along the (mainly inner) beams (fig. 3.6f).

3.4.3 Natural versus Artificial Sites

The primary structural differences between cliff swallow colonies on natural cliffs and those on artificial nesting structures were related to more uniform substrate surfaces, and greater between-year substrate stability, on artificial sites. Concrete or metal walls had fewer surface irregularities and a more regular, consistent size of overhang, which presumably allowed the birds to place their nests more uniformly across those substrates. The substrate of cliff sites tended to be too irregular, for example, for us to judge accurately whether all portions of the cliff were suitable for nest attachment and thus calculate a meaningful nest density for the cliff colonies (section 7.2.1). Artificial structures generally remained intact from year to year and did not suffer the structural collapses that befell cliff sites and often destroyed nests over the winter.

Another major difference between natural and artificial nesting substrates was the presence of parallel beams or walls on many artificial structures such as bridges and culverts. This permitted the birds to place their nests so that nests on opposite adjacent beams or walls faced each other (e.g., fig. 3.6d–f). Such a spatial array of nests was not possible in cliff colonies. Surprisingly, however, birds in nests facing each other seldom seemed to interact directly, as in trying to trespass into each other's nests (section 5.5). They interacted mostly with birds occupying nests adjacent to them on the same wall or beam.

Did cliff swallow behavior differ between birds occupying natural cliff sites and those on artificial nesting sites? If either the structural differences in substrate or other potential differences (such as degree of nest exposure) altered the birds' behavior to an extent that was outside the range of natural behavioral variation, conclusions based on studies of birds at artificial sites might be artifactual. This is potentially a problem with any study of animals occupying human-altered habitats and is an especially serious issue with hole-nesting birds that now breed primarily in birdboxes often erected by the scientists themselves (e.g., Semel and Sherman 1986; Moller 1989, 1992a; Robertson and Rendell 1990; Koenig, Gowaty, and Dickinson 1992).

We repeatedly observed cliff swallows in natural colonies and qualitatively (and in some cases quantitatively) compared those observations with ones made on birds in colonies of similar size on artificial sites. In no instance did we detect any obvious differences. For example, birds on cliffs and bridges shared foraging information (section 9.5) in the same way (we first discovered information sharing in cliff swallows occupying a cliff site); both brood parasitized neighbors' nests (section 6.3); both responded in the same way to attacks by bull snakes (section 8.2.2); both exhibited kleptoparasitism of parental care by fledged juveniles (section 6.4.2); and both suffered serious infestations of ectoparasites (section 4.3). Quantifying these similarities was not always possible owing to substantial logistical challenges of studying more than a few cliff colonies. We are confident, nevertheless, that colony type had only minor influence on cliff swallow behavior in general, and this was likely within the range of natural behavioral variation seen in the species.

Further comparisons of cliff swallows occupying colonies on different substrate types may be found in following chapters (e.g., sections 4.3.2, 7.4).

3.5 OVERVIEW OF THE STUDY POPULATION

In this section we characterize the cliff swallow population found in the study area by presenting data on population trends during the years of the study; sex ratio; arrival times of males, females, and birds of known age; dates that colonies of different sizes became active; and variation in colony size within the population.

3.5.1 Population Trends

The study population appeared to be relatively stable, at least after 1987. We assessed relative population size by summing the total number of active cliff swallow nests in eighty colony sites each year from 1984 to 1993. These sites were ones that we surveyed each year of the study. Although other colonies were present, they were discovered in later years and not included in this tabulation. The eighty colony sites used for this analysis

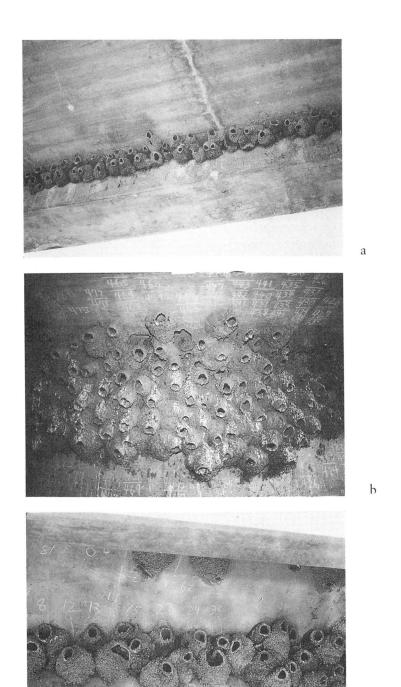

Fig. 3.6 Illustration of nest positioning in cliff swallow colonies on artificial nesting substrates. Part (*b*) courtesy of Art Gingert, Wildlands Photography.

d

e

f

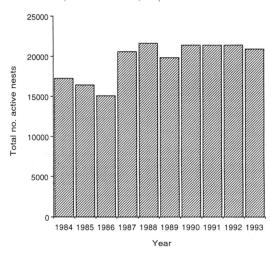

Fig. 3.7 Total number of active cliff swallow nests at eighty colony sites in the study area, 1984–93. The total number of active nests is a relative measure of population size and is based on sites that were surveyed for size and use each year.

were mostly in the primary study area, with a relatively few regularly surveyed sites from the secondary study area included. Because we were surveying the same sites and the same general area each year, this analysis presumably gave a relative measure of population size in the study area each year. Not all of the colony sites were active in a given year.

Population size seemed to increase in 1987 and remained stable after then (fig. 3.7). Fluctuations that are on the order of 1,000 to 2,000 nests may not be meaningful, because these fluctuations could result if the birds from one or two large colonies moved to an unsurveyed site in a given year. However, the large differences between 1984–86 and the remaining years (fig. 3.7) are probably meaningful. We do not know why cliff swallows apparently became more numerous in the study area in 1987. This increase followed the two driest years (1985–86) of the study (table 3.2), but why (or if) drought should lead to an increase in population size is unknown. The population was remarkably stable the last four years (fig. 3.7), varying by less than 3%. The cold, wet summer of 1992—in which some adult mortality occurred (see section 10.6)—apparently had little overall effect on population size either within that year or as measured to the next year (fig. 3.7).

We also examined population trends for 1990–93 separately, using additional colonies. For these years we surveyed colony size at 118 sites annually; this increased number of colonies presumably gave us a more comprehensive survey of the study area. We summed the total active nests in all active colonies per year: in 1990, there were 29,633 nests; in 1991, 32,101 nests; in 1992, 29,643 nests; and in 1993, 30,591 nests. Estimated population sizes for 1990 and 1992 were amazingly close. The apparent increases in 1991 and 1993 could have reflected only redistribution of birds from colonies outside the study area and hence may not have been real. The separate analyses for these later years thus support the conclusion (fig. 3.7) that population size has recently been relatively stable.

3.5.2   Sex Ratio

Several lines of evidence indicated that the sex ratio of the cliff swallow population in our study area was male biased. In mist netting, we consistently caught more males than females throughout most of each season (fig. 3.8). For this analysis we combined the total males and females caught each day at all colonies into successive five-day date intervals for each year from 1987 to 1991. Yearly percentages of males were averaged for each interval. Only the final interval, beginning on 24 July, had more females represented among the captures (and this could have reflected problematic sexing methods toward the end of the season; section 2.4). The apparent sex ratio, as based on net captures, declined significantly with date interval (fig. 3.8). The overall sex ratio, based on total captures ($N = 43,729$) across all dates and years, averaged .564 ($\pm$ .007) per year. Each year males were significantly overrepresented among the total birds caught (χ^2 tests, $p < .001$ each year). The predominance of males was especially pronounced during May (fig. 3.8), although that was probably due to older males' arriving in the study area before females (section 3.5.3).

An apparently male-biased sex ratio based on capture data at colonies could also result merely if males were more likely to be caught in nets.

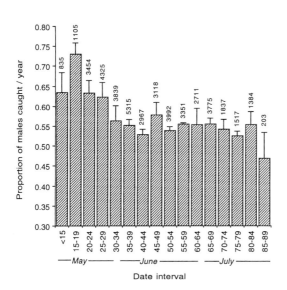

Fig. 3.8 Proportion of males among the total cliff swallows captured during five-day date intervals throughout the nesting season. Each bar illustrates the mean over five years of sampling (1987–91). Numbers above bars are the total number of birds captured in each interval across all years. Date 01 = 1 May. The proportion of males declined significantly with date interval ($r_s = -.760$, $p < .001$, $N = 16$ intervals).

Males might be more tolerant of disturbance or might simply be present at the nests more than females. This could conceivably cause a single male to be represented repeatedly in the capture totals over the season. (We excluded any birds double-caught on the same day, but the same individual could have been caught multiple times and counted on different days.) However, two kinds of evidence indicated that an approximately .56 sex ratio was probably real.

First, we found a similar sex ratio (.578) in a sample of 249 foraging cliff swallows netted in a field 2 km from the nearest colony during May and June. These birds were not in any way associated with nests or a colony site at the time and presumably represented a random sample of passersby. The sex ratio of these foragers approached a significant departure from .50 ($\chi^2 = 3.07$, $p = .080$, df = 1). Second, we dissected and sexed all cliff swallows we found dead on roads, presumably hit by cars. There was no reason to expect one sex to be more or less likely to meet such accidental deaths. (Net casualties or others we caused were excluded.) Sex ratio among sixty-eight roadkills was similar to that among net captures: .573 ($\chi^2 = 0.739$, $p = .39$, df = 1). This sex ratio was not significantly different from .50, probably because of the small sample size. Taken with the net capture data, however, it also suggested that the male bias in sex ratio was real. Apparently there was a surplus of males each year, and this surplus may have important consequences for the evolution and maintenance of extrapair copulation in particular (section 6.2.3).

3.5.3   Arrival Times of Individuals

The first cliff swallows arrived in the study area in mid-April each year. We were in southwestern Nebraska in time to observe the earliest-arriving birds on four occasions. The first arrivals appeared on 20 April in 1982, 1983, and 1989, and a few birds were already present when we arrived on 18 April 1992. The first birds in each case arrived in small groups of 4–20 birds. The population increased steadily once the first birds arrived, with new birds appearing daily (section 13.7.2) throughout May and into early June.

Analysis of when birds were first captured at colonies revealed some differences in arrival times, though not especially pronounced, among sex and age classes. We used the date when a cliff swallow was first captured during a season as a relative measure of arrival times for birds of different classes. First-capture time does not necessarily reflect actual arrival time, but it should reveal relative differences among birds if certain groups are underrepresented, especially early in the year. This analysis examined males and females divided into the age classes of yearlings and all birds older than one year. It was based on a total of 868 yearling males, 579 yearling females, 1,124 males older than one year, and 778 females older

Overview of the Study Population • 63

than one year for which first-capture dates were known in 1986–91. For each class each year, we calculated the proportion of birds that were first captured during ten-day intervals throughout the season. For graphic purposes, the yearly values for each interval were averaged and presented (fig. 3.9), although statistical tests (χ^2 tests) were done for each year sepa-

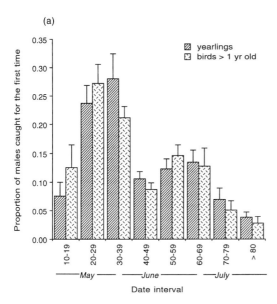

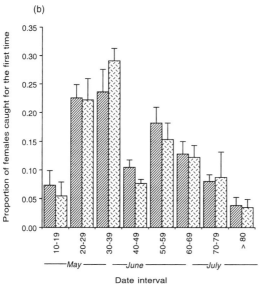

Fig. 3.9 Proportion of total males (*a*) and total females (*b*) caught for the first time that season during ten-day date intervals. Each bar shows the mean over six years of sampling (1986–91). Date 01 = 1 May.

rately. In using first-capture times to infer relative arrival times, probably only the first three date intervals (through 8 June) are meaningful. Birds captured for the first time in later intervals likely had been present earlier and simply eluded capture.

Among yearlings, there was no significant difference in the distribution of first-capture times for males versus females in any of the six years (χ^2 tests, $p > .10$ for each except 1987 where $p = .071$; relevant comparison is fig. 3.9a versus fig. 3.9b). Among birds older than one year, the male and female distributions differed significantly in three of the six years (χ^2 tests, $p < .05$ in 1989–91, $p > .10$ in 1986–88). This appeared to be largely due to more males in the first two date intervals and more females in the third one (fig. 3.9). We conclude, therefore, that the major intersexual difference is that older males arrive slightly earlier than females in some years.

Among males, the distribution of first-capture times for yearlings versus birds older than one year differed significantly in four of the six years (χ^2 tests, $p < .03$ for each, except 1988 and 1990 where $p > .10$). This was due to underrepresentation of yearlings in the first two date intervals and overrepresentation in the third (fig. 3.9a). Among females, the distributions for yearlings versus older birds (fig. 3.9b) also differed significantly in four of six years (χ^2 tests, $p < .05$ for each, except 1987–88 where $p > .10$). These differences for females may have been due largely to more yearlings in the third date interval, but when yearly proportions were averaged, the two female distributions seemed relatively similar (fig. 3.9b).

We also examined first-capture distributions for males and females that were two years old, three years old, and four or more years old. In a series of comparisons, none of the intrasexual comparisons with respect to these age classes were statistically significant (χ^2 tests, $p > .10$ for each). Thus, age beyond the first year had no apparent effect on first-capture times.

Our conclusion from the age analysis is that older males arrived somewhat earlier than yearling males in most years. However, at least with respect to the first 20 days, older females did not seem to arrive markedly earlier than yearling females. In fact, in 1986 yearling females were significantly overrepresented in the first two date intervals compared with all other intervals, relative to older females ($p = .038$).

These differences in apparent arrival times (fig. 3.9) are generally not great despite the statistical significance in some cases, and in some years there were clearly no differences among sex and age classes. One can conclude that older males tend to arrive earliest, but beyond that age and sex seem to have little consistent influence on when cliff swallows arrive in our study area. Arrival time is important, however, because it has potential bearing on (among other things) when an individual can reliably choose a colony size (section 13.7.2), its expectations of nest reuse and spatial position within a colony (section 5.2), and its seasonal breeding phenology and associated reproductive consequences (sections 4.4, 11.4).

3.5.4 Colony Start Times

Large cliff swallow colonies tended to become active earlier in the year than small ones in our study area. For 177 colonies in 1983–92, we were able to determine to within two to six days the date when birds first occupied the colony site. We assigned each colony a number from 1 to 9 denoting the week within the season when it became active, divided into six separate colony size classes (fig. 3.10). Start times declined significantly with colony size (fig. 3.10).

The pattern of large colonies' starting early in the year was striking. For example, all the colonies of 2,000 nests or larger ($N = 12$) began before 12 May. Most (83.3%) of the colonies of 500 to 1,999 nests ($N = 36$) began before 12 May, and none began after 27 May. Start times of the colonies of between 11 and 499 nests spanned the entire season, some starting in mid-April as early as the largest colonies and some as late as 3 July. We suspect that a few of these colonies began as late as 10 July in some years, although we did not have good enough information to pinpoint their start times. Solitaries and very small colonies of 10 nests or fewer also spanned most of the season, from 4 May to 19 June. Differences among colonies in start times potentially influence many aspects of cliff swallow social behavior and ecology that are explored in later chapters,

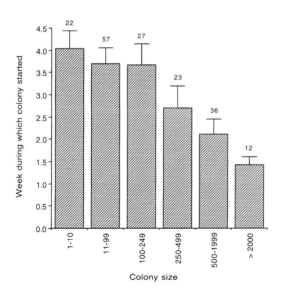

Fig. 3.10 Week during which a cliff swallow colony became active in relation to colony size. Week 1 was defined as anytime before 4 May, week 2 as 4–11 May, week 3 as 12–19 May, week 4 as 20–27 May, and so on. Numbers above bars are sample sizes (number of colonies). Mean colony start time declined significantly with colony size class ($r_s = -.99$, $p < .001$, $N = 6$ size classes).

66 • *Study Site and Study Population*

especially ectoparasitism (section 4.4), colony choice (section 13.7), and reproductive success (section 11.4).

3.5.5 Variation in Colony Size

Cliff swallow colonies in our study area varied considerably in size. The distribution of colony sizes (by size class) from a total of 726 active colonies of all substrate types, 1982–93, is shown in figure 3.11. Mean colony size was 393.0 nests ($\pm$ 24.3), with a range of 2 to 3,700 nests. Some birds also nested solitarily. Separate frequency distributions by year were similar in each case to the combined distribution (fig. 3.11). The most obvious

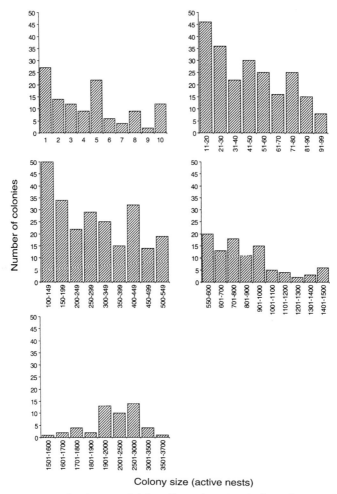

Fig. 3.11 Frequency distributions of cliff swallow colony sizes in the study area, 1982–93. Note similar scale of ordinate for each.

pattern was that despite the regular presence of several very large colonies each year, colonies larger than 1,000 nests occurred infrequently relative to the other size classes. The single most common colony size was 400 nests (23 colonies), followed by 5 nests (22 colonies), 100 nests (21 colonies), 20, 150, and 300 nests (20 colonies each), 75 nests (18 colonies), and 30 and 500 nests (17 colonies each). There were 27 cases of solitary nesting (fig. 3.11). Thus both the overall distribution and the distribution of most common sizes spanned a relatively large size range.

This natural variation in colony size within the study area forms the foundation for much of this book and presents opportunities to measure many behavioral and demographic parameters as a function of colony size. We will return repeatedly to the issue of colony size variation throughout the chapters that follow.

3.6   SUMMARY

Our study site was an area of approximately 150×50 km centered in Keith County, Nebraska, along the North and South Platte Rivers. The region is largely prairie, with rocky outcroppings and bluffs found along the south side of the North Platte River Valley. The climate is characterized by unpredictable annual rainfall, mostly occurring during the cliff swallows' summer breeding season, and relatively variable temperatures. Climate data taken during the study showed that one summer was among the coldest ever recorded in Nebraska and another was among the warmest.

Cliff swallows were known to be in the study area in the mid-nineteenth century, nesting on rocky outcroppings along the North Platte River. At present the birds nest on both outcroppings and artificial structures such as bridges, highway culverts, and buildings, and their current distribution is still mostly along the North and South Platte Rivers. The birds' behavior in colonies on natural cliff sites was similar to that seen in colonies on artificial structures.

The study population has been relatively stable in size since 1987, and unusually cold weather in 1992 apparently did not cause widespread mortality. The population was about 56% males. Older males tended to arrive in the study area earlier in some years than yearling males or females. Large cliff swallow colonies consistently became active earlier in the year than smaller ones. Colony size variation in this population was substantial. Colonies ranged from 2 to 3,700 nests, and some birds nested solitarily. Mean colony size was 393 nests.

4 Ectoparasitism

> To determine the question as to bedbugs being *brought* to houses by [cliff] swallows, I allowed about twelve pairs to raise broods under the eaves. . . . On tearing down the nests I found bugs (*Cimex*) in every one, whatever part of the roof it occupied, showing that they were brought by the birds, none having been observed in the house.
>
> J. G. Cooper (1870)

4.1 BACKGROUND

Parasites have long been recognized as important regulators of animal populations in nature, often causing mortality among their hosts (reviewed in Price 1980; May 1983; Lehmann 1993). More recently, behavioral ecologists have realized that parasites also influence their hosts' behavior, especially social behavior and mate choice (e.g., Alexander 1971, 1974; Hamilton and Zuk 1982; Barnard and Behnke 1990; Hart 1990; Keymer and Read 1991; Loye and Zuk 1991). Effective transmission of most parasites occurs when host population density exceeds a given threshold value (e.g., Black 1966; May and Anderson 1979; May 1983; Dobson 1988), and consequently population density of hosts is a critical parameter in most epidemiological models.

Many studies that have examined the relation between parasitism and host colony or social-group size have found that parasitism increases in larger host groups (e.g., Hoogland and Sherman 1976; Kunz 1976; Freeland 1979; Hoogland 1979a, 1995; Duffy 1983; Brown and Brown 1986; Moller 1987a; Shields and Crook 1987; Moore, Simberloff, and Freehling 1988; Rubenstein and Hohmann 1989; Hieber and Uetz 1990; Davies et al. 1991; Hochberg 1991; Poulin 1991a; Bennett and Whitworth 1992; Poiani 1992; Ranta 1992), although grouped hosts are not always more likely to be infested (Hieber and Uetz 1990; Poulin 1991b; Rogers, Robertson, and Stutchbury 1991; Poiani 1992; Arnold and Lichtenstein 1993; Cote and Gross 1993). In some herding animals, group formation reduces the per capita rate of parasite attacks (e.g., Duncan and Vigne 1979; Helle and Aspi 1983; Rutberg 1987; Poulin and FitzGerald 1989; Mooring and Hart 1992).

The reasons for the more typical increase in parasitism with host group size have seldom been investigated. Presumably larger groups of animals provide a greater "target area" for dispersing parasites, which are thus more likely to encounter the group (either through independent dispersal or in/on an infected host) or which aggregate in areas of high host density (e.g., Hassell 1968, 1971; Royama 1971; Kuris, Blaustein, and Alio 1980). If parasites are overdispersed on hosts, as they usually seem to be (e.g., Crofton 1971; Kennedy 1975, 1984; Anderson 1978; Anderson and May 1978; Anderson, Whitfield, and Dobson 1978; Marshall 1981; Scott 1987), and hosts settle in groups independent of their degree of parasitism, most large colonies will be colonized by relatively large numbers of parasites (by virtue of the many hosts settling there), whereas some small colonies may escape parasitism completely. The length of time hosts occupy a given colony site is also often positively correlated with parasite load, because older colonies have had longer to be found and colonized by the parasites (e.g., Danchin 1992a).

Larger colonies presumably provide better opportunities for parasites, once in a colony, to disperse successfully among (and possibly assess) hosts within the group. Within-colony dispersal of parasites is probably best promoted by close proximity of host nests and incidental or intentional physical contact among hosts during behavioral interactions. There is little or no information for colonially breeding animals about which parasite transmission mechanisms are important. Ectoparasites with direct life cycles are probably the kinds of parasites most likely to be transmitted between and within host colonies in these ways.

Cliff swallows are associated with a variety of ectoparasites throughout their range. It was obvious to us early in our research that ectoparasitism represents a major cost of coloniality for these birds (Brown and Brown 1986). In this chapter we explore ectoparasitism in detail, emphasizing in particular how its incidence varies with cliff swallow colony size and how it affects nestling swallows' growth and survivorship. We also examine various socioecological correlates of ectoparasitism, address the transmission of ectoparasites between and within cliff swallow colonies, and describe the birds' possible behavioral responses to ectoparasitism. This chapter contains some analyses updated from those originally published in Brown and Brown (1986) plus substantial new data.

4.2 ECTOPARASITES OF CLIFF SWALLOWS

Cliff swallows in southwestern Nebraska are parasitized by four types of ectoparasites: swallow bugs, fleas, ticks, and chewing lice. We focus primarily on the swallow bug and the flea, because these parasites appear to have the greatest and most easily studied effects on the birds. In this section we present the basic natural history of the parasites.

4.2.1 Swallow Bug

The wingless swallow bug *Oeciacus vicarius* (Hemiptera: Cimicidae) is a hematophagous insect that lives primarily in cliff swallow nests. The family Cimicidae contains eighty-nine species in twenty-three genera and includes species variously referred to as human bedbugs, bat bugs, chicken bugs, and pigeon bugs (Usinger 1966; Marshall 1981). All are temporary ectoparasites that feed and then return to nests or cracks in the rooms or roosts of their hosts between blood meals. They are not well adapted to cling to fur or feathers of their hosts in flight but can do so on the rare occasions when they disperse between host groups over short distances (Usinger 1966; Rannala 1995; section 4.9.1).

The genus *Oeciacus* contains two species, both parasites of swallows. *Oeciacus hirundinis* is a European species that is primarily associated with house martins but has been recorded as occasionally parasitizing other birds (Usinger 1966). *Oeciacus vicarius* is a North American species associated almost exclusively with cliff swallows. When it has been found on other birds, these alternative hosts have been either species nesting in close association with cliff swallow colonies or ones that roost transiently in cliff swallow nests (Usinger 1966; Smith and Eads 1978). Thus *O. vicarius* presumably has coevolved with the cliff swallow and exhibits various adaptations to its host's behavior and population biology (Loye and Hopla 1983). (All further references to swallow bugs pertain to *O. vicarius*.)

The swallow bug is a relatively long-lived parasite that begins to reproduce as soon as it feeds in the spring. Eggs are laid in several clutches that hatch in variable lengths of time, ranging from 3–5 days (Loye 1985) to 12–20 days (Myers 1928). Nymphs undergo five instars in about ten weeks before maturing and will mate and reproduce as long as food is available (Loye 1985). Females probably mate before winter, store sperm, and after feeding may lay eggs the following spring without remating. Adults and all five instars feed on blood of adult and nestling cliff swallows, primarily at night when the bugs are most active (George 1987). Swallow bugs possess a metathoracic scent apparatus that exudes a pungent odor (Myers 1928; Marshall 1981), which may prevent cliff swallows and other birds from eating them. When not feeding, bugs cluster on the outsides of nests, in crevices between nests, or in cracks of the substrate. Those visible during the day on the outsides of nests tend to be mostly ovipositing females (B. Rannala, pers. comm.). During the winter the bugs remain in the vacated nests or in cracks and crevices of the nesting substrate at varying distances from the nests and apparently do not travel with the cliff swallows to the birds' South American wintering range (Usinger 1966).

The swallow bug's confinement to nests and colony substrates means that cliff swallows represent an ephemeral host resource for these parasites. Not all nests or colonies are used by the birds in a given year, and an interval of several years may pass before a colony site is reused (section 7.4). Consequently the bugs seem to be adapted to withstanding long periods of host absence. Overwintering mortality of bugs can be substantial, but anecdotal reports suggest that some can survive in colonies not used by cliff swallows for up to four consecutive years (Smith and Eads 1978; Loye 1985; Loye and Carroll 1991). Although possibly in these cases alternative hosts (such as house sparrows or bats) were present in the colonies in the interim, we know that some bugs in our study area can survive in the complete absence of alternative hosts throughout the first two summers a colony site is unused (Rannala 1995). The physiological basis for this sort of dormancy is unknown.

As soon as a colony site is occupied in the spring, swallow bugs begin feeding on the adult birds. We collected newly engorged adult bugs as early as 24 April from a colony that had been active only two or three days. Bug populations at a site increase throughout the summer, reaching a peak about the time cliff swallow nestlings fledge. Swallow bugs are relatively mobile within colonies, crawling along the substrate from nest to nest. They sometimes move up to 3 m (and perhaps farther) from the nearest nest, usually hiding in crevices in the substrate when away from a nest. The frequency and mechanisms of swallow bug dispersal from colony to colony are poorly understood. The only way bugs can move between colonies is on the birds, often attached to their feet, and bugs appear to disperse between colonies relatively rarely (section 4.9.1).

4.2.2 Bird Flea

The hematophagous bird flea *Ceratophyllus celsus celsus* (Siphonaptera: Ceratophyllidae) parasitizes cliff swallows in our study area. This flea is one of approximately fifty-five species in the genus *Ceratophyllus;* all but nine species infest birds, the remainder being ectoparasites of mammals (Traub, Rothschild, and Haddow 1983). At least seven species have been recorded parasitizing cliff swallows in various parts of the bird's range, and one of these, *C. petrochelidoni,* is not known to parasitize other bird species (Traub, Rothschild, and Haddow 1983; Brown and Brown 1995). *Ceratophyllus celsus* is found primarily on cliff swallows, but there are records of its parasitizing bank swallows in Alaska (Hopla 1965). Fossil remains suggest that *C. celsus* has been associated with cliff swallows at least 4,570 years (Nelson 1972). Unless noted, all further discussion of fleas refers to *C. celsus*, and we assume all fleas in the study area belong to this species.

Like swallow bugs, fleas spend considerable time in their hosts' mud nests and do not live permanently on the birds' bodies. They appear to

72 • *Ectoparasitism*

travel on the adult birds more than do swallow bugs, and they have been found on the adult swallows throughout the breeding season (section 4.8). No native species of *Ceratophyllus* has been reported from South America (Wenzel and Tipton 1966; Traub, Rothschild, and Haddow 1983). This suggests that the cliff swallows do not carry their fleas to the wintering range, although we doubt anyone has systematically searched for *Ceratophyllus* fleas on wintering cliff swallows in South America. Large numbers of fleas winter in the birds' nests, and species of *Ceratophyllus* exhibit various physiological adaptations to cold winter temperatures (Pigage and Larson 1983; Schelhaas and Larson 1989). The fleas cluster at the entrances of the nests early in the spring and leap onto birds that pass nearby or that enter or investigate the nests in which the fleas cluster (Hopla and Loye 1983). In this way fleas find hosts and thereby move between nests and colonies (see also Bates 1962; Humphries 1969; Du Feu 1992). They do not appear to disperse within a colony by crawling on the substrate.

Once the fleas find an active nest, the females feed several times and within a week lay eggs (Hopla and Loye 1983). Larval development begins in the nest before the swallow eggs hatch and continues until after the nestlings fledge. Usually only one generation a year is produced (Hopla and Loye 1983). The larvae pupate at 7–10 days (often about the time the nestling swallows fledge). By about two weeks after the birds have departed from the colonies, the pupal cases contain teneral adult fleas, which remain quiescent and encased during the winter. The life cycle and ecology of *C. celsus* in our study area appear similar to that of *C. styx* in Britain (Bates 1962; Humphries 1969) and *C. hirundinis* in Czechoslovakia (Cyprich, Krumpal, and Hornychova 1988).

Unlike swallow bugs, fleas are unable to withstand long periods of host absence (Hopla and Loye 1983). If a colony is unused one year, most of the fleas do not survive to the following spring. Once they emerge from their pupal cases, fleas must eat immediately and thus cannot endure long periods of fasting. They compensate with their jumping ability, which enables them to leap onto hosts that may visit a nest or colony site only briefly during the spring.

4.2.3 Other Ectoparasites

A third nest-based ectoparasite of cliff swallows in southwestern Nebraska was the soft-bodied tick *Ornithodoros concanensis* (Acarina: Argasidae). These (and other) hematophagous ticks are common in cliff swallow colonies in various parts of the United States (Baerg 1944; Kohls and Ryckman 1962; Howell and Chapman 1976; Hopla and Loye 1983; Larimore 1987; Brown and Brown 1995), but they are rare in our study area. From 1982 to 1992 we examined and counted ectoparasites (section 2.6) on 12,318 nestling cliff swallows (this total includes nestlings from fumigated nests because ticks occurred in fumigated nests). Only 18 ticks on 17 nestlings (0.14%) were found, all but 3 in two colonies in Garden County in the western part of the study area. Although these ticks appeared to take large blood meals from the birds and clearly could have had major deleterious effects had they been more common (Hopla and Loye 1983; Chapman and George 1991), ticks were apparently too rare and too difficult to sample to permit us to study them in Nebraska.

Cliff swallows in our study area also were parasitized by two species of chewing lice (Mallophaga): *Machaerilaemus malleus* (Amblycera: Menoponidae) and *Brueelia longa* (Ischnocera: Philopteridae). We encountered these insects only during ectoparasite sampling of adult swallows, when they fell off the birds that were placed in the sampling jars (section 2.6). We did not begin quantifying the presence of lice until 1992. That year *Machaerilaemus* was found on 221 of 2,315 adults examined (9.5%; range 0–33 per bird) and *Brueelia* on 122 of 1,555 adults examined (7.8%; range 0–25 per bird). Mallophaga spend their entire lives on the birds, apparently feeding mostly on feathers and dead skin (although some amblycerans take blood; Marshall 1981). Because our study of chewing lice was limited to only one year (Brown, Brown, and Rannala 1995), they are not addressed in any detail in this book.

4.3 EFFECTS OF COLONY SIZE AND DENSITY

There are several reasons to expect that animals in larger colonies should be subjected to heavier infestations of ectoparasites (section 4.1). The factors causing increased ectoparasitism generally fall into two classes (section 4.9): those that enhance the introduction of new parasites into larger colonies and those that enhance the spread of existing parasites within larger colonies. In this section we describe how colony size and nest density affect ectoparasitism in cliff swallows. We assessed the extent of ectoparasitism per colony in two ways.

4.3.1 Parasite Counts on Nestlings

By counting all parasites present on the nestlings' bodies when they were 10 days old (section 2.6), we found that both swallow bugs per nestling per nest and fleas per nestling per nest increased significantly with cliff swallow colony size (fig. 4.1). Parasite counts on nestlings provided only a relative measure of ectoparasitism among colonies (section 2.6), and therefore the data reported for the fifty-one colonies in figure 4.1 do not indicate actual parasite loads. However, the relative counts (fig. 4.1a) suggest up to a tenfold increase in bug parasitism between the smallest and largest colonies. The pattern for fleas was less striking, perhaps because of generally lower overall flea numbers, but notice that the largest colony had almost five times as many fleas as the smallest (fig. 4.1b). Swallow bug and flea parasitism increased with colony size independently of each other,

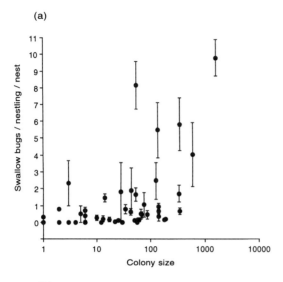

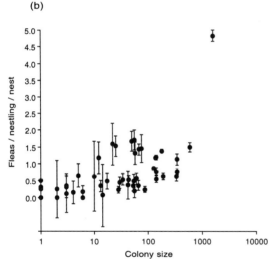

Fig. 4.1 Number of swallow bugs (*a*) and fleas (*b*) counted per nestling per nest in relation to cliff swallow colony size. Swallow bugs increased significantly with colony size ($r_s = .50$, $p < .001$, $N = 51$ colonies), as did fleas ($r_s = .66$, $p < .001$, $N = 51$ colonies). These colonies are the same as in table 4.5, and sample size for each is given there. For colonies of fewer than three nests, sample size equals colony size. In some cases several colonies with identical values are represented by a single dot.

because the number of bugs in a given nest was not positively associated with the number of fleas in the same nest ($r_s = -.11$, $p < .001$, $N = 1{,}310$ nests).

4.3.2 Parasite Counts from Nests

For a relatively small number of colonies, we assessed the swallow bug loads per nest by collecting nests after nestlings fledged and counting all bugs present in the nests (section 2.6). We collected nests from eight colonies in highway culverts and six colonies on bridges, which also enabled us to examine whether substrate type affected swallow bug parasitism. The number of nests collected per colony ranged from 10 to 50. The actual bug loads per nest were at times quite high. The most heavily infested nest we collected had 2,608 swallow bugs, and the mean number of bugs per nest for many colonies exceeded 300. The colony with the heaviest infestation averaged 707.8 ($\pm$ 67.3) bugs per nest. As with bug counts from nestlings (fig. 4.1a), counts based on total bugs in a nest increased with colony size. There was a significant positive correlation between average bugs per nest and colony size for bridge sites ($r_s = .83$, $p = .042$, $N = 6$ colonies) and a positive but not significant correlation for culvert sites ($r_s = .50$, $p = .21$, $N = 8$ colonies).

Substrate type clearly affected swallow bug numbers. Bridge colonies averaged 168.6 ($\pm$ 49.6) bugs per nest ($N = 6$ colonies), compared with culvert colonies averaging 400.2 ($\pm$ 54.2) bugs per nest ($N = 8$ colonies). The difference was significant (Wilcoxon rank sum test, $p < .001$). The increased number of bugs in culvert colonies was probably not attributable to the culvert colonies sampled being on average larger than the bridge sites, because we sampled larger colonies on bridges than on culverts. Culvert colonies sampled for this analysis ranged from 10 to 345 nests, whereas the ones on bridges ranged from 30 to 1,000 nests. The higher average numbers of bugs in culvert colonies was probably related in large part to substrate type. Nests in culverts are less exposed to the elements than those on the taller, more open highway bridges, and consequently the overwinter survival of bugs may be greater in culverts. Substrate type may account for some of the variation in bug counts on nestlings, which are presented in figure 4.1 irrespective of substrate type.

Another difference between colonies on bridges and those on culverts was that nests in bridge colonies tended to be more spread out—that is, less dense. Since the distance between nests potentially influenced swallow bug dispersal within colonies (section 4.9.2), we examined the effect of a colony's nest density (section 2.8.1) on average bug load per nest. The fourteen colonies where we collected nests for bug counts were classified by nest density, which yielded a range in densities from 0.1 to 14.9 nests per meter. Parasitism by swallow bugs increased significantly with a colony's nest density (fig. 4.2). There was an approximately sevenfold increase in swallow bug parasitism per nest between the least dense and the most dense sites. Therefore if nest density on bridges is routinely lower than in culverts, nest density alone may explain some of the difference in bug numbers between colonies of these different substrate types.

We also assessed the extent of flea parasitism using counts from nests, and in this analysis we examined how colony size the preceding year affected the number of fleas in nests during the current year. Fleas clustered

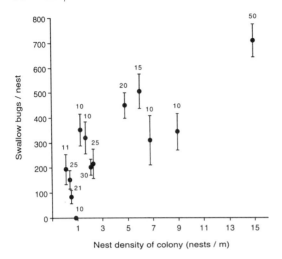

Fig. 4.2 Number of swallow bugs counted per collected nest in relation to nest density of cliff swallow colonies (number of nests per meter of substrate). Sample sizes (number of nests) are shown above error bars. Swallow bugs increased significantly with nest density ($r_s = .72$, $p = .006$, $N = 13$ colonies).

at the entrances of cliff swallow nests early in the season were sampled using a sticky black card held in front of the nest (section 2.6.2). We used data collected on two dates (9 and 22 May 1983) at six colony sites before these sites had been occupied by any cliff swallows that year. The colony sizes the previous year at these sites were 1, 2, 5, 55, 450, and 1,600 nests. No fleas were found on either date at the three smallest colonies. In order of increasing colony size, mean numbers of fleas per nest at the other sites on 9 May were 0.45 ($\pm$ 0.22, $N = 20$), 0.10 ($\pm$ 0.06, $N = 30$), and 1.80 ($\pm$ 0.24, $N = 52$) and on 22 May, 1.35 ($\pm$ 0.63, $N = 20$), 0.98 ($\pm$ 0.24, $N = 52$), and 4.38 ($\pm$ 0.70, $N = 112$). These data are presented graphically in Brown and Brown (1986). The most fleas sampled at any single nest was 39, found at the largest colony on 22 May. The mean number of fleas per nest increased significantly with colony size on each date (for each, $r_s = .88$, $p = .022$, $N = 6$ colonies).

These data were consistent with those based on flea counts from nestling swallows (fig. 4.1b); both card sampling at nest entrances and flea counts from nestlings indicated about a four- to fivefold increase in flea parasitism over the range of colony sizes observed. The consistent increase in fleas from 9 to 22 May for the three larger colonies presumably reflected increasing activity levels of fleas as the season progressed. They increased their clustering at nest entrances, apparently in attempts to disperse from nests that, in all but one colony, were never used that season.

We conclude from the analyses in this section that parasitism per nest by both swallow bugs and fleas increased substantially with cliff swallow colony size and nest density. Loye and Carroll (1991) reported a similar pattern for swallow bugs in colonies in Oklahoma. Nests in larger colonies, and in colonies where nests are more closely packed, can expect more parasites during the current breeding season. Furthermore, for fleas, and probably for swallow bugs, the number of parasites waiting in nests at a colony site at the start of a breeding season is strongly influenced by the colony size at that site the previous summer. We defer discussion of possible reasons for the increase of parasitism with colony size to section 4.9, where we address between-colony and within-colony parasite transmission.

4.4 EFFECTS OF DATE

In this section and the one that follows, we investigate potential socioecological correlates of ectoparasitism other than colony size that may have influenced parasite loads among nests and colonies. Time of year in particular may have a major effect on parasite load because swallow bug and flea populations presumably increase through reproduction over the course of a summer.

We assessed seasonal effects by using the date on which each brood of nestling swallows was examined for ectoparasites. Since each brood was examined at 10 days of age, date of parasite counting provided a relative measure of how parasite loads varied over time, standardized with respect to host nesting stage. However, parasite counts from nestlings done on the same date at different colonies could not be combined for a meaningful analysis of seasonal effects because colonies became active at different times of the year (section 3.5.4), and thus parasites were introduced into them, and began reproducing, at different times. For instance, parasite counts on 25 June from a colony that first became active on 20 April, by which time most of the nestlings would have fledged, would likely be very different from counts taken the same date at a colony that became active on 20 May, when no nestlings would yet have fledged. Therefore the most appropriate (albeit more unwieldy) way to examine the effect of date on parasite load was to analyze each colony separately.

For 25 colonies we were able to analyze how date affected swallow bug load, and we had 29 colonies for analysis of flea loads (the discrepancy in number of colonies for bugs versus fleas resulted when no parasites of a given type were found on any nestlings in a colony, preventing a within-colony correlation analysis for that parasite type at the site). We present the colonies by site and size and show the correlation (r_s) for swallow bug and flea parasite load as a function of date, the significance level, and the sample size (table 4.1).

Parasitism by swallow bugs increased with date in 21 of 25 colonies (a significant preponderance of colonies with positive correlations, binomial test, $p < .001$). In the remaining four colonies bug parasitism de-

Table 4.1 Spearman Rank Correlations between Average Number of Swallow Bugs and Fleas Counted per Nestling per Nest and the Date the Nest Was Sampled (01 = 1 May) within Each Cliff Swallow Colony

Colony Site	Colony Size	Swallow Bugs		Fleas		
		r_s	p	r_s	p	N
8227	4	—	—	.80	.20	4
8628	10	.56	.25	.22	.68	6
8709	10	−.84	.038	−.89	.018	6
8228	12	—	—	.54	.27	6
8327	13	.66	.054	−.08	.83	9
8324	17	−.39	.17	−.36	.21	14
8931	22	.63	.09	−.46	.25	8
8329	25	−.08	.81	.40	.17	13
8432	30	—	—	−.01	.98	11
8627	34	.21	.52	.44	.16	12
8431	43	.81	<.001	.06	.80	19
8207	50	.28	.14	.09	.63	30
8832	54	.25	.20	.41	.036	27
8226	56	.18	.31	.05	.78	36
8434	56	—	—	−.62	.14	7
8932	57	.41	.07	−.40	.08	20
8203	60	.29	.13	.08	.69	28
8824	66	.20	.22	.20	.20	42
8421	75	.77	<.001	−.69	.005	15
8405	125	.01	.96	.46	.003	40
8641	137	.64	<.001	−.28	.034	56
8630	140	.06	.61	−.26	.24	22
8732	140	.62	.002	−.11	.32	79
8305	180	.29	<.001	.25	.002	160
8730	340	.16	.42	−.28	.14	28
8430	345	.80	<.001	.07	.56	68
8206	350	−.20	.28	−.71	<.001	32
8202	600	.24	.017	−.12	.26	98
8201	1,600	.56	<.001	−.40	<.001	114

Note: Missing data are instances in which there was no observable variation in parasite load among the nests sampled. N = number of nests.

clined as the season progressed. In 9 cases the correlation coefficients for swallow bugs were statistically significant (table 4.1). Patterns of statistical significance in these within-colony analyses are probably not meaningful, however, because statistical significance is highly sensitive to sample size, which is determined largely by colony size. In evaluating the within-colony analyses presented here and in sections 4.5 and 4.7 and chapter 11, the focus should be on the degree to which either positive or negative correlation coefficients predominate.

The pattern for fleas was less obvious (table 4.1). In 14 colonies flea parasitism per nest increased as the season progressed, and in 15 colonies it declined. There was thus no trend across colonies for flea parasitism to vary consistently with time of year. We had earlier concluded that flea parasitism declined as the season progressed (Brown and Brown 1986), but the analysis presented here is more sophisticated and suggests that the earlier conclusion was wrong. Bug parasitism, on the other hand, clearly increased as the season advanced.

The difference in bug parasitism and flea parasitism as a function of date is consistent with these insects' life cycles. Bugs begin reproducing in a colony as soon as it is occupied, the eggs hatch relatively quickly, and instar populations continue to increase as long as the colony is occupied and the bugs are able to feed. Fleas' eggs, in contrast, produce diapausing adults that do not become active until the following spring. There is thus little or no observable flea population growth attributable to reproduction during the period within a given season when a colony site is occupied by cliff swallows.

The effect of date on swallow bug parasitism (table 4.1) suggests that it alone might explain a positive correlation between parasitism and colony size (e.g., fig. 4.1a) if large colonies routinely became active later in the year. However, large colonies became active *earlier* than smaller ones (section 3.5.4). Therefore colony initiation date cannot account for the observed increase in bug parasitism in larger colonies (fig. 4.1a).

4.5 EFFECTS OF NEST AGE, SIZE, AND SPATIAL POSITION

The ectoparasite load in a cliff swallow nest potentially can be affected by the age, size, and spatial position of the nest within a colony. Nest age is clearly important, because fleas and swallow bugs are nest-based ectoparasites that overwinter in the nests. The length of time a nest has been intact presumably affects the number of parasites present there, somewhat dependent on how often the birds use the nest or colony site. Nest size is potentially related to ectoparasite load: larger nests might accommodate (house) more parasites (more cracks and crevices in which to hide and lay eggs) or could be more likely to be encountered by substrate-dispersing parasites because they take up more space on the substrate. Finally, the number and proximity of neighboring nests may directly affect parasite dispersal and transmission (section 4.9.2), and therefore a nest's position with respect to other nests in a colony theoretically could influence ectoparasite load. For instance, Loye and Carroll (1991) reported that swallow bugs increased toward the centers of cliff swallow colonies in Oklahoma. Of principal interest to us, of course, is how the effects of these factors in turn vary with colony size and whether they help generate the increased ectoparasitism observed in larger and denser cliff swallow colonies (figs. 4.1, 4.2).

80 • *Ectoparasitism*

4.5.1 Nest Age

We examined separately how ectoparasitism varied with colony size for old and new nests (fig. 4.3). For this analysis we could use only colonies where we had information on nests' histories (section 2.8.3), and thus some of the colonies represented in figure 4.1 could not be used, especially many of those in the first year of the study. Old nests showed a significant increase in swallow bug parasitism with colony size, but there was no sig-

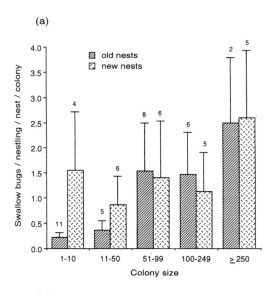

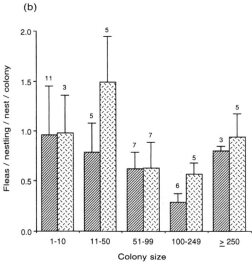

Fig. 4.3 Number of swallow bugs (*a*) and fleas (*b*) counted per nestling per nest per colony for old and new nests in colonies of different size classes. Sample sizes (number of colonies) are shown above error bars. Swallow bugs increased significantly with colony size in old nests ($r_s = .58$, $p < .001$, $N = 32$ colonies) but not in new nests ($r_s = .28$, $p = .18$, $N = 26$ colonies). Fleas did not vary significantly with colony size in either old nests ($r_s = .15$, $p = .40$, $N = 32$ colonies) or new nests ($r_s = -.23$, $p = .27$, $N = 25$ colonies).

nificant relationship for new nests (fig. 4.3a). Paradoxically, there was an indication that new nests might have on average *more* total swallow bugs than old nests, a result we did not expect. New nests averaged 0.88 ($\pm$ 0.45) more bugs per nestling per colony than old nests for a sample of 14 colonies (ranging in size from 5 to 600 nests) where we had both old and new nests in the same colony for direct comparison. The increase in bugs in new nests approached significance (Wilcoxon matched-pairs test, $p = .068$).

In contrast to the earlier result showing fleas increasing with colony size (fig. 4.1b), we found no significant effect of colony size on flea parasitism for either age of nest when old and new nests were analyzed separately (fig. 4.3b). This difference may simply reflect sampling error, in that some of the colonies in the earlier analysis (fig. 4.1b) could not be included in the analysis of nest age because nest ages there were unknown. It also suggests that the effect of colony size in figure 4.1b may not be especially strong. Fleas too seemed to be more numerous in new nests than in old nests. For the 14 colonies where we had both old and new nests for direct comparison, new nests averaged 2.30 ($\pm$ 2.18) more fleas per nestling per colony than old nests, although the difference was not significant (Wilcoxon matched-pairs test, $p = .31$). A similar noneffect of nest age was seen for *Ceratophyllus* fleas parasitizing house martins in Britain (Clark and McNeil 1981, 1991).

For swallow bug parasitism, the effect of date that we described in the previous section (table 4.1) depended on nest age. Old nests showed the same pattern of increasing bug parasitism within a colony as the season progressed. We had adequate data on old nests for 12 colonies, and in 10 of these bug parasitism increased with date. As in table 4.1, this preponderance of colonies with positive correlations was significant for old nests (binomial test, $p = .038$). However, for new nests, in only 9 of 15 colonies was the correlation between bug parasitism and date positive, and the preponderance of colonies with positive correlations was not significant for new nests (binomial test, $p = .61$). The pattern for fleas in table 4.1—little consistent effect of date on parasitism within a colony—held for both old and new nests when analyzed separately.

These results on nest age can be summarized by noting that an individual's expectation of ectoparasite load was affected somewhat by the age of the nest it occupied. Cliff swallows that used old nests could expect bug loads to increase both with colony size and as the season progressed. Birds using new nests, however, were not more likely to suffer higher bug loads as colony size increased or as the season advanced. There was no apparent difference between old and new nests in a bird's expectation of flea parasitism.

These results suggest two major questions that we cannot answer fully.

Why did new nests tend to have higher parasite loads than old nests, and why did bug loads increase with colony size in old nests but not in new ones? At present we can offer no plausible answer to the latter question, but one possible answer to the former is that swallow bugs may move out of old, inactive nests and into new nests as soon as the new ones are built. Most colonies contain old nests that are unused each year (section 4.10.3), and there is no guarantee that the birds will occupy a given old nest in a particular year. Thus ectoparasites trapped in unused nests may at times concentrate at new nests where their chances of obtaining blood meals are good. Both bugs and fleas respond to the odor and movement associated with potential hosts (Marshall 1981) and presumably use these cues to move (or jump) toward newly built nests (or other old nests that happen to be active).

Another possibility is that the difference in bug load between old and new nests reflects new nests' becoming active on average *later* in the year than old nests. Because a new nest requires substantial nest-building time, eggs are laid in most new nests after egg laying is completed in old nests. As bugs increase later in the summer, they may aggregate at new nests (section 4.9.2), which tend to still be active after nestlings in old nests have fledged. The most appropriate comparison would be between old and new nests with similar clutch initiation dates within a colony, but these occurred relatively rarely, presumably because of the additional time necessary to construct new nests.

The difference in how swallow bug parasitism in old versus new nests varied with date is more easily understood. Old nests had lasted largely intact for at least one previous breeding season, and bugs and fleas from previous seasons were presumably already there. This enabled bugs in particular to begin reproducing as soon as the nest was occupied by cliff swallows and led to increasing bug numbers in old nests as the season progressed. New nests had few or no parasites from the previous breeding season, and therefore bugs from other nests had to first find and colonize the new nests. The presumed delay in the time it takes bugs to infest a new nest may account for the lack of a consistent relation between bug parasitism and date in new nests.

4.5.2 Nest Size

Did nest size affect ectoparasite load? Nest size varied among birds, from nests with an outside maximum diameter of 9 cm up to nests of 33 cm, at least double the size of the smallest nests. Maximum outside diameter seemed to be an appropriate relative measure of nest size (section 2.8.2).

The size of a cliff swallow nest had little consistent effect on parasitism by either fleas or bugs (table 4.2). Bug parasitism increased with nest size in 15 colonies and decreased in 23. Although for fleas there was a suggestion that parasitism might have declined with nest size (23 of 36 correla-

Table 4.2 Spearman Rank Correlations between Average Number of Swallow Bugs and Fleas Counted per Nestling per Nest and a Nest's Size (as Measured by Its Diameter) within Each Cliff Swallow Colony

Colony Site	Colony Size	Swallow Bugs		Fleas		N
		r_s	p	r_s	p	
8227	4	—	—	.95	.051	4
9160	6	−.10	.87	—	—	5
8628	10	.56	.24	−.45	.37	6
8709	10	−.26	.62	−.02	.98	6
8228	12	—	—	−.25	.64	6
8327	13	−.18	.67	.06	.89	8
8825	14	−.89	.042	−.46	.43	5
8324	17	−.19	.51	−.09	.76	14
8931	22	.21	.65	.11	.80	7
8329	25	−.29	.34	.45	.12	13
8432	30	—	—	.11	.74	11
8627	34	.72	.009	.05	.88	12
9032	42	−.01	.95	.03	.91	19
8431	43	−.08	.74	.17	.49	19
8207	50	−.03	.89	−.04	.86	29
9131	53	.11	.64	.17	.47	20
8832	54	−.41	.039	.07	.74	26
8226	56	−.44	.007	−.30	.08	36
8434	56	—	—	−.62	.14	7
8932	57	.21	.40	−.21	.40	19
8203	60	−.33	.15	−.02	.95	21
8824	66	−.48	.001	−.01	.99	42
9124	68	.17	.33	−.14	.42	34
8421	75	.13	.65	−.27	.33	15
9041	86	.03	.86	−.08	.62	47
8405	125	.08	.64	.20	.23	40
8641	137	−.16	.23	−.07	.62	56
8630	140	.08	.72	−.14	.52	22
8732	140	.09	.43	.08	.46	79
9130	140	−.08	.48	.13	.25	78
8305	180	−.15	.057	−.09	.26	159
9030	190	.01	.97	−.15	.23	66
8730	340	−.03	.88	−.04	.83	28
8430	345	.17	.17	−.16	.20	66
8206	350	−.18	.34	−.05	.80	32
8202	600	.05	.66	−.05	.62	98
8201	1,600	.12	.24	−.05	.62	103

Note: See table 4.1 for explanation of missing data. N = number of nests.

tion coefficients were negative; table 4.2), there were no significant preponderances of positive or negative correlations for either fleas or bugs (binomial tests, $p > .10$ for each). When the analysis of nest size was repeated for old and new nests separately, we found no effect of nest age for flea parasitism. The correlation between fleas and nest size was negative in 13 of 20 colonies for old nests and 11 of 17 colonies for new nests; none of these differences were significant. However, in 15 of 19 colonies the correlation between bug parasitism and nest size was negative for new nests (binomial test, $p = .020$); it was negative in 11 of 18 colonies for old nests. This means that bug parasitism tended to decline in larger newly built nests but not in larger old nests. The reason for this pattern was not clear. We conclude that nest size in general had no consistent or easily explainable effect on ectoparasite load in cliff swallows. Further discussion of nest size may be found in sections 11.2.3 and 11.6.2.

4.5.3 Nest Spatial Position

Swallow bugs are relatively mobile ectoparasites, often crawling from nest to nest. Therefore a nest's proximity to other nests, some of which may be infested with swallow bugs, may be a critical factor determining the pattern of infestation within a cliff swallow colony. Although we do not know how far a swallow bug will disperse along a substrate in the absence of host nests, we have found bugs as far as 3 m from the nearest active nest, and we suspect they can move much farther than that (section 4.9.2). We showed earlier that bug loads increased in the colonies with more densely packed nests (fig. 4.2), suggesting that the closer a nest is placed to other nests, the higher its average bug load may be. Whether fleas move themselves from nest to nest within a colony is less certain, but they should at least be capable of jumping from one nest to another whenever nests are closely packed.

In analyzing how ectoparasitism varied with a nest's spatial position within a colony, we used three different but related measures of spatial position (section 2.8.2). *Distance from the center* is a measure of a nest's position relative to the center of the colony and places each nest spatially relative to all other nests. *Nearest neighbor distance* indicates the closeness of the next active nest and is useful as an index of both the packing of nests within a colony and the spatial isolation of particular nests. Finally, classifying small samples of nests as *center* and *edge* allowed comparison of nests representing the biologically relevant extremes in a spatial nest distribution that was often largely continuous within a colony (section 2.8.2; Brown and Brown 1987).

The analyses of distance from center and nearest neighbor distance could be done only within colonies. These distances could not be lumped across colonies because each was a measure of a nest's position relative to other nests in that colony. For example, in a large and spread-out colony, a distance of 400 cm from the center might place that nest only midway between the center and edge, but in a small and relatively compact colony a distance of 400 cm might place the nest at the extreme edge of the colony. Thus, again we used a series of within-colony correlation analyses.

Table 4.3 Spearman Rank Correlations between Average Number of Swallow Bugs and Fleas Counted per Nestling per Nest and a Nest's Distance from the Colony's Center within Each Cliff Swallow Colony

Colony Site	Colony Size	Swallow Bugs		Fleas		N
		r_s	p	r_s	p	
8709	10	.30	.56	.12	.82	6
8327	13	.71	.18	−.11	.86	5
8324	17	−.31	.28	.15	.60	14
8931	22	.61	.14	−.22	.63	7
8329	25	−.52	.07	.38	.20	13
8432	30	—	—	.62	.054	10
8627	34	−.15	.64	−.56	.06	12
9032	42	−.04	.90	.00	.99	14
8431	43	.56	.012	−.07	.77	19
8207	50	.13	.48	.07	.70	30
9131	53	.50	.07	.15	.60	14
8832	54	.03	.90	−.02	.91	24
8226	56	.08	.66	−.42	.010	36
8932	57	.18	.45	−.50	.029	19
8824	66	−.10	.53	−.10	.52	42
9124	68	.43	.011	.21	.23	34
8421	75	−.28	.31	.01	.97	15
9041	86	.12	.47	−.01	.94	40
8405	125	.24	.13	.05	.75	40
8641	137	.24	.08	.12	.40	56
8630	140	.06	.78	.09	.70	22
8732	140	.03	.79	.10	.37	79
9130	140	.01	.92	−.02	.89	75
8305	180	.06	.47	.08	.35	159
9030	190	.06	.61	−.04	.74	66
8730	340	.16	.41	.03	.88	28
8430	345	.05	.70	.15	.22	66
8206	350	.01	.95	.01	.97	32
8202	600	.07	.48	−.20	.054	92
8201	1,600	−.15	.10	.12	.20	114

Note: See table 4.1 for explanation of missing data. N = number of nests.

We were surprised to find that bug parasitism did not consistently decline the farther a nest was from the colony's center (table 4.3). In fact, in 22 of 29 colonies bug parasitism seemed to *increase* with a nest's distance from the center (table 4.3), a significant preponderance of positive correlations (binomial test, $p = .008$). For fleas, parasitism increased with distance from the center in 18 colonies and decreased in 12 (table 4.3). The preponderance of positive correlations for fleas was not significant (binomial test, $p = .281$).

We conclude from these analyses that linear distance from a colony's center is not a good predictor of either swallow bug or flea parasitism within most cliff swallow colonies. One possible reason is that nest density in many of the study colonies was relatively uniform (e.g., fig. 3.6), and consequently the centers of colonies were not markedly more dense than the edges. Therefore the centers were perhaps no more likely to promote ectoparasite transmission than were nests closer to the edges.

For swallow bugs, we found a different result when we analyzed parasitism as a function of a nest's nearest neighbor distance (table 4.4). In 24 of 33 colonies bug parasitism declined as nearest neighbor distance increased; the preponderance of negative correlations was significant (binomial test, $p = .013$). The magnitudes of these correlations were generally higher than those for the distance from the center (table 4.3). Because nearest neighbor distance measures the degree of a nest's isolation from others, these results (table 4.4) indicate that bug parasitism declined in nests that were farther away from other nests. These results are also consistent with those reported in figure 4.2 for a colony's nest density.

A decline in swallow bug parasitism as nests are farther apart can probably be explained by the bugs' being less likely to disperse longer distances between nests along the substrate (section 4.9.2). Flea parasitism, however, did not show a consistent relation to nearest neighbor distance (table 4.4); it increased with nearest neighbor distance in 17 colonies and decreased in 19 (binomial test, $p = .87$). That fleas would show no consistent relation to nearest neighbor distance is perhaps not surprising, because fleas apparently do not disperse by crawling on the substrate. Fleas are introduced into nests largely by riding on birds (section 4.9.1), and therefore a nest's linear distance from other nests per se is less likely to affect how many fleas reach it. Clark and McNeil (1981, 1991) similarly found no effect of a nest's degree of isolation on its level of flea infestation among house martins.

How did nest age affect these spatial measures of ectoparasitism? Bug parasitism in old nests increased with distance from the center in 13 of 15 colonies (binomial test, $p = .008$) and decreased with nearest neighbor distance in 14 of 18 ($p = .030$), mirroring the results for all nests (tables 4.3, 4.4). Bug parasitism in new nests, however, showed no consistent trends with either distance from center (increasing with distance from center in 9 of 16 colonies; $p = .80$) or nearest neighbor distance (declining with nearest neighbor distance in 13 of 19 colonies; $p = .17$). For distance from center, flea parasitism showed the same pattern in both old and new nests, increasing with distance from center in 11 of 15 colonies for old nests (binomial test, $p = .12$) and 11 of 17 colonies for new nests ($p = .33$). Flea parasitism decreased with nearest neighbor distance in 14 of 17 colonies for old nests (binomial test, $p = .012$) but in only 8 of 20 colonies for new nests ($p = .50$).

From these analyses of nest age, we conclude that there were virtually no spatial predictors of either bug or flea parasitism for newly built nests.

Table 4.4 Spearman Rank Correlations between Average Number of Swallow Bugs and Fleas Counted per Nestling per Nest and a Nest's Nearest Neighbor Distance within Each Cliff Swallow Colony

Colony Site	Colony Size	Swallow Bugs		Fleas		N
		r_s	p	r_s	p	
8227	4	—	—	.63	.37	4
9160	6	−.67	.22	—	—	5
8628	10	−.03	.95	.48	.33	6
8709	10	−.30	.56	−.12	.82	6
8228	12	—	—	.29	.57	6
8327	13	−.54	.13	−.68	.046	9
8825	14	.08	.90	−.21	.74	5
8324	17	−.47	.09	−.03	.93	14
8931	22	.62	.14	.07	.87	7
8329	25	−.28	.36	.67	.012	13
8432	30	—	—	.10	.77	11
8627	34	.06	.86	−.27	.39	12
9032	42	−.31	.21	−.43	.07	18
8431	43	.43	.07	.24	.33	19
8207	50	.05	.78	.16	.40	30
9131	53	.51	.018	−.11	.65	21
8832	54	−.61	<.001	−.17	.41	25
8226	56	−.09	.60	−.45	.006	36
8434	56	—	—	.62	.13	7
8932	57	.23	.32	−.40	.08	20
8203	60	−.28	.18	−.18	.39	24
8824	66	−.27	.08	.10	.52	42
9124	68	−.21	.23	−.33	.055	34
8421	75	−.17	.55	.19	.49	15
9041	86	−.08	.59	−.11	.94	47
8405	125	−.42	.007	.23	.14	40
8641	137	−.02	.89	−.14	.29	56
9130	140	−.13	.26	.17	.15	75
8732	140	.11	.34	−.06	.58	79
8630	140	−.09	.70	.04	.86	22
8305	180	−.20	.011	−.04	.65	159
9030	190	−.11	.37	−.01	.94	66
8730	340	−.25	.20	−.11	.57	28
8430	345	−.02	.85	.14	.26	68
8206	350	−.54	<.001	−.10	.60	32
8202	600	.02	.82	.08	.44	98
8201	1,600	−.12	.20	.05	.62	114

Note: See table 4.1 for explanation of missing data. N = number of nests.

Old nests were more likely to show a pattern in which ectoparasitism by both fleas and bugs tended to increase slightly with a nest's distance from center and decreased with a nest's nearest neighbor distance. These results underscore again the different expectations in terms of ectoparasite load for cliff swallows occupying nests of different ages. We detected no patterns suggesting that the spatial effects on ectoparasitism discussed thus far varied in any consistent manner with colony size (e.g., tables 4.3, 4.4).

Finally, we examined how the ectoparasite loads of the center nests in a colony compared with those of the edge nests. These classifications provide a direct comparison of birds that settle in nests with the most neighbors versus those settling in nests on the periphery with the fewest neighbors. Examining these discrete spatial categories enabled us to detect any spatial patterns that might apply only to birds at the two extremes of the nest distribution continuum, a revealing comparison when considering, for example, how predation varied with a nest's spatial position (section 8.7).

Center-edge comparisons for average bug and flea loads per nest per colony were done for all nests irrespective of age and also for old and new nests separately (fig. 4.4). Perhaps the most surprising result was that bug parasitism was not consistently greater in center nests than in edge nests (fig. 4.4a); it was actually higher in edge nests for all except newly built nests. For nests of all ages combined, edge nests averaged 0.90 ($\pm$ 0.67) more bugs per nestling per colony than center nests ($N = 11$ colonies). Old nests on the edge averaged 2.19 ($\pm$ 1.29) more bugs per nestling per colony ($N = 8$ colonies) than old nests in the center. New nests on the edge averaged 1.36 ($\pm$ 1.60) fewer bugs per nestling per colony ($N = 5$ colonies) than new nests in the center. None of these differences in average bug loads between center and edge nests were significant (Wilcoxon matched-pairs tests, $p > .10$ for each).

Flea parasitism also did not differ strongly between center and edge nests (fig. 4.4b). For nests of all ages combined, edge nests averaged 0.11 ($\pm$ 0.09) fewer fleas per nestling per colony ($N = 11$ colonies) than center nests. Old nests on the edge averaged 0.19 ($\pm$ 0.13) more fleas per nestling per colony ($N = 7$ colonies) than old nests in the center. New nests on the edge averaged 0.18 ($\pm$ 0.15) fewer fleas per nestling per colony ($N = 5$ colonies) than new nests in the center. None of these differences in flea loads between center and edge nests were significant (Wilcoxon matched-pairs tests, $p > .10$ for each). In each comparison of both fleas and bugs, we included only colonies in which we had ectoparasite data for both center and edge nests of the relevant ages.

The center-edge comparisons described here confirm our earlier conclusion that ectoparasite load did not vary consistently with nest position relative to the center within a colony (table 4.3). Surprisingly, even the most peripheral nests exhibited ectoparasite loads that were as high on average as those for center nests, regardless of nest age, and older edge nests often contained slightly more swallow bugs than more centrally positioned nests. One interpretation is that on a broad spatial scale, all the nests in a cliff swallow colony are sufficiently clustered that even the

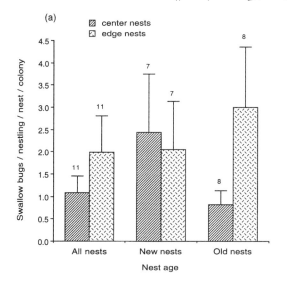
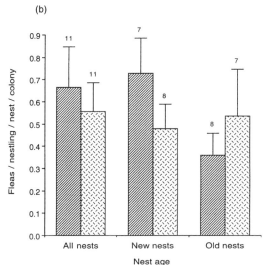

Fig. 4.4 Number of swallow bugs (*a*) and fleas (*b*) counted per nestling per nest per colony for center and edge nests of different ages. Sample sizes (number of colonies) are shown above error bars.

most peripheral are not far enough away to be beyond the parasites' effective dispersal range. On the other hand, an increasing nearest neighbor distance was associated with reduced loads of at least swallow bugs (table 4.4). Perhaps only distance to the closest neighboring nest affects parasite load, irrespective of a nest's position relative to the center. Edge nests often were somewhat clustered, meaning that although they were on the periphery of the colony, in many cases their nearest neighbor distances were small.

4.6 VARIANCE IN ECTOPARASITE LOAD

The analyses in this chapter so far have addressed the mean number of swallow bugs and fleas per cliff swallow nest. The variance in parasite load is also potentially important. If variance is high, individuals either within the same colony or in different colonies (of perhaps the same size) may not have equal expectations of being parasitized. As we emphasized in section 1.3.2, it is important to identify if possible which individuals within a group experience a given cost or benefit of coloniality and which ones do not. The first step is to measure the variance among individuals in the extent to which they are affected by the trait in question.

As a measure of variance both in this chapter and throughout this book, we used the coefficient of variation. This statistic allows comparison of variances among samples with different means (Sokal and Rohlf 1969). Coefficients of variation for parasites counted on nestlings revealed that variance among the nests within a colony increased significantly with colony size for swallow bugs (fig. 4.5a). There was no significant relationship for fleas. Fleas showed a peak in within-colony variance at colony sizes of about 100 nests, with lower variance in smaller and larger colonies (fig. 4.5b). Whether this was a real pattern is unclear. The pattern for bugs, however, was more pronounced (fig. 4.5a) and suggests that cliff swallows in larger colonies experienced greater variance in bug load within the colony than did those in smaller colonies. This increased variance was independent of the mean parasite load in the different colony sizes. Thus, some birds in large colonies occupied nests that had relatively few bugs whereas others occupied heavily infested nests that greatly exceeded the mean bug load for the colony site.

Unfortunately, we still know relatively little about how (or if) the variance in parasite load is distributed among certain classes of individuals within a colony. The earlier analyses suggest that individuals nesting later would be more likely to suffer heavy bug infestations (section 4.4, and see section 4.9.2), and differences in timing of nest initiation within a colony probably account for some of the variation seen in the larger colonies. Since large colonies are less synchronous than small ones (section 8.6.1), there is more time during the season for bugs to reproduce at sites with large colonies, potentially leading to higher internest variance as the bugs concentrate at the later nests (section 4.9.2).

Another explanation for the variance in ectoparasite load is that certain cliff swallows are better hosts than others, and ectoparasites aggregate on those superior hosts. Differences in host susceptibility could be ecological, such as one class of birds' being forced (through intraspecific competition) into nest sites that are more likely to attract parasites, or being forced to nest later in the year when parasites are more abundant. Differences in

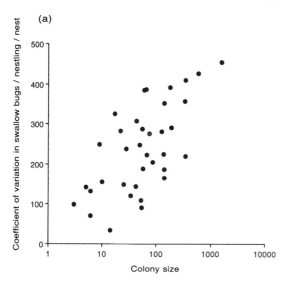

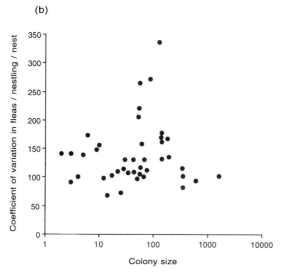

Fig. 4.5 Coefficient of variation in the number of swallow bugs (*a*) and fleas (*b*) counted per nestling per nest in relation to cliff swallow colony size. Coefficient of variation increased significantly with colony size for bugs ($r_s = .62$, $p < .001$, $N = 36$ colonies), but for fleas it did not vary significantly ($r_s = .08$, $p = .61$, $N = 41$ colonies). Sample size for each colony was the same as in figure 4.1 and table 4.5. Colonies with mean equal to zero or sample size of one could not be included.

host susceptibility could also reflect phenotypic or genetic differences among hosts that make certain ones inherently easier to parasitize (Anderson, Whitfield, and Dobson 1978; Wakelin 1978; Wakelin and Blackwell 1988; Barnard and Behnke 1990; Poulin, Rau, and Curtis 1991). As a measure of how much ectoparasitism varied among classes of individuals, we examined the extent of flea parasitism in nests occupied by adult birds of known age.

This analysis used flea counts on nestlings from nests in five colonies ranging from 140 to 1,400 nests. In each colony we identified nests occupied by both yearling cliff swallows and birds older than one year. These were the only colonies where we had known nests of both yearlings and older birds, and they had been fumigated, in large part for the parentage studies (Brown and Brown 1988b). Therefore we could not examine bug parasitism, but because the fumigant appeared only weakly effective against fleas (section 2.7, Brown and Brown 1986), we could use flea parasitism as a relative measure of ectoparasitism between these age classes of birds.

In all five colonies yearling cliff swallows occupied nests with higher average flea counts per nestling than did birds older than one year. In three of the colonies, no fleas were found on any nestlings in nests owned by parents older than one year. The average count for nests occupied by yearling parents was 0.18 ($\pm$ 0.08) fleas per nestling ($N = 48$ nests), whereas for parents older than one year it was 0.05 ($\pm$ 0.02) fleas per nestling ($N = 38$ nests). The small number of colonies ($N = 5$) prevented a statistical test (such as a matched-pairs test), but nevertheless it appeared that yearling cliff swallows occupied nests with over three times as many fleas as did older birds. (Further comparisons of yearling parents versus those older than one year may be found in section 11.8.) We do not know why yearlings were more susceptible to flea parasitism, but one possibility is that the more inexperienced yearlings visited more nests and colonies early in the year before settling than did older birds and consequently picked up more fleas clustering at nest entrances (see sections 4.8 and 13.2). Also, perhaps older birds had learned to avoid the fleas that clustered at nest entrances early in the spring (section 4.2.2). We could not identify any variables other than parental age that seemed to distinguish between birds more and less likely to be parasitized.

Another hypothesis to explain variance in ectoparasite load is that the variation among nests is the result of a stochastic process of distribution and may not correlate generally with differences in susceptibility among hosts. For example, if ectoparasites begin each breeding season uniformly distributed among the nests within a colony and their subsequent movement is at all density dependent (for example, as they seek mates or avoid competition or predation), an overdispersed distribution of parasites among nests can result from relatively simple diffusion processes (Taylor and Taylor 1977; Okubo 1980; Taylor 1981). Similarly, if parasites aggregate in nests over the winter because overwintering survivorship is perhaps density dependent, the initial overdispersed distribution could be maintained the next summer merely through random movement patterns by the parasites (Okubo 1980). In these cases some nests would suffer high parasite numbers and others would have relatively low numbers, but the distribution would be generated solely by parasite movement and have nothing to do with attributes of the hosts occupying each nest. These processes are probably more likely to apply to swallow bugs than to fleas, because bugs often move along the substrate between nests (section 4.9.2). To date we have not collected explicit data on parasite movement and how it may influence variation in parasite load within colonies of different sizes. To do so would require detailed information on parasite population biology and a series of field experiments designed to measure movement of bugs and fleas under a variety of conditions.

4.7 EFFECTS OF ECTOPARASITES ON NESTLINGS

This chapter thus far has examined the numbers of ectoparasites associated with cliff swallow nests of various attributes both within and among colonies. In this section we ask how much these parasites affect nestling cliff swallow growth and survivorship. Ectoparasitism represents a cost of coloniality only if it both increases with colony size and has a demonstrable negative effect on cliff swallow reproductive success. We assessed ectoparasitism's effects in two ways: by examining the extent of ectoparasitism in relation to nestling growth and survivorship under naturally occurring parasite loads and by manipulating ectoparasite load in the field with a fumigation experiment (Brown and Brown 1986).

4.7.1 Nestling Growth and Survivorship

Swallow bugs and fleas both feed on blood of nestling swallows. Therefore we hypothesized that these parasites represent some physiological burden on nestlings, and this burden should be reflected in lower nestling growth rates and perhaps reduced survivorship as parasite numbers increase. Chapman and George (1991) demonstrated that nestling cliff swallows parasitized primarily by swallow bugs and ticks had reduced body mass, slower feather growth, reduced blood concentrations of hemoglobin, hematocrit, and erythrocytes, increased leukocytes, and higher mortality than nestlings in nests where parasites had been removed by fumigation.

As a measure of nestling growth, we used body mass of 10-day-old nestlings. Cliff swallows follow a typical passerine pattern in which nestling body mass rises sharply for the first ten to twelve days after hatching, then reaches an asymptote and declines slightly before fledging (Stoner 1945). We selected an age of 10 days because body mass at that time should reflect how fast nestlings are growing and provides a comparative measure among nests at the time of maximal nestling growth (also see section 10.4). Relative differences in body mass of 10-day-old nestlings continued to be reflected in body mass of birds after they fledged and was positively related to survivorship to the next breeding season (section 12.5). For these reasons, 10 days of age was an appropriate time to measure both the short- and long-term effects of ectoparasites.

Factors besides ectoparasitism potentially affect body mass. The most obvious is brood size: nestlings in larger broods generally weigh less than nestlings in smaller ones (Perrins 1965; Royama 1966; Nur 1984a). This is presumably brought about by limitations in food harvest and delivery rates by parents as the number of mouths to feed increases. We removed potentially confounding effects of brood size by analyzing separately the effects of ectoparasites in cliff swallow broods of different sizes.

Average nestling body mass in a nest declined as the average number of swallow bugs per nestling in the nest increased (fig. 4.6). A decline oc-

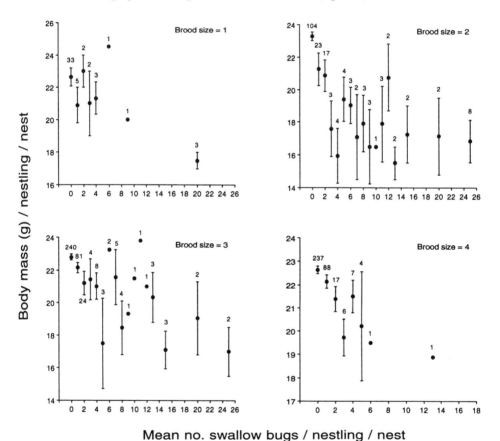

Fig. 4.6 Body mass per nestling per nest in relation to mean number of swallow bugs per nestling per nest for cliff swallow broods of one to four nestlings. Mean number of bugs for each nest was rounded to the nearest integer value. Sample sizes (number of nests) are shown above error bars. Body mass per nestling did not vary significantly with bugs per nestling for broods of one ($r_s = -.43$, $p = .29$, $N = 8$), but it declined significantly for broods of two ($r_s = -.54$, $p = .023$, $N = 17$) and four ($r_s = -.88$, $p = .004$, $N = 8$) and nearly so for broods of three ($r_s = -.45$, $p = .08$, $N = 16$).

curred in all four brood sizes, and it was significant in brood sizes of two, three, and four nestlings (fig. 4.6). Nestlings in the more heavily infested nests averaged 4–5 g less in body mass than nestlings in the lightly infested nests. This represents an 18–22% reduction in body mass for birds occupying the heavily infested nests. For this analysis and the following one (fig. 4.7), we had to combine data from different colonies to achieve sufficient sample sizes over a meaningful range of ectoparasite loads.

A different pattern emerged for fleas (fig. 4.7). Nestling body mass increased as the number of fleas increased; the increase was significant for

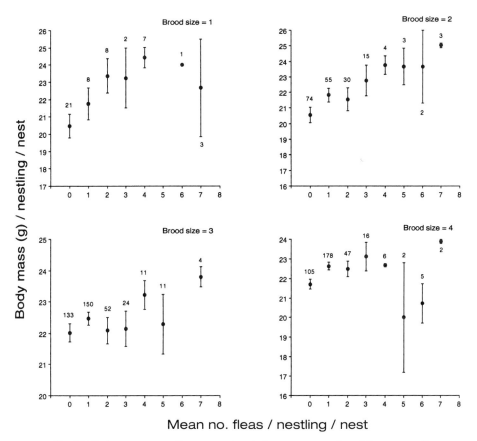

Fig. 4.7 Body mass per nestling per nest in relation to mean number of fleas per nestling per nest for cliff swallow broods of one to four nestlings. Mean number of fleas for each nest was rounded to the nearest integer value. Sample sizes (number of nests) are shown above error bars. Body mass per nestling did not vary significantly with fleas per nestling for broods of one ($r_s = .57$, $p = .18$, $N = 7$) or four ($r_s = .12$, $p = .69$, $N = 8$), but it increased significantly for broods of two ($r_s = .88$, $p = .004$, $N = 8$) and nearly so for broods of three ($r_s = .71$, $p = .07$, $N = 7$).

brood size of two and nearly so for brood size of three. This was a surprising result and suggests that fleas have no negative effect on nestling body mass. The positive correlations between mass and flea load for each brood size may simply reflect the sampling method. Heavier nestlings represented a greater surface area for flea attachment, and consequently we may have counted more fleas on the heavier nestlings because they were larger and thereby picked up more parasites. Regardless, it seems that fleas do not represent a major burden on nestling cliff swallows, at least within the range of flea loads observed in this study.

Correlations between nestling body mass and parasite load and the resulting conclusions apply only to nestling cliff swallows that survive to 10 days of age. Nestlings that died before day 10 could not be included in the analyses for figures 4.6 and 4.7. Therefore we also examined nestling survivorship as a function of ectoparasite load. We used the total number of nestlings surviving to 10 days of age per nest as a measure of survivorship, which included nests that failed before hatching. Because other variables also affected nestling survivorship between colonies (e.g., parental foraging efficiency, predation, date), the most appropriate analysis was a series of within-colony comparisons.

The number of nestlings surviving to day 10 per nest and that nest's average number of swallow bugs counted per nestling were negatively correlated in 23 of 33 colonies (binomial test, $p = .035$; table 4.5). All colonies larger than 56 nests showed negative correlations, meaning that in most colonies nestling survivorship declined (at least to some degree) as bug infestations increased. Nestling survivorship declined with increasing flea parasitism in 22 of 36 colonies (a nonsignificant preponderance of negative correlations; binomial test, $p = .24$), and there was no obvious pattern with colony size for fleas (table 4.5). In fumigated nests where the effects of bugs had been removed, nestling survivorship declined with increasing flea parasitism in 11 of 22 colonies (binomial test, $p = .99$).

These analyses indicate that increasing infestations of swallow bugs were associated with higher nestling cliff swallow mortality, but fleas had no consistent effect on nestling survivorship even in the absence of bugs. Because other factors also contributed to nest failure, we could not assign (from table 4.5) a specific fraction of the nestling mortality to swallow bugs, and thus we resorted to a fumigation experiment described below to determine how much mortality was in fact caused by the bugs.

Reductions in survivorship as a function of ectoparasite load in the natal nest were also observed for birds after they fledged. We found that swallow bug load significantly affected a nestling's probability of surviving to its first breeding season (section 12.6). However, we defer discussion of these long-term effects to chapter 12, where we address survivorship in detail.

Table 4.5 Spearman Rank Correlations between Number of Nestlings Surviving to Day 10 per Nest and Average Number of Swallow Bugs and Fleas Counted per Nestling per Nest within Each Cliff Swallow Colony

Colony Site	Colony Size	Swallow Bugs		Fleas		N
		r_s	p	r_s	p	
8227	4	—	—	−.77	.22	4
9160	6	.22	.72	—	—	5
8628	10	.36	.48	−.69	.13	6
8709	10	.10	.84	.48	.34	6
8228	12	—	—	−.49	.32	6
8327	13	−.58	.11	.06	.88	9
8825	14	.58	.30	.21	.73	5
8324	17	.06	.84	−.60	.024	14
8931	22	.08	.84	−.72	.046	8
8329	25	.06	.86	−.21	.49	13
8432	30	—	—	.11	.75	11
8627	34	−.33	.29	.38	.22	12
9032	42	−.08	.73	−.12	.64	19
8431	43	−.19	.44	.19	.42	19
8207	50	.47	.009	−.15	.43	30
9131	53	.01	.95	−.24	.30	21
8832	54	−.38	.050	.18	.38	27
8226	56	.30	.07	.02	.93	36
8434	56	—	—	−.02	.72	7
8932	57	−.09	.72	.09	.69	20
8203	60	−.13	.51	−.07	.73	28
8824	66	−.07	.65	−.60	<.001	42
9124	68	−.04	.82	−.24	.17	35
8421	75	−.30	.28	−.09	.74	15
9041	86	−.31	.035	−.13	.40	47
8405	125	−.19	.25	−.11	.52	40
8641	137	−.30	.026	.27	.048	56
9130	140	−.26	.023	−.02	.84	78
8732	140	−.04	.74	.06	.63	79
8630	140	−.19	.41	.13	.57	22
8305	180	−.13	.11	−.17	.035	160
9030	190	−.10	.43	−.10	.43	66
8730	340	−.02	.92	−.04	.84	28
8430	345	−.61	<.001	−.05	.68	68
8206	350	−.01	.95	.11	.55	32
8202	600	−.01	.94	−.09	.37	98
8201	1,600	−.22	.018	.11	.23	114

Note: See table 4.1 for explanation of missing data. N = number of nests.

Swallow bugs may also affect nestling cliff swallows by transmitting viruses to them. The swallow bug is a known vector for Fort Morgan virus (Togaviridae, *Alphavirus*) of the western equine encephalitis arbovirus complex. This virus was first isolated at a cliff swallow colony along the South Platte River at Fort Morgan in eastern Colorado (Hayes et al.

1977), about 175 km from the study area. Bugs we collected from Keith County within the study area in 1983 also carried a virus similar to the Fort Morgan type (Centers for Disease Control, Fort Collins, pers. comm.). In a study by Scott, Bowen, and Monath (1984) in Colorado, nestling cliff swallows became viremic when fed upon by bugs carrying the virus. Negative effects on the birds were not obvious in that study, although there has been no systematic study of virus transmission and its effects in colonies of different sizes. Viremia in nestling cliff swallows may compound the other deleterious effects of swallow bugs.

4.7.2 Fumigation Experiment

We used nest fumigation to measure more directly the effects of swallow bugs on nestling cliff swallows (Brown and Brown 1986). Field experiments (section 2.7) consisted of dividing colonies of various sizes in half, regularly fumigating the nests in the experimental section with a short-

Table 4.6 Clutch Size per Nest and Number of Nestlings Surviving to Day 10 per Nest in Fumigated (F) and Nonfumigated (NF) Cliff Swallow Nests

Colony Site	Colony Size		Clutch Size			Number of Nestlings Surviving		
			Mean	SE	N	Mean	SE	N
Pooled data[a]	< 10	F	3.5	0.3	8	2.3	0.5	9
		NF	3.3	0.4	11	1.1	0.6	11
8432	30	F	—	—	—	3.9	0.2	11
		NF	—	—	—	3.1	0.1	12
8431	43	F	3.7	0.1	12	2.9	0.3	23
		NF	3.4	0.2	11	3.4	0.2	19
8434	56	F	3.2	0.2	33	1.4	0.3	33
		NF	3.4	0.2	24	0.9	0.3	24
8841	68	F	3.1	0.1	37	2.5	0.2	34
		NF	2.7	0.2	32	0.4	0.2	29
8421	75	F	3.7	0.2	18	2.7	0.3	18
		NF	3.6	0.2	17	2.5	0.3	17
8405	125	F	3.0	0.1	41	2.4	0.1	41
		NF	2.9	0.1	86	1.4	0.1	86
8630	140	F	3.7	0.1	105	3.1	0.1	105
		NF	3.8	0.1	35	2.5	0.3	35
8642	163	F	3.0	0.1	116	2.5	0.1	103
		NF	2.7	0.1	47	0.1	0.1	46
8730	340	F	3.7	0.1	215	2.7	0.1	210
		NF	3.6	0.1	147	1.4	0.1	145
8430	345	F	3.8	0.1	122	3.1	0.1	96
		NF	3.9	0.0	223	1.4	0.2	154
8830	375	F	3.3	0.1	295	2.3	0.1	286
		NF	3.3	0.1	83	0.1	0.0	81

Note: N = number of nests.
[a] From sites 8425, 8426, 8427, 8428, and 8433.

Table 4.7 Brood Size per Nest and Average Nestling Body Mass per Nest in Fumigated (F) and Nonfumigated (NF) Cliff Swallow Nests

Colony Site	Colony Size		Brood Size		Nestling Body Mass (g)		
			Mean	SE	Mean	SE	N
Pooled data[a]	< 10	F	3.0	0.5	22.9	0.9	7
		NF	3.0	0.7	20.3	1.1	4
8431	43	F	3.2	0.2	23.9	0.5	21
		NF	3.4	0.2	22.1	0.8	19
8434	56	F	2.9	0.2	24.0	0.5	16
		NF	3.1	0.1	21.5	1.1	7
8421	75	F	3.1	0.2	23.8	0.3	16
		NF	2.9	0.2	21.1	0.5	15
8405	125	F	2.7	0.1	24.2	0.3	33
		NF	2.6	0.1	21.0	0.7	40
8630	140	F	3.5	0.1	21.2	0.3	77
		NF	3.4	0.1	19.7	0.7	22
8730	340	F	3.3	0.1	23.4	0.2	73
		NF	3.0	0.2	21.5	0.6	28
8430	345	F	3.4	0.1	23.7	0.2	86
		NF	3.3	0.1	20.3	0.3	69

Note: N = number of nests. Sample sizes (N) for brood size and nestling body mass were the same for each row.
[a] From sites 8425, 8426, 8427, 8428, and 8433.

lived insecticide to remove swallow bugs, and then assessing nestling body mass and survivorship at 10 days of age as described in the previous section. Because the fumigated and nonfumigated nests within each colony differed only with respect to fumigation, any differences among these nests could be attributed directly to swallow bug parasitism. Effects of fleas could not be studied because the fumigant was largely ineffective against them (section 2.7). We did this experiment in colonies of various sizes, and therefore we could measure how the relative effects of swallow bugs varied with colony size. Colonies chosen for the fumigation experiment were all culverts (to minimize substrate effects) ranging in size from 1 nest to 375 nests. Data from all colonies of fewer than 10 nests were pooled to achieve meaningful sample sizes. The analysis reported here includes results from five additional colonies studied in 1986–88 that were not available for inclusion in Brown and Brown (1986).

The average clutch size per nest did not differ significantly between fumigated and nonfumigated nests in any of the colonies studied (table 4.6; Wilcoxon rank sum tests, $p > .10$ for each colony). This allowed us to examine the average number of surviving nestlings per nest in fumigated versus nonfumigated sections for all nests pooled, regardless of initial clutch size. Similarly, the average brood size of nests with young surviving to day 10 did not differ significantly between fumigated and nonfumigated nests in any of the colonies studied (table 4.7; Wilcoxon rank sum tests,

100 • *Ectoparasitism*

$p > .10$ for each colony). This allowed us to examine the average body mass of nestlings surviving to day 10 in fumigated versus nonfumigated sections for all nests pooled, regardless of brood size.

Counting nests that failed before day 10, the average number of young surviving to day 10 was lower in nonfumigated than in fumigated nests in all but one colony, and the reductions in survivorship were relatively large for most sites (table 4.6). These reductions came about because many of the nests in the nonfumigated portions of the colonies failed completely: all nestlings were killed before day 10 or the eggs were abandoned before hatching. The fumigation treatment allowed us to attribute the nesting failures to swallow bug parasitism in these colonies. In the larger colonies (>160 nests), nestling survivorship was reduced by at least 50% and in some cases close to 100% in nonfumigated sections. In these colonies bugs clearly had major effects. In heavily infested colonies many nestlings, some as young as 8–9 days, jumped out of their nests, evidently in response to the bugs. These nestlings were of course doomed, and any on the ground were soon picked up by scavengers.

Swallow bugs also greatly reduced body mass for nestlings that survived to day 10 (table 4.7). The effects were pronounced, with nestlings in nonfumigated nests being smaller and weaker and exhibiting far less feather growth than same-aged nestlings in fumigated nests of the same

Fig. 4.8 Typical nestling cliff swallow from a nonfumigated nest (*left*) and one from a fumigated nest (*right*) at a 345 nest colony. Both were 10 days old. From Brown and Brown (1986).

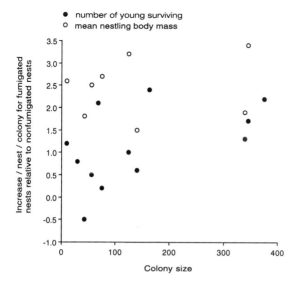

Fig. 4.9 Average increase per nest per colony for fumigated nests relative to nonfumigated nests, in number of nestlings surviving to day 10 and average nestling body mass (g), in relation to cliff swallow colony size. Increase in number of surviving nestlings rose almost significantly with colony size ($r_s = .56$, $p = .059$, $N = 12$ colonies). Increase in average nestling body mass did not vary significantly with colony size ($r_s = .21$, $p = .61$, $N = 8$).

colony (fig. 4.8). The average reductions in nestling body mass for nonfumigated nests (table 4.7) do not reflect how much healthier the fumigated nestlings seemed in general and how much greater feather development they exhibited than nonfumigated nestlings of roughly equivalent body mass (see fig. 4.8).

How did these deleterious effects of swallow bugs vary with colony size? We calculated the difference in average number of nestlings surviving and average nestling body mass between fumigated and nonfumigated nests for each colony. We predicted that the magnitude of the increase in both of these variables for fumigated nests should rise with colony size, because larger colonies have more swallow bugs and presumably the bugs' effects are greatest there.

The increase in average number of young surviving in fumigated nests relative to nonfumigated nests increased significantly with colony size (fig. 4.9). This means that nestling mortality attributed to swallow bugs increased with colony size, and thus bug parasitism represented a greater cost to cliff swallows living in larger colonies. The effect of bugs on nestling survivorship (fig. 4.9) likely would have been greater had we been able to do fumigation experiments in colonies larger than 375 nests (which, in our study area, represented only medium-sized colonies; section 3.5.5). Unfortunately, we did not have any larger colonies situated in culverts in which it was feasible to do this experiment.

The increase in average nestling body mass in fumigated nests relative to nonfumigated nests showed no significant pattern with respect to colony size (fig. 4.9). Perhaps this reflects the fact that body mass can be measured

102 • *Ectoparasitism*

only for nestlings that actually survive. If nestlings survive to 10 days of age in the larger colonies, they presumably pay about the same mass-related cost to bugs as do nestlings in smaller colonies. However, as expected, the trend was for nestlings from fumigated nests in larger colonies to weigh more than their counterparts in nonfumigated nests (fig. 4.9).

The fumigation results suggest that if the swallow bug infestation in a nest exceeded some apparent threshold, the nestling swallows were likely to be killed before day 10 or the nest to be abandoned before the eggs hatched. This represented the greatest cost of swallow bug parasitism and one that increased with colony size. If the swallow bug infestation in a nest was low enough to allow the nestlings to survive to day 10, they still suffered reduced body mass, retarded feather growth, and probably a weakened overall condition, reducing their chances for postfledging survival (section 12.6). The high variance in parasite load among nests in larger colonies (fig. 4.5) was probably what allowed some nestlings there to survive and fledge, whereas many others were killed by the bugs.

4.8 ECTOPARASITISM OF ADULTS

The previous section addressed the effects of swallow bugs and fleas on nestling cliff swallows. Adult fleas and bugs of all ages presumably feed on blood of adult birds whenever the adults are inside their nests, especially at night. However, measuring parasitism's direct effects is more difficult with adult birds than with nestlings. Unlike nestlings, adult cliff swallows cannot be reliably caught and measured at regular intervals, and scoring how many parasites may be hidden in the adults' feathers is difficult. Of course, any effect of parasitism on an adult bird's offspring is an indirect effect on that adult, but we were also interested in learning whether direct ectoparasitism of adult cliff swallows increased with colony size and thus might represent another parasite-related cost of coloniality.

We assessed parasitism of adult birds using an ectoparasite-sampling jar as described in section 2.6. We sampled adult cliff swallows at a variety of times throughout the breeding season and classified the sampling times as "early," "mid," or "late," corresponding to the stage of the nesting cycle (section 2.3.2). We sampled a total of 5,219 adult birds from 1987 to 1992.

Swallow bugs very seldom were found among the feathers of adult cliff swallows as measured by our sampling jars. Only 18 of the total 5,219 birds sampled in jars (0.34%) from 1987 to 1992 had one or more bugs. All of these bugs were found on birds in 1992. The most taken off any single bird in a jar was two, and all but one were instars. It appears that bugs (especially the instars) do not commonly travel on adult cliff swallows (and see section 4.9.1) but instead probably feed on the birds while they are confined to their nests at night (Myers 1928; George 1987) and drop off before the birds leave the nests in the morning. Most of the instar bugs found on adult swallows (principally first through fourth instars) were attached to the brood patch or lower belly and perhaps had been feeding when the bird left the nest.

However, each season we also observed a few adult and juvenile cliff swallows with one or more swallow bugs attached to their feet. Some of these birds had remarkable numbers of bugs clinging to various sides of their toes and tarsi—as many as 13 on one bird. All bugs seen on birds' feet were adults, and all observed cases occurred relatively late in the year, usually in July. We did not systematically record bugs on birds' feet until 1992–93, when we began to examine all birds caught in mist nets for them. In 1992 we found one or more bugs (total of 37) attached to the feet of 13 birds among a total of 21,272 bird captures during the season. (Each time a bird was captured and handled its feet were examined.) In 1993 we found one or more bugs (total of 59) attached to the feet of 32 birds among a total of 19,555 bird captures. As with instar bugs in the birds' feathers, adult bugs' traveling on the birds' feet was relatively rare. Swallow bugs thus occurred too infrequently on adult cliff swallows to permit any rigorous analysis of the effect of colony size, although we return to the issue of bug dispersal on adult birds in section 4.9.1, where we address between-colony transmission of ectoparasites.

Fleas were more commonly found on adult cliff swallows. In all cases, fleas that fell off the birds in the sampling jars had been hidden in the birds' feathers, and we seldom saw fleas moving on the exposed surfaces of a bird's body. The number sampled per bird ranged from 0 to 21. Flea parasitism of adult birds varied with time of year and with sex. Of 2,162 birds sampled in 1987–92 during early stages of nesting (arrival through egg laying), 440 (20.4%) had one flea or more; of 938 birds sampled during mid season (incubation), 131 (14.0%) had one or more; and of 2,028 birds sampled during late season (nestling feeding), 608 (30.0%) had one or more. The percentage of birds with fleas during each period differed significantly from each of the other periods (χ^2 tests, $p < .001$ for each pairwise comparison).

At all times of the year, male cliff swallows had more fleas than did females: early males, 0.433 ($\pm$ 0.030, $N = 1,574$) per bird, early females, 0.284 ($\pm$ 0.034, $N = 588$); mid males, 0.216 ($\pm$ 0.025, $N = 582$), mid females, 0.126 ($\pm$ 0.020, $N = 356$); late males, 0.684 ($\pm$ 0.044, $N = 1,133$), late females, 0.437 ($\pm$ 0.032, $N = 895$). The intersexual differences were significant for each time period (Wilcoxon rank sum tests, $p \leq .021$ for each). Perhaps males had more fleas than females because males may have visited more nests, both early in the season before settling (but see section 13.2.3) and later during trespassing interactions (section 5.5).

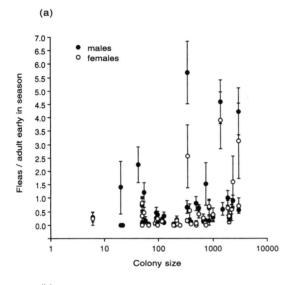
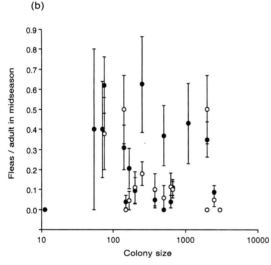

Fig. 4.10 Number of fleas per adult cliff swallow in early (a), mid- (b), and late (c) season in relation to colony size for male and female swallows. Fleas per adult early in the season increased significantly with colony size for females ($r_s = .41$, $p = .014$, $N = 35$ colonies) but not for males ($r_s = .24$, $p = .11$, $N = 47$ colonies). Fleas per adult did not vary significantly with colony size at midseason for either males ($r_s = -.28$, $p = .24$, $N = 19$ colonies) or females ($r_s = -.34$, $p = .22$, $N = 15$ colonies), or late in the season for either males ($r_s = -.04$, $p = .85$, $N = 27$ colonies) or females ($r_s = -.09$, $p = .70$, $N = 22$ colonies).

Therefore they may have encountered more fleas clustered at nest entrances than did females. Why male cliff swallows continued to have more fleas than females late in the season while feeding nestlings was not clear, however.

The number of fleas per adult bird increased with colony size for both male and female cliff swallows early in the season (fig. 4.10a), although the correlation was significant only for females. These results mirror those reported for flea loads of birds settling in colonies upon their arrival in the

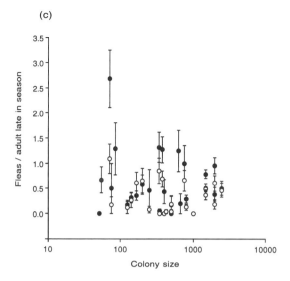

(c)

study area (section 13.3.2), but it is important to note that these data (fig. 4.10a) were taken over a longer period (extending through egg laying at some colonies) than those in section 13.3.2. That adult cliff swallows breeding in large colonies tended to have more fleas on their bodies than did birds in smaller colonies during the first part of the nesting season is consistent with the hypothesis, developed in section 13.3.2, that birds with higher initial parasite loads preferred to settle in larger colonies.

A different pattern emerged for fleas on adult birds at midseason and late in the year (fig. 4.10b,c). Fleas per adult did not vary with colony size for either sex during each period. By the end of the season, adult birds in larger colonies apparently had somehow diminished their flea loads relative to birds in smaller colonies at the same time. This could have occurred through the birds' own preening, or it may have reflected changing behavior of fleas through the season.

Was the presence of ectoparasites on adult cliff swallows costly? We examined the potential costs to adults by comparing subsequent survivorship (see chapter 12) of birds placed in the ectoparasite-sampling jars (section 2.6) versus those caught and handled but not placed in the jars at each site at the same time of the year (Brown, Brown, and Rannala 1995). Fumigation of adults in the sampling jars was similar to the nest fumigation experiment in which we studied the nest-based parasites' effects on nestlings by removing parasites and comparing those individuals with others exposed to natural levels of ectoparasitism (sections 4.7.2, 12.6).

Adult cliff swallows whose ectoparasites were removed had, on average, about a 14% greater probability of surviving to the next breeding season than did nonfumigated adults from the same colonies (Brown, Brown, and Rannala 1995). Based on this difference in survivorship and the observed frequency of ectoparasite occurrence on cliff swallows, we estimated that the presence of parasites (primarily fleas) on an adult reduced an individual's probability of annual survival by 33%. This long-term effect of ectoparasitism illustrates that the increasing flea infestations of adults in larger cliff swallow colonies (e.g., fig. 4.10a) may represent another parasite-related cost of coloniality and demonstrates a major cost of parasitism per se (see Brown, Brown, and Rannala 1995 for details).

4.9 TRANSMISSION OF ECTOPARASITES

This chapter thus far has described patterns of ectoparasitism within and among cliff swallow colonies without much explicit attention to how those patterns were generated. In this section we address transmission of ectoparasites and suggest why ectoparasitism increases with cliff swallow colony size. We should note at the start that transmission dynamics of swallow bugs and fleas have not been a primary focus of our research to date, and some of what we report here is speculation based on anecdotal observations. Ectoparasite transmission includes both that taking place between colonies and that within colonies.

4.9.1 Between-Colony Transmission

Both swallow bugs and fleas can move between cliff swallow colonies only by traveling on the adult birds themselves. Since neither parasite apparently remains with the birds during the winter, the birds presumably do not arrive in the study area in the spring carrying bugs or fleas. Transmission of parasites between colonies therefore occurs only when a bird visits one colony, picks up parasites, and then moves to another colony within the same season. Between-colony transmission of ectoparasites depends on how many (and perhaps which) cliff swallows move between colonies in a season and the number of ectoparasites those birds carry with them.

Birds that moved between colonies within a season could be classified into three groups (sections 13.6, 13.7). One group consisted of individuals early in the season that visited several colonies before settling in one. A second group included individuals whose nests failed during the season and that then moved to another colony, presumably to initiate another nesting attempt. The third group consisted of postbreeding individuals that moved between colonies late in the year immediately before migration.

We address within-year movement of birds in detail in section 13.6. Most of the movement occurred in the latter part of the season, suggesting that it was primarily by birds that had finished breeding and were simply visiting colonies before migration (section 13.6). This alone suggests that most of the ectoparasite movement between colonies probably occurred relatively late in the summer.

The overall number of cliff swallows moving between colonies was relatively small. On average about 5% of birds were caught at a second colony within the same season (section 13.6). Of those moving between colonies, relatively few probably carried swallow bugs. We detected bugs on only 0.34% of birds put in sampling jars, and only 0.11% had bugs attached to their feet (section 4.8). If we assume that the birds carrying bugs were a random sample of individuals that were not more or less likely to move between colonies, we can estimate that approximately 0.02% of cliff swallows transferred one bug or more between colonies in our study area. Thus bug movement between colonies apparently was rare.

How many bugs did the birds transfer between colonies? Among the 45 birds caught with adult bugs on their feet in 1992–93 (section 4.8), 15 (33.3%) had at least 2 bugs, with two of these birds having 6 bugs each, one bird having 10, and another having 13. (We may have slightly underestimated the total number of bugs per bird, however, since some bugs may have fallen off the feet before a swallow was removed from the mist net.) Thus, when bugs dispersed between colonies, often several did so at once. Nevertheless, the incidence of bugs observed on birds suggests that the number of swallow bugs moving between colonies is minute compared with the bug population in the study area as a whole or at each site. The population size per colony site can vary from 30 to about 100,000 bugs, depending on the history of use by cliff swallows (Rannala 1995, pers. comm.).

Another observation suggested, however, that bug movement between colonies at times could be substantial. In 1989, two colonies of 260 and 125 active cliff swallow nests (sites 8930, 8941) became active in culverts approximately 1 km apart along a highway that was under construction. Nesting progressed until approximately 20 June, when construction blocked the entrances to both culverts. The birds, most of which were feeding large nestlings at the time, abandoned their nests owing to the disturbance. Within a week many of the birds from both colonies occupied another culvert about 3.5 km away, outside the construction zone, that had been installed earlier that year (site 8960). This culvert had never before had any cliff swallows nesting in it. Eventually 245 nests were constructed in this new culvert, mostly (as banding revealed) by immigrants from the two abandoned colonies.

On 23 July, approximately four weeks after the new site was first occupied, we counted all swallow bugs that were observable on the outsides of 115 of the nests. These bugs were resting on the undersides of nests after having apparently fed on birds inside the nests, and all were adults or fifth instars. We counted 13,388 bugs! There undoubtedly were many more bugs not visible on the nests, and not all nests were sampled. Because these bugs were too old to have been reared at this colony in the relatively short time the birds had been there, all had to have been brought from elsewhere. This suggested a massive immigration of swallow bugs, unlike anything we would have predicted from our estimates of bird movement and bugs found on adult birds (also see Rannala 1995).

One could argue that this event was precipitated by human disturbance and hence was "unnatural." However, massive abandonments of colonies by cliff swallows occur in more natural contexts in our study area (at times perhaps owing to ectoparasitism but at other times for unknown reasons) and also were observed by Foster (1968) and Loye and Carroll (1991). Even if the abandonment of the two colonies was unnatural in that it was caused by humans, the number of swallow bugs observed at the new site suggests a high potential for bug dispersal between colonies. We are thus left to conclude tentatively that bug movement between colonies in general is probably rather limited but that occasionally colony abandonments may lead to extensive bug dispersal when alternative sites are colonized late in the season.

What sorts of behavioral strategies do the bugs employ to disperse? Bugs first have to make contact with birds that are likely to transport them between colonies. Among the cliff swallows found to be carrying adult bugs on their feet in 1992–93, 6 birds had been caught earlier in the same year. Of those, 4 (66.7%) had previously been caught at a different colony than where they were found carrying bugs. Among the 11 birds with instar bugs taken off them in the sampling jars that had been caught earlier in the year, 5 (45.4%) had previously been found at a different colony. Thus many of the bugs found were on cliff swallows that were moving between colonies.

Dispersal primarily on the birds' feet seems likely, based on bug behavior in unoccupied sites. If a colony is not used in a season or if some of the nests in an active colony fail, the bugs in those nests begin clustering on the entrances of the nests (fig. 4.11), much like fleas early in the year. Bug clustering began in late May and continued into July in unoccupied nests. If a cliff swallow lands even briefly at the entrance of an unoccupied nest, it is likely to pick up some of the bugs that are clustered there and carry them away on its feet. Clustering at nest entrances may be the only way bugs can effectively find a bird on which to disperse. Dispersal out of an unoccupied colony or nest is presumably advantageous to bugs because without hosts one summer they may have a poor chance of surviving another winter in the same place.

Bug dispersal may therefore depend on cliff swallows' visiting unoccupied nests or colonies, at least briefly, and would explain why adult bug

Fig. 4.11 Adult swallow bugs clustered at the entrance of an unoccupied cliff swallow nest. Clustering presumably enhances the bugs' probability of encountering an adult swallow on which they can disperse from a nest. If a cliff swallow lands at the nest entrance or brushes against it, the bugs crawl onto the bird's feet and thus are carried to another nest or colony. Photo courtesy of Art Gingert, Wildlands Photography.

movement (on feet) was most common late in the year when cliff swallows often visited several colonies (some unoccupied) before migrating (section 13.2.6). For example, we radio-tagged (section 13.2.6) one bird that was found with six bugs on its feet on 8 July 1992 and followed its subsequent movements for eight days. During this time the bird visited at least seven different colonies, entering nests and possibly picking up bugs and depositing others. On one occasion it entered a nest in an unoccupied section of a colony that had been very large the preceding year and that was presumably heavily infested with bugs trying to disperse. One bug that was paint-marked in an unoccupied colony site where bugs clustered was found several days later in an active colony about 1 km away (Rannala and Brown, unpubl.).

Between-colony transmission of fleas is not as well understood as transmission of bugs, although presumably the same general patterns pertain because fleas also can disperse between colonies only when birds move to another site during the season. Fleas apparently can and do travel on adults for longer periods than bugs do, and presence of a flea on an adult does not necessarily mean the flea will successfully disperse to a given colony or nest that the bird visits. Far more adult birds carried fleas than bugs; for example, approximately 30% of birds caught late in the season had at least one flea (section 4.8). Fleas likely were moved between colonies at all times during the summer, as suggested by the data on flea loads of adult cliff swallows (section 4.8), although they probably dispersed most often early in the season when they clustered at nest entrances (section 4.3.2). If a colony site was not used, the fleas there did not continue to cluster at nest entrances all summer. They either died or disappeared (see Humphries 1969), presumably crawling back into the nest, where they were unlikely to find a host. Where adult cliff swallows picked up fleas late in the season is not clear, unless they all came from each bird's own nest.

How can between-colony transmission of ectoparasites contribute to the observed increase in parasitism with colony size? If the cliff swallows already carrying ectoparasites are more likely to visit and settle in large colonies, then more parasites per capita will be introduced by settlers into those large colonies. There was evidence that birds carrying fleas were more likely to settle in large colonies early in the year (section 13.3.2). Preferential settlement of infested birds (for whatever reasons) can increase average parasite loads at a site and may explain, for instance, the increased numbers of fleas on adult cliff swallows in larger colonies early in the year (fig. 4.10a). Similar patterns may exist for birds carrying bugs, especially if certain colonies are founded largely by birds coming from (and importing bugs from) heavily infested colonies where their earlier nesting attempts failed, as our observations suggest.

In the absence of host sorting among colonies, if between-colony parasite transmission occurs relatively infrequently but several ectoparasites are introduced whenever a dispersal event occurs (as the data on the bugs found on the birds' feet suggest), the variation in mean parasite load between small colonies will be greater than that between large colonies. This occurs because virtually all large colonies (because of the many birds there) will contain a few birds carrying bugs, unlike small colonies in which some will happen to have some infested birds and others will have none. Whether the mean parasite load among all large versus all small colonies will be affected by between-colony transmission in the absence of host sorting is unclear, however, and may depend on how many parasites are introduced to a site relative to the timing and magnitude of the subsequent parasite population growth there.

4.9.2 Within-Colony Transmission

Swallow bugs are transmitted between nests within a cliff swallow colony when they crawl along the substrate. Fleas apparently can jump from one nest to another if nest density is high. In addition, both parasites may be moved from one nest to another by traveling on a bird that flies between two or more nests, such as during trespassing interactions early in the year (section 5.5). We know that swallow bugs often move between nests on the substrate, but we have little idea how often fleas jump between nests or how often either parasite rides between nests within a colony on adult birds. Paint-marked bugs moved up to 65 m from one nest to another within a colony over a three-day period while birds were feeding nestlings (B. Rannala, pers. comm.), although how the bugs moved was unknown.

Bug movement between nests is probably determined in part by nest spacing and in part by temporal differences in host nesting stage. If swallow bugs disperse on the substrate by random walks, a disperser is more likely to encounter another nest in large colonies or ones with densely packed nests. Consequently, more of the bugs in the large and dense colonies are likely to disperse successfully, thereby maintaining or increasing the bug population size in those sites (section 4.3). Substrate dispersal may be riskier when nests are spread out. Crossing large expanses of substrate exposes bugs to predators such as spiders (Myers 1928; Usinger 1966), or they may encounter extreme temperatures without adequate crevices in which to thermoregulate, such as when swallow bugs crawl across hot cliff faces exposed to direct sunlight (George 1987). These risks of within-colony dispersal probably prevent many bugs from reaching some of the more isolated nests (e.g., table 4.4) and may contribute to lower bug population sizes in the smaller and less dense colonies (section 4.3).

Swallow bugs clearly discriminated among cliff swallow nests based on a nest's contents. They tended to aggregate at nests where young swallows had hatched, and given an experimental choice between nests with eggs or ones with nestlings, bugs moved into the ones with nestlings (P. Walsh, pers. comm.). This was illustrated by the counts of bugs on the outsides of active nests at the 245 nest site (8960) described in section 4.9.1. These counts were all taken on the same day, and at that time we classified each nest as either being empty, having eggs, having newly hatched nestlings (less than 5 days old), or having nestlings 5 or more days old (none were older than about 12 days). Counts were 3.4 ($\pm$ 2.3) bugs for empty nests ($N = 10$), 7.1 ($\pm$ 3.2) bugs for nests with eggs ($N = 9$), 58.5 ($\pm$ 11.7) bugs for nests with newly hatched nestlings ($N = 47$), and 215.1 ($\pm$ 25.7) bugs for nests with older nestlings ($N = 49$). Nests of these different stages were often side by side. Bugs clearly concentrated on nests with older nestlings that presumably provided larger blood meals.

The general pattern apparently is for swallow bugs to prefer to parasitize nestlings up to about 12–15 days of age, when the birds begin to develop more feathers and begin to preen vigorously. Bug numbers in those nests then decline as bugs move to other nests with younger nestlings (P. Walsh, pers. comm.). Thus bugs move between nests in search of nestlings of the appropriate age, and their movement and search is probably easier when nests are close together. Their tendency to aggregate may also explain in part the increase in parasite load with date (table 4.1). Not only are more bugs present later in the season (having been reared in the colony or immigrating), but the ones there concentrate on increasingly fewer late-starting nests as the nestlings in earlier nests fledge. As a result, late cliff swallow nests are seldom successful in large colonies.

Our understanding of both between- and within-colony ectoparasite transmission dynamics in cliff swallows is still fragmentary. It is safe to conclude that both occur and that at times both may be important. The patterns of parasitism reported in this chapter can probably be best interpreted when we better understand transmission and how it varies with colony size, colony spatial structure, bird nesting stage, and seasonal factors.

4.10 BEHAVIORAL RESPONSES TO ECTOPARASITISM

It should be obvious by now that ectoparasitism in cliff swallows is a complex story. Not all the observed patterns were ones we predicted, and not all are explainable based on what we currently know about parasite or cliff swallow population biology. However, two relatively clear results were that swallow bug and flea parasitism tended to increase with cliff swallow colony size and that parasitism by swallow bugs had deleterious effects that also seemed to increase with colony size. Thus swallow bug parasitism at the least represents a major cost of cliff swallow coloniality. If so, the birds should exhibit adaptations to counter or at least ameliorate the negative effects of ectoparasites.

In this section we examine cliff swallows' behavioral responses to ectoparasitism. The potential responses we identified include direct assessment of parasite load by birds, active avoidance of nests and colonies where parasite loads are high, and building of new nests to avoid "inheriting" parasite loads from earlier nesting attempts.

4.10.1 Assessment of Parasite Load

Cliff swallows apparently assessed the parasite load of nests before settling in the spring (Brown and Brown 1986), as do other swallows (Bates 1962; Moller 1990). On arriving at a colony, cliff swallows typically hovered a few centimeters in front of the entrances of old nests, not entering them. At these times fleas and bugs were quite obvious as they clustered at the nest entrances. Birds would fly from nest to nest, hovering at each and apparently assessing parasite load by viewing the parasites. Sometimes birds would leave a colony altogether without actually entering any nests. If the cliff swallows began entering nests, it usually meant the site would become active.

The birds' responses to nest fumigation indicated that they were sensitive to parasite load. For one of the colonies (site 30) used in the nest fumigation experiment (section 4.7.2), we kept the same half fumigated and the same half nonfumigated for five consecutive years. The first year, birds settled before we began the fumigation. In that year there were 122 active nests in the portion we began fumigating and 233 nests in the portion not fumigated. The following year, after a season of fumigation, 90 nests were active in the fumigated half and no nests were occupied in the nonfumigated half, where many bugs remained from the previous summer. In the successive years, occupancy was as follows: 222 nests in the fumigated section, 82 in the nonfumigated; 262 nests in the fumigated, 147 in the nonfumigated; 295 nests in the fumigated, 80 in the nonfumigated. These differences could not have reflected merely substrate or nest availability in the respective parts of the colony, because the colony was in a culvert that permitted us to divide the substrate area into equal halves. Old nests remained intact in both sections from year to year. The birds clearly preferred to nest in the previously fumigated nests without bugs.

The same general results were obtained at another colony (site 05) that had been used for the nest fumigation experiment. The next year we discovered 39 active nests in the section fumigated the previous year and only 3 in the section that had not been fumigated. After determining the occupancy of these nests, we fumigated the entire colony. Although birds had previously shown little interest in the nonfumigated portion of the colony, within eight days large numbers began establishing nests throughout the colony. Eventually 456 additional nests were constructed or became active. Of those, 174 were in the former fumigated half and 282 were in the former nonfumigated half. The birds clearly responded to our complete removal of parasites from the colony. These experimental results indicate that the birds apparently can assess parasite load and select parasite-free nests. The same results were seen in other colonies in the study area where portions were fumigated and cliff swallows quickly moved in (P. Walsh, pers. comm.) and at a Wisconsin colony where ectoparasites were removed (Emlen 1986).

Cliff swallows may also be able to assess ectoparasite load in other contexts. We discovered that when brood parasitizing the nests around them (section 6.3.1), the birds preferentially selected neighboring nests that ultimately were to have the lowest infestations of swallow bugs and fleas among those nearby (Brown and Brown 1991). The birds were making these assessments relatively early in the season during egg laying and were therefore assessing ectoparasite load among active nests. We do not know what cues they used, but that they selected nests based on ectoparasite load suggests that there are indeed advantages to not being heavily parasitized, even by fleas.

4.10.2 Avoidance of Parasites: Abandoning Nests and Colonies

If cliff swallows can assess ectoparasite load either early in the season before settling or after nests become active, they should be able to avoid infested sites. We repeatedly saw birds abandon nests with eggs and small (usually moribund) nestlings, and this typically occurred late in the season in medium-sized to large colonies. We could not be certain what caused the abandonment, but in each case the nests were heavily infested by swallow bugs. Late nests in larger cliff swallow colonies are routinely abandoned, in all likelihood because of increasing numbers of bugs.

A more effective way for cliff swallows to avoid ectoparasites may be to avoid infested colony sites altogether. Since the bugs and fleas overwinter in the nests and cliff swallows represent their principal food source, abandoning an entire site for two or more years may effectively reduce ectoparasite populations. Various authors have suggested that cliff swallows' erratic and somewhat unpredictable annual colony site use patterns diminish parasite loads (Grinnell, Dixon, and Linsdale 1930; Earle 1985; Brown and Brown 1986; Chapman and George 1991; Loye and Carroll 1991). Annual site use is indeed unpredictable throughout most of the bird's range, some colonies being used perennially and some used only at intervals of up to five or more years (Hopla and Loye 1983; Sikes and Arnold 1984; Chapman and George 1991; Loye and Carroll 1991; Brown and Brown 1995).

We defer a detailed discussion of colony use patterns to section 7.4. However, if alternate-year site use is primarily a response to ectoparasitism, we can make one major prediction. A large colony (with more bugs) should be less likely to be used the following year than a small colony (with fewer bugs). Use of large colonies thus should be more erratic from year to year. This prediction was not supported for colonies in our study area. Large colonies were more likely to be reused in successive years, and their size was more stable relative to small colonies (section 7.4; see also Aumann and Emlen 1959). But what determines site use is a complex problem, and it is probably safe to conclude only that ectoparasitism does not solely determine whether a site will be used in a given year. Some results were in fact consistent with the hypothesis that ectoparasitism at least influences site use. For example, one or more years were more likely to elapse between successive uses of culvert sites than of bridges independent of colony size (section 7.4.2), perhaps because culverts had more swallow bugs than did bridges (section 4.3.2). Even though small colonies were less likely to be reused the next year, when reused they were more likely than large colonies to increase in size (section 7.4), perhaps because existing ectoparasite loads prevented further size increases in the large colonies. We emphasize again that alternate-year site use is probably not solely a response to ectoparasitism but may be influenced by it.

Another way of avoiding ectoparasites is by dispersing away from them between years (Brown and Brown 1992). Nestlings raised in nests infested with fleas and bugs were more likely to disperse to a nonnatal colony the next year, whereas those raised in relatively uninfested nests returned to

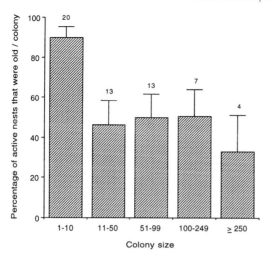

Fig. 4.12 Percentage of active nests per colony that were old in relation to cliff swallow colony size. Sample sizes (number of colonies) are shown above error bars. Percentage of nests that were old declined significantly with colony size ($r_s = -.49$, $p < .001$, $N = 57$ colonies).

their natal colony. Extent of parasitism in the natal nest may be one cue individuals use to avoid potentially heavily infested colonies the next year (Brown and Brown 1992). Natal dispersal is addressed in more detail in section 13.5.

4.10.3 Nest Reuse versus Building New Nests

Another potential tactic for avoiding ectoparasites is to build a new nest and thereby escape the parasites overwintering in an existing nest. If new nests confer any antiparasite advantage, we might predict that new nest construction should increase in larger colonies where parasite infestations are highest. In small colonies with fewer parasites, birds should reuse old nests to save the time and energy costs of nest construction. For 57 colonies ranging in size from 1 nest to 1,600 nests, we determined the percentage of active nests in the colony that were old. We averaged these percentages for colonies of different size classes (fig. 4.12). As predicted, the percentage of active nests in a colony that were old declined significantly with colony size (fig. 4.12). Cliff swallows thus were more likely to build new nests in larger colonies and seldom built new nests in the smallest ones. This result is consistent both with the finding that swallow bug parasitism of old nests increased with colony size (fig. 4.3a) and the hypothesis that building a new nest helps birds avoid parasites.

However, it is unclear how much antiparasite advantage a new nest really confers. A variety of analyses indicated no consistent parasite reduction in new nests relative to old ones in the same colony (section 4.5), and in some instances new nests seemed to have more parasites. New nests may provide at best only a temporary, initial reduction of parasites until swallow bugs from nearby old nests move into the newly built ones. Since bugs are mobile and move freely between nests (section 4.9.2), new nests built near old ones are probably infested rather quickly. Swallow bugs often hide in crevices of a colony's substrate (Foster and Olkowski 1968; Zack 1990; Brown and Rannala, pers. obs.) and may persist at colony sites even if old swallow nests are destroyed by the elements during the fall and winter. Thus, despite the increased incidence of new nest construction in larger colonies (fig. 4.12), whether this is in fact a response to ectoparasitism and whether it helps birds effectively avoid parasites is problematic. Also, building new nests may be less costly, in time and energy, in large colonies than in small ones owing to the opportunities to share walls when nest density is high (section 5.3.2). This may account in part for the increased use of new nests in the larger colonies.

4.11 SUMMARY

Theory suggests that the incidence of parasitism and disease should increase in larger groups of hosts, and several field studies of colonial animals in recent years support this prediction. Cliff swallows in southwestern Nebraska are parasitized principally by the hematophagous swallow bug (*Oeciacus vicarius*) and bird flea (*Ceratophyllus celsus*). These insects are nest-based ectoparasites that are mostly confined to occupied and unoccupied swallow nests throughout the year, disperse between colonies relatively rarely, and parasitize cliff swallows almost exclusively. Ectoparasitism of nestling cliff swallows by both bugs and fleas increased with colony size. Colonies that had been large the previous summer had more fleas per nest the next spring than did colonies that had been small. Bug parasitism increased with the nest density of a colony.

Swallow bug parasitism within a colony increased as the nesting season progressed, but flea parasitism exhibited no clear relation to date. Newly built nests averaged slightly more bugs and fleas than old nests, perhaps because parasites early in the season moved out of old, inactive nests and into new nests where they may have been more likely to find hosts. Bug parasitism increased with colony size for old nests but not for new ones. The size of a cliff swallow nest had little consistent effect on parasitism by either fleas or bugs. Nests within a colony that were more spatially isolated from other nests had fewer swallow bugs, but a nest's position in the colony relative to the centermost nest had no consistent effect on parasite load. Variance in bug load among nests was greater within large colonies than in small colonies. Variance was generated either by parasites' preferentially infesting certain classes of birds, through random movement and overwintering distributions of the parasites, or both.

Swallow bugs had severe effects on nestling cliff swallows, reducing body mass of those that survived and causing a 50–100% reduction in nestling survivorship in larger colonies. Ectoparasitism by bugs thus represented a major cost of coloniality. Fleas had little observable effect on nestlings. Up to 30% of adult cliff swallows had one or more fleas on them late in the nesting season, and flea parasitism of adults (especially of females) tended to increase with colony size early in the season.

Transmission of swallow bugs between colonies probably occurred mostly when birds visited unoccupied nests where bugs clustered. Adult bugs attached themselves to the birds' feet and traveled to new colonies, mostly late in the season when birds investigated different colonies before migrating. Occasionally mass abandonment of colonies may have introduced large numbers of bugs into late-starting colonies. Within colonies, bugs moved freely between nests, especially those close together, and aggregated at the nests with relatively young nestlings where they could feed most efficiently.

Cliff swallows assessed parasite loads at nests early in the spring and avoided those that were infested from the previous summer. There was no evidence that alternate-year use of colony sites was determined solely by ectoparasite load. Birds were more likely to build new nests in large colonies, possibly to try to avoid the overwintering parasites present in old nests.

5 Competition for Nest Sites

> Doubtless the Lord—to paraphrase Lincoln's aphorism—must love the cliff swallows, else he would not have made so many of them. Common they unquestionably are; yet I do not know that they are altogether lovable.
> William L. Dawson (1923)

5.1 BACKGROUND

Most population biologists agree that intraspecific competition generally increases with population density (e.g., Brown 1964; Buss 1981; Wittenberger 1981; Slobodchikoff and Schulz 1988; Hoogland 1995). This assumption is based on the observation that most resources are limited and therefore cannot sustain all potential individuals that have identical resource requirements. Alexander (1971, 1974) used this argument to suggest that intensified competition for resources is an inevitable consequence of group living and that this cost will increase with group size.

Resource competition within species is perhaps most easily measured by focusing on its behavioral manifestations. For example, one obvious consequence of competition is increased aggression and fighting as individuals attempt to control access to resources (e.g., Birkhead 1978; Wittenberger 1981; Christenson 1984; Rendell 1993). Another consequence may be increased reproductive interference among conspecifics as each tries to manipulate or disrupt others' activities to its own advantage (e.g., Vehrencamp 1977; Bertram 1979; Mumme, Koenig, and Pitelka 1983; Crook and Shields 1985; Chardine 1986; Brown and Brown 1988c; Schleicher, Valera, and Hoi 1993; Bensch and Hasselquist 1994; Hotta 1994; Lifjeld and Marstein 1994; Hoogland 1995; Riley et al. 1995). Even individuals that win contests for resources still suffer the costs of competition, and these costs should increase with group size and population density (Alexander 1971, 1974; Hoogland and Sherman 1976; Buss 1981).

Colonial animals compete for a variety of resources, and an important one is breeding site. An individual's ability to raise offspring can be greatly affected by the quality of its nest site or whether it can find a site at all. Most if not all colonial animals apparently compete for nesting sites (reviewed in Wittenberger and Hunt 1985). This is especially true in species that aggregate in limited nesting habitats where not all individuals can be accommodated, but competition for optimal breeding sites may also be important in colonial species that are not obviously limited by habitat (Coulson 1971; Baird and Baird 1992).

Within a colony, certain nesting sites may be better than others, and the intensity of competition for these sites should vary with the differential between the good and poor sites. This differential may often increase with colony size because the relative advantages of particular spatial positions can change with colony size. For example, in a large colony, nests in the center will have many neighbors and, for various reasons, may be less likely to suffer predation than more peripheral nests (reviewed in Burger 1981 and Wittenberger and Hunt 1985; section 8.7). Small colonies do not have as clearly defined centers and edges, and thus spatial position may have little influence on expectation of success when colony size is small. The consequence could be increased competition for center nests in larger colonies as residents jockey for position and at times try to usurp other birds' nests. Large colonies may also attract a greater fraction of the individuals that are trying to find a nest site, presumably because (all else being equal) there may be better odds of encountering a recently vacated nest site in a large colony than in a small one.

We know of no studies that have systematically evaluated nest site competition as a function of colony size. The most closely relevant studies are those of Hoogland and Sherman (1976) in which competition for nesting materials seemed to increase with colony size in bank swallows, and of Birkhead (1978) and Hoogland (1979a, 1995) in which the incidence of fighting (for unspecified reasons) increased with nest density in common guillemots and with colony size in black-tailed prairie dogs. Fights for nesting sites were greater at a dense white-tailed tropicbird colony than in a less dense colony, and the fights were sometimes violent and bloody (Schaffner 1991). In cattle egrets, loss of nesting material to thieving conspecifics was a major cause of nesting failure (Siegfried 1972), although how (or if) theft varied with colony size was not known. To the degree that nesting materials are costly to collect or limited in supply, individuals should fight to retain those they do collect and simultaneously should seek to steal from unattended nests. Opportunities to steal and be stolen from presumably both increase with colony size.

This chapter focuses primarily on competition for nest sites and associated interference among neighboring individuals. In cliff swallows this interference takes the form of furtive trespassing into neighboring nests in competition both for nesting material and for access to appropriate nests that can be brood parasitized (section 6.3). Interference also occurs as individuals trespass and seek to destroy other birds' eggs, for poorly understood reasons (Brown and Brown 1988c). These activities are presumably costly to at least some individuals, and we address the questions of whom competition and interference occur among, their reproductive consequences, and how they vary with colony size. We defer discussion of competition for mating opportunities to chapter 6 and competition for food to chapter 10.

5.2 FIGHTING FOR NEST SITES

5.2.1 Natural History of Fighting

As soon as a cliff swallow colony site became active in the spring, individuals began to defend all existing nests or nest remnants in the active sections of the colony. In most cases, these nests or nest remnants were rapidly taken over, and few if any remained unoccupied after the first two or three days a site was active. Fights developed as two or more individuals contested a nest or nest site. These fights were often violent, with birds grappling at the nest site or in flight and falling to the ground or into water below the nests. Individuals frequently chased away would-be usurpers in flights that could extend several meters from the contested nest or nest site. Individuals fought over nests in all stages of completion (section 5.2.3), especially ones ranging from partial bottoms to three-fourths present/no neck (fig. 2.2), but cliff swallows also contested positions on a cliff face or concrete wall that were devoid of mud. Fights at particular nest locations could be frequent on a given day, and although in most instances at least one of the birds involved was unmarked, it appeared that the same individuals fought each other repeatedly for several minutes to several hours.

Observations of color-marked birds at selected colonies (section 2.5) revealed that these violent fights almost invariably involved at least one bird that had recently arrived in a colony and presumably had not yet established ownership of a nest or nest site. Either both combatants were recently arrived or (more often) one bird had been present consistently at a nest site for several days, thus presumably "owning" it, and the other was a newly arrived intruder that attempted to usurp the occupied nest site. As cliff swallows established clear ownership of nests and newly arrived birds declined in abundance, violent fighting over nest sites largely ceased (at which time trespassing among neighboring birds began; section 5.5). Early-season fighting probably often involved at least one male, given that males generally tended to settle at colonies earlier than females (sections 3.5.3, 13.2.5).

The temporal progression of fighting for nest sites was studied in detail at six colonies. These sites ranged from 10 to 2,000 active nests. We quantified fights daily or every second or third day as soon as birds began showing an interest in each colony site. For two to four hours on each day of observation, we recorded all fights or defensive chases per nest at each colony, the total number of active nests, and the total number of nests under construction at that time. For small colonies we watched all active nests, but for large colonies we could watch only a sample, usually 20–60 nests. This sample was selected as randomly as possible and encompassed nests in all stages of completion. We watched the same sample each day or alternated between the same two samples on successive days.

For each colony, the average fights and chases per hour per nest each day was compared with the cumulative percentage of nests in the colony that were complete each day (fig. 5.1). In this analysis a complete nest was one with a fully constructed neck (fig. 2.2). Often the necks (or larger portions) of old nests would break over the winter, and thus most colonies had substantial numbers of incomplete nests at the start of each season, regardless of the nests' ages. With the exception of the largest colony, there was a relatively narrow window of time when fights and chases peaked in each colony (fig. 5.1). In most cases fights began at a low level at the time a site was first occupied, rapidly increased in frequency, then quickly declined. The peak period of fighting preceded the completion of most nests; as nest completion increased at a colony, fighting declined and virtually ceased when all nests were completed (fig. 5.1). This pattern is consistent with our observations that fights were mostly between birds contesting nest sites early in the season. The pattern is also consistent with the assumption that a bird's best chance of winning a nest site through fighting is before the nest is completed and the perimeter of defense becomes reduced (section 5.2.3).

5.2.2 Fighting in relation to Colony Size

The six cliff swallow colony sites studied in detail provided a comparison of fighting among birds in colonies ranging from 10 to 2,000 nests (fig. 5.1). These sites differed slightly in the temporal pattern of fighting, with some colonies showing more pronounced peaks than others, but the birds occupying nests in small colonies clearly experienced as many fights and chases per hour as did birds occupying nests in large colonies. Often fighting seemed more frequent in small colonies (fig. 5.1).

We did a more direct comparison of fighting in relation to colony size by taking an average fight rate per nest for the six colonies in figure 5.1 plus eight additional sites ranging from 2 to 350 nests. Because the degree of nest completion within a colony obviously affected the rate of fighting, for this comparison we used only data from days on which more than 50% of all active nests at the site were still incomplete. We averaged the daily mean values for fights and chases per nest per hour over all days with more than 50% incomplete nests. This gave a single mean estimate of fight rate for each colony. The eight additional colonies used in this analysis were ones studied only during the peak period of fighting.

Using the single measure of fight rate, fights and chases per hour per

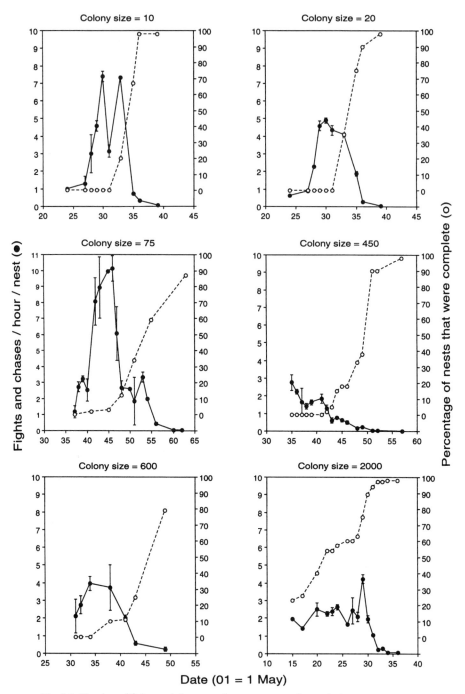

Fig. 5.1 Number of fights and chases per hour per nest and cumulative percentage of nests that were complete each day during nest establishment, construction, and egg laying periods at six cliff swallow colonies of different sizes. Note similar scales of ordinate.

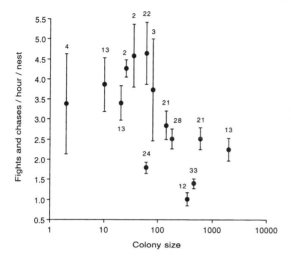

Fig. 5.2 Number of fights and chases per hour per nest in relation to cliff swallow colony size. Numbers above error bars are total hours of observation time at each colony. Fights and chases declined significantly with colony size ($r_s = -.64$, $p = .014$, $N = 14$ colonies).

nest declined significantly with colony size (fig. 5.2). This result was consistent with the pattern apparent from the colonies studied in detail (fig. 5.1). Fight rate in smaller colonies was about double that in larger colonies. It appeared that a colony size of about 100 nests was a threshold below which fighting seemed high and above which it was markedly lower (fig. 5.2). Contrary to what we predicted at the outset, cliff swallows in large colonies fought less than birds in small colonies. Possible reasons are discussed below (section 5.2.4).

5.2.3 Fighting in relation to Nest Position and Completeness

Fighting was more frequent at center nests. In five colonies ranging from 35 to 600 active nests, we directly compared fight rate for center nests versus edge nests. These colonies were ones in which clearly defined center and edge nests (section 2.8.2) were active simultaneously. We scored fights and chases simultaneously on the same days at both center and edge nests in each colony. The mean number of fights and chases per hour per nest per colony was 4.34 ($\pm$ 0.44) for center nests versus 2.36 ($\pm$ 0.64) for edge nests. The small number of colonies ($N = 5$) prevented a rigorous matched-pairs statistical comparison across all colonies, but for three we had adequate observational samples for within-colony statistical comparisons of fighting in center and edge nests (in the remaining two total observation time was short, less than 3 hours). In all three colonies, fights were significantly more numerous at center nests than at edge nests (Wilcoxon rank sum tests, $p \leq .055$ for each).

Earlier we showed that fighting for nest sites diminished as cliff swallow nests became complete (fig. 5.1). This suggests that degree of nest com-

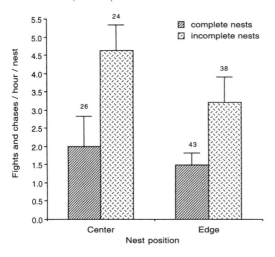

Fig. 5.3 Number of fights and chases per hour per nest in relation to nest position of complete and incomplete cliff swallow nests at a 140 nest colony. Numbers above error bars are number of nests observed. Among incomplete nests, fights and chases were significantly more numerous at center nests than edge nests (Wilcoxon rank sum test, $p = .017$), but there was no significant difference between center and edge among complete nests (Wilcoxon rank sum test, $p = .68$).

pleteness influences fight rate. Thus if completeness consistently varies with spatial position, center-edge differences might reflect only nest completeness and not necessarily spatial position per se. We evaluated this possibility by examining how fight rate varied with nest position for nests of similar degrees of completeness. This was done in one colony of 140 active nests situated on a cliff and of a circular geometry (fig. 3.4) in which center and edge differences should be greatest. There were relatively large numbers of concurrently active complete and incomplete nests both in the center and on the edge of this colony. Simultaneous observations of these nests revealed that fights increased at center nests, relative to edge nests, for both classes of nest completeness (fig. 5.3). The increase was significant for incomplete nests but not for complete nests. Fight rate was highest at incomplete center nests, followed by incomplete edge nests, and was lowest at complete edge nests.

Our conclusion is that cliff swallows fought most at incomplete nests regardless of spatial position but seemed to favor center nests among those of the same degree of completeness. Thus both nest position and degree of completeness influenced fight rate. Although completed nests are presumably a more valuable resource and for that reason one might expect more fighting for them, increased fighting at incomplete nests (figs. 5.1, 5.3) probably occurred for four reasons. First, incomplete nests were likely to be owned by more recently arrived birds that may have established less site dominance (Davies 1978) than birds owning complete nests. Consequently, the owners of incomplete nests may have been more easily evicted. Second, an incomplete nest offered a considerable perimeter for intrusion by would-be usurpers (Emlen 1954), unlike a complete nest, whose narrow entryway could be more easily defended by an owner sitting in its entrance. Third, incomplete nests still required nest construction, and thus their owners could not be constantly present at the nest and also gather mud. This may have increased a usurper's chances of gaining access to a temporarily vacated nest. Fourth, complete nests represented a large investment, so their owners fought harder to retain them. Consequently invaders probably were unlikely to acquire them, and many did not bother to try.

5.2.4   Fighting as a Cost of Coloniality

Fighting per se was presumably costly to cliff swallows in terms of both time and energy and the risk of injury or death. We had no direct way to estimate the time and energy invested in fighting (section 5.3), but we did document that it can on occasion lead to death. Many of the colonies in our study area were over water, and often we saw grappling birds tumble out of nests and hit the water. In six instances one of the combatants drowned in rapidly flowing water. In at least eight other cases a fighting bird landed in the water, became waterlogged, and swam to shore, where it was very vulnerable to predation.

However, there was no evidence that these costs of fighting increased in larger colonies or that the incidence of fighting was higher in larger colonies. If anything, fighting declined with colony size (fig. 5.2), perhaps because much of the fighting over nests was between nest owners and unestablished individuals that probably visited both large and small colonies attempting to find nest sites (e.g., section 13.2). If we assume the existence of a finite pool of wandering, unestablished birds that circulated among all colonies, the probability that a given nest will experience a fight declined in large colonies simply because more nests were present there to attract the attention of a given wanderer. Consequently, a large colony may dilute each resident's own chance of encountering a nest-seeking wanderer. Our finding that fighting did not increase in larger colonies also suggests that unestablished individuals did not recruit preferentially to large colonies at the start of the nesting season.

This pattern (fig. 5.2) may also partly be a sampling artifact. Because we scored fights only in colonies where more than 50% of nests were incomplete, any site with primarily complete nests at the beginning of the season could not have been included in our sample. If small colonies routinely had a higher percentage of complete nests, perhaps more of those sites would have essentially no fighting. This might reduce the overall average per-colony fight rate for the smaller colonies, although the observed decrease in fighting with colony size for sites with more than 50% incomplete nests (fig. 5.2) is still real.

The increased rate of fighting at center nests, relative to edge nests of the same degree of completeness (fig. 5.3), suggests that there are advantages to nesting closer to a colony's center. One such advantage is enhanced avoidance of nest predators (section 8.7). We suggested earlier (section 5.1) that the advantage of center nests should increase with colony size, because the center confers more of a benefit in a larger colony. This was supported by our center-edge comparisons in the three colonies where we had adequate simultaneous observations of fighting at both center and edge nests (section 5.2.3). These three colonies had 60, 140, and 600 active nests. The increase in average fights and chases per hour per nest in center nests (relative to edge nests) was 0.65 in the 60 nest colony, 0.96 in the 140 nest colony, and 1.56 in the 600 nest colony. Although limited, these data suggest that cliff swallows fought more over center nests as colony size increased. This may be because there are both advantages to being in the center in large colonies and strong disadvantages to being on the edge, especially if predators are attracted to large colonies and attack the edgemost nests first (sections 8.7, 8.9).

5.3 NEST BUILDING AND COLONY SIZE

Individual cliff swallows that owned nest sites often had to repel attempted intrusions and usurpations by unestablished birds. From the nest owner's perspective, a potentially serious consequence of fighting is that it interferes with nest construction and may delay completion of the nest. When wandering birds were present at a colony and attempting intrusions and takeovers, birds with incomplete nests often remained at their nests to defend them. Increased rates of fighting in certain colonies (e.g., fig. 5.2) can therefore reduce the amount of time per day that established nest owners can spend gathering mud, making it take longer to complete a nest and thus delaying the onset of egg laying. Such a delay is not trivial, given the increase in deleterious swallow bug infestations as the summer progresses (section 4.4) and the overall reduction in reproductive success in later nests (section 11.4).

5.3.1 Time Taken to Build Nests

The best analysis for this issue would be a direct comparison of total nest building time with the amount of fighting at a given nest site. Unfortunately this was not feasible, given the laborious nature of collecting meaningful fight data at nests. We would have had to watch nests for multiple hour-long periods over the entire time of nest building, and we would have had to watch each nest from the day construction first began. These conditions prevented us from obtaining good focal nest data. However, an indirect way to measure the potential effect of fighting on nest construction time was to examine in general how nest building time varied with

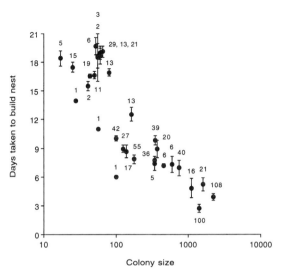

Fig. 5.4 Number of days taken to build a cliff swallow nest in relation to colony size. Sample sizes (number of nests) are shown above error bars. Number of days declined significantly with colony size ($r_s = -.82$, $p < .001$, $N = 31$ colonies).

colony size, since we had determined how fighting varied with colony size (fig. 5.2).

For this analysis we used thirty-one colonies ranging from 17 to 2,200 active nests. Colonies smaller than 17 nests could not be used, because cliff swallows seldom built new nests in those colonies, instead refurbishing old ones (fig. 4.12). In each colony we used nest checks (section 2.2) to identify new nests under construction. The date when a new nest was first classified as a "partial bottom" (fig. 2.2) was taken as the date when its construction started, and the date when it was classified as "complete" was taken as the date construction ended. We calculated the elapsed time (days) between these dates as the nest building time. Only nests that we first found as partial bottoms and that remained active until completion were included in this analysis.

Nest building time declined markedly as colony size increased (fig. 5.4). Entire nests were constructed in only 3–5 days in the largest colonies, while routinely taking 15–20 days in the smallest colonies. This result is especially striking when one considers that most large colonies became active early in the season (fig. 3.10) when days often were too cold or rainy for nest building. Despite these weather constraints, residents in large colonies nevertheless were able to build nests much more rapidly than birds in small colonies. Small colonies often began later in the summer when weather was routinely more conducive to nest building.

These results on nest building time (fig. 5.4) were consistent with our finding that fighting was more frequent in small colonies (figs. 5.1, 5.2). It is probable that some of the delay in the smaller colonies was attributable

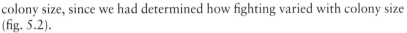

to the increased fighting and the time lost in continual nest defense, but other factors may also have contributed to the slower nest building. For example, cliff swallows that gathered mud for their nests in large groups (at large colonies) did so more efficiently than individuals in small groups, probably because they could devote less time to scanning for predators (section 8.4; Brown and Brown 1987). Consequently, more mud was gathered per unit of time in large colonies, likely hastening nest building.

5.3.2 Wall Sharing

Another reason nest building time declined with colony size (fig. 5.4) may have been that large colonies had higher nest densities (section 7.2.1), so that a nest was more likely to share walls with neighboring nests. If two nests share a wall, nest building time can be reduced because less total nest construction per bird is necessary (Gauthier and Thomas 1993). For each nest for which we knew building time, we also scored how many neighboring nests it eventually abutted (touched) on any side, providing a relative index of wall sharing potential for newly constructed nests. Nests in larger colonies averaged markedly more abutting nests than did those in small colonies (fig. 5.5).

Wall sharing probably accounted for some of the savings in nest building time in larger colonies. A partial correlation analysis showed that nest-building time decreased significantly with an increase in the number of abutting nests ($r = -.59$, $p < .001$) independent of colony size and date the nest became active. However, another partial correlation analysis showed that nest building time also decreased significantly with increasing colony size ($r = -.41$, $p = .010$) independent of the number of abutting

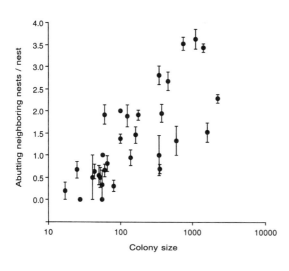

Fig. 5.5 Number of abutting (touching) neighboring nests per nest in relation to cliff swallow colony size. These are the same colonies as in figure 5.4, where sample size for each colony is shown. The number of abutting nests increased significantly with colony size ($r_s = .80$, $p < .001$, $N = 31$ colonies).

nests and date. Thus, the reduction in nest building time with colony size (fig. 5.4) occurred among nests with the same number of abutting neighbors. A similar result was seen among colonial spiders in which neighbors shared barrier web threads and thus built and repaired their webs more efficiently (Lubin 1974; Buskirk 1975; Spiller and Schoener 1989; Wickler and Seibt 1993).

5.3.3 Distance to Mud Sources

Shorter nest building times in large colonies could also result if the larger colonies were closer to mud sources, which would reduce the birds' commuting time during mud collection. In addition, availability of mud could influence colony size directly. For instance, large colonies might form near abundant mud sources, and small colonies might occur in areas with little mud. This scenario could help explain why the birds fought more for existing nests at small colonies (section 5.2.2), where it would be more costly to find and collect the mud to build their own. Limited mud availability near small colonies might also account for the cliff swallows' apparent reluctance to build new nests in small colonies (section 4.10.3).

Given that cliff swallows actively assess mud quality (section 1.4), we could not directly measure the amount of suitable mud near colonies, since we could not easily judge whether mud was appropriate. Instead, as our index of mud availability, we measured how far the birds typically traveled to collect mud. Distance traveled from the colony presumably reflects relative mud availability around different sites, with birds commuting the minimum distance required to find suitable mud. A similar rationale was used to infer food availability near colonies (section 10.8).

In 1993 we regularly visited twenty-two colonies, ranging from 3 to 2,700 nests, during the time the birds were building nests. We determined where the birds were getting mud each day and measured the linear distance from the colony. On any given day, virtually all the birds at a colony collected mud at the same site. Thus for our analysis we determined each day's distance at each colony and averaged the daily distances for each colony site.

The average one-way distance traveled to collect mud increased significantly with colony size (fig. 5.6). Cliff swallows in the smallest colonies traveled the shortest distances to collect mud, often only 1 m (dropping to the ground below their nests). Assuming that distance traveled (fig. 5.6) reflected relative mud availability among sites, it appears that large cliff swallow colonies did not form near more abundant mud sources. Hence mud availability alone cannot explain the shorter nest building times in larger colonies (fig. 5.4). It also cannot explain increased fighting in small colonies (section 5.2.2) or the variation among colonies in the use of new nests (section 4.10.3). The shorter nest building times in larger colonies

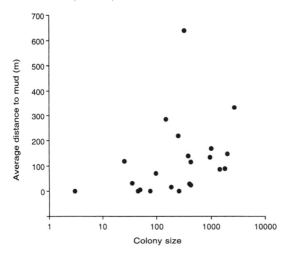

Fig. 5.6 Average one-way distance (m) traveled to collect mud in relation to cliff swallow colony size. Average distance increased significantly with colony size ($r_s = .51$, $p = .015$, $N = 22$ colonies).

are all the more remarkable given the increased travel necessary to collect mud in the larger colonies.

The decrease in nest building time in larger colonies, whether brought about by reduced fighting, increased mud gathering efficiency, or wall sharing, is probably a major advantage of coloniality. This enables cliff swallows in large colonies to begin egg laying earlier in the summer, before swallow bug infestations become severe, and to reduce their energy investment in nest building, which can be substantial (Withers 1977; Gauthier and Thomas 1993). Birds in smaller colonies can partly compensate for the time cost by initiating egg laying before nests are complete (section 11.4), but even so it seems likely that their egg laying still must be delayed several days and they must pay a higher energy cost than individuals starting nest construction at the same time in a large colony. Perhaps the best way to compensate for increased nest building times in smaller colonies is simply to reuse old nests, which birds in small colonies were more likely to do (section 4.10.3).

5.4 DISADVANTAGES OF HIGH NEST DENSITY: BEING BURIED ALIVE

The previous section noted that one potential advantage of having close neighboring nests is that walls can be shared and nest building time reduced. However, we discovered a related potential disadvantage of having close neighbors: another bird may build a nest that blocks the entrance to an already completed nest, sometimes entombing the owners alive.

When cliff swallows first arrived at a colony in the spring, they defended the entire perimeter of their incomplete nests and (later) the entrance holes of their completed nests by lunging at any intruding swallow

Fig. 5.7 Two separate cases (*arrows*) of a cliff swallow nest constructed so that if completed it will block the entrance of an upper nest. In these cases the incomplete nests were abandoned, probably because of interference from the birds whose entrances were threatened.

that approached. Because the birds defended their nests so assiduously (section 5.5.2), they usually succeeded in keeping any other birds from starting to build a nest near the entrance to their own. Although later-arriving birds often began building along the sides or above the necks of existing active nests (e.g., figs. 2.2, 3.6) without any interference from the existing nests' owners, prolonged fights resulted when newcomers tried to build nests directly *below* an existing nest's entrance (fig. 5.7). Newcomers building below existing nests usually offset their nests slightly so that they avoided direct interference with the entrances of residents above them (fig. 3.6; this also achieved a hexagonal pattern of nest placement, perhaps contributing to nest stability). However, we sometimes saw birds begin to build directly below the entrance of a completed nest (fig. 5.7), usually in large colonies with high nest densities. In some cases, as the newcomers' nest progressed upward and came closer to the entrance of the upper nest, fighting between the newcomers and the owners of the upper nest became so constant that the newcomers abandoned their incomplete nest.

In other cases, however, the newcomers persisted and eventually succeeded in constructing their nest so that its top partially or completely blocked the entrance to the nest above it. We observed 32 cases in which a nest entrance was completely blocked and 4 in which it was partially blocked. Of the latter, two of the owners deserted their nests, but the other two persisted with nesting despite their partially blocked entrances. Of the 36 cases of sealed-up nests, 9 occurred before egg laying; 16 occurred while the nest had eggs; and 11 while it had nestlings. The nestlings in these nests perished after the entrance was blocked. In two instances of nests with eggs, we discovered an adult cliff swallow alive in a nest whose entrance had been completely sealed. These birds presumably had been incubating and allowed their neighbors to entomb them alive in their own nests! Why these birds could not have broken out before the mud dried is unclear, but they obviously would have died had we not intervened.

Sealing up of nest entrances was observed in eight colonies ranging from 345 to 2,200 active nests. Of the 36 cases, 14 (38.9%) occurred in one colony of 1,400 nests. These eight colonies ranged in nest density from 14.9 to 78.6 nests per meter. We observed at least one instance of a nest's being sealed up in every colony (where we checked nests) with nest densities greater than 16 nests/m ($N = 6$) but saw none in any colony with a nest density less than 14 nests/m ($N = 64$). Birds nesting in dense colonies obviously had a greater risk of having a neighbor block their entrance, although the overall frequency was low. The colonies where we saw nests being sealed up contained a total of 4,896 nests, and only 0.7% suffered blockage. Nevertheless, this illustrates at least a measurable cost of nesting in a large, dense colony and having close neighbors. Avoiding having one's nest blocked by a newcomer is perhaps one reason residents at times fought so hard with newly arrived birds that were apparently looking for nest sites in a colony.

A related problem occurred later in the nesting season, when nestlings were occasionally entombed by their own excrement (Stoddard 1983). At about 8 days of age nestling cliff swallows begin to back up to the nest entrance and defecate so that the fecal sac falls out of the nest. When nests were underneath others with nestlings, fecal sacs sometimes dropped onto the tops of the lower nests and began to pile up. When the pile of excrement became high enough, it blocked the entrance of the upper nest it came from. We observed three instances of this, all again in a colony with a very high nest density (50 nests/m). The nestlings in each case died, since the parents could not reach them to feed them. Stoddard (1983) observed a similar case of nestlings' being entombed by their own feces. Although these events were rare, they illustrate again that high nest density can be costly for cliff swallows.

5.5 TRESPASSING AMONG NEIGHBORS

Another form of nest site-based competition among cliff swallows occurred as established neighboring birds attempted to trespass into each other's nests. This interference among residents at times had serious consequences for an individual's reproductive success, but the reasons for such activity were not entirely clear. Most likely, trespassing was related to attempts to steal nesting material, seek forced copulation, toss out eggs, brood parasitize a nest, and possibly assess ectoparasite load, nesting stage, and parental quality of one's neighbors.

5.5.1 Natural History of Trespassing

A typical trespassing sequence would begin as one nest owner sat in the entrance of its own nest, apparently closely monitoring the activity of nearby residents. Suddenly the nest owner would fly to a neighboring nest and attempt to enter it. Most often the intruding bird was repelled by the owner of the invaded nest, whereupon the intruder either attempted to enter yet another nest nearby or immediately returned to its own nest. Fights were rarely violent, with most owners defending their nests simply by sitting in the entrance and lunging at birds attempting to enter.

Occasionally a trespasser succeeded in entering a nest while an owner was present, and in almost all these cases the trespasser was immediately ousted in a brief fight during which the owner did not leave its nest. Often trespassers attempted to enter a neighboring nest just as its owner did so, following the owner in. Some individuals repeatedly tried in vain to enter the same nest within a few minutes, even though the owner was constantly present. However, occasionally a trespasser did manage to enter a neighboring nest when no owner was present (section 5.5.2).

Trespassing increased in frequency as nests became completed and reached a peak during egg laying, usually seven to ten days *after* the peak in fighting for nest sites (e.g., fig. 5.1). At one colony of 180 nests, approximately 82% of trespassing (all occurring after egg laying began) was between known color-marked residents of the colony that maintained nests of their own. The remaining 18% probably also included some unmarked colony residents. Trespassing in a colony ceased almost completely as most of the birds began incubating. We emphasize that there were important distinctions between trespassing behavior and the fighting for nests described in section 5.1: trespassing was less violent and more frequent, was not apparently related to nest usurpation attempts, occurred mainly after nests were completed, and usually involved established neighbors.

Male cliff swallows engaged in trespassing more often than females. We knew the sex of all birds observed at two colonies in 1986–87 (see section 2.4). At these colonies we recorded 1,491 total incidents of trespassing, and 1,128 (75.6%) were by males.

Cliff swallows primarily directed their trespassing at nests nearby. For a total of 3,727 trespassing incidents observed in six different colonies, we knew the identity of the trespasser and thus its "home" nest. In each case

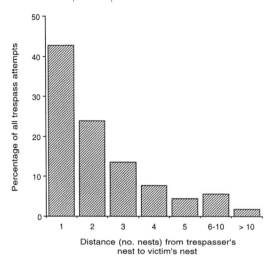

Fig. 5.8 Percentage of observed trespass attempts ($N = 3{,}727$) directed at cliff swallow nests at various distances from the trespasser's own nest. A one-nest distance represented a victim's nest that was adjacent to the trespasser's, a two-nest distance was for a victim's nest separated by one intervening nest, and so on.

we determined the distance between the trespasser's own nest and the nest it attempted to enter and expressed this in terms of nest distance (fig. 5.8). For example, a victim's nest adjacent to the trespasser's was one nest away, a victim's nest separated by only one intervening nest was two nests away, and so on. The frequency of trespassing directed at a nest declined the farther a nest was from a given trespasser (fig. 5.8). Individuals on average directed two-thirds (66.7%) of their trespassing to nests either adjacent to their own (on any side) or only two nests away. Cliff swallows seldom went to another part of the colony (that is, more than ten nests away) to attempt trespassing.

5.5.2 Costs and Benefits of Trespassing

Despite the prevalence with which cliff swallows attempted trespassing, they rarely succeeded in gaining entrance to an active nest. We observed a total of 5,405 trespassing attempts in the six colonies studied; only 895 of these (16.5%) were classified as "successful." We considered a trespass attempt successful if the trespasser gained entrance to an unattended nest or if it was able to interact with a nest owner to its own obvious advantage—for example, by copulating with a female owner. We predicted that the payoff to the occasional successful trespasser must be large to compensate for the presumed costs of an activity that often seemed energetically wasteful and unproductive.

We were surprised to find, however, that cliff swallows often apparently did nothing when they managed to gain entrance to an unattended neighboring nest. In 572 successful attempts (63.9%, $N = 895$), the trespassing bird merely sat in the nest for a few seconds (sometimes up to a minute or more) and then returned to its own nest. Eggs or nesting material often were present in the nest, which the trespasser could have either tossed out or stolen. We do not know why individuals would apparently try so hard to get into a neighbor's nest and then do nothing once they did gain entrance, though perhaps some of these birds were males seeking to forcibly copulate with a female nest owner they then discovered was not there.

Another possibility might be that gaining entrance enabled a bird to assess its neighbor's nesting stage or its ectoparasite load. Information on nesting stage could be useful to a female trespasser that might later attempt to brood parasitize its neighbor by laying an egg or transferring an egg to that nest (section 6.3). There might be other cues about the parental quality of neighbors, gleaned from trespassing, that potential brood parasites could use in selecting which nests to parasitize. A male trespasser might also benefit from information on nesting stage, to gauge whether its female neighbor was still receptive to forced extrapair copulations. Assuming that birds can assess ectoparasite load of active nests (section 4.10.1; Brown and Brown 1991), trespassers of either sex possibly could use information on their neighbors' ectoparasite loads to predict local patterns of parasitism within the colony and thus determine whether abandonment of their own nests owing to swallow bugs might eventually be necessary.

Not all birds benignly sat in their neighbor's unattended nest. Of the 895 successful trespassing attempts, 128 (14.3%) resulted in the trespasser's stealing some of the grass its neighbor had used to line its nest. Trespassers pulled the grass out of the bottom of the neighbor's nest and flew back to their own nests with it. In 86 cases (9.6%) a trespasser attempted a forced copulation with the female owner of a neighboring nest. In these cases the trespasser gained entrance to the nest and copulated with the owner before the owner could evict it. (We saw no cases where a female trespasser entered a nest with the male owner present and solicited or "allowed" extrapair copulation with him.) In 64 cases (7.2%) a trespasser picked off some still-wet mud from the neighbor's nest and stole it. In 31 cases (3.5%) a trespasser brood parasitized an unattended neighbor's nest by laying a parasitic egg (28 cases) or transferring an egg from its own nest to the neighbor's (3 cases); these are discussed in section 6.3. In 11 cases (1.2%) a trespasser tossed out one of the neighbor's eggs (section 5.6). And in 3 cases (0.3%), a trespasser actually evicted the nest's owner. In these instances the trespasser eventually returned to its own nest and the evicted owner regained its nest, but the dynamics of the interaction were unclear.

Thus, in approximately 36% of successful trespassing attempts, the trespasser presumably gained some benefit and the victim suffered some cost. Pilfering of grass stems or mud may not represent a major cost to a victim or benefit to a perpetrator, especially since this kind of thievery was relatively rare. But having an egg tossed out was clearly a major cost to a victim (the benefit to a trespasser was not so clear; section 5.6.3), and brood parasitism probably had a substantial effect on the fitness of both perpetrator and victim (section 6.3.3). The fitness consequences of forced extrapair copulation depended on how often it led to fertilization, which was unknown in these cases.

Besides being costly to victims, trespassing may also have been costly to those that perpetrated it. We found a positive correlation between how often the birds at a nest engaged in trespassing and how often their own nests suffered trespassing ($r = .41$, $p < .001$, $N = 440$ nests). This was probably because frequent trespassing elsewhere meant that one's own nest was often left unattended and thus vulnerable to other potential trespassers in the colony. The same pattern was seen in birds that often broodparasitized other nests and left their own nests unattended; they suffered a high incidence of brood parasitism themselves (Brown and Brown 1989; section 6.3.3).

Given the potential costs of leaving one's nest unattended and vulnerable to trespassing by neighbors, one would expect cliff swallows to exhibit defensive countermeasures. The most obvious such measure was intensive nest guarding. Most trespass attempts were unsuccessful because one nest owner or both was present to repel the intruder. During nest building both the male and the female constructed a nest, working out of phase with each other. Consequently one bird was usually present at the nest, either guarding or applying mud, while its mate was away gathering mud. This arrangement allowed nearly constant nest defense against nest usurpers early in the year and continued after a nest was completed. In complete nests both nest owners were often present simultaneously, but whenever one left to forage, the other usually remained at the nest. Males even guarded nests in preference to guarding their mates (section 6.2.4).

We quantified nest guarding at a 345 nest colony and at a 456 nest colony. Observations were done during two to three days immediately preceding egg laying and during laying, a time when trespassing was high. An observer watched single nests for 30 minute periods and with a stopwatch recorded the cumulative time when at least one of the (colormarked) owners was present at the nest. Intense concentration was required to be sure that owners did not depart or arrive undetected, and it was possible to watch only one nest during each 30 minute period. Results of 44 watches at 20 nests at the 345 nest colony revealed that owners nest guarded a mean 80.2% ($\pm$ 2.5) of the time, whereas 20 watches at 15 nests at the 456 nest colony showed that owners nest guarded a mean 92.0% ($\pm$ 2.7) of the time. Birds at the larger colony guarded for a significantly greater percentage of the time (Wilcoxon rank sum test, $p = .002$).

During much of the remaining time that nests were unguarded, *all* birds at the colony were absent because of alarms or for other reasons. The increased nest guarding time in the 456 nest colony, relative to the 345 nest colony, suggests that nest guarding may vary with colony size. Perhaps it is more essential in large colonies where birds potentially have more trespassing neighbors (but see section 5.5.4). Unfortunately, we did not have nest guarding data from smaller or larger cliff swallow colonies to test this prediction.

In contrast, more solitary barn swallows exhibited much less nest guarding than cliff swallows. We did similar 30 minute watches at barn swallow nests in a 4 nest colony while the birds were laying eggs. Barn swallows nest guarded a mean 29.3% ($\pm$ 4.0) of the time ($N = 33$ watches), significantly less than cliff swallows (all cliff swallow data combined; Wilcoxon rank sum test, $p < .001$). We did not observe any trespassing at nests in this barn swallow colony. The interspecific difference in nest guarding suggests that the high levels in cliff swallows were a direct response to their larger colonies and the more frequent opportunities for trespassing.

Similar trespassing behavior has been reported for colonial house martins (Lifjeld and Marstein 1994; Riley et al. 1995). Although some birds stole nesting material, trespassing incidents in house martins were believed to be primarily attempts to secure extrapair copulation with neighboring females, because they occurred mostly during the female's "fertile period." However, Lifjeld and Marstein (1994) and Riley et al. (1995) had no direct information on when female house martins are fertile, and the fertile period also is unknown for cliff swallows. The predominant occurrence of trespassing in cliff swallows before and during egg laying (section 5.5.1) suggests that some may be related to extrapair copulation. But that only 9.6% of successful trespass attempts actually led to extrapair copulation, that females also trespassed, and that there were other payoffs from trespassing suggest that the interpretation of Lifjeld and Marstein and Riley et al. is not the entire story, at least for cliff swallows.

5.5.3 Variation in Trespassing

Cliff swallow nests within the same colony exhibited great variation in how many times their owners tried to trespass elsewhere and in how many trespass attempts were directed at them by other birds. For all nests from all colonies, we examined the frequency distributions of nests whose owners trespassed elsewhere to various degrees (fig. 5.9a) and of nests that suffered trespass attempts (fig. 5.9b). These relatively overdispersed distributions meant that owners of many nests virtually never tried to trespass elsewhere, whereas owners of a relatively few nests often tried, and that a substantial number of nests had relatively few (or no) trespass attempts directed at them, whereas a few attracted extensive trespassing.

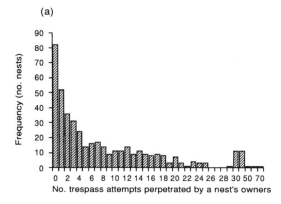
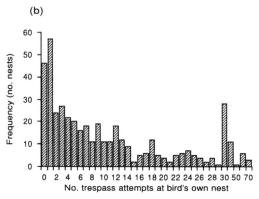

Fig. 5.9 Frequency distributions for the number of trespass attempts perpetrated by a nest's owners (*a*) and the number attempted at their nest by other birds (*b*). The bars illustrate the total number of nests with each level of trespassing, combined across all colonies.

These results (fig. 5.9) show that the costs and benefits of trespassing, whatever their magnitude, are borne unequally by the individuals in a colony. Unfortunately, we could not identify any obvious phenotypic attributes of birds that were more or less likely to engage in, or suffer, trespassing. Most of our observations were conducted in the early years of the study before we had banded enough birds in the population to be able to examine age or history of trespassers. There appeared to be no difference in trespassing frequency among nests of different spatial positions (center-edge). Nest initiation date also did not seem to affect trespassing frequency, at least in the relatively narrow period during a colony's peak egg laying when we conducted our observations. We did not systematically observe trespassing in late-nesting birds, although we would predict that late nesters might have fewer trespassing attempts directed at them (after most of the earlier-settling birds have begun incubation). Further work is needed to determine the basis for the variation among individuals in propensity to trespass and attract trespassing.

5.5.4 Trespassing in relation to Colony Size

So far we have described trespassing behavior and its outcomes without regard to any effect of colony size. In a qualitative sense, trespassing seemed to be similar among cliff swallows living in all six study colonies, ranging from 125 to 1,100 nests. The various outcomes of successful trespassing attempts (section 5.5.2) were distributed roughly equally among all the colony sizes. The one exception was mud stealing, which was not observed at all in the two smallest colonies (125 and 180 nests) but occurred on 4.8–14.3% of successful trespassing attempts in colonies of 340 nests or more. Whether this reflects anything but sampling error is unclear, but the increased average distances to mud sources in large colonies (section 5.3.3) may have promoted mud stealing there.

Colony size did affect the frequency of observed trespassing (fig. 5.10). For each colony we computed the average observed trespassing incidents per nest per hour. We were not trying to count *all* trespassing incidents at each nest. This is important, because it means that the trespassing frequencies reported in figure 5.10 are not absolute measures of how often trespassing occurred. However, these data provide a relative measure of trespassing frequency among colonies, because in focal nest watches we

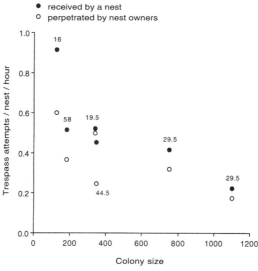

Fig. 5.10 Average number of trespass attempts received per nest per hour and average number of trespass attempts made by nest owners per nest per hour in relation to cliff swallow colony size. Numbers above dots are total observation time (hr) at each colony. Number of trespass attempts suffered per nest decreased significantly with colony size ($r_s = -.94$, $p = .006$, $N = 6$ colonies), as did the number of trespass attempts made by nest owners ($r_s = -.89$, $p = .020$, $N = 6$ colonies).

presumably observed approximately the same percentage of total trespasses occurring at each colony. (Our impression was that we saw and recorded virtually all trespassing that occurred among the focal nests at each colony.) Both the per capita number of trespasses perpetrated by residents and those suffered per nest declined significantly with colony size (fig. 5.10).

These results indicated that cliff swallows in large colonies did not engage in, or suffer, more trespassing than birds in small colonies. If anything, trespassing incidents per nest were more frequent in small colonies. Cliff swallows in large colonies have more close neighbors (e.g., section 7.2), and thus they potentially have more nests to trespass on, and to suffer trespassing from, within a given radius from their own nests. Most trespassing, however, involved residents in adjacent or nearly adjacent nests (fig. 5.8), and therefore increases in total colony size should have little influence on the per nest trespassing rate once birds have a relatively constant number of close neighbors. This could explain why trespassing did not increase in larger colonies (fig. 5.10).

Why birds in smaller colonies seemed to engage in more trespassing than birds in larger colonies (fig. 5.10) is unknown. One possibility is that the potential payoffs to a trespasser (section 5.5.2) are greater in small colonies. This could be so if, for example, collecting nesting material was riskier (in terms of being attacked by a predator) in small colonies where birds gathered mud and grass in smaller, less efficient groups (section 8.4.2) and nests consequently took longer to build (section 5.3.1).

Another possibility is that decreased foraging efficiency in smaller colonies (chapter 10) meant that cliff swallows there had to spend more time foraging and necessarily left their nests unattended more often. As a result, birds that did not happen to be foraging at a particular time may have had greater access to unattended nests in small colonies, leading to more trespassing. If so, one would predict that a greater percentage of trespassing would be successful (that is, the owner would not be present in its nest) in small colonies. This possibility was not supported for the six colonies studied (fig. 5.11). The percentage of trespasses that were successful did not vary significantly with colony size, and the percentage of successful trespassing was markedly lower in the two smallest colonies than in the two largest (fig. 5.11).

A final possible explanation for the increased trespassing in smaller colonies (fig. 5.10) is simply that the observed pattern is not real but reflects sampling error. Although substantial observation time was invested at each of these sites, we had only six colonies with relevant data. These observations were laborious and did not permit us to study additional colony sites. Partial support for the possibility of sampling error came from a limited set of observations on trespassing at a 5 nest colony in

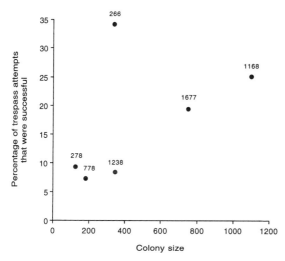

Fig. 5.11 Percentage of trespass attempts that were successful in relation to cliff swallow colony size. Numbers above dots are total number of trespass attempts observed at each colony. Percentage of successful attempts did not vary significantly with colony size ($r_s = .43$, $p = .40$, $N = 6$ colonies).

1984. We observed the nests in this colony for a total of six hours on two days while the colony residents were laying eggs. During this time we saw *no* trespassing. The cliff swallows in this very small colony showed no apparent interest in other residents' nests; birds came and went independently and never tried to interact in any obvious way. We did not include this colony among those in figure 5.10 because observation time was so limited, but the results, if true, would not support the trend of trespassing's being more frequent in smaller colonies. In extremely small colonies, it is perhaps not surprising that individuals would show no interest in trespassing. A trespasser presumably would find few opportunities to trespass successfully in a tiny colony with spread-out nests that could not be monitored easily.

We conclude that the potential costs and benefits of trespassing may be substantial (section 5.5.2), and thus they can appropriately be regarded as costs and benefits of group living per se. Solitary nesters obviously did not experience them, and birds in tiny colonies probably also did not. Among colonies larger than about 125 nests, there was no compelling observational evidence that trespassing represented a colony-size-dependent cost or benefit, but that egg tossing (section 5.6.2) and brood parasitism (section 6.3.2) both varied directly with colony size suggests that colony size can affect trespassing and its outcomes to some degree.

5.6 EGG DESTRUCTION BY CONSPECIFICS

One of the least understood and (to a victim) most costly outcomes of trespassing was a trespasser's destroying an egg in an unattended nest. Egg destruction in cliff swallows was particularly perplexing because it was perpetrated by colony residents that maintained nests of their own and was therefore not related to nest usurpation attempts. In many cases a trespasser tossed only one egg out of a nest, even though other eggs were present and it clearly had the opportunity to destroy more. Here we summarize the important points; for a more thorough treatment we refer readers to Brown and Brown (1988c). In this section we also update our earlier analysis of the effect of colony size on the frequency of egg destruction.

5.6.1 Natural History of Egg Destruction

Our data on the behavioral aspects of conspecific egg destruction were based on eleven observations of egg tossing by birds of known identity and another twelve cases in which we saw an egg thrown from a nest by a bird of unknown identity. Egg destruction was an outcome of successful trespassing, and thus egg destruction occurred in the same contexts. That is, it was directed primarily at nests within one to five nests of the perpetrator's, occurred when nests were left momentarily unattended, and peaked in frequency during egg laying. Among birds of known sex, nine incidents of egg tossing were perpetrated by males and only one by a female. Of these males, one individual accounted for two separate cases of egg tossing and another individual for three. The female laid a parasitic egg at the same time she destroyed an egg (section 6.3.1). All egg tossing except two cases appeared to be deliberate. Birds rolled an egg up to the nest entrance by repeatedly flicking it with the bill and then either flicked it out the entrance or speared it with the bill and dropped it out of the nest. In two cases an egg was knocked out of a nest, perhaps inadvertently, during a fight inside the nest between the owner and an intruding neighbor. We did not see any color-marked nest owner toss an egg out of its own nest, so we assume all egg loss was caused by intruding conspecifics.

Five cases of egg tossing (21.7%, $N = 23$) occurred during colony alarm responses when large numbers of cliff swallows flushed from their nests and remained away for 15–120 seconds. These alarms were sometimes in response to observed predators, but more often there was no apparent stimulus. Not all individuals left their nests during the alarms, and the ones remaining behind often took advantage of the momentary absence of some of their neighbors to trespass into nests and sometimes to destroy eggs. We were unable to determine which individuals gave the alarm calls, but if the trespassers were doing so in the absence of a predator to flush neighbors from their nests, this could be a case of deceptive alarm calling, as suggested by Munn (1986) for mixed-species bird flocks. Individuals that, for whatever reasons, could not gain access to unattended nests during normal colony activity may have resorted to remaining at the nests during alarm responses, a potentially risky option. The contexts in which alarm calls are used in cliff swallow colonies need to be studied, with particular attention to why alarms are sometimes given when no apparent predator is present and why only some residents respond.

We were able to infer instances of egg destruction by conspecifics based on nest check data. Egg tossing was inferred whenever part of a clutch disappeared between successive nest checks and the nest remained unaltered and still active after the egg(s) disappeared (Brown and Brown 1988c). There were no known nest predators in our study area that would take only part of a clutch, and therefore partial clutch losses were likely caused by other cliff swallows. Inferences of egg destruction based on partial clutch loss were conservative and underestimated the true extent of conspecific-caused egg destruction. This was because inferences could not detect full clutch loss if caused by other swallows (we assumed all such cases were attributable to predation), especially if another cliff swallow removed an egg from a single-egg clutch. Also, egg losses to conspecifics during laying (before clutch size was definitive) were masked by our not knowing precisely when egg laying began or ended in a nest and by irregular laying caused by brood parasitism (section 6.3). Thus all inferences about egg destruction made from nest check data were based on nests in which the loss came during incubation, after egg laying presumably had ceased (see Brown and Brown 1988c for further details). The ability to infer egg loss to conspecifics from nest checks enabled us to examine egg destruction more thoroughly than most of the other outcomes of trespassing.

Nests started early in the year were more likely to suffer egg loss to conspecifics than later-starting nests (Brown and Brown 1988c). Among nests with losses during laying (known from direct observations of color-marked birds), 64.3% occurred either on the day the victim began laying or the next day. There appeared to be little pattern in when cliff swallows destroyed others' eggs relative to the perpetrators' own stage of nesting: birds destroyed others' eggs from 15 days before to 5 days after laying began in their own nests. Within-colony laying synchrony had no effect on the likelihood of losing an egg to a conspecific, with synchronous nests being as likely as asynchronous nests to lose eggs (Brown and Brown 1988c).

A relatively pronounced pattern was one in which nests with clutch sizes of four eggs or more were more likely to suffer an egg loss to a conspecific than were nests with smaller clutches (Brown and Brown 1988c). Either trespassers sought out nests with larger clutches to destroy eggs, or they simply succeeded more often at nests where the female had laid a larger clutch and thus perhaps had to leave more often to forage and recoup the energy costs of producing a large clutch.

Among 479 nests with inferred egg losses in all colonies, there were 517 separate instances of a partial clutch loss attributed to conspecifics in which 655 eggs were destroyed. In 407 instances (78.7%) a single egg was destroyed; in 83 instances (16.0%) two eggs were destroyed; in 26 instances (5.0%) three eggs were destroyed; and in one instance (0.2%) four eggs were destroyed. In all these cases other eggs remained in the nest after the egg losses. The mean number of eggs destroyed per egg-tossing event ($N = 517$) was 1.27 ($\pm$ 0.04). For all nests suffering losses ($N = 479$), the mean number of eggs destroyed per nest was 1.37 ($\pm$ 0.03). Although it was not detectable by nest checks, trespassers probably also often removed single eggs from nests that had only a single egg (e.g., during the victim's egg laying period). In observed cases involving color-marked birds, 52.9% of trespassers that tossed eggs removed the only egg present in the nest. In the remaining cases the intruder tossed out a single egg but ignored the additional eggs in the nest.

Cliff swallows also occasionally tossed small nestlings out of nests, although we did not systematically study this sort of infanticide. We twice saw birds push a nestling out of a neighboring nest in what appeared to be a rare trespassing sequence among birds that were then feeding nestlings. When collecting data on nestlings at 10 days of age (section 2.3.1), we noticed wounds on some birds, suggesting that they had survived an attack by a conspecific. One cliff swallow usurped a Say's phoebe nest; it chased away the parent phoebes and then killed all five nestlings (about 7 days old), pecking the young birds to death before shoving them out of the nest.

5.6.2 Egg Destruction in relation to Colony Size

Data inferred from nest checks at fifty-nine colonies ranging from 1 nest to 2,200 nests indicated that the percentage of nests in a colony that lost at least one egg to a conspecific increased significantly with colony size (fig. 5.12). Most colonies smaller than 10 nests had no detectable egg losses attributed to conspecifics, a result consistent with our observations that cliff swallows in a 5 nest colony never interacted with each other (section 5.5.4). The percentage of nests per colony with egg losses ranged from 0 to 33%.

Why egg loss increased with colony size (fig. 5.12) is not clear, however. Egg losses resulted from trespassing among close neighbors, and there was no indication that the number of total trespassing attempts (fig. 5.10) or those that were successful (fig. 5.11) increased with colony size. Perhaps the increased egg destruction in larger colonies (fig. 5.12) occurred in part because egg losses sometimes occurred during colony alarm responses (section 5.6.1), and there were more alarms in the large colonies that both attracted more predators and detected those predators more efficiently (chapter 8).

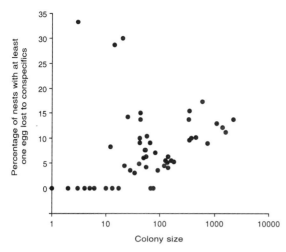

Fig. 5.12 Percentage of nests that lost at least one egg to a trespassing conspecific in relation to cliff swallow colony size. Percentage of nests with egg loss increased significantly with colony size ($r_s = .59$, $p < .001$, $N = 59$ colonies). Sample size (number of nests) for each colony is the same as in figure 6.9.

5.6.3 Costs and Benefits of Egg Destruction

Whatever the reasons, the risk of losing part of one's clutch to an intruding neighbor represented a clear cost of coloniality that increased with colony size (fig. 5.12). The average egg loss per nest for those nests victimized was more than one egg, a major reduction in reproductive success for an animal that usually lays only three or four eggs per season. A nontrivial percentage of the total cliff swallow nests in our study colonies lost eggs: we estimated from nest check data and observations of color-marked birds that 14.7–19.6% of all nests, combined across colonies, lost at least one egg to a conspecific (Brown and Brown 1988c). The relative cost of egg loss was partly ameliorated in that many of the victims had laid larger than normal clutches to begin with. After the egg losses, these birds eventually raised about the same number of offspring as (color-marked) egg tossers and birds not known to be either victims or perpetrators (Brown and Brown 1988c). On the other hand, nests with egg loss to conspecifics were more likely to fail completely before the nestlings reached 10 days of age (Brown and Brown 1988c), underscoring the potentially costly nature of egg destruction for victims.

As with trespassing in general, there was a cost to birds tossing eggs in that they were more likely to lose eggs from their own nests. Finding an unattended neighboring nest presumably entails some cost in leaving one's own nest unguarded and thus vulnerable to successful trespassing. Of eight color-marked perpetrators for which we had data on reproductive success, three (37.5%) had eggs destroyed in their own nests by other birds. Unfortunately, our sample of known perpetrators was too small to determine whether they consistently suffered greater egg loss than the population as a whole, although it seems likely. If so, whatever advantage is gained from destroying eggs must be important.

How could an individual benefit from destroying part of its neighbor's clutch? One possibility is spite (sensu Pierotti 1980), in that the behavior reduces a neighbor's fitness at some cost to the perpetrator. This interpretation is unlikely on theoretical grounds, however (Pleasants and Pleasants 1979; Wittenberger and Hunt 1985), and we rejected it for cliff swallows (Brown and Brown 1988c). Another unlikely possibility is egg cannibalism, as reported in a variety of birds, primarily gulls (reviewed in Wittenberger and Hunt 1985). Cliff swallows are strictly insectivorous and were never observed trying to eat the yolk or shells of eggs they destroyed. Unlike purple martins (Bent 1942), parent cliff swallows do not even eat the shells of their own eggs after hatching.

A more likely benefit of destroying a neighbor's eggs is that it reduces the clutch size and therefore enhances the survivorship of the egg the perpetrator may later add to the nest by brood parasitism (section 6.3.1; Lombardo et al. 1989). We directly observed one instance of a female's destroying a neighbor's egg and then laying a parasitic egg, and if this occurs routinely our nest checks would not have detected it. Egg losses during incubation may often be a prelude to physical transfer of parasitic eggs. This was suggested by a link between nests suffering egg losses and those with parasitic eggs physically transferred into them. Of those nests with egg loss attributed to conspecifics, 19.4% ($N = 479$) had an extra egg or eggs added one to four days *after* the egg loss and during incubation, in all likelihood by physical transfer. The comparable percentage of nests with egg transfers for all nests combined was 6.3% ($N = 4,821$; section 6.3). The difference was significant ($\chi^2 = 106.8$, df = 1, $p < .001$). This means that nests suffering egg loss were over three times more likely than nests in general to have a parasitic egg added later by transfer.

Thus one interpretation of egg loss is that it "prepared" a nest for later brood parasitism. Yet not all nests that lost eggs later had eggs added to them by physical transfer. This was perhaps because gaining access to an unattended nest was difficult and the odds of finding the same nest unattended twice—once to remove an egg and later to add a parasitic egg—were small. As a result, some nests may lose eggs to conspecifics, but through nest guarding their owners may be able to prevent the perpetrators from returning with parasitic eggs. If this link between egg destruction and physical transfer of eggs is real, it implies that male cliff swallows may be active participants in brood parasitism (section 6.3), because most (nine out of ten) cases of egg tossing by known birds were by males.

Another possible benefit of egg destruction for males is that destroying a neighboring female's eggs may cause that female to continue to lay eggs and thus remain sexually receptive to forced copulation attempts. Cliff swallows may be partially indeterminate layers, because adding an egg during laying can cause a female to cease laying (Brown 1984). Thus removing eggs could be a way to prolong a female's egg laying period and extend the time a male could seek extrapair copulation. This could help explain why egg tossing most often occurred among neighbors. Monitoring a female's activities and thus knowing when to seek forced copulations is probably most efficient when a male lives near a female.

The timing of egg loss, and the number of eggs lost from nests, was consistent with males' destroying eggs to keep neighboring females sexually receptive. Losses probably often occurred during laying (Brown and Brown 1988c), the time to destroy eggs from the standpoint of a male seeking to keep a laying female receptive. (Presumably egg destruction during incubation is related mostly to brood parasitism.) Egg losses tended to occur early in the season when females were likely to have the time and energy reserves to replace lost eggs. Males did not destroy entire clutches of their neighbors, presumably because full clutch loss causes cliff swallows to desert their nests (removing opportunities to copulate with the neighboring female) and because it is to an individual's apparent advantage to have neighbors (for example, with which to forage socially once eggs hatch; section 9.5; Brown 1986).

We conclude by emphasizing that we are unsure why cliff swallows destroy their neighbors' eggs. The behavior clearly is costly for the victim and probably for the perpetrator, but the possible benefits to a perpetrator are unclear. Regardless, however, the risk of losing eggs to a neighboring conspecific represents a cost of coloniality. Whether the benefits of egg destruction, whatever they may be, also increase with colony size, and therefore represent a benefit of coloniality, is unknown.

5.7 SUMMARY

Increased competition for resources such as nesting sites is probably an inevitable cost of coloniality for social animals. Fights for nesting sites peaked before most nests in a cliff swallow colony were complete and declined with colony size. Individuals fought more intensely for incomplete nests, probably because the odds of a successful takeover were greater there. Birds contested center nests more than edge nests and fought for center nests more strongly in larger colonies. Fighting for nest sites was costly, since birds sometimes died during fights, but this cost apparently did not increase in larger colonies.

Cliff swallows were able to construct their nests significantly faster in large colonies than in small ones, even though the average distance traveled to collect mud tended to increase with colony size. Shorter nest building times in larger colonies may have occurred partly because individuals in small colonies fought more during nest construction, because residents of large colonies showed increased efficiency at gathering mud, and because nests in large colonies had more abutting nests to share walls with. Shorter nest building times in larger colonies likely represented an important benefit of coloniality.

A disadvantage of living in colonies with high nest density was a greater probability of having one's nest entrance blocked when a neighboring nest was subsequently built nearby. Nests whose entrances were blocked failed in every case. Nestlings in a few nests were entombed by their own excrement, which piled up on a neighboring nest below until it blocked their own entrance.

Neighboring cliff swallows often tried to trespass into each other's nests. Trespassing occurred almost entirely among established neighbors whose nests were five or fewer nests apart. Males were more prone than females to try to trespass. Trespassing was rarely successful, because residents guarded their nests up to 92% of the time. A successful trespassing attempt usually occurred when a nest was left unattended momentarily. Upon gaining entrance to an unattended nest, over half of the trespassers sat in the nest and did nothing. These birds may have been assessing their neighbor's nesting stage or ectoparasite load. Other trespassers stole grass or fresh mud from their neighbor, attempted a forced copulation with a female neighbor, tossed an egg out of the nest, or brood parasitized the neighbor's nest.

Frequent trespassers probably suffered costs in that their own nests were more likely to suffer trespassing when often left unattended. There was substantial variation among nests in propensity to attract trespassing and for owners to engage in trespassing, but the basis for this variation was unknown. The incidence of trespassing apparently declined with colony size, for unknown reasons.

The destruction of eggs in a neighbor's nest occurred in the same general context as trespassing, was directed at nests with relatively large clutches, and often happened during a victim's egg laying or early incubation period. Perpetrators generally destroyed only one egg at a time. The percentage of nests with egg loss to conspecifics increased significantly with colony size and thus represented a cost of coloniality for potential victims. One possible benefit to a perpetrator is that egg destruction removes some eggs in a nest that the perpetrator or its mate may later brood parasitize by physical egg transfer. Another possible benefit for males is that egg destruction prolongs a neighboring female's egg laying period and thus keeps her sexually receptive to forced extrapair copulation by the egg destroyer himself.

6 Misdirected Parental Care

Extrapair Copulation, Brood Parasitism, and Mixing of Offspring

> The ordinary [cliff swallow] clutch is from four to six eggs. When a larger number occurs, it is attributed to the laying of two females in the same nest—a thing very likely to occur now and then among birds so communistic in their notions.
> Ernest Ingersoll, 1889
> (Sharpe and Wyatt 1885–94)

6.1 BACKGROUND

There is increasing evidence that animals sometimes invest parental care in unrelated offspring. The two principal ways this can occur are through cuckoldry, in which a male invests in offspring of his mate that were sired by another male, and (in egg laying species) brood parasitism, in which a female lays an egg in the nest of an individual of the same species (or in some cases a different species) to be cared for by the unsuspecting "host." Among birds, evidence is proliferating that both cuckoldry and conspecific brood parasitism occur frequently, even in "monogamous" species (Yom-Tov 1980; Rohwer and Freeman 1989; Westneat, Sherman, and Morton 1990; Birkhead and Moller 1992).

Cuckoldry is a result of extrapair copulation, in which females are forced to copulate by males other than their mates or in which females may overtly or covertly solicit copulation from other males. That extrapair copulation at times leads to fertilization and resulting offspring has been documented in a variety of species using molecular methods (references in Pena et al. 1993). In many other species extrapair copulation has been observed in the field (reviewed in Westneat, Sherman, and Morton 1990 and in Birkhead and Moller 1992).

Brood parasitism among conspecifics historically has been difficult to study because host eggs and parasitic eggs often look alike. Recently, however, several studies have overcome this difficulty and have suggested that a surprisingly high percentage of nests may in fact contain eggs from parasitic females (Brown 1984; Earle 1986; Emlen and Wrege 1986; Gibbons 1986; Semel and Sherman 1986; Moller 1987b; Eadie, Kehoe, and Nudds 1988; Hoffenberg et al. 1988; Brown and Brown 1989; Davies and Baggott 1989; Lank et al. 1989; Power et al. 1989; Eadie 1991; Gowaty and Bridges 1991; Sorensen 1991; Weigmann and Lamprecht 1991; Lyon 1991, 1993; Jackson 1992, 1993; Bjorn and Erikstad 1994; McRae 1995). Extrapair copulation and brood parasitism are alternative reproductive options that individuals may employ either as their primary mode of reproduction, depending on circumstances, or as tactics to supplement their reproductive output.

As emphasized in section 1.3.2, our primary interest in this book is to what degree these alternative reproductive options increase with group size and unequally affect individuals within a group. Although the presumably greater opportunities to seek extrapair copulation in larger colonies have led some to predict that the incidence of cuckoldry should increase with group size (Gladstone 1979; Hoogland and Sherman 1976; Westneat, Sherman, and Morton 1990; Moller and Birkhead 1993a; Wagner 1993), only a few studies have shown a greater incidence of extrapair copulation in larger (or denser) colonies within the same species or among closely related species (MacRoberts 1973; Moller 1985; Hatchwell 1988; Birkhead et al. 1992; Ramo 1993; Hill et al. 1994; cf. Dunn, Whittingham, et al. 1994). This is in part because, among those colonial species in which extrapair copulation has been studied, surprisingly little effort has been made to sample individuals in different-sized colonies. Similarly, little is known about how conspecific brood parasitism generally varies with colony size, although colonial species seem more likely to engage in it than solitary species (Hamilton and Orians 1965; Rohwer and Freeman 1989). The best evidence that alternative reproductive options such as extrapair copulation and brood parasitism affect individuals within a colony unequally comes from Morton, Forman, and Braun's (1990) study of purple martins (section 1.3.2).

The net result of both extrapair copulation (assuming that it sometimes leads to fertilization) and conspecific brood parasitism is that individuals may misdirect parental care toward nestlings to which they are unrelated. To the degree that limited resources intended for one's genetic relatives (offspring) are diverted to nonrelatives, misdirection of parental care represents a potentially huge cost (Riedman 1982; Carter and Spear 1986; Rohwer 1986; Pierotti 1991; Brown, Woulfe, and Morris 1995; Redondo, Tortosa, and de Reyna 1995; cf. Boness 1990). This cost should be especially great in species with a relatively small average annual reproductive success (such as cliff swallows). If the incidence of misdirected parental care routinely increases in larger colonies (e.g., Riedman and Le Boeuf 1982) or for certain classes of individuals within a colony (e.g., younger birds; Morton, Forman, and Braun 1990), potential victims must pay both the cost of misdirected parental care per se and also the cost of any defensive countermeasures. On the other hand, coloniality represents a net benefit to those colony residents that are better able, by living in a group, to foist off their offspring on others (Morton, Forman, and Braun 1990; section 1.3.2).

Another (perhaps less often observed) way parental care may be misdirected is through the mixing of mobile offspring, such that parents cannot easily recognize or locate their own young. Consequently parents may care for the "wrong" offspring. This is potentially a problem in any species that continues to provision young once they leave the nest and breeds in a large enough group that offspring from different nests may mix. The opportunities to misdirect parental care among mobile offspring presumably have led to the development of relatively sophisticated forms of parent-offspring recognition in most colonial species (Hoogland and Sherman 1976; Burtt 1977; Beecher, Beecher, and Lumpkin 1981; Beecher, Beecher, and Hahn 1981; Beecher 1988, 1991; Medvin, Stoddard, and Beecher 1993). Solitary species, which seldom encounter opportunities for mixing of young, generally show little or no ability to recognize their own offspring.

The existence of parent-offspring recognition in colonial species, however, does not necessarily prevent individuals from misdirecting their parental care. There may be constraints on how large a group can be and still let parents discriminate their own young from other parents' (Medvin, Stoddard, and Beecher 1993). If group size exceeds this threshold, parents may make mistakes in identification. In large groups, parents may have difficulty finding and feeding their own mobile offspring among hordes of similar-age conspecifics. The opportunity also exists for older chicks to invade nests containing smaller offspring of other parents and steal food before the latter's parents have learned to recognize their own young. All these opportunities to misdirect parental care presumably increase in larger colonies, although no studies have explicitly addressed how the incidence of chick mixing and parents' efficiency at locating young vary with colony size.

In this chapter we explore to what degree cliff swallows invest in the offspring of unrelated individuals. We focus on the three ways this may happen: extrapair copulation (cuckoldry), conspecific brood parasitism, and mixing of mobile offspring. Our emphasis is on how these forms of misdirected parental care vary with colony size and to what extent they represent either a cost or benefit of group living.

6.2 EXTRAPAIR COPULATION

6.2.1 Natural History of Extrapair Copulation

Extrapair copulation in cliff swallows occurred in two contexts: among birds at their nests during trespassing by close neighbors (section 5.5) and among birds gathering mud or grass for their nests at variable distances from a colony site. Extrapair copulation attempts were far more frequent at mudholes than at nests, and this chapter focuses on incidents among mud gathering birds. The activity at cliff swallow mudholes resembled in many respects the extracolony "leks" described for razorbills (Wagner 1992a), where males congregated to seek extrapair copulation with unescorted females.

Cliff swallows traveled to mudholes that, in our study area, varied from 1 m to over 0.5 km from the colony site (section 5.3.3). The shortest distances were at several culvert sites where birds simply dropped directly below their nests to collect mud from the wet ground underneath, but at most colonies cliff swallows commuted to mudholes that were not within direct sight of the active nests. Swallows formed groups of variable sizes when collecting mud (sections 6.2.2, 8.4.2); we observed some groups as large as 200 or more birds, although some gathered mud as solitaries. Mud gathering birds usually crowded closely together, brushing against each other, and often fluttered their wings and elevated their tails (section 6.2.4; fig. 6.1). At any given time, they would cluster at one particular spot along a creek bank or other mud source, even though areas containing mud of apparently identical quality were widespread. Birds flying to a mudhole from a colony seemed to be attracted to other birds that were already on the ground and attempted to land as close as possible to them. Others at a colony may have located mud sources by following the birds streaming between the mud site and the colony.

Fig. 6.1 Cliff swallows gathering mud. A stuffed model with wings in the extended position is shown at far left.

Fig. 6.2 An extrapair copulation attempt in cliff swallows. The apparent male attacked the apparent female from above while she was on the ground gathering mud. The female appeared to resist.

Extrapair copulation attempts (fig. 6.2) occurred as flying birds circled low over the mud gatherers and "pounced" from above onto individuals on the ground. Other birds would alight, walk toward a mud gathering bird, and jump on its back from the side or rear. These copulation attempts were in many instances forced, as the victim struggled and often a fight ensued with both birds flailing in the mud (fig. 6.2). Perpetrators attempted cloacal contact and often seemed to be successful. Of 843 copulation attempts observed at mudholes in which we noted whether cloacal contact occurred, 726 (86.1%) appeared to achieve some contact. The remaining unsuccessful instances might have represented cases in which males attacked males (section 6.2.4). Some perpetrators of copulation attempts appeared to be birds that had traveled to the mudhole to gather mud themselves and encountered a bird that for some reason elicited a forced copulation. Other perpetrators seemed not to be involved in collecting mud at all. Forced copulation attempts at mudholes were extremely common at times, with some individuals each repeatedly attacking up to six or eight birds during a five- to ten-minute period. Copulation attempts also occurred in similar contexts while birds were on the ground gathering grass stems. Grass gathering was an equally social activity, with up to 100 birds sometimes collecting grass together.

6.2.2 Perpetrators of Extrapair Copulation

Our efforts to learn the identities of birds engaging in copulation attempts at mudholes were often frustrated by the cliff swallows' unwillingness to tolerate close approach by a human observer during the times they gathered mud. At several colonies where we color-marked birds for the observations on trespassing (section 5.5), we attempted to approach mud gatherers closely enough to read color marks, but the birds usually stopped work when we appeared. In general, they were much less tolerant of people near mudholes than at their nests. However, at two colonies (sites 8730, 8430) of virtually identical size (340 and 345 nests), birds gathered mud where we could hide in nearby brush and observe them at close enough range to see color marks.

At the 345 nest colony we observed 102 extrapair copulation attempts by 38 color-marked males at the mudhole over a ten-day period (28 May to 6 June). None of these were between individuals known to be paired (that shared a nest in the colony). Thus, it was in fact accurate to refer to copulation between mud gathering birds as extrapair copulation. Among the 38 birds, there was (as with trespassing; section 5.5.3) substantial variation in the frequency with which they attempted extrapair copulation (fig. 6.3). One male committed twelve observed copulation attempts, another committed eleven, and a third committed eight. These three birds accounted for 30.4% of all extrapair copulation attempts among color-marked males (fig. 6.3). The sample of color-marked perpetrators with relevant data was too small for meaningful analysis of potential differences in age, history, or own nest status among birds prone to attempt extrapair copulation versus those unlikely to do so. The conclusion is simply that a few males engaged in this behavior commonly, whereas most did it casually or not at all.

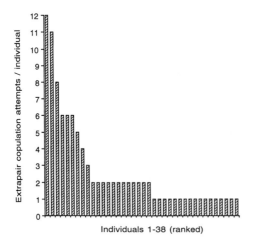

Fig. 6.3 Number of observed extrapair copulation attempts at a mudhole by 38 color-marked male cliff swallows of a 345 nest colony.

The color-marked males attempting extrapair copulation were residents at the colony, and at the times they were engaging in such behavior, their own mates either had not yet begun egg laying, were in various stages of laying, or were incubating full clutches. By noting the precise times that marked individuals left and returned to their nests and were also observed at the mudhole, we verified that these males often went directly to the mudhole, remained there for several minutes either sitting on the ground or flying nearby, and apparently sought females to copulate with. On these occasions the birds did not try to gather any mud.

Although the resident males of a colony at times attempted extrapair copulation at mudholes, it appeared that most extrapair copulation attempts were perpetrated by *nonresident* males. One suggestion of this came from the ratio of extrapair copulation attempts perpetrated by color-marked versus unmarked birds at both study colonies. At each colony we knew the approximate percentage of residents that had been color-marked (94.3% and 51.6% at colonies 8430 and 8730, respectively). If colony residents primarily committed the extrapair copulation attempts and assuming that marked and unmarked males were equally likely to engage in them, the percentage of observed extrapair copulation attempts committed by unmarked birds should roughly equal the percentage of unmarked residents in the colony. This enabled us to calculate the expected number of extrapair copulation attempts committed by color-marked and unmarked males and to compare these expected frequencies with those observed for each colony (fig. 6.4).

Significantly more unmarked males committed extrapair copulation attempts at the mudhole than expected if they were perpetrated solely by colony residents (fig. 6.4). Unless color-marking radically affected the birds' tendencies to attempt extrapair copulation (and we saw no evidence of this), these results (fig. 6.4) mean that a disproportionate number of extrapair copulation attempts were committed by unmarked nonresidents. Nonresident birds were unlikely to be caught and color-marked, because an individual had to repeatedly enter and exit these culverts in order to have any likelihood of being caught in our mist nets. Only resident birds that had active nests came and went enough to be reliably caught. The unmarked males attempting extrapair copulation probably were individuals that rarely approached the colony site itself.

Further insight into the identities of the birds engaging in extrapair copulation attempts came from observations of a radio-tagged cliff swallow (section 13.2.1) early in the season before it had established ownership of a nest. This male was captured and radio-tagged on 4 May 1992 as it was investigating nests at a colony site that eventually contained 85 active nests. This bird's activities were monitored at repeated intervals throughout each day until 16 May, when its transmitter fell off. From 4 to 6 May this bird visited at least five colonies, investigating nests at each,

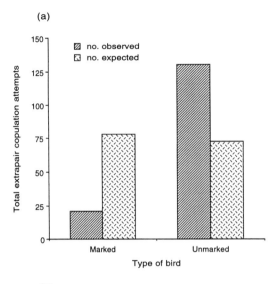
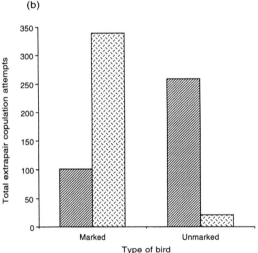

Fig. 6.4 Total number of extrapair copulation attempts observed to be committed by marked and unmarked male cliff swallows versus the total number expected if extrapair copulation attempts were committed exclusively by colony residents, in a 340 nest colony (*a*) and a 345 nest colony (*b*). The observed distribution differed significantly from the expected distribution at the 340 nest colony ($\chi^2 = 48.8$, df = 1, $p < .001$) and at the 345 nest colony ($\chi^2 = 330.4$, df = 1, $p < .001$).

but was clearly not resident at any. On 7 May he began confining most of his activities to one site that eventually contained 1,500 active nests. On that date he had still not established ownership of a nest, however, because we located him in a different part of the colony on each of his occasional visits to nests. On the morning of 7 May, for a continuous twenty-minute period, the male remained at a grass gathering site near the colony and repeatedly attempted forced copulation with birds collecting grass there.

This male began gathering mud at the 1,500 nest colony on 8 May, but he did not clearly establish ownership of a nest there until 9 or possibly 10 May. Thus the radio-tagged bird attempted extrapair copulation before he had either a mate or a nest site and perhaps before he had fully decided what colony to live in. His history and status were consistent with the inference (from fig. 6.4) that extrapair copulation at mudholes or grass-collecting sites was often attempted by birds that had not yet (or ever) settled in the colony or obtained a nest or mate of their own. He presumably eventually attracted a mate and bred at the 1,500 nest colony, since he was caught twice at that colony later in the season.

A historical note is in order here. Most of our observational studies of extrapair copulation were done before the important role of females in extrapair mating strategies was generally appreciated (e.g., Wagner 1992b, 1993; Dunn, Robertson, et al. 1994; Mills 1994). Consequently we concentrated almost exclusively on identifying which males engaged in extrapair copulation attempts and in most instances did not collect data on females. It is now obvious that additional observations are needed to determine which females within a colony typically engage in extrapair copulation and whether nonresident females might also recruit to colonies to seek extrapair mating (and see section 14.1.2). To date we have not had the opportunity to make these types of observations.

6.2.3 Extrapair Copulation in relation to Colony Size

A large colony presumably presents more opportunities for a male to engage in extrapair copulation, because larger numbers of sexually receptive females are potentially available at any given time. The many males present in large colonies also present more opportunities for females to seek extrapair copulation with other colony residents. The costly consequence of each alternative is that resident males in large colonies may stand a greater chance of being cuckolded.

Measuring the frequency of extrapair copulation at mudholes as a direct function of cliff swallow colony size was not straightforward. The frequency with which birds attempted extrapair copulation was strongly affected by the size of a mud gathering group at any given time (below), but group size at mudholes varied enormously from minute to minute at each colony. Because group sizes changed so often, we could not meaningfully measure the average size of groups gathering mud (or grass) at different colonies. We therefore made the assumption that the average size of mud gathering groups increased in larger colonies, simply because smaller colonies did not contain enough residents to form large groups. Although birds in large colonies sometimes gathered mud solitarily or in small groups, the huge flocks (100–200 birds) always occurred at the large colonies. Rather than trying to measure the dynamic variation in the size of groups gathering mud at different colonies and the associated incidence of extrapair copulation, we focused instead on a relatively few colonies where we could study extrapair copulation as a direct function of mud gathering group size.

We chose three colonies of approximately 450, 1,400, and 1,500 active nests (sites 8316, 8416, 8410) at which the residents commonly gathered mud near roads. Mud gathering cliff swallows were tolerant of vehicles and allowed them to approach closely, so we could conduct observations from the car. Whenever a solitary mud gatherer or several birds landed and began collecting mud, they were considered a separate group (group size = 1 for solitaries) until other birds arrived or some of the original ones departed. At that point the group size changed and the birds were considered to represent another group. Individuals were constantly traveling to and from the mudhole, so group size and composition changed often. As a result we considered each group statistically independent. With stopwatches we recorded how long each group was present, its size, and how many times extrapair copulation attempts were directed at group members.

A greater fraction of the larger mud gathering groups attracted extrapair copulation attempts than smaller groups (fig. 6.5). The percentage of groups with at least one extrapair copulation attempt increased significantly with group size; over 50% of groups with 20 or more mud gatherers experienced extrapair copulation attempts (fig. 6.5). Whether an *individual* had a greater likelihood of experiencing an extrapair copulation attempt, however, depended on the number of copulation attempts occurring per unit of time (fig. 6.6). The mean number of extrapair copulation attempts received per individual per minute increased significantly with

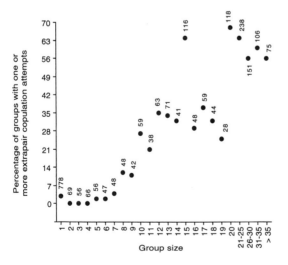

Fig. 6.5 Percentage of all mud gathering groups of cliff swallows that had at least one observed extrapair copulation attempt directed at the group in relation to group size. Numbers by dots are sample sizes (total number of groups of each size). Percentage of groups with at least one extrapair copulation attempt increased significantly with group size ($r_s = .88$, $p < .001$, $N = 24$ group sizes).

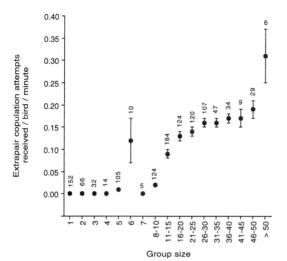

Fig. 6.6 Number of extrapair copulation attempts per minute directed at an individual cliff swallow while gathering mud in relation to group size. Numbers by dots are total number of groups of each size that were observed. Extrapair copulation attempts received per bird per minute increased significantly with group size ($r_s = .96$, $p < .001$, $N = 17$ group sizes).

colony size (fig. 6.6). Thus a female cliff swallow gathering mud in a large group was more likely to experience an extrapair copulation attempt than was a female in a small group.

Assuming that mud gathering group size varied directly with cliff swallow colony size, these results (figs. 6.5, 6.6) suggest that extrapair copulation attempts at mudholes increased with colony size. A likely reason is that the increased conspicuousness of larger mud gathering groups, consisting of birds fluttering their wings in a very visible way (fig. 6.1, section 6.2.4), attracted the attention of the nonresident males that frequently perpetrated extrapair copulation attempts (section 6.2.2). Another possibility is that both resident and nonresident males, and females, were more prone to attempt extrapair copulation when mud gathering groups were large and overall vigilance of the group against predators was consequently increased (section 8.4.2; Brown and Brown 1987).

If large colonies represent a greater pool of sexually receptive females than can be found in small colonies, unmated males should recruit to large colonies where they might be more likely both to achieve copulation with other males' mates and to perhaps find an unpaired female. The sex ratio of the cliff swallow population in southwestern Nebraska was male biased (section 3.5.2), meaning that unmated males probably were available. We examined whether unmated males recruited to colonies of different sizes by computing sex ratios for total bird captures during netting throughout the season (section 3.5.2). This assumes that nonresidents remain at a colony site long enough so that eventually they are caught in a mist net. If nonresident males prefer large colonies, sex ratios of birds caught should show increasing male bias with colony size.

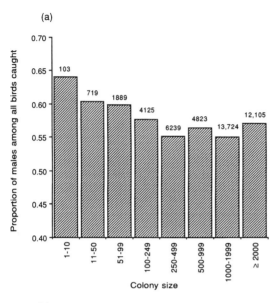

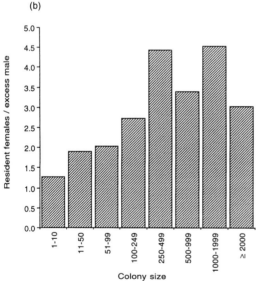

Fig. 6.7 Proportion of males among the total cliff swallows captured throughout the season in 1987–91 (a) and the average number of female colony residents per excess male (b) in colonies of different sizes. Total captures in (a) are shown above bars. The proportion of males declined significantly with colony size ($r_s = -.83$, $p = .010$, $N = 8$ size classes). The number of resident females per excess male increased significantly with colony size ($r_s = .83$, $p = .010$, $N = 8$ size classes).

Surprisingly, the opposite pattern was found (fig. 6.7a). Overall sex ratios of captures from 1987 to 1991 were more male biased in smaller colonies; sex ratio declined significantly with increasing colony size (fig. 6.7a), suggesting there was no preferential recruitment of nonresident males to larger colonies. Surplus males, if anything, were overrepresented in smaller colonies. We predicted the reverse and can offer no reasonable hypothesis for the observed pattern. One possibility, of course, is that the results are artifactual, which could happen if there was a systematic bias in when colonies of different sizes were sampled. Males seemed to arrive earlier than females in some years (section 3.5.3), and thus these results (fig. 6.7a) could reflect differences in male arrival times if small colonies were consistently sampled earlier in the year. That was not the case, however, because large colonies tended to start earlier than small colonies (section 3.5.4). If anything, arrival time differences per se should have caused more male-biased sex ratios in *large* colonies.

In considering the effects of colony size, the ratio of available nonresident males to resident females with which they might engage in extrapair copulation is important. We divided the total number of females by the number of excess males for each colony size (fig. 6.7b) to determine how many resident females a nonresident male could expect to encounter, per capita, in colonies of different sizes. The number of resident females per nonresident male increased significantly with colony size (fig. 6.7b). Each nonresident male in colonies of 250 nests or larger could expect, in a numerical sense, access to between 3 and 4.5 females, whereas males in smaller colonies had access to 1 or 2. However, each female had to contend with relatively fewer nonresident males as colony size increased. For example, the number of excess males per female ranged from 0.78 in colonies of 1–10 nests to 0.22 in colonies of 1,000–1,999 nests ($r_s = -.81$, $p = .014$, $N = 8$). Thus, although there was increased overall frequency of extrapair copulation attempts in larger colonies (figs. 6.5, 6.6), those colonies also had fewer nonresident males per capita that a female had to try to avoid mating with or, depending on the benefits to her, with which she could seek extrapair copulation.

6.2.4 Defenses against Extrapair Copulation

Extrapair copulation, if successful, represents a large cost to a resident male that may thus invest in another male's offspring. Males can take a variety of countermeasures to avoid being cuckolded: guarding their mates and frequent intrapair copulation, presumably to swamp foreign sperm, are the two most common (reviewed in Birkhead and Moller 1992). Extrapair copulation may also be costly to a female if engaging in it either jeopardizes parental assistance from her own mate or is physically dangerous (e.g., Le Boeuf and Mesnick 1990; Mesnick and Le Boeuf 1991; Hiruki, Gilmartin, et al. 1993; Hiruki, Stirling, et al. 1993). Females might therefore resist extrapair copulation in some circumstances (Wittenberger and Hunt 1985; Westneat, Sherman, and Morton 1990).

Given the frequency with which extrapair copulation attempts at mudholes occurred among cliff swallows (fig. 6.6), we were surprised to find that males did not guard their mates during mud or grass gathering or indeed apparently at any other time. As emphasized in section 5.5.2, members of a cliff swallow pair guarded their nest almost continuously, gathering mud at different times so that either the male or the female was always there. This arrangement did not allow males to escort their mates to the mudholes.

To verify quantitatively that males did not guard their mates, we observed the departures of color-marked birds from 20 focal nests at a 345 nest colony (site 8430) over a six-day period from 27 May to 1 June. The residents of these nests were laying eggs during this time, which meant that the females presumably were at their maximum sexual receptivity, and extrapair copulation attempts were common at the mud gathering sites. If males guarded their mates, departures from nests should be paired, as the male followed the female away (Beecher and Beecher 1979; Lifjeld and Marstein 1994). We observed a total of 386 departures at these nests, and 385 (99.7%) were of single birds, even though both members of a pair were often present in the nest simultaneously.

Males clearly allowed their mates to travel unescorted to mudholes, where the females were likely to be subjected to, or could solicit, extrapair copulation with nonresident and other resident males. We suspect that males dealt with the threat of cuckoldry by frequently copulating with their mates, as observed in tree swallows (Venier and Robertson 1991; Chek and Robertson 1994) and other species (Moller and Birkhead 1989; Birkhead and Moller 1992). We could not quantify how often, and precisely when, intrapair copulation occurred, because copulation between members of a pair apparently happened inside nests that were often completed and was thus not easily visible to us. However, in a few instances in which birds under observation began laying eggs in incomplete nests (section 11.4), the male copulated with his mate virtually each time she returned to the nest (and before he left for the mudhole). In these instances there likely were dozens of intrapair copulation incidents per pair during a single morning, and we suspect the same pattern occurred among birds with complete nests.

Cliff swallows exhibited another apparent defense against extrapair copulation: wing fluttering. When gathering mud (and to a lesser extent, grass), both sexes raised their wings above their backs and fluttered them and elevated their tails (fig. 6.1). Butler (1982) suggested that wing fluttering prevented birds from pouncing from above onto the backs of mud gatherers. When mud gathering cliff swallows were presented with stuffed models mounted with their wings closed or wings extended above the back, Butler (1982) found that models with wings closed were more likely to be attacked. We repeated Butler's experiment at a mudhole used by birds from a 1,500 nest colony (8416), setting out two stuffed models (see fig. 6.1) with wings completely closed and two with wings completely extended. All four models were presented simultaneously, and their spatial position with respect to each other was varied randomly over the course of the nine days on which we conducted observations. We recorded 77 attempts to copulate with the models, and 68 (88.3%) were directed at the models with closed wings ($\chi^2 = 26.5$, df = 1, $p < .001$). Copulating birds would typically grasp the nape of a model's neck, and birds copulated so often and so violently with the closed-wing models that they eventually tore the head off one of them! They then continued to copulate with the headless model.

If wing fluttering primarily serves to deter extrapair copulation, its incidence should increase in larger mud gathering groups where extrapair copulation attempts are more frequent. In observing birds at mudholes (section 6.2.3), we recorded the number of birds in each group that fluttered their wings during the time the group was present. We calculated the overall percentage of wing flutterers for each group size (fig. 6.8). The percentage of birds fluttering their wings increased significantly with group size (fig. 6.8). Virtually all group members fluttered their wings as mud gathering group size approached 20 birds, the point at which most groups had extrapair copulation attempts directed at them (fig. 6.5).

The results of the model experiments and the wing fluttering analysis supported Butler's (1982) hypothesis that wing fluttering served to deter extrapair copulation attempts. However, both males and females fluttered their wings when gathering mud. This may have been because males sometimes attacked mud gathering males in apparent copulation attempts. In making the stuffed models, we used two males and two females, each sex

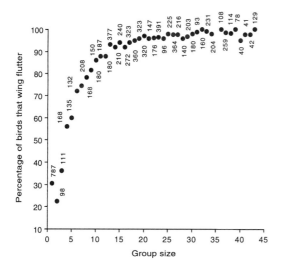

Fig. 6.8 Percentage of total cliff swallows that wing fluttered while gathering mud in groups of different sizes. Numbers by dots are sample sizes (total birds observed). Percentage of birds wing fluttering increased significantly with group size ($r_s = .91$, $p < .001$, $N = 42$ group sizes).

mounted in both wing configurations. Of the 77 observed copulation attempts with these models, 50 (64.9%) were directed at models made from males; this preponderance of attacks on males was almost significant ($\chi^2 = 3.51$, df = 1, $p = .061$). Possibly cliff swallows were unable to distinguish the sex of stuffed birds, but these results and the fact that all birds wing fluttered suggested that males sometimes made mistakes and attacked other males. In such cases the copulation attempt turned into a fight in the mud.

6.2.5 Costs and Benefits of Extrapair Copulation

Measuring the costs of extrapair copulation for males that may be cuckolded, and the benefits for perpetrators, requires knowing how often these copulation attempts lead to fertilization and offspring. This is also necessary in evaluating potential costs and benefits for females that willingly or unwillingly copulate with males other than their mates. And ideally, we should know how the success of extrapair copulation varies with colony size. Unfortunately, we do not at present know with certainty how often extrapair copulation attempts in cliff swallows resulted in offspring. We do know that sperm was probably transferred during some of these interactions, since we found wet substances near the cloacas of the stuffed models.

In an earlier allozyme study of parentage in cliff swallow broods (Brown and Brown 1988b), we analyzed seven polymorphic loci resolvable from blood among 105 complete cliff swallow families (both parents plus all nestlings) from three colonies. We excluded one or both putative parents as biological parents for 35 nestlings from 22 different families. In most cases the electrophoretic phenotypes did not permit unambiguous exclusion of either the mother or father, and thus it was impossible to conclude from the data directly whether these cases of mixed parentage within a brood resulted from extrapair copulation (cuckoldry) or conspecific brood parasitism (section 6.3). However, based on the likelihood of detecting a "nonkin" individual given the degree of polymorphism in the loci studied, it was possible to partition the detection probabilities into specific exclusion types (such as ones in which only the male, only the female, or both parents were excluded, or the exclusion was ambiguous). A different distribution of expected exclusion types can be generated depending on whether exclusions putatively result from extrapair copulation or brood parasitism (Westneat, Frederick, and Wiley 1987). The observed distribution of exclusion types did not differ significantly from that expected if exclusions resulted from brood parasitism ($\chi^2 = 3.21$, df = 3, $p = .36$), but it differed significantly from that expected if exclusions resulted from extrapair copulation ($\chi^2 = 6.34$, df = 2, $p = .042$). We thus concluded that the instances of multiple parentage detectable by allozyme analysis

analysis more likely resulted from brood parasitism (Brown and Brown 1988b; see section 6.3.3).

Although the allozyme data did not reveal widespread evidence of cuckoldry, this technique has various limitations that prevent it from detecting multiple parentage in many instances (e.g., Mumme et al. 1985; Wrege and Emlen 1987; Hoffenberg et al. 1988; Romagnano, McGuire, and Power 1989). Higher-resolution DNA-based techniques now exist for accurate assignment of parentage in natural populations and are increasingly used in studies of reproductive success (e.g., Jeffreys, Wilson, and Thein 1985; Burke and Bruford 1987; Burke 1989; Burke et al. 1991; Hadrys, Balick, and Schierwater 1992; Pena et al. 1993). Applying these methods to cliff swallows is necessary if we are both to understand fully the costs and benefits of extrapair copulation for males and females and to interpret observed patterns of reproductive success in colonies of different sizes (section 14.2.3). At present we have begun DNA profile analyses of cliff swallow parentage ("fingerprinting"), and it is perhaps no surprise that preliminary results suggest that cuckoldry is more frequent than the cruder allozyme results indicated.

A preliminary study of three cliff swallow families that were used in the earlier analysis (Brown and Brown 1988b) confirmed the presence of one nestling arising from an extrapair fertilization in two of the families. In each case the similarity (sensu Lynch 1988, 1990) between the excluded nestling's DNA "fingerprint" and those of its putative parents suggested that it was unrelated to the male nest owner but had approximately the same degree of relatedness as the other brood members to the female nest owner (Brown and Brown, unpubl.). One of these families had exhibited an ambiguous allozyme exclusion in the earlier study (Brown and Brown 1988b), and the DNA-based exclusion in the other family had not been detected at all in the allozyme analysis.

Although extensive DNA profile testing on cliff swallows has not been completed, these initial results suggest that on occasion extrapair paternity does result from extrapair copulation. That males attempt extrapair copulation so often suggests they must be successful at times. Until better data are available, however, we do not feel confident in trying to estimate either the overall frequency of multiple paternity or its incidence in different-sized colonies. The potential for cuckoldry must represent a cost of coloniality for resident males, whose mates have increased chances of copulating with other males as colony size increases. Increased opportunities to engage in extrapair copulation, and access to more females per capita, must represent a benefit of coloniality for those nonresident males that concentrate their activities in larger colonies. Therefore extrapair copulation on balance is either potentially costly or beneficial to different classes of males, depending on their residence status in a colony.

The costs and benefits of extrapair copulation to females are less clear in a general sense (Westneat, Sherman, and Morton 1990; Wagner 1992b; Dunn, Robertson, et al. 1994) and for cliff swallows in particular. That females struggled and resisted suggested that extrapair copulation was not always in their interest. Females copulating at a mudhole often soiled their feathers in the mud, either by being forced down by the weight of the males or in fighting them off. Whether extrapair copulation represents a cost to female cliff swallows by increasing the likelihood of their mates' witholding parental assistance (Morton, Forman, and Braun 1990; Moller and Birkhead 1993b; cf. Whittingham, Taylor, and Robertson 1992; Whittingham, Dunn, and Robertson 1993) is unknown.

That females might at times benefit from extrapair copulation—for example, by increasing the genetic diversity of their offspring—was suggested by our observation that females did not always appear to struggle with the males attempting to copulate with them at a mudhole. Some females clearly allowed successful cloacal contact with little or no resistance. Although we did not systematically try to score degree of resistance by females, our impression was that the likelihood of resistance declined as the season progressed. Struggling by females was commonplace during the early phases of mud gathering, but later (primarily in June) most females seemed not to resist. A possibility is that these later-nesting females may have been forced into nest sites occupied by younger or somehow less fit males (see section 14.1.2), and extrapair copulation may have been a way to enhance the genetic quality of their offspring. (However, some of the males at mudholes did not own a nest at all, so copulating with them may not have provided much competitive advantage to offspring.)

The evolution of extrapair copulation as a reproductive strategy in cliff swallows is perhaps not surprising, given the many animals in which extrapair copulation is known to occur (reviewed in Birkhead and Moller 1992). It seems especially likely in this Nebraska population of cliff swallows, which showed a consistent surplus of males (section 3.5.2). Copulation with other males' mates may be the only means through which the excess males can achieve any reproductive success. Given the high density of nests and the many conspecific females in colonies, it is also not surprising that resident males engage in extrapair copulation, thereby pursuing a classic example of a mixed reproductive strategy (e.g., Trivers 1972; Beecher and Beecher 1979; Morton 1987; Morton, Forman, and Braun 1990).

Many questions about extrapair copulation in cliff swallows remain, but the most important center on the precise frequency with which it results in offspring, to what degree nonresident versus resident males within a colony achieve success, and the role of females. Answers to these questions await results of DNA profile analyses of paternity. In the meantime, perhaps the safest conclusions are that the potential for cuckoldry via extrapair copulation is high in cliff swallows in general and probably increases with colony size, and that the costs and benefits associated with extrapair copulation likely differ for males versus females and especially for nonresident versus resident males. The potential influence of extrapair copulation on the evolution of cliff swallow coloniality is explored further in section 14.1.2.

6.3 BROOD PARASITISM

A second major way parent cliff swallows invest in unrelated offspring is conspecific brood parasitism. As soon as we began studying cliff swallows in 1982, we discovered that laying eggs in other individuals' nests was relatively common in these birds. We have explored brood parasitism in cliff swallows in detail in a series of papers (Brown 1984; Brown and Brown 1988a, 1989, 1990, 1991). Here we summarize the important conclusions of the past work, analyze the effect of colony size on the frequency of brood parasitism, and speculate on the costs and benefits of this alternative reproductive tactic.

6.3.1 Natural History of Brood Parasitism

Cliff swallows brood parasitized nests in two ways: by laying eggs in neighbors' nests and by physically transferring eggs laid in their own nests to other nests by carrying them in their bills (Brown and Brown 1988a, 1995). Both forms of brood parasitism were outcomes of successful trespassing attempts and therefore occurred in the same contexts as trespassing and egg tossing (section 5.5, 5.6). We directly observed color-marked birds laying eggs in, or transferring eggs to, neighboring nests within five different colonies ranging from 125 to 1,100 active nests. In addition to parasitizing other nests, color-marked parasitic females all maintained nests of their own where they raised offspring themselves. No case of an unmarked nonresident female parasitizing a nest was observed. As with trespassing and egg tossing, birds mostly parasitized their close neighbors: over 80% of parasitism by known birds occurred within five nests of the perpetrator's (Brown and Brown 1988a, 1989). However, we recorded one instance in which an egg physically transferred into a nest came from another colony. This egg hatched so soon after appearing in the nest that it clearly was added by transfer, but there were no other nests in that colony of the appropriate stage where it could have originated. Thus these birds must occasionally carry eggs over relatively long distances. Parasitism typically occurred when a host's nest was left momentarily unattended, although in four cases a male nest owner allowed a neighboring female to enter his nest and lay an egg there while he was present. Perhaps these males had earlier copulated with these particular females; if so, these events would represent quasi-parasitism sensu Wrege and Emlen (1987).

Parasitic females and those serving as their hosts tended to be relatively well synchronized in nesting stage, with only a few days separating their clutch initiation dates (Brown and Brown 1989). Parasitism often occurred early in a host's laying period or one to three days before the host started laying eggs. Hosts accepted any egg(s) added to their nests from four days before they themselves began laying until the end of incubation. Earlier than four days before laying, however, hosts rejected eggs placed in their nests either by us or by parasitic females (Brown and Brown 1989). Parasites seldom wasted eggs by laying them in a host nest before the hosts would accept them, suggesting that parasitic females accurately assessed potential hosts' nesting stage, probably through trespassing (section 5.5.2).

Parasitized nests had significantly smaller clutches than nests not known to have been parasitized (Brown and Brown 1989). This was probably in part because adding a parasitic egg caused the host female to stop laying before she had laid a full clutch (Brown 1984) and also because parasitic females may have assessed their neighbors and preferentially selected as hosts those that were to lay smaller clutches, perhaps to minimize later competition between offspring of parasite and host. Brood parasitism in general seemed relatively successful, with only about a quarter of parasitic eggs known not to survive.

Cliff swallows rarely removed one of the host's eggs at the same time they laid or transferred a parasitic egg. In twenty-nine cases we directly saw a bird lay or transfer an egg, and only once did the parasite remove a host's egg at that time (section 5.6.1). However, brood parasitism via egg transfer was significantly more likely to occur in nests that had earlier had an egg tossed out by a conspecific (section 5.6.3). Perhaps parasites assessed nests and removed eggs in those they would later try to parasitize. Cliff swallows apparently assessed host nests using several cues to select ultimately the more successful and less ectoparasite-infested nests in which to place their parasitic eggs (Brown and Brown 1991). Nest age was one cue they used—old nests were preferred—but how the brood parasites, remarkably, predicted nest success and eventual degree of ectoparasitism at the relatively early dates when nests were brood parasitized was not clear.

Cliff swallows also occasionally transfer nestlings between nests. In collecting data on nestling survivorship at 10 days of age (e.g., section 4.7), we found some instances in which the brood in a nest was larger than the clutch laid there. The extra nestlings had to have been transferred in, because they were too small and helpless to have reached the nest on their own. This represents another variation on how to brood parasitize nests (Brown and Brown, in prep.).

6.3.2 Brood Parasitism in relation to Colony Size

Brood parasitism in cliff swallows was similar to egg tossing in that we could infer its occurrence based on nest checks. This permitted us to examine its frequency among colonies of various sizes in which we did not conduct intensive observations of color-marked birds. The perpetrators of parasitism inferred from nest checks were, of course, not known, but it can be reasonably presumed from the observations of color-marked birds (section 6.3.1; Brown and Brown 1989) that in most instances the parasites were colony residents living nearby. Criteria for inferring that a nest had been parasitized via egg laying were the appearance of more than one egg per day during laying or the appearance of a single egg in a nest three days or more before additional eggs appeared (Brown and Brown 1989). The criterion for inferring that a nest had been parasitized via egg transfer was the appearance in a nest after incubation began of an egg that hatched at the same time as the rest of the clutch (Brown and Brown 1988a). The data in this section on the effect of colony size are based on inferred instances of brood parasitism (and do not include any parasitism via nestling transfer).

We calculated the percentage of cliff swallow nests with at least one parasitic egg in sixty-three colonies ranging from 1 to 2,200 active nests (fig. 6.9). The percentage of brood parasitized nests increased significantly with colony size. This was in part because brood parasitism rarely occurred in the smallest colonies, although it still increased significantly with colony size when only colonies of more than 10 nests were considered (fig. 6.9). Virtually no cliff swallows living in colonies of fewer than

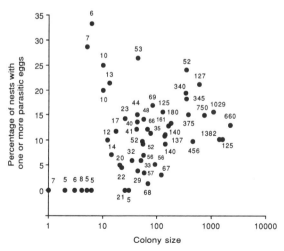

Fig. 6.9 Percentage of cliff swallow nests in a colony with at least one parasitic egg of a conspecific in relation to colony size. Parasitism here includes eggs both laid in and transferred into host nests. Numbers by dots are sample sizes (number of nests) for each colony. Percentage of nests with brood parasitism increased significantly with colony size for all colonies ($r_s = .55$, $p < .001$, $N = 63$) and for colonies greater than 10 nests only ($r_s = .34$, $p = .023$, $N = 44$).

10 nests were parasitized (or parasitized others), whereas in almost all colonies of more than 100 nests at least 10% of the nests contained one or more parasitic eggs.

In a sense these results (fig. 6.9) were surprising, because brood parasitism (like trespassing and egg tossing; chapter 5) mostly occurred among close neighbors. We therefore initially predicted that once colony size exceeded some relatively low size threshold where all birds had roughly equivalent numbers of neighbors within a five-nest radius, the frequency of brood parasitism should not vary with colony size. However, availability of potential host nests was affected not only by their spatial position relative to the parasites, but also by whether they were temporally available. A neighboring nest whose owner began laying many days before or after the parasite did was effectively unavailable to be parasitized and did not exist as a potential host nest. Thus the nests surrounding a parasite's nest represented a mosaic of spatially and temporally suitable and unsuitable potential host nests (Brown and Brown 1991).

We introduced the term "sphere of choice" to describe the set of nests within a five-nest radius of each parasite's nest, representing the maximum number of potential host nests (Brown and Brown 1991). The number of nests within each sphere that were actually suitable as host nests obviously varied, owing to the degree of nesting synchrony within the colony or the extent of prelaying nest failure and abandonment. Thus, in analyzing how often cliff swallows in different colonies brood parasitized nests, it was important to know how colony size might have affected the total number of *suitable* host nests available within each parasite's sphere of choice. (In designating spheres of choice, we assumed the respective parasite lived adjacent or nearly adjacent to each parasitized nest and thus used a sphere centered on the host nest; Brown and Brown 1991.)

For this analysis we used thirty-two colonies that were large enough to designate meaningful spheres for each parasitized nest (colonies of more than ten nests) and that had at least three nests with inferred brood parasitism. We calculated the number of suitable nests within each sphere of choice in each colony. The mean number of potential host nests that were suitable for brood parasitism per sphere increased significantly with colony size (fig. 6.10). This meant that parasitic birds in larger colonies indeed had more potential nests of appropriate temporal stages within a five-nest radius to choose from when selecting host nests than did parasites in smaller colonies. This pattern probably reflected high localized nesting synchrony within large colonies, so that many neighbors initiated egg laying on the same day. With more suitable host nests available in larger colonies (fig. 6.10), potential parasites there had greater opportunities. Perhaps as a consequence, the frequency of brood parasitism increased with colony size (fig. 6.9).

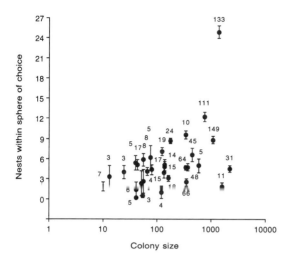

Fig. 6.10 Number of temporally suitable potential host nests within a parasitic cliff swallow's sphere of choice in relation to colony size. Numbers by dots are sample sizes (number of spheres) for each colony. The mean number of suitable host nests per sphere increased significantly with colony size ($r_s = .45$, $p = .010$, $N = 32$).

6.3.3 Costs and Benefits of Brood Parasitism

The first step in assessing the potential costs and benefits of brood parasitism, as with those of extrapair copulation (section 6.2.5), was to determine how often it occurred. Detecting brood parasitism was much easier than detecting cuckoldry: if an egg appeared in a nest in an irregular sequence or if a bird was observed to lay an egg in its neighbor's nest, brood parasitism obviously occurred. Most parasitic eggs survived (Brown and Brown 1989), and therefore most individuals that were parasitized in fact invested in other birds' offspring. However, our methods of inferring brood parasitism based on laying sequences were conservative and clearly overlooked many instances. For example, a parasitic egg added to a host nest the day before the host began laying, or the day after the host stopped laying, would be scored as laying by the host. Therefore we attempted to estimate how often parasitism was overlooked on nest checks by using data on observations of parasitism by color-marked birds (Brown and Brown 1989) and parental exclusion based on allozyme data (section 6.2.5; Brown and Brown 1988c).

Nests with parasitism by color-marked birds and those used in the allozyme study were checked in the same manner as all other nests (section 2.2). This enabled us to estimate the percentages of nests in which brood parasitism was known to have occurred from either observation of color-marked birds or allozyme data but in which parasitism would not have been detected by nest checks. This provided a relative measure of the efficiency with which nest checks detected known parasitism. We concluded that, minimally, at least 22% of nests, overall, in our population contained one or more parasitic eggs, and that perhaps up to 43% of nests were parasitized (Brown and Brown 1989). The higher figure is based on the assumption that all allozyme mismatches resulted from brood parasitism (section 6.2.5), and this estimate would decline insofar as any of the mismatches resulted from extrapair copulation. If we consider extrapair copulation and brood parasitism collectively, however, we can conclude that at least 43% of all cliff swallow nests contained at least one nestling unrelated to either the mother or the father. This is an unusually high level of parental uncertainty that may rise further once DNA profile analyses (section 6.2.5) are completed. The 43% average value probably varied somewhat among colonies (e.g., fig. 6.9). Almost all cliff swallows in this population were at relatively high risk of investing parental resources in other individuals' offspring.

Brood parasitism was costly for both hosts and parasites. Hosts paid the obvious cost of raising unrelated offspring plus a numerical reduction in their own clutch and brood sizes. This reduction was both direct, as when future egg transferrers removed one of the host's own eggs (section 5.6.3), and indirect, as when the host stopped laying after a parasitic egg was added. Color-marked hosts on average laid 0.71 fewer total eggs than known parasites and eventually raised on average 0.86 fewer of their own (putative) young than did parasites and birds not known to be either hosts or parasites (Brown and Brown 1989). Certain host individuals were repeatedly parasitized; 7 of 21 color-marked hosts (33.3%) were parasitized multiply (one individual four separate times in a season), compounding the costs for them.

The costs of parasitism for the parasites themselves were the same as for those birds that perpetrated trespassing and egg tossing (section 5.5, 5.6): parasites often left their own nests unattended and consequently were likely to suffer more brood parasitism themselves (Brown and Brown 1989). This was illustrated well when a male nest owner left his nest momentarily and tried to enter a neighboring nest. While he was fighting with its owner, another neighbor parasitized his unattended nest. Of 24 color-marked females known to be parasites, 13 (54.2%) suffered observed or inferred parasitism in their own nests. Of these 13 females, 6 (46.1%) were parasitized more than once. And of 19 parasitic eggs laid in the nests of the known parasites, 12 (63.2%) were added on the same day the parasites parasitized another nest (Brown and Brown 1989). There was thus an obvious cost of being a parasite, and this cost may have been enough to ameliorate, in a relative sense, the high cost for hosts. Parasites themselves may have realized little relative advantage from brood parasitism if, as a consequence, they were also likely to be parasitized.

The outcome of brood parasitism in cliff swallows appeared to be a scrambling of eggs among different nests as neighbors parasitized, and were parasitized by, each other. This led us (Brown and Brown 1988a, 1989) to suggest that perhaps the benefit of brood parasitism was to disperse an individual's eggs such that it insures itself against total reproductive failure; that is, it spreads the risk (Gillespie 1974, 1977; Payne 1977; Rubenstein 1982). Although questioned on theoretical grounds (Bulmer 1984; Sorenson 1992), risk spreading as a possible cause of brood parasitism in cliff swallows was suggested by the negative correlation between average probability of successful reproduction at a colony and the incidence of brood parasitism there (Brown and Brown 1989). As odds of successful reproduction declined in a colony, birds apparently resorted more to brood parasitism, perhaps to offset the risk of total nest failure.

We updated the earlier analysis of how brood parasitism varied with success at a site by considering twelve additional colonies from 1988 to 1991. We used only colonies where we had followed nests throughout the entire season to know whether they had been successful. We did not use sites where nest failures were caused by recently introduced house sparrows that competed for cliff swallow nests (section 8.2.3) or by other unnatural events such as road construction that blocked culvert entrances (section 4.9.1). We had data for forty colonies, as opposed to twenty-eight in the earlier analysis (Brown and Brown 1989). The same general pattern was evident in the updated analysis (fig. 6.11): the frequency of brood parasitism increased at a site as successful reproduction (defined as whether a nest produced at least one nestling alive to day 10) declined. However, the

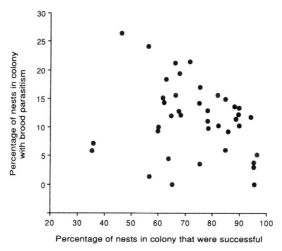

Fig. 6.11 Percentage of cliff swallow nests with brood parasitism per colony in relation to the percentage of nests in the colony that were successful in producing nestlings to day 10. All colonies of 10 nests or smaller were pooled for this analysis. Frequency of brood parasitism per colony did not vary significantly with nest success ($r_s = -.24$, $p = .13$, $N = 40$), but when the two colonies with the lowest nest success were excluded, the correlation was significant ($r_s = -.36$, $p = .027$, $N = 38$).

pattern was not as pronounced as earlier, and the negative correlation coefficient was not significant (fig. 6.11). This may have been largely due to two colonies (8627, 8825) in which reproductive success was very low (about 35% of nests being successful) and brood parasitism rare. Although there was no particular reason to consider these sites anomalous, when they were removed, the negative correlation coefficient for the remaining thirty-eight colonies was significant (fig. 6.11).

Since the publication of Brown and Brown (1989) and our earlier arguments for risk spreading, we have discovered parasitic cliff swallows' ability to assess potential hosts and direct parasitism toward those nests where their parasitic young would be most likely both to survive and fledge in presumably good condition in the absence of ectoparasites (Brown and Brown 1991). This ability to place parasitic young in the better host nests within a colony means that those young probably have higher rates of survivorship than the average nonparasitic nestling. The resulting advantage of brood parasitism may be that it increases a parasitic individual's mean reproductive success. A risk spreading advantage need not be invoked.

There was no evidence that brood parasitism in cliff swallows was a reproductive strategy practiced by birds without nests of their own, enabling them to avoid the costs of parental care (cf. Davies 1988). Our observations of color-marked birds revealed no parasitism by either a "professional parasite" or even a resident bird that suddenly and accidentally lost its own nest. Although the latter might be expected to occur occasionally given the high incidence of weather-related nest loss at times (Brown and Brown 1989), there was no evidence that parasitism regularly represented an avoidance of parental care. Parasites raised as many offspring in their own nests as hosts did (Brown and Brown 1989).

The incidence of brood parasitism varied considerably between colonies (fig. 6.9) and probably among individuals within colonies. The observations of color-marked birds showed that some birds appeared to engage in brood parasitism consistently, whereas others did not. For example, we estimated that 29.2% of known parasitic females each parasitized at least two nests (Brown and Brown 1989). Unfortunately, the sample of known parasitic females (based on observations of color-marked birds) was too small to reveal meaningful phenotypic differences between parasites and hosts, and therefore the basis for the apparent variation among individuals in their propensity to brood parasitize nests is unknown. Variation among individuals would best be addressed by applying DNA-based methods to accurately assign maternity in nests throughout two or three colonies of different sizes.

Although the absolute percentages of nests with brood parasitism may be difficult to know in the absence of molecular analysis, the pattern of increased relative rates of parasitism with colony size (fig. 6.9) is probably real, based on the large sample of colonies ($N = 63$). The average odds of being parasitized increased with colony size and represented a cost of coloniality. The average odds of finding a nest to parasitize also increased with colony size (fig. 6.10) and represented a benefit of coloniality. That in many cases the same individuals were both parasites and hosts suggests that this cost and benefit, for some birds, may have canceled each other out.

Conspecific brood parasitism is apparently a sophisticated alternative (supplemental) reproductive strategy in cliff swallows. We conclude that it may both increase an individual's mean reproductive success and insure against total reproductive failure, and its incidence increases with colony size. The challenge ahead is to identify the basis for the apparent variation among individuals in their propensity to engage in brood parasitism and to determine how colony size affects this variation.

A recent study of brood parasitism in the cliff swallows of the Sierra Nevada in California (Smyth, Orr, and Fleischer 1993), using electrophoresis of egg white proteins, revealed a very low incidence of parasitism in that population (3.7% of nests). Why brood parasitism should be so much lower in these California birds is not clear. The results suggest substantial variation between populations in this life history trait and illustrate that studies on other populations of cliff swallows are needed before we can fully understand the adaptive context promoting brood parasitism and perhaps other forms of social behavior discussed in this book.

6.4 MIXING OF MOBILE OFFSPRING

The third way cliff swallows misdirected parental care was by provisioning unrelated juveniles that joined a parent's own offspring after fledging. This was a consequence of the high breeding synchrony within cliff swallow colonies (section 8.6.1), which resulted in many nestlings' fledging nearly simultaneously. After fledging, some juveniles gathered in large flocks or crèches, sometimes at considerable distances from the colony site, whereas others returned repeatedly to a colony where they entered both their natal nest and nests of other parents. This section addresses two important questions stemming from the postfledging activities of juveniles: How easily can parents find their own offspring in large crèches? And to what extent do fledged juveniles steal food from, or kleptoparasitize (sensu Brockmann and Barnard 1979), smaller birds still in the nest?

6.4.1 Efficiency of Locating Offspring in Crèches

Cliff swallows in southwestern Nebraska formed crèches of recently fledged juveniles that varied from 2 to over 1,000 juveniles. Parents of the juveniles were usually in attendance, typically foraging nearby and visiting the crèche intermittently to feed their young, usually remaining perched with the juveniles for only a few seconds after delivering food. Crèches assembled on wires, fences, trees, and along the ledges of steep cliffs. Birds sometimes traveled surprising distances from colonies to form crèches. We observed banded juveniles in crèches approximately 2.5 km from their natal colony. Some colonies were completely deserted by juveniles once they fledged, and we could not find any within several kilometers of the colony site. Juveniles from different colonies almost certainly combined to form some of the larger crèches. Crèches were often in the same place for two to three weeks, although membership was continually changing from day to day as older individuals departed and more recently fledged juveniles replaced them.

Crèches consisted of juveniles of several apparent ages. Some obviously were recently fledged, flew little, and still seemed completely dependent on their parents for food. Others seemed older, flew often, and received little or no food from parents. Our guess is that juvenile cliff swallows are probably largely dependent on their parents for food for about three to five days after fledging and may take food occasionally for several days after that. Thus each juvenile was probably in a crèche for up to a week. While in a crèche, virtually all nestlings vigorously begged for food by fluttering their wings, opening their mouths, and calling loudly and frequently whenever any adult cliff swallow approached them or flew nearby. Frequent alarms at unknown stimuli would momentarily flush most of the birds, then in a few seconds all would resettle on their perches, although almost certainly not in the same spots as before. Parent cliff swallows were thus faced with locating their own offspring among the hordes of begging juveniles, with positions often changing in these crèche "reshufflings."

Cliff swallow parents apparently recognize their offspring through the juveniles' begging calls (Stoddard and Beecher 1983; Beecher 1988; Medvin, Stoddard, and Beecher 1993). These calls are individually distinctive enough that they represent a "signature" system through which parents can identify their own calling chicks when the young are cross-fostered among nests (Stoddard and Beecher 1983; Loesche et al. 1991; Medvin, Stoddard, and Beecher 1992, 1993) and presumably also within crèches. Parents learn these calls by gradual exposure to them as the chicks mature, so that their ability to recognize offspring has developed by the time of fledging, when potential mixing of mobile young occurs. Parents also call to their young, and there is some experimental evidence that offspring may recognize their parents by the adults' own signature calls (Beecher et al. 1985). At present it is unknown whether the individually distinctive facial plumage of the juveniles (illustrated in Stoddard and Beecher 1983) is used in parent-offspring recognition, although this seems likely.

We hypothesized that parents, though presumably able to recognize their own young, might have more trouble simply finding their chicks in

Fig. 6.12 Parent cliff swallow feeding a begging juvenile in a crèche. The parent paused with food near the juvenile on the left but did not feed it.

larger crèches. If so, chicks in larger crèches might receive less food as parents waste time in locating them. A parent cliff swallow approaching a large crèche, greeted by scores of begging juveniles, would fly to each perched juvenile, briefly hovering above or beside it but transferring no food, until it eventually found and fed (apparently) its own offspring (fig. 6.12). In small crèches, parents typically flew directly to a juvenile and fed it.

We examined the efficiency of parents in locating offspring by observing crèches and scoring how many juveniles a parent visited and hovered by before finding and feeding one, presumably its own (fig. 6.12). No birds in the crèches were color-marked, so we assumed that parents fed the first chick of their own that they encountered. Because juveniles were constantly arriving and departing and changing positions within the crèche, crèches were considered to be statistically independent whenever their size changed (as in the observations of mud gathering groups; section 6.2.3). All observations were made on wires where all birds were clearly visible. We used a regularly occurring crèche near the Cedar Point Biological Station that was about 2 km from the nearest colony and that varied in size over an eleven-day period in 1983 as birds presumably from several nearby colonies arrived and departed from it.

The number of visits (or feeding attempts) parents made before finding their own chicks increased significantly with crèche size (fig. 6.13).

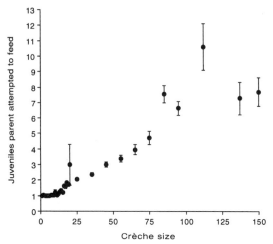

Fig. 6.13 Number of juveniles a parent cliff swallow attempted to feed on returning to a crèche in relation to crèche size. Birds that presumably located their own chick immediately and fed it without visiting other chicks were scored as 1. Crèches of between 21 and 100 birds were divided into intervals of 10, and those between 100 and 150 into intervals of 25, and the midpoints for each were plotted. Crèche size of 150 includes a few crèches larger than 150 birds. The mean number of juveniles a parent attempted to feed increased significantly with crèche size ($r_s = .98$, $p < .001$, $N = 31$ size classes).

Parents almost immediately located their offspring in crèches of fewer than 20 birds but showed a marked increase in the total number of juveniles visited in crèches larger than that. Parents seemed least efficient at locating their young in crèches of more than about 80 birds (fig. 6.13).

These data suggest that one potential cost of large crèches is parents' reduced efficiency in locating their offspring on returning to the crèche with food. This cost may have been greatest for crèches assembled in trees, where juveniles were less visible in the foliage. How this reduced efficiency translated into energy disadvantages for both parents and young is unknown, and addressing this issue would require having color-marked birds from known families within the crèche. Unfortunately, there were substantial technical problems with getting fresh color marks on parents and fledglings and then reliably locating them in a given crèche. Nevertheless, whatever the energy costs, parents did waste more time when their chicks were in large crèches (fig. 6.13).

A related issue is how often parents in large crèches mistakenly fed other birds' chicks (McCracken 1984; Beecher 1991). The studies illustrating parents' recognition of offspring's voices were done in small colonies of about 50 nests (Stoddard and Beecher 1983). Information analyses of cliff swallow begging calls indicate that parents can identify their own chicks in groups up to about 80 birds but probably begin to lose that ability as crèches continue to increase (Medvin, Stoddard, and Beecher 1993). For example, by the time a cliff swallow crèche reaches about 500 birds, Medvin et al.'s analyses suggest that parents may correctly identify their own young only about 50% of the time. In these larger groups amid the many similar signals from other chicks, parents lose some of their discriminatory ability because of structural limitations on the information capacity of the signature calls (Medvin, Stoddard, and Beecher 1993). Thus, in large crèches parents may more often misidentify chicks and feed unrelated young. It is worth noting that parents' efficiency in locating young decreased markedly in groups larger than about 80 birds (fig. 6.13), suggesting that 80 may be a threshold crèche size above which discrimination based on call information content begins to decline (Medvin, Stoddard, and Beecher 1993).

To what degree does efficiency in locating offspring in crèches vary with colony size? It is probable that crèche size tends to vary directly with colony size. Although birds from different colonies did use the same crèche on occasion, we saw no fragmenting of crèches into small groups or individual broods. Juveniles from large colonies thus had no apparent option of joining a small crèche, in contrast to birds from some small colonies that could choose either large crèches (by traveling up to several kilometers) or small ones (by staying near the colony). Presumably, the main advantage of crèches is predator avoidance (section 8.4), which may in fact compensate for disadvantages in locating offspring, depending on crèche size.

We conclude that mixing of mobile offspring in crèches is potentially costly for parents, both in the time taken to find their young and in the possibility of feeding unrelated juveniles, and costly for juveniles in reduced food intake as parents waste time. These costs increase with crèche size. The major unanswered questions focus on how often parents in large crèches make mistakes in feeding young, to what degree the benefits of crèches outweigh the costs for certain group sizes, and why individuals, when given a choice, choose crèches of particular sizes.

6.4.2 Kleptoparasitism of Parental Care

Not all juvenile cliff swallows traveled to and remained in a crèche after fledging. Some birds returned to the colony and apparently spent long periods in active nests that often contained much younger birds. In the process, these fledged juveniles stole food intended for nestlings that were almost certainly unrelated to them. Attempted kleptoparasitism of food by fledglings at nests other than their own probably occurs in a variety of species (Stoner 1942; Graves and Whiten 1980; Bitterbaum and Brown 1981; Poole 1982; Morton and Patterson 1983; Lombardo 1986; Bustamante and Hiraldo 1990; Donazar and Ceballos 1990; Ferrer 1993;

Kenward, Marcstrom, and Karlbom 1993; Frumkin 1994; Redondo, Tortosa, and de Reyna 1995) and may be common among birds nesting at relatively high densities. In this section we examine how often kleptoparasitism occurred, in what colonies, and by which juveniles.

Juvenile cliff swallows at times returned to their natal nests. This usually occurred on the day of fledging (sometimes several times), and always as a parent led them back to the nest. The juvenile followed closely behind its vocalizing parent, which would fly directly to the nest and enter it, as described for purple martins (Brown 1978a). The only times juveniles seemed capable of returning to their natal nests was when led by a parent. These comings and goings in the company of parents rarely occurred more than two days after fledging, when the juveniles began to fly more frequently.

Older fledglings, however, often returned to colonies and entered nests in the absence of parents. We mist netted hundreds of juvenile cliff swallows that were going in and out of colonies, and many of these birds had been fledged for a week or more and were in all likelihood independent of their parents. We discovered kleptoparasitism when banding and scoring survivorship of broods at 10 days of age (section 2.3.1). At these times we often found juveniles, many of which had obviously been flying at least a week, in the nests with the much smaller 10-day-old nestlings. By observing nests, we verified that these kleptoparasites sat in the nest entrance, blocking the parents' access to the smaller nestlings, and were willingly fed by the nest's owners. Kleptoparasitic juveniles took virtually all the food delivered while they were in the nest. We could not determine how long a kleptoparasite generally spent in a given nest, but our frequent net captures of juveniles suggested that they moved in and out of colonies frequently. Usually we found only one kleptoparasite in a single nest, but occasionally we recorded up to four there simultaneously.

Perhaps kleptoparasites were actually trying to return to their natal nests (to rest, preen, or spend idle time) and "missed" by a nest or two. If so, one might expect the kleptoparasites to be found in nests clustered near the natal nest. At three colonies, ranging from 750 to 1,400 nests, where we frequently found kleptoparasites, we calculated the distance between nests where banded kleptoparasites were found and their own natal nests. For this analysis we could use only birds that were banded as 10-day-old nestlings and later found as kleptoparasites in the same colony. We divided distances from the natal nest into ten-nest increments and examined the frequency with which birds were found at the various distances from their natal nests (fig. 6.14). If the natal nest had no influence, birds should have been found at equal frequencies in all distance categories. There was no significant departure from this expectation (fig. 6.14), although kleptoparasites did seem slightly overrepresented in

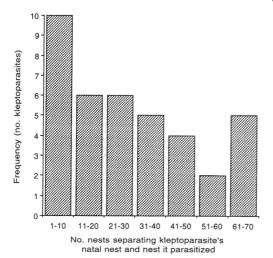

Fig. 6.14 Number of kleptoparasitic juvenile cliff swallows found in nests at various distances from their natal nest. If kleptoparasites entered other nests randomly with respect to their natal nest, the frequency expected for each distance class would be 5.4. The observed distribution did not differ significantly from the expected distribution ($\chi^2 = 3.24$, df = 6, $p = .78$).

the closest distance category. This suggests that a few of these birds may have been trying to return to their natal nests (or at least to a familiar part of the colony), but kleptoparasites generally appeared to enter nests independent of the location of their own natal nests.

How often did kleptoparasitism occur? We examined its incidence as a function of colony size by calculating the percentage of nests per colony with at least one kleptoparasite at the time nestlings were banded. The percentage increased significantly with colony size (fig. 6.15). No kleptoparasitism was observed in any colony smaller than 61 nests, whereas kleptoparasitism occurred in all colonies of 375 nests or more. The chances of having one's nest kleptoparasitized by older juveniles represented a cost of coloniality (fig. 6.15).

Although these data gave an accurate relative measure of kleptoparasitism among colonies (fig. 6.15), they probably drastically underestimated the absolute frequency of this behavior, because we checked each nest for kleptoparasites only once (at 10 days of age), and kleptoparasites in most instances entered nests undetected by us. They also often flew out of nests as we entered a colony. Of 4,678 broods examined, 166 (3.5%) contained kleptoparasites. This overall percentage is misleading both because the method assessed each nest only once and, especially, because some nests had no opportunity to be kleptoparasitized owing to timing.

For example, the early nests in a colony were unlikely to be kleptoparasitized because no juveniles had fledged by then. To account for this, we reexamined the incidence of kleptoparasitism in the three largest colonies where it was most frequent. Based on total nests examined, these colonies had 5.7%, 5.2%, and 7.1% of the nests kleptoparasitized. However, if we

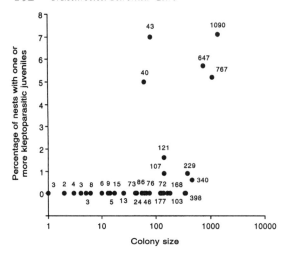

Fig. 6.15 Percentage of cliff swallow nests with at least one kleptoparasitic juvenile when owners' young were 10 days old in relation to colony size. Numbers by dots are sample sizes (number of nests) for each colony. Percentage of nests with kleptoparasites increased significantly with colony size ($r_s = .59$, $p < .001$, $N = 39$).

excluded all nests with 10-day-old nestlings before the date the first kleptoparasite was found—that is, the early nests that had little or no chance to be kleptoparasitized (and the same nests that probably produced most of the kleptoparasites that plagued the later nests)—the remaining nests suffered a much higher percentage of kleptoparasitism: 15.4%, 10.2%, and 12.2%. Considering the infrequent sampling scheme, these percentages seem remarkably high and probably indicate frequent kleptoparasitism of later nests. We can conclude that within a colony, later and more asynchronous broods are likely to suffer higher kleptoparasitism from earlier fledging broods, while later broods probably also experience fewer opportunities to kleptoparasitize other nests (because few with young nestlings remain at the later dates).

A large cliff swallow colony represents more nests that may potentially be kleptoparasitized than a small colony. Thus juveniles might recruit to larger colonies to find nests with smaller chicks that may be victimized. Juveniles move between colonies after fledging. For example, we caught fifteen previously banded juveniles in mist nets at a colony to which they had moved after fledging. These birds were caught going in and out of the colony and, with no natal nests there, were likely (though not proved to be) kleptoparasites. The juveniles were caught at the second colony between one and fourteen days (mean 5.7 ± 0.9) after they had fledged from their natal colony . All but five were found at the second colony three days or more after fledging, at which time they were presumably obtaining most of their food independent of their parents. These birds moved to colonies 1.1, 6.9, 42.1, 42.7, and 59.1 km from their natal sites. The most remarkable movements were by two juveniles found only three days after fledging, one at a colony 59.1 km from its natal colony and the other 42.1 km away. (Only two of the five juveniles moving more than 40 km after fledging were recaught in subsequent years as breeders, but both had then moved back to their natal colonies!) Five of the fifteen juveniles were found in colonies much larger than the natal colony (size changes of 137 to 750 nests, 140 to 500 nests, 140 to 1,700 nests, and two cases of 375 to 1,400 nests), two were in a colony slightly larger (163 nests) than the natal one (137 nests), and the remaining eight were found in a colony that was essentially the same size as the natal colony (size changes of 63 to 61 nests and 140 to 137 nests). None was found in a markedly smaller colony. In California, Robertson (1926) also observed juveniles moving between colonies after fledging.

The movements of the banded birds suggested that juveniles moved to certain colonies to kleptoparasitize. To establish this more directly, we analyzed the ratio of marked to unmarked kleptoparasites in the three large colonies where kleptoparasites were relatively common. This analysis was similar to the one in which we inferred recruitment of excess males to colonies by an overrepresentation of unmarked birds engaging in extra-pair copulation (section 6.2.2). At these three colonies we knew nestling survivorship for all nests, and consequently we knew what percentage of the total nestlings in the colony had been banded at 10 days of age. If kleptoparasites were primarily juveniles raised at that same colony, the percentages of banded and unbanded kleptoparasites should approximate the same percentages for 10-day-old nestlings.

In these three colonies (of 750, 1,100, and 1,400 nests), the percentages of nestlings in the colony that were banded were 72.6 ($N = 1,802$), 46.6 ($N = 2,197$), and 47.0 ($N = 3,041$), respectively. These percentages are based on all nestlings hatched and banded in each colony ten days or more before the last recorded capture of a kleptoparasite in a nest that year; nestlings banded or unbanded after that date would not have been available to be caught as kleptoparasites (because no nests were still active once they became juveniles). We tabulated the observed numbers of banded and unbanded kleptoparasites caught in nests at day 10 and compared those with the expected numbers based on the percentage of nestlings banded in each colony (fig. 6.16). The observed frequencies differed significantly from the expected; unbanded juveniles were significantly overrepresented among the kleptoparasites. Thus juveniles from other colonies apparently recruited to these three large colonies, where the bulk of kleptoparasitism in general occurred (fig. 6.15).

This analysis (fig. 6.16) indicated that residents in large colonies not only had to contend with kleptoparasites from the nests in their own colony but also experienced kleptoparasitism by juveniles from other colonies. Although we could not quantify the costs of kleptoparasitism to

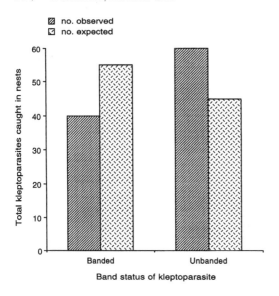

Fig. 6.16 Total numbers of banded and unbanded kleptoparasitic juveniles found in cliff swallow nests versus those expected if kleptoparasites consisted exclusively of juveniles from that same colony. Data from three colonies of 750, 1,100, and 1,400 nests were combined. The observed distribution differed significantly from the expected distribution ($\chi^2 = 4.51$, df = 1, $p = .034$).

younger nestlings and their parents, our qualitative observations suggested that kleptoparasites took all the food delivered to a nest while they were in it. Depending on how long kleptoparasites are routinely present, they could cause a substantial reduction in food intake for the smaller nestlings. Why do parent swallows tolerate kleptoparasites? They presumably could easily evict them, as they routinely do trespassing adults (section 5.5). Adult cliff swallows were observed to evict juveniles of other broods from nests in small colonies in Washington State (P. Stoddard, pers. comm.), but in twelve years we never saw an adult bird evict a juvenile from a nest in Nebraska. A possible reason is that our observations of kleptoparasitism all occurred in relatively large colonies (fig. 6.15) in which parent-offspring recognition is probably diminished (Medvin, Stoddard, and Beecher 1993). Parents may not be able to efficiently recognize their own chicks amid the inevitable call similarity in the larger colonies (section 6.4.1), perhaps explaining both why kleptoparasites were tolerated and why we did not see evictions in the large Nebraska colonies. This might also explain the apparent absence of kleptoparasitism in small colonies (fig. 6.15), where chicks can be more readily recognized.

Another possible reason kleptoparasites may have been tolerated is that parents of 10-day-old nestlings had not yet learned to recognize their own chicks' signature calls. These calls develop throughout the nestling period (Stoddard and Beecher 1983) and probably are only partially developed and learned by day 10. Consequently one might predict that kleptoparasitism would decline in nests with older chicks, at least in smaller colonies, as parents gain the ability to recognize their own offspring (and as nestlings become larger and less easily dominated by kleptoparasites).

Kleptoparasitism probably represents a cheap way for fledged juveniles to find food, and it may be an unavoidable cost of coloniality, especially for birds nesting late in larger colonies. Juveniles that engage in this behavior may be able to improve their postfledging survivorship (section 12.5.2) by increasing food intake during, and immediately after, the weaning period and also perhaps by better avoiding the predators that may attack crèches (section 8.2.1). In addition, visiting colonies might help juveniles assess colony sites to which they may return the following year (sections 13.2.6, 13.5; Brown and Bitterbaum 1980; Lombardo 1987; Danchin et al. 1991; Baker 1993; Boulinier and Danchin 1994). On the other hand, returning to kleptoparasitize a nest or assess the site also means continuing to suffer the costs of nest-based ectoparasitism (chapter 4). As with so many aspects of cliff swallow social behavior, kleptoparasitism probably represents a complex trade-off between different costs and benefits.

6.5 SUMMARY

Cliff swallows sometimes invested parental care in the unrelated offspring of other individuals. This came about through cuckoldry, conspecific brood parasitism, and mixing of mobile offspring after fledging. All of these potentially increased with colony size and thus represented costs of coloniality for most individuals, while simultaneously representing benefits for those birds able to exploit others within a colony.

Extrapair copulation often occurred at mudholes or grass gathering sites where females traveled to collect nesting material. Both males resident in a colony and unmated, nonresident males engaged in extrapair copulation; nonresidents recruited to colonies and apparently committed more extrapair copulation attempts than residents. Larger groups of mud gathering birds were more likely to attract extrapair copulation attempts than were smaller groups. The incidence of extrapair copulation attempts per female increased with mud gathering group size, meaning attempts per capita were more frequent in larger colonies where birds gathered mud in larger groups.

Male cliff swallows did not guard their mates, and they probably resorted to frequent intrapair copulation as a defense against cuckoldry. Birds of both sexes fluttered their wings above their backs while gathering mud as an apparent defense against being attacked in extrapair copulation attempts. Allozyme parental exclusion analyses did not reveal widespread evidence of multiple paternity in cliff swallow broods, although allozymes did not have high resolution and preliminary DNA profile analyses showed clearly that extrapair copulation at times led to cuckoldry. The evolution of extrapair copulation behavior in cliff swallows has probably been promoted by a male-biased sex ratio.

Resident females within a colony laid eggs in (brood parasitized) the nests of their close neighbors, usually when a neighboring nest was left unattended. Parasitic females preferentially parasitized nests likely to successfully fledge young and nests in which infestations of ectoparasites were lowest. The incidence of brood parasitism increased with colony size and probably reflected more temporally suitable host nests available to each parasite locally in the larger colonies. At least 22% of cliff swallow nests were estimated to be brood parasitized, and up to 43% of nests had one nestling or more resulting from brood parasitism and extrapair copulation collectively.

Brood parasitism was costly for hosts in that it reduced their output of young and forced them to invest in unrelated offspring. Brood parasitism was costly for parasites in that they were often parasitized themselves when their own nests were unattended. Parasitic individuals supplemented their mean reproductive success by having one or more of their young raised in some of the better, less ectoparasite-infested nests within the colony.

After fledging, juvenile cliff swallows assembled in large crèches sometimes containing hundreds of same-aged juveniles from different broods and colonies. The efficiency of parents in finding their offspring declined in larger crèches. Problems with locating mobile offspring represented a cost of coloniality.

Older juveniles returned to colonies, entered nests, and stole food that was delivered by the parents of smaller birds. Kleptoparasitic juveniles appeared to come to these nests specifically to steal food, since there was no evidence that they were simply trying to get back to their own natal nests. The incidence of kleptoparasitism increased with colony size. Kleptoparasites traveled among colonies and recruited to the larger ones, where later-starting nests had the greatest chances of being kleptoparasitized. Kleptoparasitism probably is a cheap way for juveniles to obtain food.

7 Shortage of Suitable Nesting Sites

> There are things that grow on human accoutrements and structures: molds, mosses, fungi, mildews, and cliff swallows.
> John Janovy Jr. (1978)

In the previous three chapters we described a variety of costly consequences of coloniality for cliff swallows, especially for individuals living in the larger colonies. Without compensating advantages, these costs seem substantial enough to prevent colonial nesting. In this chapter, and in chapters 8–10, we address possible selective pressures that potentially provide enough benefits to have caused and maintained coloniality in these birds.

7.1 BACKGROUND

Animals may form colonies for several reasons, including to better avoid predators and to enhance their ability to find food. Another potential cause of coloniality is a shortage of suitable breeding habitat that forces individuals into localized areas of high density. If breeding can occur at only a relatively few sites, coloniality may be inevitable. In such cases the automatic costs of group living still exist and may be severe, but individuals gain no benefit from the presence of conspecifics (Alexander 1971, 1974). In general, there is surprisingly little good evidence that coloniality is or is not a direct result of habitat limitation. The issue is critical, however, because arguments positing habitat shortage as a cause of coloniality are perhaps the most parsimonious and thus should be evaluated before considering predator avoidance and social foraging.

There are at least three major types of nesting site limitation. Some animals may require specialized breeding substrates that are relatively uncommon, and individuals are forced into loose colonies at the suitable sites that exist (e.g., Snapp 1976; Farr 1977; Muldal, Gibbs, and Robertson 1985; Shields and Crook 1987; Stutchbury 1991; Szep 1991). Other animals may use relatively common substrates, but only a few areas are safe from predators, and individuals must aggregate in the safe sites to be successful (e.g., Lack 1967, 1968; Schmutz, Robertson, and Cooke 1983; Robinson 1985; Post 1994). Finally, some species, especially pelagic ones, occur in areas with few (if any) nesting substrates of any type, and consequently many individuals must aggregate to breed on isolated islands and coastlines (e.g., Lack 1967, 1968; Wittenberger 1981).

Habitat limitation is not easy to address directly in the field for two reasons. First, we usually do not know what constitutes "suitable" habitat. Breeding sites in use are obviously suitable, but unused sites can rarely be evaluated. Unused sites may appear to us to be identical to used sites, but it is possible that subtle differences undetectable by humans may render them unacceptable. There is obvious circularity inherent in defining sites as suitable only if occupied and then concluding that the animals are colonial because only a few sites exist. Second, most studies are done over such a small spatial scale that it would be difficult to conclude anything about nesting site availability for the population as a whole even if suitability of sites could be determined and defined.

Perhaps as a result of these difficulties, conclusions that habitat limitation does or does not lead to coloniality have been based on indirect evidence (Hoogland and Sherman 1976), assertions without data (Lack 1967, 1968), a default alternative when no other benefits of coloniality were found (Coulson 1971; Snapp 1976; Farr 1977; Schmutz, Robertson, and Cooke 1983; Shields and Crook 1987; Shields et al. 1988; Post 1994), or somewhat contradictory evidence (Robinson 1985). In the last case, for example, Robinson concluded that coloniality in yellow-rumped caciques evolved because safe nesting sites were limited, yet he found a substantial number of birds still nesting in presumably unsafe sites, and effective mobbing of predators represented a clear benefit for colonial individuals. Thus, although avoidance of predators was probably important generally in the evolution of cacique coloniality, whether coloniality was a direct result of limited nesting sites per se was not so clear.

Another problem with the hypothesis that coloniality results from limited nesting sites is that in most species, even ones that appear to be limited by breeding habitat, nests are more closely clumped in space than would be required by the amount of substrate available (e.g., Waltz 1981; Schmutz, Robertson, and Cooke 1983; Wittenberger and Hunt 1985; Szep 1991; Burger and Gochfeld 1993). If individuals gain no benefits from the presence of conspecifics, nests should be spread out as much as possible to reduce the unavoidable costs of group living. For example, nest-site-limited tree swallows, which sometimes nest in loose "colonies," tend to maximize nearest neighbor distances when given experimental arrays of plentiful nest boxes (Muldal, Gibbs, and Robertson 1985). Nest clustering is perhaps the best evidence against habitat limitation as a cause of coloniality (Hoogland and Sherman 1976), although some animals that are indeed site limited might also group their nests to maximize the secondary advantages of group living (Waltz 1981; Shields et al. 1988; Winkler and Sheldon 1993).

In this chapter we address to what extent coloniality in cliff swallows results directly from limited nesting sites. We do this by considering nest spacing within colonies, the importance of total substrate area, and colony reuse patterns. We also compare nest spacing in cliff swallows with that of sympatric barn swallows, a semicolonial species believed to be limited by nesting sites. This chapter addresses nesting substrate limitation both within colonies and between different colony sites, because both types of limitations may occur among colonial animals (Coulson 1971).

7.2 NEST SPACING WITHIN COLONIES

The way nests are spaced within a colony may suggest to what extent animals are limited by nesting substrates. For example, if space is limited, individuals may be forced into high density in the few suitable sites, and consequently nest density should increase with colony size. Increased nest densities in larger colonies do not conclusively demonstrate a shortage of nesting sites or substrates, because there may be active advantages to crowding with conspecifics (e.g., section 9.5), but no relationship (or an inverse relationship) between nest density and colony size would provide evidence against there being a shortage of nesting substrates.

7.2.1 Nest Density in relation to Colony Size

We measured nest densities for 254 cliff swallow colonies ranging from 2 to 2,350 nests. Nest density was measured as described in section 2.8.1 and expressed as nests per meter of substrate. We calculated an average density for each colony, although sections of some colonies had higher nest densities than other sections. However, because we had no good way to evaluate substrate suitability at the microhabitat level within colonies and because we were interested in how density varied with colony size, we judged average nest density per site as an acceptable relative measure. Nest density in this analysis was based on the total substrate area included only within the span of the edgemost nests in a colony (section 2.8.1), which resulted in a more realistic measure of nest spacing within the occupied portions of a site. Only colonies on bridges and culverts were included in this analysis. Cliff sites exhibited too irregular a surface to allow confident measurement of the substrate available for nest attachment.

Nest density increased significantly with colony size (fig. 7.1). Average densities were especially high in the largest colonies; those larger than 1,000 nests typically had nests packed in multiple tiers. These results indicate that cliff swallows packed their nests more closely as colony size

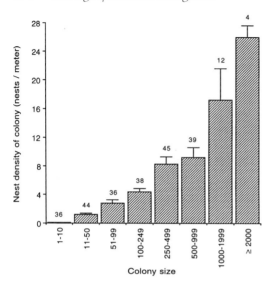

Fig. 7.1 Nest density (nests per meter of substrate) of cliff swallow colonies of different size classes. Sample sizes (number of colonies) are shown above the error bars. Nest density of a colony increased significantly with colony size ($r_s = .78$, $p < .001$, $N = 254$ colonies).

increased, a result consistent with the hypothesis that nesting substrates were somewhat limited. The direct association between density and colony size also suggests that analyses throughout this book based on colony size would yield the same general conclusions if nest density was substituted for colony size at each site.

7.2.2 Nearest Neighbor Distance in relation to Colony Size

Another measure of nest spacing within a colony is nearest neighbor distance. Nearest neighbor distance is an especially useful measure of nest dispersion in loosely colonial species that occur in varying densities over a relatively wide area and in which colonies per se are ill defined (e.g., Davis and Dunn 1976; Parsons 1976; Veen 1977; P. Buckley and F. Buckley 1980a; Emms and Verbeek 1989; Clarke and Fitz-Gerald 1994). In addition to its implications for assessing nesting site availability, nearest neighbor distance may directly affect how successfully ectoparasites can move from nest to nest (section 4.5.3, 4.9.2), how well potential trespassers can assess neighbors (section 5.5), and the amount of foraging information readily obtainable from neighbors (section 9.5). An advantage of using nearest neighbor distance in assessing nest spacing is that it can be objectively measured and does not require making assumptions about where suitable substrate starts and stops.

We defined nearest neighbor distance as the distance to the nearest active nest (section 2.8.2). We calculated these distances only for colonies where we had checked nest contents (and thus knew the status of each nest) and had mapped all nest positions. Data were available for sixty-one

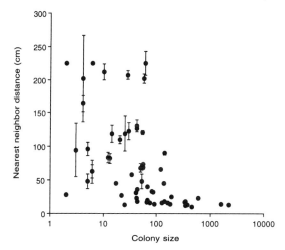

Fig. 7.2 Nearest neighbor distance (cm) per cliff swallow colony in relation to colony size. Nearest neighbor distance decreased significantly with colony size ($r_s = -.65$, $p < .001$, $N = 61$ colonies).

colonies ranging from 2 to 2,200 nests. Nearest neighbor distance declined significantly with colony size (fig. 7.2). This result illustrated generally the same conclusion derived from figure 7.1: cliff swallows packed their nests more closely in larger colonies. However, the analysis of nearest neighbor distance (fig. 7.2) revealed an apparent threshold at colony sizes of about 100 nests. Colonies larger than this, regardless of size, seemed to show a similar and perhaps irreducible minimum in nearest neighbor distance (about 10–16 cm), presumably as nests approached the maximum degree of nest packing. In colonies smaller than 100 nests, average nearest neighbor distance varied widely, with birds in some colonies grouping their nests closely and others spreading out (fig. 7.2). The small colonies with relatively small nearest neighbor distances are especially instructive, because in those the few birds present still clustered their nests almost as closely as did birds in large colonies. This presumably indicates a willingness to group together even when space on the substrate was available to spread out.

The analyses in this section, showing increased nest density (fig. 7.1) and decreased nearest neighbor distance (fig. 7.2) with increasing colony size, were consistent with the hypothesis that cliff swallows are forced into colonies by a shortage of nesting substrates. Other interpretations are possible, however, including the possibility that individuals crowd together for antipredator (chapter 8) or foraging (chapter 10) benefits that may be especially great in large colonies. Birds in small colonies also crowded together at times. Thus increased packing of nests within larger colonies (e.g., fig. 7.1) alone is not sufficient support for the nesting site limitation hypothesis. As we show in the following sections of this chapter, all the other evidence and analyses suggested that cliff swallows were not limited by local availability of nesting sites or substrates.

7.3 TOTAL SUBSTRATE SIZE OF COLONY SITES

The preceding section used the actual positions of nests to infer substrate that was presumably suitable for nest attachment. This method of calculating nest density (section 2.8.1), although reasonable for the active portions of a colony, was not adequate to assess the total substrate at each site. This was so because many colonies had large unused sections in which no cliff swallows nested in a given year but that represented at least grossly suitable substrate, as evinced by the presence of old, inactive nests. Our objective in this section is to evaluate the total substrate available at each colony site and determine how that area may have affected colony size, nest spacing as measured by nearest neighbor distance, and probability of site use from year to year. If cliff swallows are forced into colonies primarily by limited habitat, sites with a large total substrate and room for many individuals should consistently be occupied by large colonies, and small colonies should not regularly occur at sites with a large total substrate. Colonial bank swallows in Hungary conformed to this prediction (Szep 1991).

The strength of the analyses and resulting conclusions in this section rely on our ability to measure total substrate size at each colony. They are thus potentially tainted by the uncertainty that we are measuring truly *suitable* substrate (section 7.1). However, cliff swallows present perhaps a better opportunity than most species to define (at least broadly) what constitutes suitable substrate for nest attachment, because of their affinity for nesting on artificial structures such as bridges and culverts. These mostly concrete and metal structures presented surprising uniformity in both vertical substrate and sheltering overhang (section 3.4.3). Artificial sites thus provided an experiment of sorts in which colony sites with different amounts of roughly uniform nesting substrate were presented to the birds. (Possible problems with conclusions about nesting site availability based on artificial sites are addressed in section 7.6.1.)

At sixty-three bridge and culvert colony sites we measured (in square meters) the total amount of vertical substrate with an overhang both in sections of the site that had existing cliff swallow nests and in those that did not but were structurally identical to sections with nests. We measured all areas of the site that contained substrate on which nests could obviously have been attached, and we did not make subjective judgments about a substrate's relative suitability or desirability. The only parts of sites we excluded were the ends of some bridges where limbs from trees growing in the river below had reached the level of the bridge and blocked the birds' access. Because on occasion cliff swallows stacked their nests in layers (e.g., fig. 3.6b), we assumed that the entire vertical span of a beam or wall was suitable substrate and could have accommodated nests.

7.3.1 Colony Size in relation to Substrate Size

Total substrate size among the sites surveyed ranged from 19 m² at a small and low culvert (the ceiling being only about 1.7 m above the ground) to 5,040 m² at a bridge spanning the South Platte River. Total substrate area was a relative measure of how large a colony site was in a structural sense. The sites with the most substrate tended to be the relatively long bridges over rivers (e.g., fig. 3.5c), whereas those with the least substrate tended to be the shorter, single-tunnel culverts (e.g., fig. 3.6d). We calculated the average colony size at each site (using only the years it was active) and compared this average with the total substrate size for the colony site (fig. 7.3).

Average colony size per site did not vary significantly with total substrate per site (fig. 7.3). There was a slight trend for larger colonies to be at sites with greater substrate area, but on balance there was no evidence that sites with large substrates consistently had the largest colonies. In fact, some sites with relatively large substrates had quite small colonies (fig. 7.3). For example, the three colony sites with the largest substrates (over 4,000 m²) had average colony sizes of only 150, 299, and 578 nests, which constituted relatively small to medium-sized colonies in our study area (fig. 3.11). Cliff swallows obviously did not crowd into these sites, despite the extensive substrate. Even more instructive was the colony with the fourth largest substrate (3,900 m²), which averaged only 30 nests and

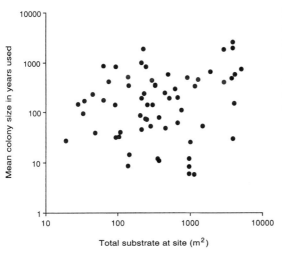

Fig. 7.3 Mean colony size in the years a site was used in relation to total nesting substrate (m²) per cliff swallow colony site. Mean colony size did not vary significantly with total substrate at a site ($r_s = .15$, $p = .23$, $N = 63$ sites).

was not used at all in some years! Conversely, one site with a substrate of only 211 m² had an average colony size of 984 nests, the fifth largest colony size among those sites in which substrate was measured. These extremes illustrate well that the birds were probably not forced into colonies by substrate availability: small colonies formed at sites with large substrates, and large colonies formed at sites with small substrates (fig. 7.3).

7.3.2 Nearest Neighbor Distance in relation to Substrate Size

We used nearest neighbor distance (section 7.2.2) as a measure of nest packing in colonies with different substrate sizes. We had data on nearest neighbor distances for seventeen of the colony sites at which we measured substrate size. Most of these sites were occupied in multiple years, and therefore we calculated an average nearest neighbor distance for each colony site based on the yearly mean values. Average nearest neighbor distance per site did not vary with total substrate per site (fig. 7.4). There was a nonsignificant trend for average nearest neighbor distances to increase with substrate size, but these data do not suggest that nest spacing within colonies was greatly affected by total substrate size. If cliff swallows are forced into colonies by limited nesting substrates, one might predict the greatest crowding of nests in colony sites with relatively small substrates. This prediction was not well supported (fig. 7.4). For example, the site with the largest substrate in our study area (5,040 m²) exhibited a high degree of nest packing, as evinced by a relatively small average nearest neighbor distance (fig. 7.4).

The analyses of how total substrate affected average colony size (fig. 7.3) and average nearest neighbor distance (fig. 7.4) at a site both suggested that cliff swallows were not limited by nesting substrates, at

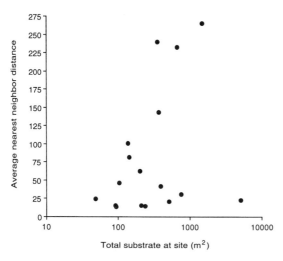

Fig. 7.4 Average nearest neighbor distance (cm) per year a cliff swallow colony was used in relation to the total nesting substrate (m²) at a site. Average nearest neighbor distance did not vary significantly with total substrate ($r_s = .31$, $p = .22$, $N = 17$ sites).

least within colonies. Cliff swallows clearly preferred to cluster their nests even in the presence of essentially unlimited substrate in some of the larger bridge sites that presumably would have permitted the birds to spread out into low nest densities. Sites with large sections of unused substrate sometimes contained only small or medium-sized colonies. Assuming this unused substrate was suitable for nest attachment (and there was no obvious physical reason it should not have been), there was thus no compelling evidence that these birds were limited by availability of nesting substrates.

7.3.3 Annual Site Use in relation to Substrate Size

Not all portions of a colony site's substrate may, of course, be suitable for nesting in a given year. This most likely would result from ectoparasites' infesting certain nests or sections of a colony too heavily for the birds to use them (section 4.10.1). Thus, if nesting substrates are routinely limited either in overall availability or suitability owing to ectoparasite infestations, colony sites with large total substrates might be more likely to be used perennially, since birds there have more "room" to move around and find presumably suitable substrates within the site. Annual site use therefore might increase with substrate size.

We classified the sixty-three sites where we measured substrate into ten size categories. For each colony site we determined the percentage of years during the study when it was used (at least one active nest was present). The mean percentage of years a site was used was calculated for each substrate size class (fig. 7.5). Contrary to the expectation if substrates were limited, the percentage of years a site was used did not vary significantly

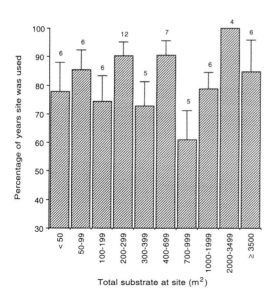

Fig. 7.5 Percentage of years a cliff swallow colony site was used in relation to the total substrate (m^2) at a site. Sample sizes (number of colonies) are shown above the error bars. Mean percentage years used did not vary significantly with substrate size class ($r_s = .21$, $p = .56$, $N = 10$ size classes).

with substrate size (fig. 7.5). Sites with small total substrates generally were as likely to be used in a given year as sites with large total substrates. Thus, once again, we found no evidence for cliff swallows' being limited by nesting substrate availability within colonies.

7.4 ANNUAL USE OF COLONY SITES

In this section we examine patterns in annual colony site use by cliff swallows. Site use is relevant to considerations of nesting site availability in this chapter and to other issues such as avoidance of ectoparasites (section 4.10.2) and local resource differences among sites (sections 7.6.2, 10.7).

The previous two sections on nest spacing and total substrate size focused primarily on the availability of nesting substrate *within* colonies. Addressing site use patterns expands the analysis to account for availability of nesting substrate between sites. Whether birds reuse a colony site the following year presumably depends both on how many alternative sites are available at other locations and on the suitability and quantity of the substrates in the previously used sites. For example, some workers have suggested that ectoparasite infestations were severe enough in nests and substrates of certain colonies to cause cliff swallows to avoid those sites entirely the following year (section 4.10.2). If this is true, the ectoparasites would render nesting substrates unsuitable and hence unavailable to be chosen by the birds.

The data in this section are based on colony site use over an eleven-year period from 1982 to 1992. Some sites were discovered or first became active in later years, however, and therefore use patterns are based on fewer years for some sites (see table 7.1).

Table 7.1 Total Cliff Swallow Colony Sites Surveyed for Use in the Study Area Each Year and Number and Percentage Not Used by the Birds Each Year

Year	Total Colony Sites	Number Not Used	Percentage Not Used
1982	51	18	35.3
1983	61	18	29.5
1984	76	27	35.5
1985	79	24	30.4
1986	77	23	29.9
1987	83	24	28.9
1988	91	24	26.4
1989	107	45	42.1
1990	123	43	35.0
1991	128	40	31.2
1992	134	54	40.3

7.4.1 Total Numbers of Used and Unused Colony Sites

We surveyed colony use in the 150 × 50 km study area annually (section 2.1). For each year we tabulated the total number of cliff swallow colony sites that were unused (table 7.1). This tabulation is based only on the sites known to us in a given year, which tended to increase each successive year as we gradually discovered more colonies and as new ones were established. Therefore year-by-year comparisons in total number of sites are not meaningful; what is relevant is the percentage of unused colony sites each year (table 7.1). Unused sites in this analysis were ones known to be broadly suitable for cliff swallows, as evinced by birds' having nested there at some time, in either previous or subsequent years. If a site obviously became unsuitable in a later year, such as when an overhang fell off a cliff, the site was dropped from the tabulation at that point.

Each year a substantial number of cliff swallow colony sites in the study area were not used (table 7.1). Averaged across years, 33.1% of sites were unused annually, this figure varying from 26.4 to 42.1% per year. This is perhaps the strongest direct evidence that cliff swallow nesting sites were not limited when the study area as a whole was considered. Our estimate of how many sites were not used in a given year was extremely conservative, because we included only sites that cliff swallows actually used at some time during the study. Each year the birds occupied sites that had never before been used; these sites had previously been present but never chosen (and therefore not included to that point in our definition of "sites"). This was especially evident when cliff swallows suddenly occupied the side of a building. The birds seemingly chose these sites at random, even though hundreds of identical structures were present and not used. Similar scenarios occurred at cliff nesting sites. The conclusion from this analysis (table 7.1) is that at least 33% of colony sites, on average, were available but unused annually, and additional suitable sites (unidentified by us) that had not yet been used were also available.

7.4.2 Site Use in relation to Colony Size

It could be argued, of course, that the unused sites in table 7.1 were somehow unsuitable in a particular year, and therefore cliff swallows might really have occupied 100% of the colony sites that were in fact suitable. This argument hinges on what might make a colony site unsuitable in a given year. One candidate is ectoparasitism (section 4.10.2; also see section 7.6.2). Because ectoparasitism varied directly with colony size (section 4.3), larger colonies more likely might be unsuitable the following year owing to ectoparasites and thus less likely to be used (unless the larger substrates at large colonies allow the birds to move around more within a site; section 7.3.3).

We examined the percentage of cliff swallow colony sites that were re-

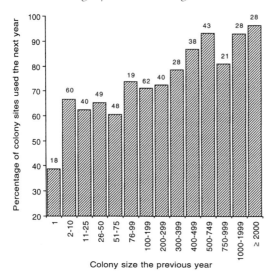

Fig. 7.6 Percentage of cliff swallow colony sites that were used the next year in relation to their colony size the previous year. Sample sizes (total number of colonies) are shown above the error bars. Percentage of sites used the next year increased significantly with colony size ($r_s = .92$, $p < .001$, $N = 14$ size classes).

used in a given year as a function of the colony size at the site the previous year (fig. 7.6; the large sample size made possible finer size classes here than in most analyses in which we used colony size classes). There was no evidence that sites with larger colonies were less likely to be reused the next year. In fact, the opposite pattern was found: probability of site reuse increased significantly with colony size (fig. 7.6). This result did not support the hypothesis that ectoparasites make colony sites unsuitable and suggests if anything that a site with a large colony one year is perhaps more suitable the next year. A similar result was reported by Aumann and Emlen (1959) for Wisconsin colonies.

In scoring site use (fig. 7.6), we classified a site as used if at least one active nest was present. The colony size at a site reused the next year is also potentially instructive. Among such sites, we examined what percentages either increased or stayed the same size (fig. 7.7), and the magnitude of those size changes (fig. 7.8), as a function of the colony size at the site the previous year. Colonies that were large the previous year were significantly less likely to increase the next year (fig. 7.7). The very small colonies when reused the next year usually increased. As might be expected, most of these small colonies experienced a huge percentage change in size, on average (fig. 7.8). Interestingly, the sites with large colonies, though usually decreasing in overall size the next year when reused, tended to decrease only slightly in percentage. At sites of about 400 nests or larger, colony size tended to remain relatively similar whenever the site was reused the next year.

The scenario that can be pieced together from these analyses is that the average cliff swallow colony site might experience cyclical size changes.

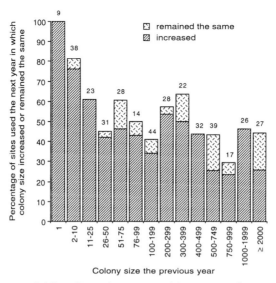

Fig. 7.7 Percentage of cliff swallow colony sites used the next year that increased in size or stayed the same in relation to their colony size the previous year. Sample sizes (total number of colonies) are shown above the error bars. Percentage of colonies increasing in size declined significantly with colony size the previous year ($r_s = -.68$, $p = .0076$, $N = 14$ size classes); percentage of colonies staying the same size did not vary significantly with colony size the previous year ($r_s = .35$, $p = .22$, $N = 14$ size classes).

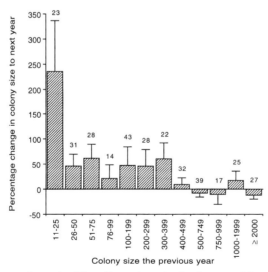

Fig. 7.8 Percentage change in cliff swallow colony size for sites reused the next year in relation to their colony size the previous year. For reasons of scale, colony size classes of 1 nest—mean 11,600.4% ($\pm$ 3,881.5, $N = 9$)—and 2–10 nests—mean 1,096.5% ($\pm$ 571.7, $N = 38$)—are not shown. Sample sizes (total number of colonies) are shown above the error bars. Percentage change in colony size decreased significantly with colony size the previous year ($r_s = -.88$, $p < .001$, $N = 14$ size classes).

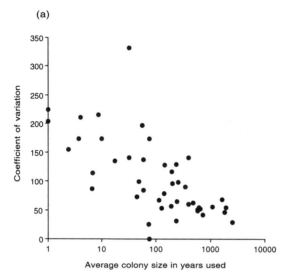

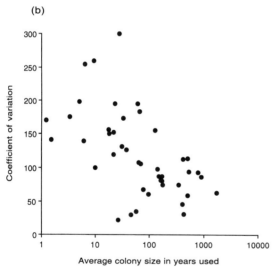

Fig. 7.9 Coefficient of variation in cliff swallow colony size and use at a site (see text) in relation to average colony size at a site in years it was used, for bridge sites (a), culvert sites (b), and cliff sites (c). Coefficient of variation declined significantly with average colony size for bridges ($r_s = -.71, p < .001$, $N = 45$ sites), culverts ($r_s = -.63, p < .001$, $N = 44$ sites), and cliffs ($r_s = -.65, p < .001$, $N = 34$ sites).

When large it was likely to be reused, but it gradually declined in size until it became small enough to be less likely to be reused; then it eventually increased until it once more became large. If this is the typical pattern, however, it probably occurs for any given site over a longer time scale than the eleven years of our study. We saw tremendous variation from site to site when examining use patterns at particular colony sites. There were no obvious trends apparent from each colony site's individual use and size history, although early in our research we saw what appeared to be

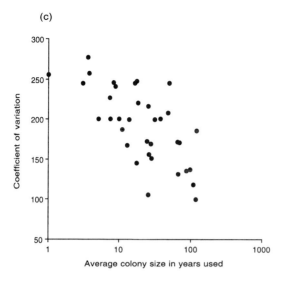

clear patterns of alternate-year use of certain colonies, with birds switching back and forth between sites (Brown 1985a). As we added data for more colonies and more years, the patterns based on a few years at a few sites broke down. Conclusions based on only three, or even five or six, years of site use data would have been wrong. In particular, it does not seem that ectoparasitism can "explain" the long-term patterns of cliff swallow colony site use.

Although some sites varied annually in whether they were used (fig. 7.7) and in their colony sizes (fig. 7.8), other sites were remarkably consistent in size and use from year to year. For example, thirty-six colony sites were used every year they were known to us ($N \geq 3$ years in which use was known), although they sometimes varied greatly in size. We wanted a single measure of each colony site's predictability in terms of both its probability of being reused and its colony size from year to year. For that measure we used the coefficient of variation (section 4.6) in colony size, and to factor in colony use we used a colony size of zero for any year a site was unused. The coefficients of variation were plotted against each site's average colony size in the years it was active (fig. 7.9).

Coefficients of variation declined significantly with average colony size for colonies on bridges, culverts, and cliffs, the three major substrate types (fig. 7.9). Sites typically supporting large colonies were consistently more predictable in both use and colony size, whereas sites with small colonies fluctuated widely in size and annual probability of being used, and hence were somewhat ephemeral (fig. 7.9).

Colony sites of different substrate types differed in predictability of size and use. Mean coefficient of variation for bridges was 104.9 ($\pm$ 10.0), for culverts 120.1 ($\pm$ 9.5), and for cliffs 194.8 ($\pm$ 7.6). These means differed significantly (Kruskal-Wallis ANOVA, $p < .001$), with the coefficient of variation for cliffs significantly higher than that of either bridges or culverts (Wilcoxon rank sum tests, $p \leq .001$ for each); those for bridges versus culverts did not differ significantly (Wilcoxon rank sum test, $p = .17$). Therefore natural colony sites were more unpredictable in use and size than artificial sites.

We conclude that there was no strong evidence that cliff swallows were limited by nesting sites when the study area as a whole was considered. Unused sites existed each year, and annual site use patterns did not suggest that large colonies were made unsuitable by ectoparasites or other factors and were thus unavailable to the birds the next year.

7.5 NEST SPACING COMPARISONS WITH BARN SWALLOWS

Another way to address the potential importance of nesting site limitation in cliff swallows is through a comparison with the congeneric and ecologically similar barn swallow. Barn swallows are solitary to loosely colonial, and colonies appear to be a direct result of nesting site limitation (Snapp 1976; Shields and Crook 1987; Shields et al. 1988). We compared nest spacing among cliff swallow and barn swallow nests within the same colony, to determine if cliff swallows exhibited the same pattern of nest spacing shown by the putatively nest-site-limited barn swallow.

For our comparison we chose a culvert site in 1987, which contained 64 active barn swallow nests and 129 active cliff swallow nests. This was the largest barn swallow colony found in the study area, perennially containing 50 or more nests. Most of the other barn swallows in southwestern Nebraska nested solitarily or in colonies of 10 nests or fewer. The culvert was a double tunnel, yielding four separate walls where the birds built nests. We mapped the locations of all nests and determined the total length of substrate on each wall. If the birds gained no benefits from living together (as reported for barn swallows by Snapp 1976 and Shields and Crook 1987), their nests should be distributed so as to maximize nearest neighbor distances and thus minimize the costs of crowding. An even distribution of nests across each wall with equal internest distances would yield the maximum nearest neighbor distances for the greatest number of birds.

Based on the number of nests on each wall and the total length of the wall, we calculated an expected internest distance if nests were evenly distributed. We used the entire length of the wall in calculations for cliff swallows, because the substrate was uniform and cliff swallows were so thoroughly behaviorally dominant over barn swallows that they could build

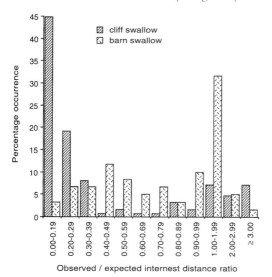

Fig. 7.10 Percentage occurrence of observed/expected internest distance ratios for cliff swallow nests ($N = 125$) and barn swallow nests ($N = 60$). Expected distances assumed the birds spread their nests uniformly to maximize nearest neighbor distances. The distributions for cliff swallows versus barn swallows differed significantly ($\chi^2 = 88.6$, df = 11, $p < .001$).

their nests anywhere they chose, sometimes usurping active barn swallow nests. For barn swallows, in calculating the total length of wall available, we did not include sections that had rows of active cliff swallow nests, because barn swallows avoided these areas, which therefore were unavailable to them as nesting substrates.

We expressed each observed internest distance as a ratio of the observed to the expected distance (fig. 7.10). Ratios of 1.0 indicate an essentially uniform dispersion of nests with little clumping and suggest little advantage of nesting close to conspecifics. In contrast, ratios close to zero indicate significant clumping and suggest benefits to having close neighbors. We elected to use ratios of observed to expected distances given that we were comparing two species, although an alternative method to evaluate clumping within species (using MacArthur's broken-stick model) was presented by Pruett-Jones and Pruett-Jones (1982).

The distributions of ratios for cliff versus barn swallows were significantly different (fig. 7.10). Most cliff swallow internest distances were much smaller than expected if nests were uniformly dispersed (indicating a high degree of nest clumping), whereas barn swallow internest distances were closer to that expected if nests were uniformly dispersed across the substrate (fig. 7.10). Over 40% of barn swallow internest distances were in the 0.90–1.99 observed-to-expected categories, meaning these nests matched relatively closely the predicted nest spacing for a species receiving no benefits from living together. Although most cliff swallow nests were highly clustered, a few were spaced far apart as if the owners were trying to avoid having close neighbors.

We conclude from this analysis (fig. 7.10) that cliff swallows did not space their nests uniformly within a colony. When presented with ample substrate, they crowded their nests together more closely than did the nest-site-limited barn swallows, suggesting that cliff swallows received more benefits from close nest spacing. That the two species differed so strongly is further evidence against substrate limitation as a cause of cliff swallow coloniality.

7.6 ARE CLIFF SWALLOW NESTING SITES LIMITED?

The bulk of the evidence presented in this chapter suggests that cliff swallows were not limited by availability of nesting substrates and thus forced into colonies. We acknowledge that each analysis per se is not an unequivocal rejection of nesting site limitation, but across analyses we found no consistent support of predictions made by the nesting site limitation hypothesis. In particular, the finding of small colonies at sites with large total substrates (fig. 7.3), the relative abundance of apparently suitable but unused nesting sites each year (table 7.1), and the crowding together of nests (e.g., figs. 7.2, 7.10) even in small colonies are probably the most convincing evidence that coloniality is not a direct result of substrate limitation.

7.6.1 Assessing Nesting Site Limitation on Artificial Structures

The colony sites used in most of the analyses in this chapter were bridges and culverts, because these sites were commonly used by cliff swallows in the study area and especially because they presented a uniform and easily measured substrate. However, perhaps these artificial sites, which are quite recent in evolutionary time, have become so common at present as to reduce or remove any nesting site limitation experienced by cliff swallows throughout most of their past. If so, conclusions based on data from bridges and culverts may be suspect. This is a difficult issue to address rigorously, regardless of species, but two sorts of observations suggest that our conclusions for cliff swallows based on artificial sites also pertain to birds on natural cliff sites.

There appeared to be an abundance of broadly suitable nesting substrate on the cliffs along the south side of the North Platte River Valley (see section 3.4.1). Portions of cliffs were unused for many years before the birds suddenly occupied them, attesting to their suitability. Although the protecting overhangs would often crack and fall off, limiting the number of years a given site could be used, *to us* there appeared to be available many more unoccupied sites with adequate overhangs. Natural cliff faces certainly were more irregular in substrate characteristics than bridges, and not all portions of the substrate within a cliff colony allowed nest attachment, but all cliff colonies we studied appeared to be in areas with ample suitable substrate for further colony expansion. Substrate did not limit the size of colonies on cliffs.

If artificial sites represent a sudden superabundance of nesting sites and coloniality historically was caused solely by substrate limitation, cliff swallows now on bridges and culverts should spread out to minimize the substantial costs of group living. As emphasized earlier (sections 7.2, 7.5), there was no evidence that the birds routinely dispersed their nests when the substrate permitted it. Nests were closely packed even on bridges and culverts. Providing the birds with an abundance of artificial colony sites gave them the opportunity to spread their nests out if they so chose, but they did not.

Although we conclude that cliff swallows on both artificial nesting structures and cliff sites were not limited by nesting site availability in southwestern Nebraska, there were differences between artificial and natural colony sites (section 3.4.3). One difference was the greater unpredictability in annual size and use of cliff sites (section 7.4.2). Perhaps the biggest difference, however, was in mean colony size, which was 85.8 nests for cliffs versus 604.6 nests for bridges and 309.3 for culverts (section 3.4). The largest colony seen on a cliff was only 735 nests. If smaller colony sizes on cliffs were not caused by substrate limitations (see above), other differences between natural and artificial sites must have accounted for this result.

Perhaps colonies on bridges and culverts were larger because artificial sites were safer from predators than natural cliff sites (section 7.1; Robinson 1985; Wittenberger and Hunt 1985). This seems unlikely, however, given the apparent ease with which bull snakes reached nests on metal and concrete bridges and that all observed instances of avian predation on cliff swallows occurred at artificial colony sites (section 8.2.1). There were clear differences among sites in probability of attack from predators (section 8.10), but there was no indication that artificial sites with large colonies were any more challenging for predators to reach. Thus there were no structural differences among sites that made certain ones particularly safe from predators and consequently might have led more birds to nest there.

7.6.2 The Influence of Local Food Resources

We have so far addressed the suitability of potential cliff swallow nesting sites primarily in terms of structural characteristics that allow nest attachment and protect the nests from the elements, but other factors may also affect whether a colony site is suitable. Infestations of ectoparasites from the previous summer are one such factor, which did not seem to affect site use in a consistent pattern (e.g., section 7.3.3). Another potential influence on site suitability was food availability near the colony site. In herons, for instance, larger colonies typically are found in areas with more food (Fasola and Barbieri 1978; Gibbs et al. 1987; Gibbs 1991). If food was scarce near a cliff swallow colony site in a given year, that site could be functionally unsuitable for nesting even though it might be structurally capable of supporting nests and devoid of ectoparasites.

Measuring food availability around colony sites was virtually impossible for cliff swallows, given the diversity of insects they fed on (section 9.4), so we could not rule out the possibility that it influenced suitability of sites. However, it seems unlikely that food resources varied enough among sites to make some colony sites unusable in some years. For instance, an indirect measure of local food availability suggested no differences among active sites (section 10.7). In a qualitative sense, the foraging habitat near colony sites that were used one year and not used the next did not seem to change between those years. Land use patterns around colonies remained similar from year to year, as did water levels in the rivers and creeks (which can influence insect populations) with which many cliff swallow colonies were associated. Creeks in the study area tend to be spring fed and hence consistent in flow from year to year (Bentall 1989), and Platte River levels in the study area are now heavily regulated via upstream dams that yield similar yearly flows.

The only indication that food resources might differ enough to influence site use came from our finding a reduction in the nearby foraging areas of the cliff swallows occupying cliff colonies (section 9.4.3). Most of the cliff colonies in the study area were along the southern shore of Lake McConaughy. This large expanse of water substantially reduced the area near each colony that was suitable for foraging. Cliff swallows in the study area typically did not forage over water (except during severe weather), instead preferring fields and grasslands where thermals originating from warm earth concentrated flying insects (section 9.4.4). Consequently, birds living in cliff colonies along the shore of the lake commuted longer distances to forage (section 10.8.2).

Possibly the increased travel costs for the cliff sites could have caused those colonies to be smaller (section 7.6.1) if fewer birds were willing to pay the foraging costs. It is important to recognize that variation in local resource availability among sites may influence both whether a site is used and the size of the colony if it is used. The effect of local resources on colony formation should be explored further, but at present we have little evidence that differences in resources among sites influence colony use or size (section 13.8; Brown and Rannala 1995).

7.6.3 The Traditional Aggregation Hypothesis

Based on their work with barn swallows, Shields et al. (1988) proposed the "traditional aggregation" hypothesis to explain the evolution of barn swallow coloniality. Because they suggested that this hypothesis may apply to all swallows, including cliff swallows, some consideration of it is warranted here. This hypothesis is based on nesting site limitation. The idea is that animals annually aggregate at traditional, high-quality colony sites, using the presence of existing nests as information that the site is likely to be successful. One implicit assumption of the hypothesis is that such sites are somehow limited, leading to aggregation in the relatively few that are available. Another assumption is that the quality of a colony site remains the same from year to year (Brown and Rannala 1995). Some data for barn swallows support this hypothesis (Shields et al. 1988), but others do not (Barclay 1988).

The traditional aggregation hypothesis makes four testable predictions (Shields et al. 1988), none of which was supported for cliff swallows.

1. Colonies should be found *only* at traditionally used sites; novel sites should initially be settled by single pairs. Only after these pioneers are successful and have left evidence of their success (e.g., a nest or nest remnant) should other birds move in. New colony sites in cliff swallows, however, were routinely settled by large numbers. For instance, in 1992 three new concrete bridges built in the study area the previous winter were settled by approximately 250, 500, and 675 pairs of cliff swallows. There were obviously no old nests existing on these virgin substrates. None of the solitarily breeding cliff swallows in the study area used a completely new colony site. Among sites unused the previous year, the average colony size was 190.6 ($\pm$ 28.9) nests and ranged up to 2,000, demonstrating that birds often moved into vacated sites en masse.

2. Colony size should increase as total substrate increases (Shields et al. 1988). This was not the case for cliff swallows (fig. 7.3).

3. Birds reusing old nests should in general experience greater reproductive success than those building new nests either within the same colony or at novel colony sites. Although we did not have systematic data on reproductive success of birds occupying new nests at novel colony sites, there was no evidence that reproductive success consistently differed for birds occupying old and new nests within the same (traditional) colony site (section 11.6.1).

4. The largest colonies should occur at sites with the maximum number of old nests that survive the winter. According to Shields et al. (1988), these existing nests should signal the site's likely success to potential settlers. In cliff swallows, however, colony size grew most rapidly when all nests were removed from a substrate over the winter; when old nests were left standing, colony size declined (Buss 1942; Emlen 1986). It is not surprising that this prediction in particular does not hold for cliff swallows, because old nests at a colony site probably signal the presence of ectoparasites from the previous summer, which the birds avoid (see section 4.10.1). Old, infested nests might indicate, if anything, the

*un*likelihood of settlers' having success there. We therefore conclude that Shields et al.'s (1988) hypothesis does not apply to cliff swallows, and its explicit rejection is further evidence against nesting site limitation in cliff swallows.

Recently a similar idea was proposed by Boulinier and Danchin (1994, pers. comm.), who argued that coloniality is maintained through individuals' assessing habitats by visiting active colonies and in later years aggregating at the sites where breeding success has been the highest. Although their hypothesis makes the same problematic assumptions as the traditional aggregation hypothesis (a limited supply of desirable breeding sites and unchanging site quality between years), it differs from that of Shields et al. (1988) in that the animals directly observe the reproductive success of conspecifics in one year and use that information in selecting a site the next year, rather than relying on the presence of old nests. Boulinier and Danchin's hypothesis could be tested by determining whether reproductive success at a colony site in one year varies directly with population size at that site the next year and whether the same nonbreeding birds that visited a successful colony one year returned to nest there the next year. At this time we have not analyzed our cliff swallow data to address these predictions specifically. However, often the birds failed to reuse a colony site even after successful reproduction there the previous year (especially if the colony had been small; section 7.4), and thus Boulinier and Danchin's hypothesis probably does not apply to cliff swallows.

7.7 SUMMARY

One possible cause of coloniality in cliff swallows is a shortage of suitable nesting sites, forcing birds to aggregate in high densities and pay the costs of crowding. There are challenges in rigorously evaluating whether any species is limited by nesting sites, primarily because we often cannot measure what constitutes a "suitable" substrate or whether sites are in fact limited at the population level. The cliff swallow's use of artificial colony sites such as bridges and culverts permitted us to measure suitable substrate more objectively than for most species and to completely survey the relatively large study area for nesting site availability. The bulk of our analyses found no support for the hypothesis that cliff swallows are limited by available nesting substrates.

Nest density within cliff swallow colonies increased significantly with colony size. Nearest neighbor distances declined with increasing colony size, suggesting greater crowding in larger colonies. Cliff swallows in some small colonies clumped their nests tightly, suggesting benefits of nest crowding. We found no significant relation between average colony size at a site and total substrate size at the site. Some sites with large substrates routinely had small colonies, providing perhaps the best evidence against the nesting site limitation hypothesis. Nearest neighbor distance did not vary significantly with the size of a colony site's substrate, and the birds clustered their nests at sites with essentially unlimited substrate that would have allowed them to spread out. Total substrate size did not significantly affect the probability that a colony site would be used in a given year.

About 33% of cliff swallow colony sites known to be suitable were unused in a given year, on average. Annual colony use patterns showed that sites with large colonies were more likely to be reused the next year. When reused, sites with large colonies the previous year tended to decline in size, and sites with small colonies increased. Colony sites that typically supported large colonies were generally more predictable in size and probability of annual use than were sites that typically supported small colonies. Comparison with the nest-site-limited barn swallow revealed that cliff swallows clustered their nests more closely on the substrate than did barn swallows nesting in the same colony. Internest distances in barn swallows were closer to those predicted if the birds spread their nests to minimize the costs of group living. Conclusions based on data collected at artificial colony sites such as bridges and culverts probably also applied to cliff swallows nesting on natural cliff faces. There was no evidence that suitability of colony sites was affected by local food resources. We found no support for the "traditional aggregation" hypothesis, a variant of the more general nesting site limitation hypothesis.

8 Avoidance of Predators

> When one approaches from below, an alarm is sounded ... and then the air becomes filled with flying [cliff] swallows, charging about the head of the intruder in bewildering mazes, and raising a babble of strange frangible cries, as though a thousand sets of toy dishes were being broken.
> William L. Dawson (1923)

8.1 BACKGROUND

One major advantage of living in a colony is the enhanced opportunity for some or all residents to avoid predators. This avoidance can occur in several ways. Predators may be detected sooner if group vigilance increases with colony size. Many studies, primarily on noncolonial species that flock when feeding, have shown that overall vigilance increases in larger groups even though each individual may reduce the time it spends alert (reviewed by Bertram 1980; Pulliam and Millikan 1982; Elgar 1989; Lima and Dill 1990). With more "eyes," breeding colonies should detect approaching predators at greater distances than can individuals nesting alone. This was perhaps demonstrated best in prairie dogs, which detected predators sooner in larger colonies, although individual alertness declined with increasing colony size (Hoogland 1979b, 1981).

Predators, once detected, may be effectively deterred through group mobbing and defense. Mobbing has been observed in many colonial animals (e.g., Burton and Thurston 1959; Kruuk 1964; Horn 1968; Burger 1974a; Hoogland and Sherman 1976; Andersson and Wiklund 1978; Smith and Graves 1978; Dominey 1981; Halpin 1983; Rood 1983; Shields 1984a; Elliot 1985; Wittenberger and Hunt 1985; Burger and Gochfeld 1990, 1991; Stutchbury 1991; Wiklund and Andersson 1994). The effectiveness of mobbing can be increased, and an individual's risk of falling victim during mobbing decreased, as the number of mobbers increases (section 8.5). Thus a major antipredator benefit of coloniality could be more potential mobbers.

A third way predators may be avoided is when colonial individuals synchronize their reproduction closely in time and thereby swamp the ability of predators to exploit them. In large colonies the probability that an individual's nest will be attacked can be reduced during the peak reproductive period because a predator will not continue to eat once satiated. Nesting synchrony affords a benefit if predators are relatively rare or territorial and do not recruit to colonies of their prey. In a variety of species, the percentage of successful nests and individual survival rise during periods of breeding synchrony (e.g., Darling 1938; Patterson 1965; Veen 1977; Emlen and Demong 1975; Estes 1976; Arnold and Wasserug 1978; Gochfeld 1980; Gross and MacMillan 1981; Wittenberger and Hunt 1985; Hagan and Walters 1990; Burger and Gochfeld 1991; Westneat 1992; cf. Ims 1990). The decreased statistical likelihood within a group of falling victim to a predator, whether reproduction is synchronized or not, is sometimes called the "dilution" effect (e.g., Bertram 1978; Foster and Treherne 1981) and is most effective when a predator is less likely to detect a single group of its prey than one of many single, scattered individuals (Turner and Pitcher 1986; Inman and Krebs 1987).

Another way of avoiding predators is clustering nests in space to create the "selfish herd" effect. Hamilton (1971) pointed out that animals may group to increase the probability that predators will attack other group members, attempting to maximize the number of conspecifics between themselves and the predator. In nesting colonies (Tenaza 1971; Vine 1971), the selfish herd is achieved by positioning nests as close as possible to the geometric center of each colony, assuming that predators are equally likely to approach from any side. Terrestrial predators encounter edge nests first, resulting in greater reproductive success of individuals breeding toward the center. Numerous studies have examined position effects on reproductive success, and many have concluded that edge nests do suffer increased predation (e.g., Taylor 1962; Coulson 1968, 1971; Siegfried 1972; Feare 1976; Buckley and Buckley 1977; Coulson and Dixon 1979; Burger 1981; Gross and MacMillan 1981; Wittenberger and Hunt 1985; Simpson, Smith, and Kelsall 1987; Kharitonov and Siegel-Causey 1988; Hagan and Walters 1990; Rayor and Uetz 1990; Anderson and Hodum 1993; cf. Knopf 1979). However, where aerial predators can directly attack the center of groups first (where their odds of success may be greatest), the risk of predation may be higher in the center (Parrish 1989).

These forms of predator avoidance have often been cited as potential or actual benefits of group living (e.g., Lack 1968; Alexander 1971, 1974; Bertram 1978; Buskirk 1976; Krebs 1978; Veen 1980; Wiklund and Andersson 1980, 1994; Burger 1981; Gross and MacMillan 1981; Post 1982; Van Schaik 1983; Haas 1985; Wittenberger and Hunt 1985; Godin 1986; Stacey 1986; Terborgh and Janson 1986; Turner and Pitcher 1986; da Silva and Terhune 1988; Forbes 1989; Hagan and Walters 1990; Magurran 1990; Siegel-Causey and Kharitonov 1990; Baird and Baird 1992; Van Schaik and Horstermann 1994; Hoogland 1995; Tyler 1995; cf. Spiller and Schoener 1989; Clode 1993). There is substantial empirical evidence favoring predator avoidance as a common advantage of animal coloniality, although this may be in part because predation (or its absence) is operationally easier to document than other potential benefits of coloniality such as nesting site limitation (chapter 7) or social foraging (chapter 10). Furthermore, the body of general theory developed for how foraging and nonbreeding animals may better avoid predators by flocking (e.g., Pulliam 1973; Bertram 1980; Pulliam and Millikan 1982; Caraco and Pulliam 1984; Pulliam and Caraco 1984; Elgar 1989) is directly applicable to animals breeding in colonies and has perhaps promoted the view that the antipredator advantages of coloniality are widespread.

In this chapter we examine the suite of potential antipredator advantages of coloniality for cliff swallows. We first describe the cliff swallow's predators in Nebraska and the contexts in which observed predations occurred and then address to what degree the birds avoided these predators through enhanced detection, mobbing, temporal synchrony, and the selfish herd effect. In evaluating the detection of predators, we also examine what antipredator advantages flocking away from the colony sites affords. After addressing these benefits, we return to the costs of coloniality by investigating how nest clustering may enhance the predators' physical access to nests within colonies and to what extent predators are attracted to cliff swallow colonies. This chapter contains analyses first reported in Brown and Brown (1987), some of which have been updated and modified, and it also contains extensive new material.

8.2 NATURAL HISTORY OF PREDATION ON CLIFF SWALLOWS

During eleven years, we documented successful predation on cliff swallows by eight kinds of predators. Seven of these were birds, and the eighth was the bull snake. Three other animals occasionally competed for existing cliff swallow nests, in the process destroying swallow eggs and nestlings.

8.2.1 Avian Predators

The avian predators that attacked cliff swallows in southwestern Nebraska included sharp-shinned hawks, American kestrels, barn owls, great horned owls, black-billed magpies, loggerhead shrikes, and common grackles.

We observed only one successful, and one unsuccessful, predation attempt by sharp-shinned hawks. Both occurred near dusk, as cliff swallows were entering a colony to roost in their nests. The hawks apparently were perching either in low shrubs or on the ground, relatively close to the colony, and made passes at flying birds. In one case the hawk caught a flying adult swallow; in the other instance the birds detected the hawk and responded by exiting the colony and swirling overhead. Both predation attempts happened early in the season (22 April and 10 May). Sharp-shinned hawks are winter residents in the study area and generally migrate north before many cliff swallows arrive in the spring (Rosche and Johnsgard 1984). Because they overlap with the swallows only briefly, they are probably not important predators.

The loggerhead shrike was another relatively unimportant predator. We observed only one attempt at predation by a shrike. This occurred at a 2,000 nest colony (8401) where cliff swallows continually streamed between the colony and the foraging sites while feeding nestlings. The shrike flew directly into the stream of birds before they could give an alarm and collided with a swallow, driving it to the ground. Shrikes formerly may have been more important predators in the study area, but their populations in many parts of North America have declined in recent years (Morrison 1981; Tate 1986). We saw few, if any, in the study area from 1986 to 1992, although noticeably more appeared in 1993.

Black-billed magpies were both predators and scavengers of cliff swallows. We often saw magpies at colony sites, usually on the ground underneath nests or sitting on top of the cliff or bridge above the nests. They scavenged both eggs tossed out of nests (section 5.6) and nestlings that either fell out of broken nests or jumped out in response to ectoparasites (section 4.7). During the cold weather in 1992 in which some cliff swallows starved (section 10.6), magpies both scavenged dead adults found on the ground and killed weakened and moribund adults, primarily at one large colony (9245) where mortality was especially high. In these cases they walked up to grounded swallows and killed them by several jabs with their beaks.

Occasionally, however, magpies hunted healthy adult cliff swallows by perching on the top of a bridge containing nests and flying out toward birds that were entering and exiting the nests. Magpies were slow and clumsy in these attempts and rarely caught a bird. They presumably succeeded occasionally, however, since some individual magpies repeatedly hunted in this manner at certain colonies. One bird was seen hunting almost daily at a colony (8601) from 25 May to 28 June! Cliff swallows always responded vigorously whenever a magpie passed near a colony, exiting their nests and swirling above, alarm calling. Sometimes a loose group of swallows would follow a magpie as it flew by, and occasionally a cliff swallow would dart down at the magpie's back as if to chase it (cf. section 8.5).

American kestrels were relatively common breeders in the study area and often passed over or near colonies while foraging. In most cases they did not even pause or seem in any way directly interested in the swallows, especially early in the season. Later in the season, as young cliff swallows began to fledge and the kestrels themselves had young to feed, they became more interested. Their usual hunting method was to hover above or near a colony and dive downward in a classic falcon stoop at juveniles flying beneath them. Their success rate on these stoops was well below 50%, and even juvenile swallows could often elude them. Only once did we see a kestrel chase (unsuccessfully) an adult cliff swallow. They were successful in catching mostly what seemed to be recently fledged juveniles that were not flying strongly or that did not respond quickly to alarm calls. At one colony a kestrel repeatedly flew underneath a bridge in attempts to catch swallows coming to and from their nests, and once we saw a kestrel fly directly to a cliff swallow nest and try to pull a bird out (see Bonnot 1921; Wilkinson and English-Loeb 1982). The earliest in the season that kestrels were seen hunting swallows was 28 June, and most of their predation attempts occurred in July.

Cliff swallows responded more strongly to the presence of kestrels than to any other predator in the study area. Whenever one passed near a colony, the birds would flush out of their nests, quickly fly in a very coordinated and synchronized group up to the altitude of the kestrel, and then spread out in a loose flock, giving continual barrages of alarm calls. Once at or above the altitude of the kestrel, cliff swallows often milled near it, following it and continuing to alarm call as if signaling to the predator that they knew it was there. The birds' antipredator behavior made sense, because a kestrel hunts by diving downward and thus represents no threat to swallows at or above its own altitude. During alarm responses at kestrels or other predators, the colony site became eerily quiet, even though some birds always remained behind in their nests, usually peering out (and sometimes using the opportunity to enter their neighbors' nests; section 5.6.1).

Surprisingly, the most destructive avian predator of cliff swallows in southwestern Nebraska was the omnivorous common grackle. Grackles were common breeders in the study area, themselves nesting in small, loose colonies. As with kestrels, most of the grackles that passed near cliff swallow colonies showed no obvious interest in the swallows. The relatively few grackles that were interested attacked in several contexts and used several methods. Grackle predation most often occurred when groups of adult cliff swallows were gathering mud. A grackle, apparently ignored by the swallows, would slowly stroll toward such a group and pounce on a bird, killing it by pecking it several times. At one culvert colony, where birds dropped directly beneath their nests to gather mud inside the culvert, a grackle (recognizable by a missing wing feather) hunted swallows by waiting beneath the nests and jumping on birds that landed on the ground. Inexplicably, the swallows frequently did not respond to the grackle as a predator. Over a two-day period this grackle killed at least 50 mud gathering swallows, often eating only their brains. This same grackle also learned to hunt grass gathering cliff swallows in a pasture near the colony. The grackle would fly over the pasture and drop onto swallows on the ground. This particular grackle became proficient at hunting cliff swallows, not eating some of those it killed. It killed at least 70 birds during a twelve-day period before our presence, and the swallows' cessation of nest building, apparently caused it to leave. Of the 24 banded cliff swallows of known age that the grackle killed, 21 (87.5%) were yearlings. Overall, this colony (8905) consisted of 49.7% yearlings (section 13.3.1). Thus the grackle killed disproportionately more first-year birds, perhaps illustrating their inexperience in dealing with predators.

Grackles also hunted cliff swallows by flying toward birds that were perched on wires and fences, trying to collide with ones that did not fly. They were occasionally successful. Grackles would sit on a fence near a colony and chase juvenile cliff swallows flying from nests. Despite their large size and ungainly long tails, grackles could catch recently fledged juvenile swallows. Occasionally grackles flew into a colony and perched on the tops or sides of nests and tried (usually unsuccessfully) to pull nestlings out of adjacent nests (section 8.8). Grackles also scavenged dead or moribund nestlings that fell out of nests and road-killed adults. All cases of grackle predation seemed to involve a relatively few individual grackles that had learned to attack cliff swallows and began to specialize on them. Grackle predation occurred throughout the nesting season. At colonies where grackles seldom or never attacked them, the swallows ignored all grackles and never responded to them as potential predators. At sites where grackle predation attempts were relatively common, the birds eventually began to respond to them, flying out of their nests and alarm calling, but they never reacted to grackles as strongly as to kestrels or magpies.

Great horned owls at least occasionally preyed on cliff swallows, although these owls' nocturnal and crepuscular habits meant we were seldom at colonies during the times they were likely to attack. In 1984 personnel at the Cedar Point Biological Station found cliff swallow remains in a great horned owl pellet; how the owl obtained the bird(s) was unknown. During the cold weather in 1992, we observed two great horned owls sitting on the top of an irrigation structure (fig. 3.5f), unsuccessfully hunting cliff swallows by flying toward those coming into their nests to roost at dusk. Once we saw a barn owl fly underneath a bridge at dusk, apparently trying to catch birds coming to roost. Both horned owls and barn owls occasionally passed near colonies during daylight and elicited a strong reaction from cliff swallows. We doubt they took many swallows in general, however, because cliff swallows are not active at night, and we saw no evidence that owls attacked nests at night.

Other species that passed near cliff swallow colonies and occasionally elicited alarm calls from colony residents included turkey vultures, red-tailed hawks, Swainson's hawks (see section 8.9), ospreys, and prairie falcons. None of these species ever tried to attack cliff swallows. Although prairie falcons are known to prey on cliff swallows in some areas (Brown and Brown 1995), they are relatively rare in our study area during the cliff swallow's nesting season (Rosche and Johnsgard 1984). The few we observed all appeared to be migrants passing high overhead and not interested in hunting.

No mammal predators were known for cliff swallows in southwestern Nebraska. The only potential ones we saw near colonies were long-tailed weasels, raccoons, and badgers, and none of these seemed interested in cliff swallows. At colonies near towns, house cats probably scavenged nestlings that fell out of nests.

8.2.2 Bull Snakes

Even though seven kinds of avian predators attacked cliff swallows, all were relatively insignificant compared with the bull snake. We observed several bull snake predations each year of the study, and the snakes preyed on far more individuals on each successful predation attempt than did most of the avian predators. The bull snake and the related rat snake are known predators in other parts of the cliff swallow's range (Bent 1942; Ganier 1962; Bullard 1963; Sutton 1967, 1986; Oliver 1970; Thompson and Turner 1980; Hopla and Loye 1983; Brown and Brown 1995).

We saw bull snakes attacking colonies of all substrate types except buildings; they easily reached colonies on cliffs, culverts, and bridges. Remarkably, they reached nests even on metal bridges. Their usual (and most successful) method of attacking a colony was to approach from above the nests, crawl down and underneath the horizontal overhang while still holding on to a flat surface above, and, on reaching the nests, cling to the nests' exteriors. They moved within a colony by crawling on the sides and tops of nests. They often went inside, and even large adult snakes could completely conceal themselves inside a nest. Bull snakes had less success when trying to scale a vertical wall or cliff face from the ground below the nests, usually failing to gain a strong enough grip on the substrate to go all the way up.

Once they gained access to the nests in a colony, bull snakes preyed on anything in them: eggs, nestlings of all ages, and adult swallows. They captured adult cliff swallows by coiling out of sight inside a nest and waiting for the owners to return. Most snakes remained in a colony for at least 45 minutes per attack. One bull snake stayed almost three full days, at times concealed inside nests, and ate the contents of at least 50 nests (about 150 cliff swallow eggs). At other colonies one snake cleaned out 19 nests, another ate 35 eggs, one ate 8 eggs and 2 adult swallows, and another ate 12 nestlings. When satiated, bull snakes simply dropped off a nest onto the ground or into the water below the colony.

Cliff swallows responded to bull snakes at colonies, but usually only if the reptiles were actively trying to reach nests. When a snake reached a nest, the swallows would swirl near it alarm calling, again as if to signal to it that it had been detected. An entire colony would seldom respond to a bull snake, however, even if the snake was successfully entering nests. Birds at the opposite end of the colony would carry on their activities, apparently oblivious to the snake. Once a snake entered a nest and disappeared from view, the alarm response ceased, and birds began entering nests even adjacent to the one containing the snake. This enabled snakes to catch unsuspecting nest owners returning to their nests.

The bull snake was the only reptilian predator we observed, although snapping turtles occasionally scavenged doomed nestlings. At one colony a turtle lived in murky standing water below the nests. It preyed on juvenile cliff swallows that fell out of nests and landed in the water. We also saw something below the water surface, perhaps a snapping turtle or a fish, grab a juvenile cliff swallow that was bathing on the wing in a lake as part of a large communal bathing group. A garter snake was seen scavenging a dead nestling at one colony.

8.2.3 Nesting Site Competitors

Not all losses of cliff swallow nest contents were due to bull snakes. House wrens, introduced house sparrows, and deer mice competed for existing cliff swallow nests in some colonies and at times usurped active swallow nests. These species could oust the owners and then would destroy the eggs or kill the nestlings. Nest failures in these instances resembled predation events. House sparrows were particularly destructive; in a 100 nest colony (8322) sparrows destroyed all the eggs in all the nests (also see Bent 1942; Samuel 1969; Krapu 1986; Silver 1993). Often a single house sparrow cleaned out 10 or 15 nests before selecting one as its own. Sparrows were generally present only at cliff swallow colonies near towns, and most of our data in this chapter on presumed nest predation came from colonies that contained few or no sparrows. Known cases in which sparrows were responsible for apparent "predation" were excluded from our analyses, because interaction between cliff swallows and house sparrows is an artifact of humans' having introduced these sparrows to North America.

Wrens probably destroyed cliff swallow eggs infrequently, as they seemed to prefer to use abandoned swallow nests if available (see Elliott 1983; Gutzwiller and Anderson 1986). Mice, which could climb the concrete walls of culverts, usurped cliff swallow nests in colonies where the nests were over dry ground, particularly in culverts with low ceilings. Mice chewed through the walls of adjacent nests and destroyed any cliff swallow eggs or small nestlings they encountered. Mice also chewed on the wing and tail feathers of older nestlings, and though it did not kill them, the damage to their feathers at times seriously impaired the juveniles' ability to fly. Even though deer mice were natural nest competitors, their effects were too infrequent and localized to be important.

8.3 DETECTION OF PREDATORS IN RELATION TO COLONY SIZE

The probability that an approaching predator will be detected by at least one group member should increase with group or colony size. This is especially likely for those predators that hunt in relatively open habitats and do not use cryptic, ambush strategies (Pulliam 1973; Wittenberger 1981; Van Schaik, Van Noordwijk, Warsono, and Sutriono 1983). The various predators of cliff swallows all tended either to approach a colony quickly from fairly great distances (most of the avian predators) or to approach slowly from shorter distances over open ground (bull snakes). Thus cliff swallows might be expected to routinely detect incoming predators, and their detection times and distances should increase with colony size. We investigated predator detection in colonies of different sizes by presenting a model predator under standardized conditions at each site (Hoogland 1981; Wilkinson and English-Loeb 1982) and measuring how far from the colony the birds detected it.

8.3.1 Model Predator Presentation Procedures

The model predator we used was a lifelike inflatable snake that resembled a real bull snake (fig. 8.1). We placed the snake model in a small wooden box (snake blind) 100 m from each colony. This seemed to be beyond the distance at which real predators were detected in even the largest colonies. An observer's blind was placed approximately 10 m from an edgemost nest in each colony. Because most predators presumably approach from the edge of a colony, we made our predator presentations on the edge, rather than at the center, of each colony (cf. Wilkinson and English-Loeb 1982). The model snake was towed by monofilament fishing line from the snake blind to the observer's blind at about 0.1 m per second across sand or dirt. The model was constantly in full view for the birds and not obstructed by grass or debris. Tow rate was fixed by the spool diameter of the tow line take-up reel, and our estimates indicated that 0.1 m per second was a good approximation of the velocity of an undisturbed foraging bull snake. Presentations were always done when at least half of the active nests in a cliff swallow colony contained eggs or nestlings.

Fig. 8.1 Inflatable model snake used to simulate bull snake attacks against cliff swallow colonies, to measure the distance at which predators were detected.

The birds virtually ignored our blinds and ceased responding to us immediately after we entered the observer's blind. To avoid habituation to the snake model, we did no more than three presentations on which the snake was detected, or a total of six presentations, on any single day at a colony, and we did not present it on consecutive days at the same colony. Between successive presentations, we waited at least ten minutes after the last alarm call (whether elicited by us or not) before starting the next presentation, to ensure that the swallows were not already in a state of alarm.

Cliff swallows responded to the snake by alarm calling, hovering above it, and occasionally diving at it. Either a single swallow or up to ten or more would detect the snake and respond. We were interested in *initial* detection distance at each colony, so as soon as one bird detected it, we stopped the tow and measured the snake's distance from the colony. The swallows' responses to the model snake were similar to their behavior when real bull snakes appeared. Six times we towed a stick of about the same size as the model snake at the colony where responses were most pronounced, but the birds always ignored the stick and never responded to it. This further suggested that the cliff swallows perceived our model as a predator.

8.3.2 Detection of Predators

We presented the model snake predator to cliff swallows 205 times at nineteen colonies ranging from 1 to 2,000 nests. Based on detection dis-

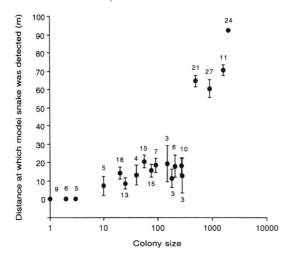

Fig. 8.2 Distance (m) at which the model snake predator was detected in relation to cliff swallow colony size. Numbers by dots are presentations at each site (sample size). Detection distance increased significantly with colony size ($r_s = .83$, $p < .001$, $N = 19$ colonies).

tances, three classes of colonies could be distinguished (fig. 8.2): small colonies of fewer than 10 nests that never detected the approaching predator; colonies containing between 10 and 275 nests that detected the predator on average at about the same distance; and large colonies of 500 or more nests that detected the predator at great distances. The distance at which the approaching predator was detected increased significantly with colony size (fig. 8.2). In the largest colony, the model snake was detected almost as soon as it emerged from its box. We detected no qualitative differences in *how* birds in different-sized colonies responded to the predator.

These results clearly showed that birds nesting in large colonies detected snake predators at greater distances than did birds in small ones. We do not know whether the same pattern held for avian predators, though it seems likely. We briefly explored using a model avian predator (a stuffed kestrel), as in Wilkinson and English-Loeb (1982), but we were deterred by the logistics of realistically presenting an incoming flying model over a distance of 100 m.

8.4 INDIVIDUAL ALERTNESS AND GROUP VIGILANCE

The model snake presentations suggested that overall colony vigilance was increased when more individuals were present (fig. 8.2). Did this result because each individual in a large colony was more vigilant (perhaps in response to a large colony's increased conspicuousness to predators; see section 8.9; Birkhead 1977), or simply because the large numbers of birds present ensured that at least a few were scanning for predators at any given instant? We attempted to learn if the time individual cliff swallows spent alert at colonies varied with colony size, as Hoogland (1979b, 1981) found for prairie dogs. Hoogland showed that each individual's alertness declined with colony size, thereby presumably freeing it to engage more often in foraging and other activities. Measuring individual alertness at cliff swallow colonies proved difficult, however, because often the birds were not readily visible to us as they sat in their nests. Furthermore, because there is considerable interaction between birds in adjacent nests (sections 5.5, 6.3), it was often impossible to know if "alert" individuals were scanning for predators or monitoring the activities of neighbors. Because of these difficulties, we examined individual alertness and group vigilance in cliff swallows in flocks away from the colony sites.

Throughout the nesting season in Nebraska, the birds assembled in "loafing" flocks, sometimes up to 2 km from the nearest colony site, where they rested, preened, and sunbathed. Loafing flocks assembled on wires, rock ledges, trees, and the ground. Cliff swallows also gathered mud in flocks away from the colony sites (section 6.2). Birds in these situations presumably showed vigilance patterns similar to those at colony sites and permitted us to evaluate vigilance-related antipredator advantages of group living. By studying these groups in which relatively few intraspecific interactions (e.g., fights) occurred, we were more likely to measure alertness truly directed toward predators.

8.4.1 Measuring Vigilance

The percentage of time each cliff swallow spent preening and sunbathing was used to examine individual vigilance. Cliff swallows sunbathed by rolling over to one side, ruffling the feathers, drooping the wings, fanning the tail upward, opening the bill, and pointing one eye toward the sun (Barlow, Klaas, and Lenz 1963). Sunbathers appeared rigid, did not look around, and seemed largely unaware of their surroundings. Whenever birds were not preening or sunbathing, they were alertly looking around. Thus time spent preening and sunbathing was an inverse measure of how much time the birds spent scanning, presumably for predators. The number of cliff swallows in loafing flocks with their heads up at any given time was used to examine overall group vigilance.

We selected focal birds as randomly as possible from a flock, noted flock size, and with stopwatches recorded the total time each bird was observed and the time it spent preening or sunbathing. We watched focal birds for as long as possible, usually between one and eight minutes. An observation was terminated if the flock was disturbed by either apparent or observed predators or if the flock size changed. Because flock size and composition were constantly changing, each observation was treated independently. These data were taken from the start of the nesting season until juveniles appeared in the loafing flocks. For flocks containing at least 15 birds, where possible we distinguished focal birds on the *edge* of the flock, defined as ones with no other birds adjacent on at least one side, and *nonedge* birds, defined as individuals with at least five (and usually many more) other birds surrounding them. If a focal bird's position changed, the observation was terminated. Using the same procedures and criteria, we also quantified the percentage of time focal individuals spent gathering mud at a mudhole.

We measured group vigilance in cliff swallow loafing flocks by scan sampling each flock, recording at that instant the number of alert individuals, defined as birds with their "heads up" that were not preening or sunbathing. Scan samples were done at five-minute intervals if flock size persisted unchanged or more often if flock size changed in the interim. After each scan, we recorded flock size.

8.4.2 Vigilance in relation to Group Size

Individual and group vigilance was measured for flocks ranging from 2 to almost 400 birds and for solitary individuals. The percentage of cliff swallows that were alert at any instant declined significantly with flock size, but the absolute number of vigilant birds at any instant increased significantly with flock size (fig. 8.3). This means that overall group vigilance in larger flocks was enhanced simply because more birds were present to be alert at any one time.

Because the percentage of birds that were alert at any given time declined with flock size, each individual in a large flock spent less time alert and was freed to engage in other activities such as preening and sun-

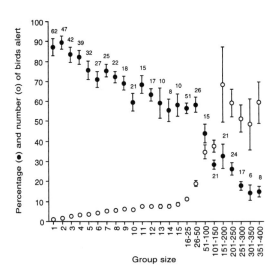

Fig. 8.3 Percentage and number of cliff swallows that were alert on a given scan sample in relation to group size. Numbers by dots are groups of each size class observed (sample size). Percentage of birds alert declined significantly with group size ($r_s = -.98$, $p < .001$, $N = 24$ group sizes). Number of birds alert increased significantly with group size ($r_s = .98$, $p < .001$, $N = 24$ group sizes).

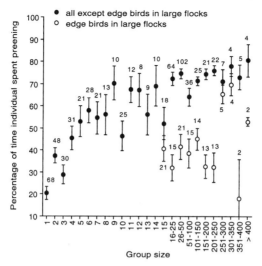

Fig. 8.4 Percentage of time a cliff swallow spent preening in relation to group size for all individuals except edge birds in large flocks and for edge birds in large flocks. Numbers by dots are birds in each size class observed (sample size). The percentage of time spent preening increased significantly with group size for all birds exclusive of edge birds in large flocks ($r_s = .87$, $p < .001$, $N = 25$ group sizes). Percentage of time spent preening did not vary with group size for edge birds in large flocks ($r_s = .25$, $p = .45$, $N = 11$ group sizes).

bathing. When all individuals were considered (except edge birds in flocks larger than 15), the percentage of time spent preening and sunbathing per individual increased significantly with flock size (fig. 8.4). There was thus a corresponding decrease in time spent alert. These individuals probably benefited greatly from flocking because their time available for preening and sunbathing increased. However, edge birds probably realized less benefit from flocking, because their preening did not increase with flock size (fig. 8.4). Instead, they spent much of their time scanning. Because individuals at the edges of a flock presumably are closest to a potential predator's approach, their increased scanning supports our assumption that alertness in these flocks was indeed directed toward predators (cf. section 9.7.2). The individuals on the edge of a flock presumably change with time; otherwise consistently peripheral birds might gain nothing by flocking and in fact might be at greater personal risk in large groups that attract predators (section 8.9). Unfortunately, since birds in loafing flocks were unmarked, we could not determine which individuals occupied the edges.

Similar results were obtained for mud gathering flocks of cliff swallows. When individuals were on the ground to gather mud, they often scanned, presumably for both aerial and terrestrial predators. When actually gath-

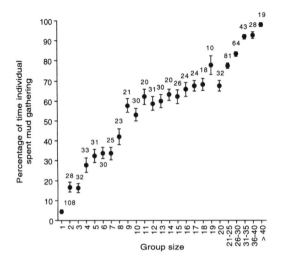

Fig. 8.5 Percentage of time a cliff swallow spent mud gathering in relation to group size. Numbers by dots are birds in each size class observed (sample size). Percentage of time spent mud gathering increased significantly with group size ($r_s = .96$, $p < .001$, $N = 25$ group sizes).

ering mud, however, their heads were bent close to the ground (fig. 6.1), and they were then probably vulnerable to predators such as kestrels and, especially, common grackles (section 8.2.1). Individuals that gathered mud in large flocks did so more efficiently than individuals in small flocks or alone (fig. 8.5). By relying on enhanced group vigilance, cliff swallows in large flocks spent almost their entire time at a mudhole actually gathering mud and relatively little time alert. Individuals in small flocks, however, spent most of their time at a mudhole scanning for predators (fig. 8.5). Efficient mud gathering in large flocks likely reduced the cumulative time birds were vulnerable to predation, and over the season it might have hastened the building of nests (section 5.3.1). These results also illustrate the trade-off inherent in gathering mud in groups: as flock size increases, so does mud gathering efficiency and the probable odds of avoiding predation, but the likelihood (at least for females) of being attacked and forced to copulate by an extrapair male also increases (section 6.2.3).

The results on group vigilance and individual alertness (figs. 8.3–8.5) suggest that cliff swallows received the classical benefits of flocking—enhanced group vigilance and decreased individual alertness. These are presumably important antipredator advantages, although we do not know how they directly translate into mean reproductive success. We assume that the observed patterns of vigilance among birds in loafing and mud gathering flocks also apply to cliff swallows when at their colonies. These antipredator advantages may have led in part to the gregariousness these birds exhibit when away from their colonies and during the nonbreeding season.

8.5 MOBBING OF PREDATORS

The enhanced detection of predators in larger cliff swallow colonies (fig. 8.2) presumably gave the birds there longer to take evasive or defensive action. Defense against predators can take the form of mobbing. In some cases a predator may be deterred from attacking by a seemingly coordinated, conspicuous mob of colony residents that swirl near it or (in rare instances) physically assault it. Mobbing presumably distracts or confuses an approaching predator and may inform it that it has lost the element of surprise and is thus unlikely to succeed in its attack. The effectiveness of mobbing presumably increases with mob size (Hoogland and Sherman 1976; Elliot 1985; Robinson 1985; Burger and Gochfeld 1991; Poiani 1991; Wiklund and Andersson 1994; cf. Lemmetyinen 1971) and the distance from the colony at which the predator is detected.

Cliff swallows' mobbing responses to predators were relatively weak. A successful bull snake attack at a colony elicited reaction from only a small fraction of the colony residents (section 8.2.2), and relatively few birds typically responded to approaching snakes not yet at a colony. There were no coordinated, colonywide mobbing displays directed toward bull snakes, and these snakes were unlikely to have been deterred even had there been. Bull snakes are bold and not easily intimidated, and they ignored even humans when attacking colonies.

The birds responded more vigorously to avian predators, but not in ways that seemed to deter them effectively. The typical response by most colony residents was to fly to the same altitude as the predator or higher and then mill in a loose group, alarm calling heavily (section 8.2.1). This behavior never seemed to frighten the predator or otherwise directly cause it to leave, unless it in fact signaled that it had been seen and deterred it from hunting there for that reason (Woodland, Jaafar, and Knight 1980; Bildstein 1983; Hasson 1991; FitzGibbon 1994; Flasskamp 1994). Because actual predator attacks were too infrequent at small colonies (section 8.9) and too unpredictable in time at large colonies, we did not try to collect quantitative data on the swallows' mobbing responses to predators. Our qualitative observations, however, showed that they could not deter any of their predators through mobbing. Therefore how mob size and effectiveness varied with colony size was moot, although obviously mobs were larger in larger colonies where more individuals had eggs or nestlings at any given time.

Whether effective or not, the mobbing behavior of cliff swallows—swirling and milling near a predator—presumably carried some risk. The cost of mobbing in general has been illustrated in cases where a predator seized a mobber that came too close (reviewed by Sordahl 1990). We have not seen mobbing cliff swallows preyed on, though getting too close to a hunting sharp-shinned hawk or American kestrel clearly seems risky. Brown and Hoogland (1986) suggested that one antipredator benefit of coloniality is the opportunity to reduce one's own risk of predation by mobbing in large groups. In an interspecific comparison of cliff, barn, bank, and northern rough-winged swallows, they showed that the colonial species took fewer risks in mobbing than the solitarily nesting species. When presented with rubber models of great horned owls or stuffed long-tailed weasels, solitary barn and rough-winged swallows typically dived at the predator or otherwise approached very closely, whereas the colonial cliff and bank swallows did not make these presumably risky dives or close approaches. Brown and Hoogland (1986) suggested that species such as cliff swallows benefited from mobbing in groups by reducing their own probability of being the predator's victim simply because of the many individuals present, and by using the confusion created by a large mob to forgo diving altogether. Solitary nesters, without conspecifics nearby to help mob, presumably have to engage in high-risk tactics such as diving to have any hope of deterring the predator.

The experiments in Brown and Hoogland (1986), many of them done in our Nebraska study area, showed that cliff swallows rarely dived at predators. Occasionally we saw a swallow very briefly chase a flying predator such as a magpie (section 8.2.1), but the birds generally avoided any close contact with a predator. This alone suggests that cliff swallows seldom deterred predators. We did not investigate the effect of colony size in the study of mobbing risk (Brown and Hoogland 1986), but we predict that the two should vary inversely in a colonial species like the cliff swallow. Pairs nesting solitarily or in small colonies should take greater risks to deter predators, and the two cases of diving by cliff swallows that we observed in the earlier study occurred in relatively small colonies of 6 and 56 nests. A similar result in which individuals in larger colonies invested less in mobbing and dived less at predators was reported for common terns (Burger and Gochfeld 1991).

8.6 BREEDING SYNCHRONY AND THE SWAMPING OF PREDATORS

Synchronizing reproduction within colonies presents predators with a brief period of superabundance of potential prey individuals and may therefore reduce the per capita risk of predation for individuals nesting during the peak period of reproduction (reviewed in Gochfeld 1980; Wittenberger and Hunt 1985; Ims 1990). In this section we examine breeding synchrony in cliff swallows and assess whether synchrony confers any antipredator advantages.

8.6.1 Synchrony in relation to Colony Size

Cliff swallow colonies exhibited relatively high breeding synchrony (figs. 8.6, 8.7). The percentages of nests in which egg laying began on various dates are illustrated for representative small and medium-sized colonies in figure 8.6 and for relatively large colonies in figure 8.7. With two exceptions, the colonies shown are ones in which we knew exact clutch initiation dates (from nest checking; section 2.2) for over 90% of all active nests in the colony (the exceptions were ones with 78.5% and 85.6% of

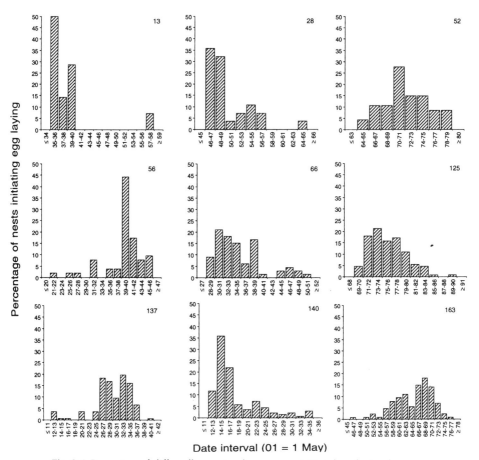

Fig. 8.6 Percentage of cliff swallow nests whose owners initiated egg laying during two-day date intervals (01 = 1 May) for nine small and medium-sized colonies. Colony size is shown in the upper right corner for each. Note similar scale of ordinate for all. Sample sizes were, from left: *upper row*, 13, 28, 47 nests; *middle row*, 52, 66, 125 nests; *lower row*, 137, 137, 128 nests.

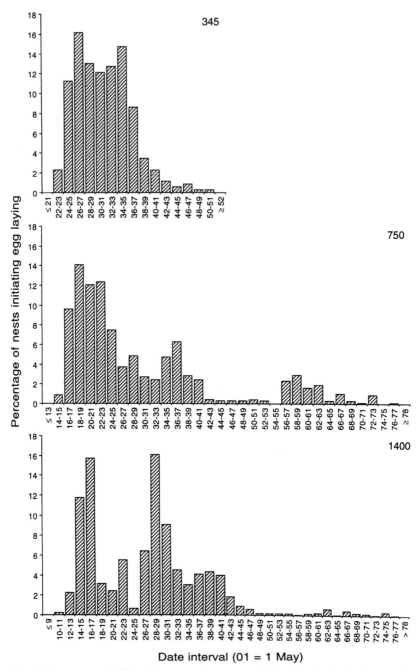

Fig. 8.7 Percentage of cliff swallow nests whose owners initiated egg laying during two-day date intervals (01 = 1 May) for three relatively large colonies. Colony size is shown in the upper right corner for each. Note similar scale of ordinate for all. Sample sizes (*upper to lower*) were 345, 696, and 1,199 nests.

nests checked). Thus each of these is a profile for virtually an entire colony and does not represent a spatial "pocket" of synchrony in an otherwise more asynchronous group. Profiles for other colonies not illustrated were similar.

Several general patterns were apparent. In most colonies clutch initiation dates were skewed toward the earlier part of the season. A relatively few cliff swallows laid eggs first, followed very closely by large numbers of colony residents. Egg laying would taper off more gradually after the peak of laying (figs. 8.6, 8.7). The two smallest colonies showed more birds initiating laying in the first date interval than in any of the others, leading to a very truncated distribution (fig. 8.6). The largest colony (1,400 nests) showed an interesting bimodal distribution (fig. 8.7), probably resulting from two separate waves of birds colonizing the site. The sites with 52, 125, and 163 nests were late colonies, first becoming active in late June and July. Two of these showed a more bell-shaped pattern in which laying gradually increased, reached a peak, and then declined (fig. 8.6). This was the typical distribution in the late-starting colonies we observed, in contrast to the early skew in laying dates in most other colonies.

Breeding synchrony varied with colony size. Approximately 75% of the nests in the smallest colonies initiated laying during periods of 6 days or less (fig. 8.6), whereas in larger colonies the comparable time interval was 20–21 days (fig. 8.7). We used the standard deviation of clutch initiation date as a comparable measure of within-colony synchrony among different colonies (section 2.9). We calculated the standard deviation for forty-five colonies ranging from 3 to 2,200 nests (including many not shown in figs. 8.6 and 8.7). The standard deviation increased significantly with colony size (fig. 8.8), emphasizing that larger cliff swallow colonies tended to be relatively less synchronous than smaller ones. Two small colonies showed substantial asynchrony (fig. 8.8), but most were as synchronous as those illustrated in figure 8.6.

8.6.2 Reproductive Success in relation to Synchrony

Because we were interested in the potential antipredator advantages of breeding synchrony, we used the percentage of nests that were successful as a measure of reproductive success and as a potential index of predation. When predators such as bull snakes attacked nests, they invariably took all the eggs or nestlings. Thus presence or absence of nestlings in an intact nest was probably the most accurate indication of whether predation had occurred, although complete nest failure was also caused by other factors such as ectoparasitism (section 4.7).

As a measure of each nest's relative degree of synchrony within a colony, we used its standard deviation from the colony's modal laying or hatching date (section 2.9). The advantage of this measure is that it per-

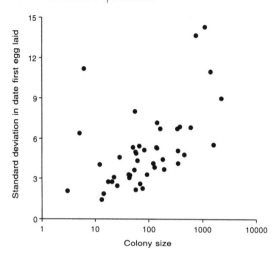

Fig. 8.8 Standard deviation in the date the first egg was laid in the nests of each cliff swallow colony in relation to colony size. Standard deviation is a measure of breeding synchrony. Standard deviation increased significantly with colony size ($r_s = .57$, $p < .001$, $N = 45$ colonies).

mitted us to pool data from different colonies to examine the overall effect of synchrony on reproductive success (Brown and Brown 1987). We examined both laying and hatching synchrony, the latter important especially for colonies where substantial numbers of nests failed before hatching (section 2.9). We compared nests in fumigated colonies with nests at sites exposed to natural levels of ectoparasitism. The data on nest failures for fumigated nests presumably reflect largely predation, because the effects of swallow bugs (section 4.7) were removed.

Nesting success in general varied with a nest's degree of within-colony synchrony (fig. 8.9). Success of both fumigated and nonfumigated nests differed significantly across laying synchrony classes (fig. 8.9a). Not surprisingly (section 4.7), fumigated nests had higher success than did nonfumigated nests for each synchrony class, and the fumigated and nonfumigated distributions of nest success across classes differed significantly from each other (fig. 8.9a). Nonfumigated nests showed a pattern of declining success for the later classes in particular, emphasizing that nests begun after the peak period of egg laying in a colony faced poor prospects of success (see sections 4.4, 11.4). Similar patterns were seen for fumigated and nonfumigated nests when analyzed with respect to within-colony hatching synchrony (fig. 8.9b).

8.6.3 Antipredator Benefits of Synchrony?

These patterns (fig. 8.9) provided no clear support for the hypothesis that cliff swallows escaped predation by nesting synchronously within a colony. In none of the distributions did the most synchronous nests (classes −1, 0, and 1) have markedly greater success than the others. It appeared that the early asynchronous nests generally did best, especially

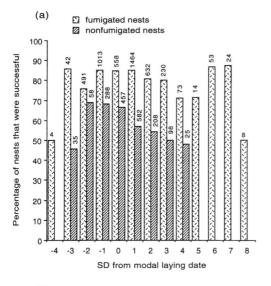
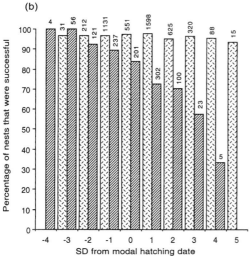

Fig. 8.9 Percentage of successful cliff swallow nests of different degrees of egg laying (*a*) and hatching (*b*) synchrony (see text). Total nests of each class (sample size) are shown above bars. Success of nonfumigated nests varied significantly with laying synchrony ($\chi^2 = 41.8$, df = 7, $p < .001$) and with hatching synchrony ($\chi^2 = 100.3$, df = 8, $p < .001$). Success of fumigated nests varied significantly with laying synchrony ($\chi^2 = 39.9$, df = 10, $p < .001$) and almost significantly with hatching synchrony ($\chi^2 = 13.6$, df = 8, $p = .09$). The distributions for fumigated versus nonfumigated nests differed significantly for laying synchrony ($\chi^2 = 201.4$, df = 7, $p < .001$) and for hatching synchrony ($\chi^2 = 287.4$, df = 7, $p < .001$).

with respect to hatching date among the nonfumigated nests (fig. 8.9b). The early asynchronous nests should have suffered higher nest loss relative to synchronous nests if synchrony effectively swamped predators.

The differences between fumigated and nonfumigated nests (fig. 8.9) can be attributed to swallow bug parasitism. The number of swallow bugs increased as the nesting season advanced (section 4.4), apparently causing increased nestling mortality. Among nests exposed to bugs, the earliest nests were generally the most successful (fig. 8.9). Synchrony per se was therefore probably less important in determining nesting success than when in the season, relative to other colony residents, a bird could breed (see section 11.4). It would be interesting to know more about the relatively few individuals that, in almost all colonies (figs. 8.6, 8.7), were able to begin breeding before the rest of the colony residents. The factors (such as energy constraints) that kept most colony residents breeding in a fairly narrow window of time (figs. 8.6, 8.7) must not have applied to the few birds that started before everybody else.

The pattern for fumigated nests is perhaps the strongest evidence against synchrony's serving to swamp predators. In the absence of ectoparasites, the differences among synchrony classes in nesting success presumably reflected largely predation, and there was no evidence for consistently higher predation rates among asynchronous fumigated nests (fig. 8.9).

Finding little evidence for predator swamping was in a sense surprising, given that the principal predator on cliff swallow nests was the bull snake. Snakes, though capable of eating many eggs and nestlings on a single visit to a colony (section 8.2.2), are satiable predators because they eat infrequently (e.g., Fitch 1982; Shine 1986). A bull snake might need to attack a colony only two or three times during the colony's peak period of nesting, assuming that it was satiated on each attack. If so, the per capita risk of predation during that period would be lower than for early or late nests, which a snake might have to attack repeatedly for smaller meals. However, perhaps swamping of predators is not important for cliff swallows simply because nest predation attempts are relatively rare. If most colonies are successfully attacked by a bull snake only once or twice a season, or often not at all (section 8.9), nest losses to predators would not be frequent enough to be distinguishable (or selectively important) among other causes of nest failure (such as falling from the substrate, accidental death of parents, weather, or conspecific interference).

Because there was no evidence that synchronous nesting swamped predators, how this potential antipredator advantage varied with colony size was moot. It is worth noting, however, that although smaller colonies (fig. 8.6) were relatively more synchronized than larger ones (fig. 8.7), the absolute *number* of nests active at any given time in large colonies was usually much greater. Therefore, if predator swamping is important in rare circumstances, it should be much more effective in large colonies, where each individual can more likely "hide" its nest, statistically speaking, among the many others active simultaneously.

If cliff swallows do not synchronize their reproduction within a colony to swamp predators, why are colonies so synchronous? Several other factors probably lead to synchrony, ectoparasitism being chief among them.

If each individual seeks to initiate nesting as early as it possibly can to avoid the ectoparasites that begin reproducing and thus increasing in a colony as soon as it is occupied (section 4.4), one would expect the pattern of nesting synchrony seen in most colonies (figs. 8.6, 8.7). A few birds have the ability to start slightly ahead of the pack, the majority of birds come next (as early as they can), and the stragglers, which are perhaps young and inexperienced (section 11.8) or inferior in some respect, bring up the rear. The observed nesting "synchrony" may simply be a consequence of each individual's nesting as early as it can.

Another possible explanation for the observed synchrony is a seasonal peak in food abundance, to which many of the birds respond by nesting during that peak. We had no evidence for this in Nebraska, although cliff swallows in southeastern Arizona synchronize their nesting with the onset of the summer rains, which presumably affect food resources (Brown and Brown 1995). Birds might also actively synchronize nesting so as to maximize the opportunities to gain foraging information from neighbors (section 9.5), although this seems more likely to be an incidental consequence of synchrony than a cause. Various factors potentially influence breeding synchrony in birds (Gochfeld 1980; Wittenberger and Hunt 1985), and the causes of temporal nest clustering in most species are not clear.

8.7 PREDATION IN RELATION TO NEST POSITION

If cliff swallow colonies represent a selfish herd (Hamilton 1971; Tenaza 1971; Vine 1971), birds should jockey for nest position early in the season, and the birds forced to the edge should suffer higher nest predation. We had no way to test the former prediction, but we examined the latter by comparing nest success of center and edge nests, as defined in section 2.8.2 (also see section 4.5.3). The fate of the edge nests is likely to be the most instructive, because predation events appeared to be so rare that they probably affect primarily the nests on the edges of each colony. In most of the successful predations by bull snakes we observed ($N = 32$), the snake began preying on the first nest it encountered along a colony's edge, progressively moving inward. Only once did we find a bull snake that had, inexplicably, reached a nest in the colony's center without apparently attacking the edge nests first.

Among nonfumigated nests of all colonies (in which one egg or more was laid), 65.9% of those on the edge ($N = 258$) were successful in producing at least one nestling alive to day 10, versus 73.0% of those in the center ($N = 226$). The difference approached statistical significance ($\chi^2 = 2.86$, df = 1, $p = .09$). Among fumigated nests of all colonies, 72.9% of edge nests ($N = 647$) were successful, versus 84.3% of center nests ($N = 645$); the difference was highly significant ($\chi^2 = 24.9$, df = 1, $p < .001$).

Thus edge nests were in general less successful than center nests, and this difference was probably attributable to predators, at least for the fumigated nests where the effects of swallow bugs were removed. Not only were edge nests in fumigated colonies significantly less successful than center nests, they were significantly less successful than all other nests (all nests except edge: 81.8% successful ($N = 5{,}578$); $\chi^2 = 29.7$, df = 1, $p < .001$). This suggested that predation was likely to occur mostly among the 10 or so nests closest to the colony's edge, and beyond that, predation rate probably was not influenced by nest position. Success of center fumigated nests, for instance, did not differ significantly from that of all other nonedge nests ($N = 4{,}286$), of which 82.8% were successful ($\chi^2 = 0.94$, df = 1, $p = .33$). These results are consistent with the apparent infrequency with which predators attacked cliff swallow colonies; any given predator probably seldom needed to venture beyond the edge nests to find a meal. We did not attempt these analyses separately in colonies of different sizes, because only relatively large colonies had clearly defined center and edge nests. Spatial comparisons of nesting success are probably meaningless for small colonies without obvious centers and edges.

Reduced nesting success of peripheral individuals may have been caused in part by factors other than predation. In some species younger, inexperienced birds are forced to the edges of a colony, and reduced success of edge nests may be partly attributable to these birds' inherent lower quality (e.g., Coulson 1968, 1971; Coulson and Dixon 1979; reviewed in Wittenberger and Hunt 1985 and Kharitonov and Siegel-Causey 1988). Cliff swallows occupying the edges of colonies tended to be younger and settled later than the birds that lived closer to the center (section 11.8). We can probably safely conclude only that the lower expectation of success for edge nesters was due in part to higher predation rates on the edges of colonies. We suggested earlier (Brown and Brown 1987) that nest positioning represented a trade-off between predation that increased toward the edges of colonies and ectoparasitism that increased toward the center. However, our most recent analyses indicated no consistent pattern of increased ectoparasitism in center nests (section 4.5.3), so such a trade-off is unlikely (and see section 14.4.1).

8.8 NEST PACKING AND VULNERABILITY TO PREDATORS

In this section and the next, we examine two predator-related costs of cliff swallow coloniality: nest packing that enhances predators' access to nests and increased attractiveness of colonies to predators. These costs must be considered in evaluating whether coloniality provides any net antipredator advantage for cliff swallows (section 8.10).

Fig. 8.10 Bull snake attacking a cliff swallow nest by clinging to an adjacent nest. High nest density helped snakes enter nests within colonies.

Cliff swallows usually placed their nests against one or more adjacent nests, often sharing walls, especially in larger colonies (section 5.3.2). Although this may have reduced the time spent nest building (section 5.3.1), it appeared to be costly in that predators used some nests as a perch from which to attack adjacent ones. This was well illustrated with both bull snakes and common grackles (section 8.2). Snakes typically held on to one nest while entering another (fig. 8.10). These reptiles moved easily from nest to nest, to which they could cling firmly, but they had substantially more difficulty traversing sections of substrate within a colony where either there were no nests or nests were widely spaced. Usually whenever a snake reached a break in a continuous row of nests in a colony, it either fell off and left the colony or turned back to the nests it had just raided. Isolated cliff swallow nests, especially in culverts or on bridges, proved extremely difficult for bull snakes to reach, and only once did we see a snake get to an isolated nest. In this case the nest was built only 1 m below the top of a cliff in a very anomalous location (we could almost touch it by reaching over the top of the cliff), and a bull snake successfully raided it after the eggs hatched.

Grackles sat on the tops of nests and tried to reach nestlings in adjacent nests. Without a place to perch, the grackles could not have attacked these nests because common grackles are too large and clumsy to hover in front of an isolated cliff swallow nest. We conclude that a neighbor's nest can enhance both grackles' and bull snakes' access to a bird's own (e.g., fig. 8.10), representing a cost of nest clustering. Actual predator attacks were too infrequent and too uncontrolled to allow us to measure this cost among colonies in any quantitative way, but clearly the cost was greater in larger colonies, where nests tended to share walls (section 5.3.2) and overall nest density was higher (section 7.2).

8.9 ATTRACTION OF PREDATORS TO COLONIES

One potential disadvantage of living in a large colony is that the conspicuousness of the many residents attracts the attention of predators. As a result, predators may recruit to large colonies of their prey, where the per capita risk of predation for colony residents potentially may increase relative to that for birds living in small colonies or solitarily. Most workers have assumed that predators (or brood parasites) are more likely to be attracted to large breeding groups (e.g., Lack 1968; Wiley and Wiley 1980; Wittenberger 1981; Wittenberger and Hunt 1985; Phillips 1990; Sillen-Tullberg 1990; Szep and Barta 1992; cf. Turner and Pitcher 1986), but there are few field data from natural populations to support this assumption. The only support we are aware of comes from Munro and Bedard's (1977) finding that herring gulls attacked large common eider crèches more often than small ones, Rypstra's (1979) results showing increased avian predation in larger *Cyrtophora citricola* spider colonies, Pienkowski and Evans's (1982a,b) study in which herring gulls were attracted to areas of high shelduck nest density, the report by Ferguson (1987) of a slight tendency for larger groups of white-browed sparrow-weavers to be attacked more often by a variety of predators, and Uetz and Hieber's (1994) demonstration of more wasp attacks at larger colonies of *Metepeira incrassata* spiders. Wilkinson and English-Loeb (1982) found no evidence for predator recruitment to colonies of cliff swallows in California, although they had data for only two colonies of relatively (by Nebraska standards) similar size (147 and 320 nests). A marked rat snake twice attacked an Oklahoma cliff swallow colony, with its visits separated by twenty-nine days (Oliver 1970).

During eleven years, we witnessed a total of 157 successful and 113 unsuccessful avian predation attempts, and 32 successful and 37 unsuccessful predation attempts by bull snakes. These were observed while we were present at colonies either checking nests, mist netting, or observing birds, and in a few instances as we were passing by colonies. During these eleven years, there were 310 colonies of various sizes that we visited often enough to have a reasonable chance of seeing predation. We partitioned the observed predation attempts by colony size and calculated the number of predator attacks per colony over the course of the study (fig. 8.11). This measure does not represent the actual predator attack rate, but it provides a relative index of predation among colonies of different sizes. Avian predators in particular were probably deterred in many cases by our presence at colonies, so these values (fig. 8.11) underestimated avian predation rates. Bull snakes, however, seemed unaffected by the presence of humans. We spent roughly equivalent amounts of time at colonies of all sizes over

SSs

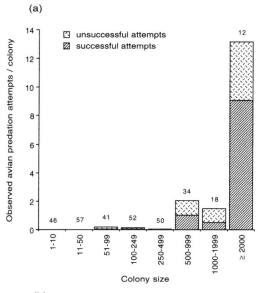

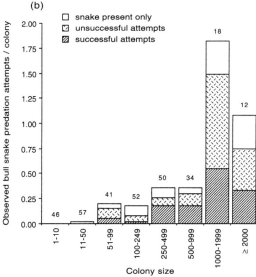

Fig. 8.11 Number of observed avian (*a*) and bull snake (*b*) predation attempts per cliff swallow colony (1982–92) in relation to colony size. Numbers above bars are total number of colonies observed. Total avian predation attempts increased significantly with colony size ($r_s = .87$, $p = .004$, $N = 8$ colony size classes), as did successful avian predation attempts ($r_s = .83$, $p = .011$, $N = 8$). Total and successful bull snake predation attempts both increased significantly with colony size (for each: $r_s = .94$, $p < .001$, $N = 8$).

the course of the study (except for a few colonies where we conducted intensive observational or mist netting studies in some years). We thus had approximately equal opportunities to see predation attempts at each site.

Both the total avian predation attempts per colony and the number that were successful per colony increased significantly with cliff swallow colony size (fig. 8.11a). This same pattern held for bull snake predation (fig. 8.11b). For snakes we also scored how often they were "present" at colonies without actually trying to reach nests. Often a bull snake appeared at a colony, seemed to be aware of the nests as it crawled under or above them investigating, but did not try to scale the wall to reach them. These events were included in the totals. Avian predators too often passed by colonies without obvious interest in the cliff swallows for us to meaningfully score them as "present only." Avian predation attempts (fig. 8.11a) referred only to cases where a predator made an attempt to get a bird.

The results of this analysis were striking (fig. 8.11). For avian predation attempts, 94% occurred in colonies of 500 nests or larger, and 58.5% occurred in colonies of 2,000 or larger. We saw no avian predator attacks in any colony of 50 nests or smaller. For bull snake predation attempts, 81% occurred in colonies of 250 nests or larger, and 49% occurred in colonies of 1,000 or larger. Only one bull snake was ever seen at a colony of 50 nests or smaller, and it was "present only," not trying to reach the nests. These attack rates also translated into an increasing per capita risk with colony size. For example, the risk of avian predation per cliff swallow was .007 in a 2,000 nest colony, .004 in a 500 nest colony, .003 in a 51 nest colony, and .000 in a 10 nest colony. Consequently predator attacks were frequent enough to overcome the "dilution" effect of large group size (section 8.1; cf. Uetz and Hieber 1994).

Visually hunting avian predators such as common grackles and American kestrels seem to be the kinds of predators most likely to be attracted by the conspicuousness of a large cliff swallow colony. A Swainson's hawk, for example, was clearly attracted to one large cliff swallow colony, where it was present almost constantly for the entire nesting season, usually sitting on a telephone pole above the culvert in which the swallows nested. This hawk did not attack the cliff swallows but instead scavenged the many road-killed birds there. A Swainson's hawk—presumably the same individual—returned to this same colony in two successive years. There was evidence that less visually oriented predators such as bull snakes also were attracted to the larger colonies (fig. 8.11; see Eichholz and Koenig 1992).

It could be, however, that avian predators and snakes did not necessarily recruit to larger colonies, but rather that local predators simply attacked larger colonies more often because the greater nest densities there enhanced their access (section 8.8) and might thus have increased their success rate. Among actual bull snake predation attempts, 52.2% of those ($N = 23$) in colonies of fewer than 500 nests were successful, versus 43.5% of those ($N = 46$) in colonies of 500 nests or larger; the difference was not significant ($\chi^2 = 0.47$, df $= 1$, $p = .51$). This indicated that the success rate of bull snake predation attempts did not vary with colony size, leading back to the suggestion that increased attack rates in larger colonies (fig. 8.11b) were due principally to additional snakes' being attracted there.

These data (fig. 8.11) provide one of the few demonstrations to date that predators are attracted to larger breeding colonies. Cliff swallows living in larger colonies have to contend with more frequent predator attacks, and these attacks are frequent enough to increase the birds' per capita risk of predation despite the large number of potential prey individuals available. Increased attraction of predators thus represents another cost of cliff swallow coloniality.

8.10 DO CLIFF SWALLOWS AVOID PREDATORS BY NESTING COLONIALLY?

The evidence on whether coloniality helps cliff swallows avoid predators more effectively was mixed. These birds detected approaching predators at greater distances in larger colonies (fig. 8.2), and flocking and mud gathering in groups away from colony sites conferred some benefits by allowing individuals to reduce the time they spent alert although overall group vigilance remained high (section 8.4). These appeared to be the only antipredator advantages of coloniality, however. There was no evidence that approaching predators were more effectively deterred through mobbing (section 8.5) or that within-colony breeding synchrony swamped predators enough to reduce per capita predation risk during peak nesting (section 8.6). If cliff swallow colonies represent selfish herds, it is only because the relatively few individuals on the colony edges were likely to suffer most of the predation (section 8.7); most birds probably gained relatively little from a selfish herd because predator attacks were comparatively rare. On the other hand, there was evidence for two predator-related costs of coloniality (sections 8.8, 8.9).

If the net result of these predator-related costs, benefits, and presumed noneffects is to enable cliff swallows to avoid predators more effectively, we predicted that predation rate (as measured by nesting success) should decrease with increasing colony size. A variety of studies on colonial birds have shown that nesting success increases with colony size (section 11.1), and this is usually interpreted to be because predation is reduced in larger colonies (Brown, Stutchbury, and Walsh 1990). We again used the per-

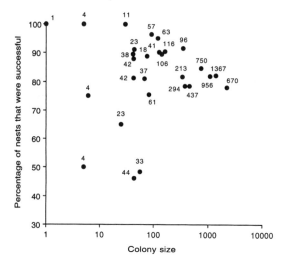

Fig. 8.12 Percentage of successful cliff swallow nests per colony in relation to colony size for fumigated sites only. Numbers by dots are sample sizes (number of nests) for each colony. Percentage of successful nests did not vary significantly with colony size ($r_s = -.16$, $p = .41$, $N = 29$ colonies).

centage of successful nests (those with nestlings alive to day 10) as an index of predation (section 8.6) and compared colonies of various sizes. Because ectoparasites also increased with colony size (section 4.3) and affected nesting success (section 4.7), for this analysis we used only fumigated colonies where the confounding effects of swallow bugs had been removed. Nesting success in fumigated colonies presumably more accurately reflected the effects of predation (section 8.6; Brown and Brown 1987).

The percentage of successful nests did not vary significantly with colony size for fumigated colonies ranging from 1 to 2,200 nests (fig. 8.12). If anything, success declined slightly with colony size, although the correlation coefficient was not significant. There appeared to be more variation among small colonies, with nesting success high in some and low in others (fig. 8.12). Assuming that these results reflected predation to some degree, there was no evidence for a net antipredator advantage in larger cliff swallow colonies. This conclusion was also supported by the data on predator attack frequencies in colonies of various sizes (fig. 8.11). The early warning in larger colonies (section 8.3) was not sufficient to reduce the per colony rate at which predators attacked these sites.

We conclude that coloniality does not afford cliff swallows major antipredator advantages. The early detection of predators in large colonies is most likely an incidental consequence of the presence of many individuals. Enhanced detection per se probably does not represent a strong enough selective factor to cause coloniality, because it does not seem to afford the birds in larger colonies any real benefit over those in smaller colonies. The lack of strong antipredator effects is perhaps not surprising, given the relative rarity of predator attacks against cliff swallow colonies in Nebraska. Although we observed more attempted predation than in most field studies of colonial birds, these data (fig. 8.11) represented cumulative totals over eleven years. We saw no predators attack most of the colonies each year. Cliff swallow nests, situated under overhangs and usually high above the ground or water, were relatively well protected from most kinds of predators. Bull snakes were the only ones that commonly could reach nests, and though they could have serious effects at certain colonies (section 8.2.2), their overall impact on cliff swallow nesting success across all colonies was probably minor. Wilkinson and English-Loeb (1982) also found no clear evidence that cliff swallows routinely avoided predators more effectively by living in larger colonies.

Perhaps only these birds' flocking away from colony sites represents a direct response to predators and leads to important antipredator benefits (section 8.4), but even these advantages may be only incidental consequences of the presence of many individuals. Enhanced group vigilance as a cause of flocking might be most likely if cliff swallows are often attacked by avian predators while migrating and in South America during the nonbreeding season. Unfortunately, we have no information on their predators outside Nebraska.

8.11 SUMMARY

Coloniality potentially enables breeding animals to avoid predators more effectively than if they nested solitarily. Cliff swallows at breeding colonies in our study area were attacked by sharp-shinned hawks, American kestrels, barn owls, great horned owls, black-billed magpies, loggerhead shrikes, common grackles, and bull snakes. Only snakes frequently gained access to nests in colonies. Cliff swallows usually reacted to these predators by alarm calling and milling near the predator in a loose group, perhaps to signal to it that it had been detected.

We presented a snake model to cliff swallow colonies and measured the distance from the colony at which the birds detected it. Detection distance increased with colony size. Birds flocking and gathering mud away from colony sites reduced the per capita time they spent scanning for predators as flock size increased. Overall group vigilance increased with flock size because many individuals were present. Cliff swallows thus received the classical benefits of flocking when away from their colonies. These birds did not obviously deter predators from attacking colonies by mobbing, but mobbing in a large group may have reduced each individual's own risk of falling victim to the predator during mobbing, a potential advantage of living in a large colony where mobs are large.

Cliff swallows exhibited a high degree of within-colony breeding synchrony. Small colonies were relatively more synchronized than large ones. The earliest-starting, most asynchronous nests tended to be the most successful, and the most synchronous nests did not have the highest success. There was little evidence that breeding synchrony enhanced nest survival by swamping and satiating predators. Cliff swallow nests at the extreme edges of colonies suffered the highest rates of failure, probably in part caused by predation. Predation rate did not seem to differ among other spatial positions within a colony.

Large cliff swallow colonies, where nests were packed more closely and often shared walls, enhanced predators' access to nests by providing a convenient perch from which adjacent nests could be attacked. Relative attack rates by both avian predators and bull snakes increased with colony size, with most of the attacks occurring in colonies of 500 nests or larger. This led to an increasing per capita risk of predation with colony size. Higher attack rates in larger colonies probably occurred because predators were attracted to the more conspicuous large colonies. There was no net antipredator advantage for cliff swallows living in large colonies. Nesting success in fumigated colonies, presumably reflecting the effects of predators, did not vary with colony size. We conclude that coloniality does not afford cliff swallows major antipredator advantages, and these birds probably do not breed in colonies to escape predators.

9 Social Foraging 1

Natural History, Food Distribution, and
Mechanisms of Information Transfer

> For as man or beast moves about in the luscious grasses, swarms of insects arise, and these the avaricious eye of the swallow eagerly notes. Bird-study—of anything but the cliff swallow—is difficult in such distracting circumstances.
> William L. Dawson (1923)

9.1 BACKGROUND

Living in a group can at times enhance an individual's efficiency at finding or harvesting food. This may result because a social animal can safely reduce the percentage of time it spends scanning for predators (section 8.1), automatically yielding more time for foraging. Studies on various nonbreeding flocking species have suggested this sort of enhanced foraging efficiency (reviewed in Pulliam and Millikan 1982; Elgar 1989), but it is unlikely to be a generally relevant advantage of colonial nesting unless individuals forage at or very near their colony as, for example, in prairie dogs (Hoogland 1981, 1995).

A perhaps more likely way for colonial animals to improve their foraging efficiency is through active or passive transfer of information on the whereabouts of food (e.g., von Frisch 1967; Wilson 1971; Ward and Zahavi 1973; Waltz 1982; Wittenberger and Hunt 1985). A colony may represent an "information center" to which individuals unsuccessful at finding food can return to learn where food may be located at that instant. Information may be actively transmitted to unsuccessful foragers by other colony residents through displays, as occurs in several social insects (e.g., von Frisch 1967; Wilson 1971; Seeley 1985) and a few colonial vertebrates (e.g., Greene 1987; Stoddard 1988, Brown, Brown, and Shaffer 1991), or an unsuccesssful forager may simply watch other colony residents and follow a successful one to food when it next leaves the colony (Ward and Zahavi 1973).

Whether bird colonies and winter roosts serve as information centers has been controversial (Wittenberger 1981; Bayer 1982; Waltz 1983; Mock, Lamey, and Thompson 1988; Danchin 1990; Richner and Marclay 1991; Clode 1994; Heeb and Richner 1994; Richner and Heeb 1995), despite continuing theoretical interest in how the process might work (Waltz 1982; Beauchamp and Lefebvre 1988; Allchin 1992; Barta and Szep 1992, 1995). Most of the empirical evidence for information centers in birds is indirect and open to alternative interpretations, in part because of the logistical challenges of studying information transfer. To make any progress, one must have distinctively marked individuals and be able to track the foraging movements of colony residents. Only recently have several studies provided relatively strong evidence for information centers in colonial birds (Brown 1986; Greene 1987; Rabenold 1987; Waltz 1987; Gori 1988) and the evening bat (Wilkinson 1992).

In those species that transfer foraging information at colony sites, an individual's foraging efficiency can presumably be improved because it wastes less time in unprofitable searching (Brown 1988a; Wilkinson 1992) or focuses its efforts on high-quality food patches (Visscher and Seeley 1982). We know little about how information transfer varies with colony size, but unsuccessful foragers in larger colonies, with more neighbors, presumably can more rapidly locate a successful individual to follow (Hoogland and Sherman 1976; Brown 1988a). As a result, birds nesting in large colonies may increase food intake rates for both themselves and their dependent offspring and thus improve their survivorship and reproductive success. The potential benefits of enhanced foraging efficiency may be great enough to cause or maintain coloniality (Alexander 1971, 1974; Hoogland and Sherman 1976; Krebs 1978; Rypstra 1979; Waltz 1982; Wittenberger and Hunt 1985; Brown 1988a; Binford and Rypstra 1992; Spiller 1992; Clode 1993; Caraco et al. 1995; cf. Richner and Heeb 1995).

Another potential benefit of coloniality is the increased opportunity to locate foraging sites via local enhancement. Local enhancement (Thorpe 1956; Hinde 1961; Poysa 1992) refers to animals' observing others actively feeding and thereby learning the present locations of food. The consequence is generally an aggregation of foragers in areas of high prey density, and the net result may be to increase each forager's feeding efficiency. Local enhancement often occurs among nonbreeding animals and does not require that participants nest together (Hoogland and Sherman 1976; Bayer 1982; Wittenberger and Hunt 1985; Mock, Lamey, and Thompson 1988). Individuals nesting separately could theoretically commute to the same general foraging area and regularly find food by observing other foragers, and thus local enhancement is probably not a primary selective factor leading to colonial nesting. Nevertheless, coloniality fosters opportunities for local enhancement by increasing the number of potential foragers in the general vicinity of a colony (Krebs 1974, 1978; Brown 1988b; Brown, Brown, and Shaffer 1991).

In the next two chapters we investigate social foraging and information transfer in cliff swallows and assess how colony size affects an individual's feeding efficiency. We begin in this chapter by describing how cliff swallows forage and characterizing their food resources. We examine in detail the spatiotemporal variability in food locations, because the degree of variability dictates the potential importance of information transfer. We illustrate how information transfer occurs at colony sites and on the foraging grounds and how social foraging influences the mean and variance in prey capture rates. In chapter 10 we examine how colony size affects both the benefits of information transfer and the potential foraging-related costs of coloniality and social foraging per se. In chapter 10 we also summarize our conclusions (Brown, Brown, and Ives 1992) about why Horn's (1968) geometrical model—which is often invoked to explain the evolution of avian coloniality—is unlikely to apply to cliff swallows. Together, chapters 9 and 10 present an integrated look at social foraging in cliff swallows, aspects of which were reported earlier (Brown 1986, 1988a,b; Brown, Brown, and Shaffer 1991; Brown, Brown, and Ives 1992). These chapters also contain substantial new data and analyses.

9.2 NATURAL HISTORY OF FORAGING BEHAVIOR

Cliff swallow foraging can be divided broadly into activities occurring when the birds do not have eggs or nestlings to care for, so that frequent departure and return to a fixed location (colony) is unnecessary, and those occurring while the birds are incubating eggs and feeding nestlings and thus often move to and from a colony site. These two periods represent fundamentally different situations for the birds, because information centers may be important in the latter period but not in the former.

9.2.1 Foraging Behavior Early and Late in the Season

Both before the birds laid eggs and after nestlings had fledged, cliff swallows tended to forage continuously for extended periods (up to several hours) each day during which they seldom returned to their colony sites. Their partitioning of foraging and nonforaging activities into distinct and extended periods meant that generally there was little traffic between the colony and foraging areas during these times of the year. Because there also was little if any way birds could presumably gauge a returning forager's success (insects were not being brought back; section 9.5.1), cliff swallow colonies in all likelihood did not serve as information centers early and late in the season.

When not caring for eggs or nestlings, cliff swallows often engaged in "network" foraging (sensu Wittenberger and Hunt 1985). The birds would spread out over a relatively wide area in loosely defined groups or as solitaries and presumably within sight of each other. When some group members discovered insects the other birds would almost instantly converge on that location. The swallows apparently used the characteristic prey capture behavior of the foragers that discovered an insect source as a visual cue. These convergences consisted of 30 to 1,000 or more cliff swallows that fed over areas often not greater than 10 m^2. Foragers were sometimes distributed in "towers" of birds from 0.5 m to more than 30 m above the ground. Cliff swallows converged on areas where insects concentrated, often along stream banks late in the afternoon, on the leeward side of road cuts and bluffs during windy weather, and among trees in canyons during cool and rainy weather. Convergence of birds at a given site lasted up to thirty minutes, after which the foragers dispersed or reformed at another site 100 to 500 m away. These were clear examples of local enhancement (Brown 1988b). Foraging birds often used squeak calls to signal the presence of insects during network foraging (section 9.6).

Cliff swallows' foraging behavior varied with weather conditions, presumably because they affected insect activity. During cold, rainy, and windy weather (usually early in the year), the birds foraged low and in poorly defined groups, often just above the grass tops or water surfaces. Early in the season on cool days, up to 5,000 cliff swallows regularly foraged low over a lake and adjacent bluffs near the Cedar Point Biological Station; these foragers probably consisted of both migrants and local residents. In contrast, during warm, sunny conditions the birds fed much higher (often to 25 m or more) and in clearly defined groups, were dispersed over wider areas, and generally foraged over open fields and pastures. Cliff swallows resorted to feeding over lakes and ponds or among trees and in canyons only during poor weather or early and late in the day when temperatures were routinely cool.

The most striking characteristic of cliff swallows' foraging throughout the breeding season was their propensity to forage in groups. Except during the relatively rare periods of bad weather, foraging was generally done in groups ranging from 2 to over 2,000 birds. Groups tended to be well enough defined that we had no difficulty operationally distinguishing one group from another for observation (Brown 1988b; Brown, Brown, and Ives 1992). Foraging groups tended to be largest in the afternoons, when temperatures were warmest and insects most abundant, but there were always at least a few birds foraging solitarily (for reasons that were not clear), even during the hottest weather when group foraging was most pronounced.

9.2.2 Foraging Behavior during Incubation and Nestling Feeding

Once cliff swallows started to incubate eggs, they began foraging in shorter, more frequent bursts, often commuting between the colony sites and surrounding fields where most foraging occurred. Arrivals and departures at a colony increased as nestlings hatched and the birds began to devote most of their time to provisioning the young. Consequently, during these periods there were many opportunities for information transfer at the colony (section 9.5).

Group foraging was especially pronounced once birds began incubating and feeding nestlings, reflecting both the generally warmer, sunnier weather at that time of the year and perhaps also a behavioral shift by the birds to increase their feeding rate when nestlings had to be fed. Cliff swallows commuted to foraging locations up to 1 km away, flying in straight, direct flight and rarely pausing to catch prey en route to either the foraging site or the colony. Foraging groups formed, dissipated, and reformed continually, being maintained by individuals arriving on the foraging grounds from the colony.

Foraging locations typically changed frequently (section 9.4.3), and birds rarely used the same general area for longer than an hour at a time. Occasionally, however, cliff swallows exploited large, long-lasting concentrations of insects near colonies. These usually occurred as a result of mowing or haying by farmers in nearby pastures, which flushed many insects, or when herds of horses and cattle walked through tall grass, stirring up insects and attracting large numbers of flies. The birds would associate continuously for four to five hours with the mowers or the grazing animals, hawking insects above them. The large numbers of swallows there at any one time continually tracked the mowers' or grazing animals' movements over the course of a day. Historically, cliff swallows probably foraged similarly around bison herds, before mowers or farm animals appeared in the study area. The closely related South African cliff swallow also feeds in association with grazing animals (Skead 1979).

While incubating and feeding nestlings, cliff swallows maintained their decided preference for foraging over land, generally avoiding lakes and ponds. Grassy pastures were the most popular foraging areas, but the birds also often foraged above trees in the floodplain of the North and South Platte Rivers.

9.3 CLIFF SWALLOW FOOD SOURCES

A key determinant of the potential importance of information sharing in a species is spatiotemporal variability in food location (Bayer 1982; Waltz 1982; Erwin 1983; Clark and Mangel 1984; Wittenberger and Hunt 1985; Mock, Lamey, and Thompson 1988; Allchin 1992; Barta 1992; Barta and Szep 1992, 1995; Richner and Heeb 1995). Information on food locations is of little value if they are stable and predictable over time and space. If food locations often change, however, information transfer may benefit the animals greatly. The first step in assessing spatiotemporal variability is identifying with some accuracy the food items a colonial animal exploits. Knowing the food base per se can alone yield some insight into whether, and perhaps how often, animals might be expected to share information on the whereabouts of food (Wilkinson 1992).

Cliff swallows feed exclusively on flying insects. Before our study, the only specific information on the birds' diet came from early reports by Beal (1907, 1918) based on stomach content analysis of individuals from unspecified parts of North America. To learn what cliff swallows in our study area ate, we systematically sampled the food delivered to nestlings and opportunistically collected miscellaneous food items that adults dropped while feeding their young.

Food samples from nestlings were collected using the ring-collaring technique of Orians and Horn (1969), in which a pipe cleaner was placed around a nestling's neck to prevent it from swallowing food boluses. Prey samples were thus collected intact. Ring collaring does not harm nestlings if the collars are adjusted correctly, and it does not normally affect their growth (Henry 1982). We left nestlings collared for about twenty minutes in most cases and never collected more than one sample a day from any one nestling. Like other swallows (Turner 1982), cliff swallows delivered tightly compressed boluses, always to only one nestling per visit. Boluses seldom came apart, allowing us to collect all food delivered at a visit. Parents did not feed any nestling that already had a bolus lodged in its throat. We collected boluses from the nestlings' throats with forceps, placed them in 70% alcohol, and identified all insects they contained to family. We ring collared nestlings in colonies ranging from 10 to 2,200 nests in 1983, 1984, 1987, and 1988.

Food items that adults dropped were primarily grasshoppers, on which cliff swallows at times fed heavily. When the parent birds returned to their nests, grasshoppers often fell or jumped out of the adults' mouths as they attempted to transfer them to the young. We collected those we saw fall into the water beneath nests at two colonies.

Cliff swallows fed on many kinds of insects (table 9.1). We identified 84 insect families among the 144 food boluses we collected. Most of these families were relatively rare, however, and only 17 families (primarily dipteran, hymenopteran, and homopteran) were represented by more than 20 individual prey items (table 9.1). The grasshoppers taken ($N = 117$) were of 12 species (A. Joern, pers. comm.), but predominantly *Melanoplus sanguinipes* (75.2%). The next most common was *Ageneotettix deorum* (6.0%), followed by *Hesperotettix viridis*, *Amphitornus coloradus*, *Aulocara elliotti*, *M. augustipennis*, *Trachyrhachys kiowa*, *Sphagemon cellare*, *M. foedus*, *M. confusus*, *Cordillacris occipitalis*, and *Phoetaliotes nebrascensis*. (The dropped grasshoppers were not part of complete food boluses and were not included in table 9.1.) The data in table 9.1 and the

Table 9.1 Insect Families Represented in Diet Samples of Cliff Swallows, in Order of Decreasing Abundance

Order	Family	Total Individuals	Individuals per Bolus	Swarming[a]
Homoptera	Cicadellidae	539	12.53	No
Diptera	Dolichopodidae	500	12.82	Yes
Diptera	Simuliidae	389	11.44	Yes
Hymenoptera	Formicidae	325	5.24	Yes
Diptera	Empididae	220	8.46	Yes
Diptera	Chironomidae	176	7.33	Yes
Diptera	Muscidae	135	3.46	Yes
Diptera	Culicidae	92	23.00	Yes
Hymenoptera	Argidae	89	3.87	No
Homoptera	Delphacidae	68	4.00	No
Hymenoptera	Halictidae	52	13.00	No
Diptera	Phoridae	38	6.33	Yes?
Diptera	Tipulidae	34	3.78	Yes
Diptera	Syrphidae	25	2.27	Yes
Coleoptera	Scarabaeidae	25	1.47	Yes
Neuroptera	Chrysopidae	24	3.00	No
Hemiptera	Lygaeidae	22	2.20	No

Note: Others (<20 total individuals): Coleoptera: Cerambycidae, Tenebrionidae, Coccinellidae, Chrysomelidae, Colydiidae, Curculionidae, Dytiscidae, Haliplidae, Histeridae, Hydrophilidae, Meloidae, Phalacridae, Pselaphidae, Scaphidiidae, Staphylinidae, Carabidae, Conthasidae; Diptera: Stratiomyidae, Sarcophagidae, Drosophilidae, Asilidae, Anthomiidae, Calliphoridae, Chamaemyiidae, Lauxaniidae, Mycetophilidae, Pipunculidae, Psychodidae, Rhagionidae, Tabanidae, Tachinidae, Tephritidae, Trupaneidae, Bibionidae, Thereridae; Ephemeroptera: Ephemeridae, Baetidae; Hemiptera: Alydidae, Miridae, Notonectidae, Berytidae, Enicocephalidae, Nabidae, Pentatomidae, Scutelleridae, Tingidae; Homoptera: Fulgoridae, Cercopidae, Aphididae, Copromorphidae, Dictyopharidae, Psyllidae; Hymenoptera: Braconidae, Ichneumonidae, Tenthredinidae, Siricidae, Apidae, Pteromalidae, Vespidae; Lepidoptera: Pyralidae; Neuroptera: Hemerobiidae, Myrmeleontidae; Odonata: Coenagrionidae, Libellulidae; Orthoptera: Gryllidae, Tetrigidae, Acrididae.
[a] From Edelmann 1990.

grasshoppers were lumped across years because we did not have enough samples each year to rigorously investigate yearly differences in prey selection, although such differences probably existed. For example, Aphididae were found in 1983 only, Pyralidae and Halictidae in 1984 only, and Delphacidae in 1987 only.

Patchiness in the cliff swallows' food supply conceivably could be generated through swarming by the prey species. Swarming, for mating, feeding, or unknown reasons (Chapman 1954; Downes 1969; Sullivan 1981; Svensson and Petersson 1994), creates localized and temporary areas of high insect density. If a prey species is predisposed to swarm, patchiness of the birds' food sources may be maintained even in the absence of wind, thermals, or other abiotic factors that at times concentrate the insects involuntarily (section 9.4.4).

We assessed general swarming tendencies of the cliff swallows' insect prey by examining the entomological literature for references to swarming for each of the families represented in the diet (Edelmann 1990). We consulted both general references (such as Comstock 1940; Imms 1951; Borror, Delong, and Triplehorn 1981) and more specific studies of swarming on particular insect groups (such as Van Dyke 1919; Gibson 1945; Nielsen and Greve 1950; Southwood 1957; Peterson 1959; Downes 1969, 1970; Oliver 1971; Alcock 1973; Brodskiy 1973; Moorhouse and Colbo 1973; Waloff 1973; Eberhard 1978; Allan and Flecker 1989; Svensson and Petersson 1994). We found specific references to swarming for 30 of the 84 families represented in the cliff swallow's diet. Of the 15 families most frequently taken (table 9.1), 10 (66.7%) were swarmers, versus only 20 of the remaining 69 families (29.0%), suggesting that cliff swallows primarily exploited swarming insects among those taxa taken. Unfortunately, we could find no information on swarming among the grasshopper species taken by cliff swallows, although the most common one (*M. sanguinipes*) is a widespread species that in some years is abundant in southwestern Nebraska (A. Joern, pers. comm.).

The birds' reliance on swarming insects was also illustrated by the number of individuals of each family represented per food bolus collected (table 9.1). The most common families were typically represented by multiple individuals in each sample, meaning that the birds had encountered an abundance of these prey (probably a swarm) on the foraging trip when the bolus was collected.

In cliff swallows most food boluses contained multiple insects (except when the birds took grasshoppers or lepidopterans), in marked contrast to solitarily foraging barn swallows. In 1987 we simultaneously ring collared nestling barn and cliff swallows at a colony occupied by both species (the one described in section 7.5) to compare diets. Parents of these nestlings fed in the same general habitat (nearby pastures), although barn swallows typically foraged low and solitarily and cliff swallows usually in groups at high altitudes. Barn swallows delivered significantly fewer individual insects per bolus (3.50 ± 0.86, $N = 26$) than did cliff swallows at the same colony (18.70 ± 2.99, $N = 53$; Wilcoxon rank sum test, $p < .001$). Of 26 barn swallow boluses, 10 (38.5%) consisted of a single insect, versus 6 of 53 cliff swallow boluses (11.3%); this difference was significant ($\chi^2 = 7.95$, df $= 1$, $p = .005$). (Although barn swallows delivered fewer prey items, the insects they caught tended to be larger than those collected by cliff swallows.) These differences in diet likely reflected the species' different foraging tactics, further suggesting that cliff swallows went after relatively small, swarming insects and consequently often fed on patchy food sources.

We were surprised at some of the rarer insect families represented in the cliff swallows' diet. Some of these were not reported to swarm and not known to occur voluntarily at the relatively high altitudes where cliff swallows foraged. Most of the coleopterans, hemipterans, and homopterans are not known to swarm or regularly fly at high altitudes, including the family most commonly taken, the Cicadellidae (table 9.1). How a cliff swallow encountered, for example, a haliplid water beetle at the altitudes at which the birds usually foraged was perplexing. The most likely explanation is that thermals of warm air concentrated some of the nonswarming insects into localized areas and transported them aloft (section 9.4.4) where the cliff swallows encountered them.

9.4 SPATIOTEMPORAL VARIABILITY IN FORAGING LOCATIONS

We showed in the previous section that cliff swallows apparently preferred to forage on swarming insects, meaning that the birds' food probably often varied in location depending on movements and activity patterns of the swarms. In this section we address how these patchy insect swarms affected spatiotemporal variability in cliff swallow foraging locations. Only if food is spatiotemporally variable can individuals gain important benefits from information transfer (section 9.3).

Spatiotemporal variability in prey is best addressed through direct, systematic sampling of food resources. Although direct sampling has been attempted for insectivorous birds and bats (Orians 1980; Wilkinson 1992), and swallows in particular (Bryant 1975; Moller 1987a; Earle and Underhill 1991), for two reasons we doubt that direct sampling would yield meaningful data for cliff swallows. Cliff swallows feed on such a wide diversity of insects that probably no single method can accurately sample them all (see Cooper and Whitmore 1990). Furthermore, although flying insects can be sampled from airplanes (Glick 1939), there are no feasible ground-based methods of sampling insects at the altitudes at which cliff swallows regularly feed or on a large enough spatial scale to examine spatiotemporal variability. Hunt and Schneider (1987) point out related problems with sampling food resources of colonial seabirds.

However, alternative ways of assessing spatiotemporal variability exist. Wilkinson (1992) pointed out that most studies on information transfer have relied on the behavior of foragers to infer predictability and location of prey patches. This approach may be valid if the foragers consistently distribute themselves in an ideal free way (sensu Fretwell and Lucas 1970; Milinski 1988) such that the number of foragers at each site maps directly to the number of potential prey. But problems develop if, for example, foragers aggregate for other reasons (such as to avoid their own predators; section 8.1), in which case the number or presence of foragers at a site may not necessarily reflect the richness or even the presence of a prey patch. Therefore, to use the foragers' behavior to infer spatiotemporal variability in prey abundance, one must first examine at least on a limited scale how the presence of foragers correlates with the presence and abundance of food at a given site (see Leighton and Leighton 1982).

9.4.1 Spatiotemporal Associations between Foraging Birds and Flying Insects

To examine the spatiotemporal associations between the number of cliff swallows foraging at a site and the insects present there, we selected as a study site a section of the shoreline of an irrigation canal and adjacent lake over which cliff swallows fed. We often saw insects, especially midges (Chironomidae), emerging from the grass and rising upward toward the foraging swallows. Ground-based insect traps at this site probably accurately reflected the insects available to the birds, for the following reasons. First, during the sampling periods some birds fed quite low, at the level of the ground-based traps. Second, the chironomids were plainly moving from the grass to higher altitudes. Third, we sampled on a limited basis the insects at 10 m above the ground at this site, using sticky traps suspended from helium balloons. Insects caught 10 m above the ground were taxonomically identical to those captured at ground level.

We selected a 50 m sampling area along this shoreline (we moved the sampling area along the shoreline to a slightly different location each day) and placed five 135×110 mm sticky traps 1 m above the ground, at 12.5 m intervals. The sticky traps were overhead transparencies cut and coated with Tree Tanglefoot and mounted on wire sticks. Every ten minutes we examined each trap, counting all insects that had been caught since the last check, and we also counted all foraging swallows present up to about 10 m above, or 5 m on either side of, the length of the sampling area. The birds were usually in a single group over the traps, making it easy to determine if foragers were in the designated airspace.

Sampling results from six representative days are illustrated in figure 9.1. The number of cliff swallows foraging over the traps on each ten-minute count tended to match directly the total number of insects caught at that time (fig. 9.1). Whether a foraging group was present there reliably indicated the presence or absence of emerging insects, and the size of a foraging group varied directly with the number of insects present. Most important, there was no evidence that large numbers of insects were present in the *absence* of foraging cliff swallows, or vice versa. These results (fig. 9.1) indicate that the birds tracked the presence and abundance of insects at this site. Therefore, assuming that these results also applied to cliff swallows foraging in other contexts and habitats, we can justify using the locations and number of foragers to infer spatiotemporal variability in prey locations.

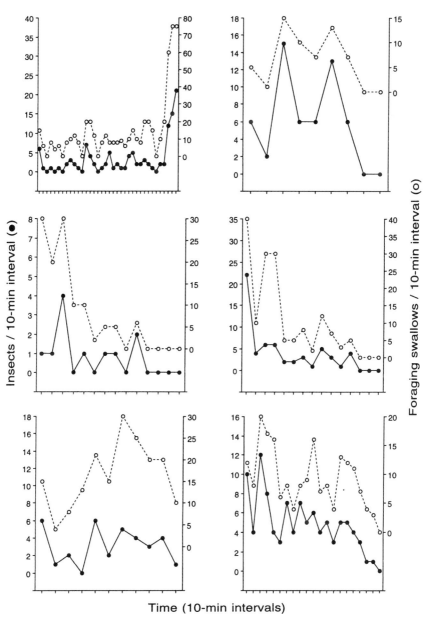

Fig. 9.1 Total number of emerging insects caught on all sticky traps in the sampling area and total number of foraging cliff swallows counted foraging above the sampling area at ten-minute intervals for six separate days in May and early June.

9.4.2 Methods for Determining Foraging Locations

To assess spatiotemporal variability in bird foraging locations, we had to measure where the cliff swallows fed. Generally we did this by observing the birds and noting their approximate positions relative to known landmarks at variable distances from a colony (Brown, Brown, and Ives 1992). These data were collected only at colonies in open areas with clear views of the surrounding terrain on all sides. Because cliff swallows feed in well-defined groups, usually high above the ground (section 9.2), we could see them with binoculars at virtually any location (Brown 1988b).

We surveyed foraging birds' positions at ten-minute intervals for continuous periods of one and a half to four hours a day, in both mornings and afternoons. The observer was positioned at the colony site, usually sitting above the nests either on a road (in the case of culvert colonies) or at the top of a cliff. The surrounding fields were scanned 360° around the colony. When a foraging group or solitarily foraging bird was spotted, the compass direction, approximate foraging group size, and the birds' distance from a nearby landmark of known distance from the colony were recorded. Landmarks consisted of a wide variety of stationary objects for which linear distances from the colony could be measured (e.g., stop signs, telephone poles, road intersections, irregularities of a creek bank, sandbars). All observations at all sites were made by one person only (MBB). To check the accuracy of the estimated group positions, at one site another observer (CRB) independently estimated the positions of all groups. Estimates of the two observers did not differ by more than 20 m; thus we considered group positions accurate to 20 m. If a foraging group was large and relatively spread out, we used the compass heading and distance of the group's center as its position. Surveys of all foragers around the colony could be accomplished within two to three minutes, so each survey was almost an instant record of where all foraging was occurring at that time. Total days on which observations were made at each colony ranged from seven to eleven (Brown, Brown, and Ives 1992).

Cliff swallows were scored as foragers only if they were actively feeding at the time of the scan. Foraging birds can be easily identified by their characteristic twisting and turning movements as they pursue prey (Brown 1988b). Birds that were commuting between the colony and the foraging sites were not included, because cliff swallows did not forage while commuting (section 9.2.2).

We made all observations during relatively warm, sunny weather when cliff swallows fed at altitudes of at least 5 m, where we could see them easily. We did not conduct these observations during occasional cloudy, cool, or rainy weather when cliff swallows fed low over the grass and were difficult to see. (This type of weather occurred on fewer than 10% of days during the period when we observed colonies.) Observations were made at each site on days relatively late in the nesting season (mid-June to mid-July) when over half of all birds in the colony were feeding nestlings and foraging activity was at a peak.

Using binoculars, we could detect birds up to a radius of at least 1.5 km from each site, but cliff swallows rarely were seen foraging more than 1 km from their colony. Because of the high visibility and openness of the terrain, we are confident we did not routinely overlook birds farther from the colony. The total number of foraging cliff swallows on any given scan was always consistent with the number of birds present in the colony as evinced by colony size (number of nests). Thus our surveys of foragers did not include large numbers of birds that did not live at the colony from which they were seen (see Brown, Brown, and Ives 1992).

We constructed maps of each colony's foraging arena by plotting the average number of birds recorded at each location around the colony. To do this, we divided the surrounding area into a grid of points placed at 20 m intervals, because 20 m was the apparent resolution of the observations on bird foraging locations. For notation in later mathematical equations, each point on the grid was denoted (x_i, y_i) where x increases from west to east and y increases from south to north. The colony was assigned the coordinate $(0, 0)$. For each day of sampling at a given colony, the scans of the foraging arena were pooled to give the number of birds seen at each point on the 20 m grid. The daily average number of birds seen per scan at the point (x_i, y_i) was denoted b_i. Note that b_i will equal zero if no bird is seen at (x_i, y_i), as is frequently the case (Brown, Brown, and Ives 1992).

9.4.3 Observed Spatiotemporal Variability

Foraging arenas were mapped for sixteen colonies ranging from 10 to 3,000 nests in 1985–91. To illustrate the characteristic spatiotemporal variability in cliff swallow foraging locations, we present data for five representative days from four colonies of 10, 75, 750, and 3,000 nests (fig. 9.2). Foraging locations are shown on a 2 × 2 km plot, with the colony location at the center. For this graphic presentation, we used a grid with points at 100 m intervals, with the b_is from the 20 m grid grouped accordingly and dot size indicating for each colony the relative number of birds at each foraging location (fig. 9.2). The total area of the foraging arena is also shown for later reference (section 10.8).

These data (fig. 9.2) indicate that cliff swallow foraging locations around a colony often changed from day to day. The numbers of birds using a particular foraging site varied considerably, with some locations not used at all on certain days. The same patterns were seen for other colonies not illustrated. Foraging locations also varied in similar ways when we plotted consecutive ten-minute periods within a single day. The

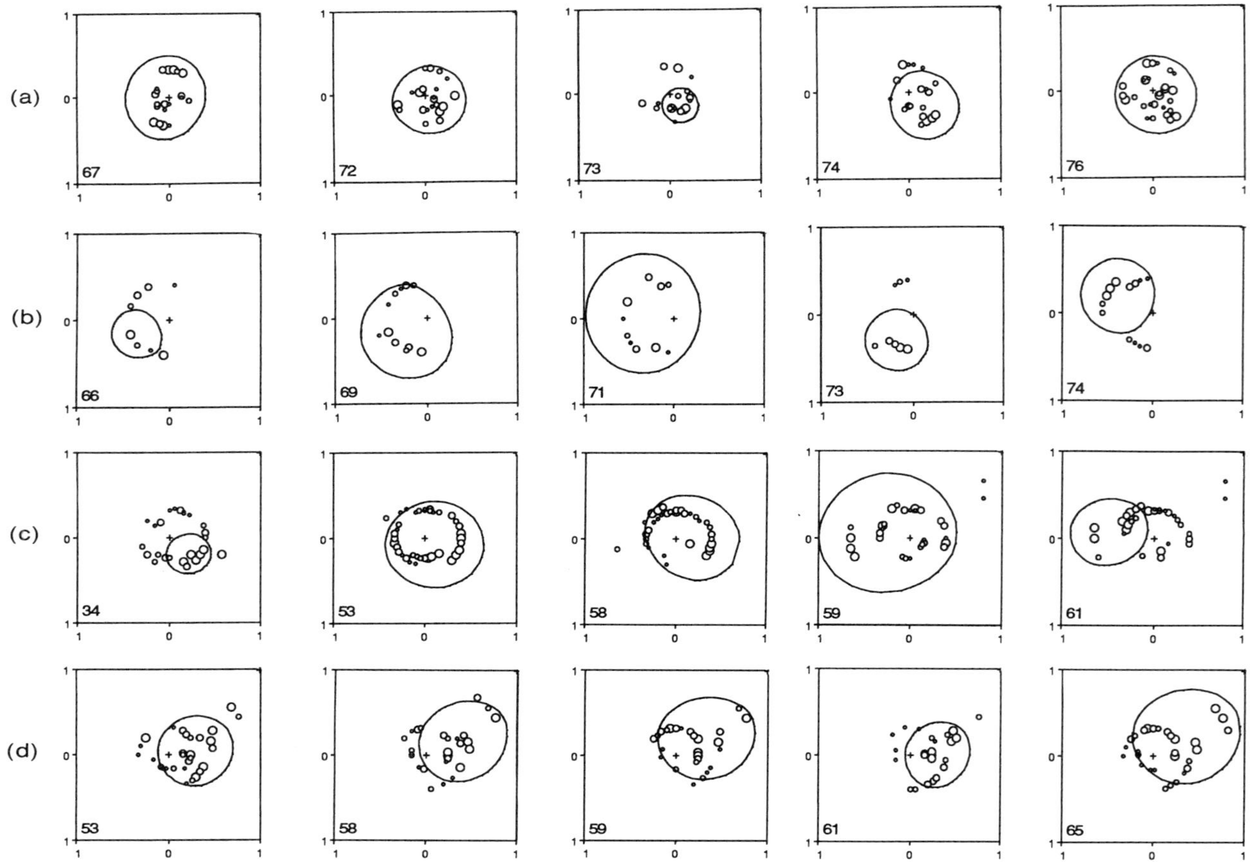

Fig. 9.2 Distribution of foraging cliff swallows around the colony site on five separate days (date shown in lower left corner of each; 01 = 1 May) for colonies of 10 nests (*a*), 75 nests (*b*), 750 nests (*c*), and 3,000 nests (*d*). Colony depicted in (*b*) was on a cliff with a lake adjacent. These maps were constructed as described in section 9.4.2; each plot is 2 × 2 km. Each colony has equal numbers of four sizes of dots. The largest dot size corresponds to the largest 25% of the b_i's for that colony, and the three smaller dot sizes are assigned to the smaller quartile ranges of b_i's. Circled region is colony's foraging area as defined in section 10.8.1. Colony's location is denoted by +.

maps (fig. 9.2) illustrate generally what seems to be high spatiotemporal variability in use of foraging locations and thus prey availability, although in the absence of similar studies for other colonial and solitary species we cannot be sure what really constitutes "high" variability. We are unaware of similar data for any other bird species at present, although several studies have documented relatively high spatiotemporal variability in honeybee foraging locations (Visscher and Seeley 1982; Schneider 1989; Waddington et al. 1994).

Maps of foraging locations (fig. 9.2) do not allow easy comparisons in spatiotemporal variability between colonies of different sizes. Such a comparison would be potentially insightful. For example, colony size might reflect local food availability and, in particular, food stability (Orians 1961; section 7.6.2). If so, large colonies might be more likely to form in areas with abundant, stable food sources, which in turn could diminish the importance of information transfer at those sites.

We examined spatiotemporal variability both within and between days for nineteen colonies ranging from 10 to 3,000 nests. This was done by first determining the mean compass bearing (e.g., 10°, 120°, 300°) from the colony for all foraging birds on each scan. In calculating the mean angle per scan, we weighted each bearing by its total number of observed foragers (group size). The mean angle per scan was computed using standard circular statistical techniques (Batschelet 1965). For each day, we then calculated the angular deviation (the circular equivalent of the standard deviation) among all scan mean angles. This yielded a single measure of how much foraging locations varied during that day. The numerical average of these daily angular deviations represented the mean within-day variability in bird foraging locations. To determine between-day variability, the mean angle among all scan means was computed for each day using circular techniques. The angular deviation of these daily means represented between-day variability in foraging locations for a given colony.

Neither measure of spatiotemporal variability varied significantly with colony size (fig. 9.3). Within-day variability in foraging locations seemed to decline with colony size, but the correlation was not significant (fig. 9.3a). Between-day variability showed no trend with respect to colony size (fig. 9.3b). In figure 9.3 we denoted separately two colonies situated on cliffs. These cliffs were adjacent to the relatively extensive Lake McConaughy (see fig. 3.1), whose surface created a large region of unsuitable foraging habitat. Because cliff swallows did not feed over water except during poor weather, the residents of these cliff colonies generally did not have as large a potential foraging arena within a given radius as did the birds living in colonies surrounded by dry land. Partly because of this, perhaps, the birds in the cliff colonies foraged in slightly more predictable locations (the few pastures nearby) than did birds at other sites of similar size.

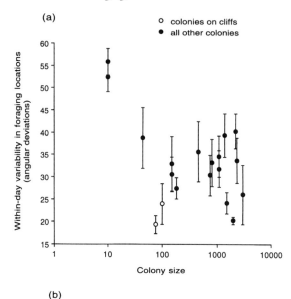
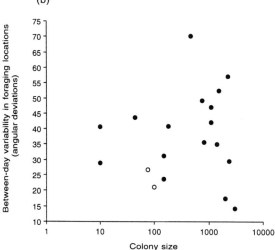
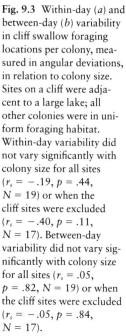

Fig. 9.3 Within-day (*a*) and between-day (*b*) variability in cliff swallow foraging locations per colony, measured in angular deviations, in relation to colony size. Sites on a cliff were adjacent to a large lake; all other colonies were in uniform foraging habitat. Within-day variability did not vary significantly with colony size for all sites ($r_s = -.19$, $p = .44$, $N = 19$) or when the cliff sites were excluded ($r_s = -.40$, $p = .11$, $N = 17$). Between-day variability did not vary significantly with colony size for all sites ($r_s = .05$, $p = .82$, $N = 19$) or when the cliff sites were excluded ($r_s = -.05$, $p = .84$, $N = 17$).

On balance, there was no evidence that cliff swallows in large colonies foraged in more spatiotemporally predictable locations from day to day than did birds in small colonies (fig. 9.3b), although within a day foraging locations in large colonies may have been slightly more stable (fig. 9.3a). This latter trend could have resulted from more efficient search in large colonies, where the greater numbers of foragers at any given time may have more completely tracked an insect swarm's movements once it was located.

9.4.4 Generating and Maintaining Spatiotemporal Variability

The observations on where cliff swallows fed (e.g., fig. 9.2) showed that the locations of their insect prey often varied in space and time. That large numbers of birds fed in certain places suggested that concentrations of insects formed, lasting for variable lengths of time. The insects' own tendencies to swarm (section 9.3) accounted for, at minimum, some spatial patchiness and may have also caused temporal patchiness, since swarming in some species varies with time of day (Sullivan 1981). In instances where cliff swallows fed in the vicinity of rivers and ponds, localized and ephemeral mass emergences of aquatic insects (Riggs 1947; Lewis and Taylor 1964) may have caused patchiness. Wind can concentrate insects on the leeward side of windbreaks (Lewis 1965), such as (in our study area) low hills, road cuts, buildings, and isolated stands of trees. Flights of aerial insects tend to cycle during the course of a day, and different taxa fly at different times (Lewis and Taylor 1964; Peng, Sutton, and Fletcher 1992), conceivably generating some patchiness.

Perhaps the most important mechanism generating and maintaining spatiotemporal patchiness in insect distribution was localized convection. The generally open nature of the habitats where most cliff swallow colonies were situated promotes frequent formation of thermals, and convection-supported thunderstorms are a characteristic feature of the study area during summer afternoons. Rising warm air can transport insects aloft and concentrate them in relatively well defined areas (Glick 1939; Freeman 1945; Hardy and Ottersten 1969; Schaefer 1976; Drake and Farrow 1988). Convection currents were probably responsible for the presence of many ground-dwelling insects in the cliff swallow diet samples, because otherwise the birds would not have encountered them (section 9.2). There have been relatively few studies of convection on small spatial scales, but there is evidence, for instance, that localized thermals may be only 180 to 550 m wide (Wallington 1961), persist for only twenty to thirty minutes, and move horizontally at speeds of 3 m a second (Hardy and Ottersten 1969), characteristics sufficient to generate spatiotemporal variability in insect locations. Most convection cells in clear weather are doughnut shaped, with insects concentrated in the outer walls of rising warm air and fewer in the cooler interior of the cells (Hardy and Ottersten 1969; Schaefer 1976). This probably causes spatial variation in insect abundance even within a convection current.

Several aspects of the birds' foraging behavior suggested that they in fact used thermals to find insect concentrations. Cliff swallows preferred to feed over open fields, where convection currents are most likely to form (Wallington 1961; Drake and Farrow 1988), and the birds mostly avoided water, where convection is least likely. The birds fed higher and in better-defined groups as the day warmed up; thermals also increase in frequency

260 • *Social Foraging 1*

and strength as the temperature rises (Drake and Farrow 1988). Finally, some direct observations suggested that cliff swallows cue on thermals. When we sampled insects to verify their movement from the ground (section 9.4.1), our tethered helium balloons often rose and fell with air currents. Each time the balloon rose, swallows converged near it; they dispersed as it fell. They appeared to be actively feeding whenever they converged on the balloon, as revealed by their characteristic behavior (section 9.4.2). There was no indication that they were converging specifically on the balloon through an alarm response (no alarm calls were ever heard) or for other reasons. The cliff swallows appeared to detect the localized thermal that lifted the balloon and moved to its vicinity to feed.

9.5 INFORMATION TRANSFER AT THE COLONY SITE

Information on whereabouts of food must be transferred among individuals at the fixed location of a colony if information sharing is to be an important advantage of colonial nesting (Hoogland and Sherman 1976; Wittenberger and Hunt 1985; Mock, Lamey, and Thompson 1988; Richner and Heeb 1995). In this section we address foraging information that is available to cliff swallows at the colony site, emphasizing how information transfer at the colony occurs.

Once incubation began, and especially after nestlings hatched, cliff swallows frequently came and went from a colony, commuting to spatiotemporally changing foraging sites in surrounding fields. There were three ways a cliff swallow might learn from other colony residents where food was at a given time: (1) successful foragers could actively signal the location on returning to a colony; (2) unsuccessful foragers could observe individuals returning with food in their mouths to feed nestlings and then follow those successful individuals to food on their next trip to the foraging site; and (3) unsuccessful foragers could simply join the streams of commuting birds traveling to foraging sites, perhaps cueing on their direct, "purposeful" flight (sensu Ward and Zahavi 1973).

There was no evidence that cliff swallows used food-finding calls at the colonies to signal the presence or location of food (Brown, Brown, and Shaffer 1991), although food signals were used on the foraging grounds (section 9.6). Thus we focus here on the birds' gaining information from other colony residents through watching and following presumably successful foragers at the colony site. This represents passive information exchange or information parasitism (Wilkinson 1992).

9.5.1 Foraging by Following Successful Foragers

When feeding nestlings, cliff swallows returned with food that was often obvious as it protruded from their bills or bulged in their throats (fig. 9.4).

Fig. 9.4 A cliff swallow upon its return to its nest to feed nestlings after having foraged successfully. The bird has insects in its throat and mouth; note the slightly open bill, indicating that it has a load of insects.

Other cliff swallows returned to the colony without food. These presumably unsuccessful individuals typically clung to the outside of their nests, apparently monitoring the activities of their neighbors. To test whether unsuccessful cliff swallows gained information from those returning with food, we examined whether they were likely to follow the successful birds on their next trip away from the colony (Brown 1986).

These observations were conducted at two colonies of 450 and 800 nests (8410 and 8449). We recorded whether the nest owners at a sample of about 100 nests in each colony arrived at their nests with food and fed nestlings or whether they arrived without food. We confined our observations to the period when cliff swallows were feeding relatively large nestlings that effectively blocked the entrance to a nest and forced the parents to feed them without entering, thus making it easy for us to observe them. During the time these observations were made, birds were foraging continuously, and there was no indication that nest owners scored as unsuccessful were instead engaged in any nonforaging activities.

After scoring whether a bird had been unsuccessful on its previous foraging trip, we observed it until it next left the colony. At that time we determined whether it followed another bird, was followed by another, or left singly. The criterion for whether a bird followed or was followed by another was a departure within five seconds of the other individual and in

Table 9.2 How Success on Previous Foraging Trip Influenced Whether Cliff Swallows Followed Others on Subsequent Trip and Whether Birds Were Followed by Others

	Previous Trip	
Subsequent Trip	Successful	Unsuccessful
Followed	524	1,355
Did not follow	2,610	454
	$\chi^2 = 1,647.6$	
	df = 1	
	$p < .001$	
Was followed	1,378	172
Was not followed	1,756	1,637
	$\chi^2 = 632.8$	
	df = 1	
	$p < .001$	

Source: Brown 1986.

the same direction, traveling together for at least 50 m from the colony (Brown 1986). Individuals that left the colony behind others usually stayed behind them until they reached the foraging locations; that is, leader and follower status on leaving the colony was maintained. Other methodological details can be found in Brown (1986).

We recorded data for 4,943 departures of cliff swallows whose recent foraging success we knew. Individuals that had been unsuccessful on a previous foraging trip were significantly more likely to follow other birds than were ones that had been successful (table 9.2). The relatively few returning without food that did not follow others may have been birds whose nestlings at that time were satiated; hence perhaps they had not tried to collect food and therefore had not actually been unsuccessful. Individuals that had been successful on their previous foraging trip were significantly more likely to be followed on their next trip than were birds that had been unsuccessful on their previous foraging trip. The number of successful birds not followed probably reflected an absence of unsuccessful birds at nearby nests at the time these successful individuals departed. An unsuccessful individual was unlikely to be followed at any time.

These observations suggested that cliff swallows used the same cue as did humans—a mouthful of insects—to determine foraging success and decide which birds would presumably be a good bet to follow. Carrying food is a reliable signal that a bird knows the location of a concentration of prey. Therefore cliff swallows might more easily observe and follow their closest neighbors rather than more distant ones. We tested this prediction using 46 focal nests in a 165 nest colony (8305). Each time a color-marked nest owner followed another bird from the colony, we recorded

Table 9.3 Number of Times Owners of Cliff Swallow Nests Followed Others during Feeding of Nestlings

Action	Observed	Expected[a]
Nest owner followed mate	88	29
Nest owner followed owner 1–5 nests away	676	191
Nest owner followed owner 6–10 nests away	294	162
Nest owner followed owner > 10 nests away	2,088	2,941
	$\chi^2 = 1{,}706.5$	
	df = 3	
	$p < .001$	

Source: Brown 1986.
Note: N = 46 nests.
[a] If followings are directed toward all birds with equal probability.

the identity of the follower and the bird being followed. We observed 3,146 followings and divided them into ones directed at a nest owner's mate and at nest owners living 1–5, 6–10, and more than 10 nests away. We calculated the number of times those owners should have followed mates and neighbors if all birds in the colony were equally likely to be followed (see Brown 1986 for details).

Individuals clearly preferred to follow neighbors one to ten nests away and especially those within a five-nest distance (table 9.3). However, followings were not directed exclusively at close neighbors; distant neighbors were also followed (section 9.5.2). Mates followed mates more often than expected if followings were random (table 9.3), in marked contrast to earlier in the year, when members of a pair virtually never left the nest together (section 6.2.4). We conclude from these data that cliff swallows probably more easily monitored their neighbors (and mates) owing to physical proximity, as we also found earlier in the year in other contexts (e.g., sections 5.5.1, 6.3.1). Consequently, birds knew their neighbors' foraging success better and often fed with them.

The cliff swallows that followed others or were followed by others fed together (Brown 1986). Because of the open terrain around our study colonies, we were able to visually track birds that left the colony together. We kept them in sight until they began foraging. Almost all of the birds leaving a colony together foraged together (see Brown 1986).

For an information center to evolve in the absence of other benefits of coloniality, the relative foraging success of different individuals must change regularly (Rubenstein et al. 1977; Clark and Mangel 1984; Wittenberger and Hunt 1985; Galef 1991; Wilkinson 1992; Richner and

Heeb 1995). That is, each individual must at times be both a "producer" and a "scrounger" (sensu Barnard and Sibly 1981; Barnard 1984), although recent models suggest that this may not always be necessary (Caraco and Giraldeau 1991; Vickery et al. 1991; Nowak and May 1992; Barta and Szep 1995). If individuals' success does not change regularly, the consistently successful birds gain nothing by living in a colony and unintentionally informing others on the whereabouts of food. We examined whether tendencies to follow other birds varied among the residents of a colony. Not surprisingly, we found that the propensity to follow others—that is, to be unsuccessful—was similar among the residents of a colony (Brown 1986). Virtually all birds were likely to follow others about 40% of the time. Thus no birds were consistently unsuccessful (or successful) and consistently took advantage of (or were exploited by) other colony residents. Further details are in Brown (1986).

9.5.2 Foraging by Joining Streams of Departing Birds

Another way cliff swallows appeared to gain information at the colony site on food locations was to join continuous streams of foragers commuting between the colony and the foraging grounds. This foraging behavior was not easily quantified but was obvious and qualitatively quite different from that described in section 9.5.1.

We discovered the importance of "stream foraging" when we attempted to conduct observations on following behavior of color-marked birds in a 2,200 nest colony (8905). It immediately became apparent that cliff swallows in extremely large colonies did not usually monitor and follow specific neighbors. This colony was so large that dozens of birds were leaving the colony to forage each second (section 10.1), creating a constant stream of birds between the colony and the foraging sites. Residents of this colony seldom returned to their nest without food (perhaps because of enhanced foraging efficiency in large colonies; e.g., section 10.3). Only 12% of arriving birds had no visible insects in their mouths ($N = 182$). Unsuccessful foragers typically scanned first (section 10.1.2) but then invariably joined the stream of outgoing foragers. Although it was difficult for us to resolve amid the bedlam whether these birds were in fact following neighbors, in most cases they did not seem to be following specific birds. The stream clearly indicated to us where food was at a given time. By tracking where the stream was going with binoculars, we could find the colony's current foraging location(s). At virtually any time during the day, a cliff swallow that was resident in a very large colony could presumably find the general location of food simply by joining the stream of foragers. The streams leaving a colony sometimes bi- or trifurcated a short distance from the colony, with some birds going to different locations.

Streaming to foraging sites was most pronounced at large colonies (section 10.1), but bursts of birds would often leave together at smaller ones. These departing groups were probably responsible for many of the cases of birds following distant neighbors to food (table 9.3). If a successful neighbor was not available to follow at a particular time, especially in the smaller colonies, birds often followed other residents elsewhere in the colony. There may have been less information available to an unsuccessful forager about these distant neighbors, but their flying out of the colony was perhaps a reliable indication that one of the birds in the group knew the whereabouts of food. Following these individuals may have been ultimately more rewarding than waiting for neighbors, depending on colony size. Unsuccessful foragers possibly cued on the "purposeful" flight of these groups or streams of departing birds, as suggested by Ward and Zahavi (1973). Purposeful individuals presumably are those flying out straight and direct. This might also represent the only way for an unsuccessful forager to assess which colony residents to follow during periods of incubation before food is brought back to nestlings and during the early periods of nestling feeding when cliff swallows completely enter their enclosed nests to feed small young.

9.5.3 Disguising Foraging Success?

The data and observations reported above show that unsuccessful cliff swallows have various opportunities to observe other colony residents and use them to find food. Did the birds do anything to prevent others from following them? Discoverers of a food source presumably gain little from sharing the resource with other colony residents and would suffer a major cost of being followed if recruitment of more cliff swallows depleted the food source. We saw no evidence, however, that cliff swallows returning to a colony with food tried to disguise their success. They seemed willing to perch on the outside of their nests, in full view of neighbors, once the nestlings began sitting in the nest entrance, although parents were capable of pushing the nestlings aside and entering the nest when they wished. Parents did typically feed their young immediately upon arrival, minimizing the time they were visible with food, but this may have been primarily to deposit the insects—some of which were still alive—into the nestlings' mouths before they escaped. Successful birds also did not apparently try to disguise the foraging location they had just visited by taking circuitous, indirect routes on their next trip. Successful foragers commuted directly back to foraging sites, even when other cliff swallows followed them.

Disguising foraging success perhaps was not necessary because sharing the food with other swallows may not have been costly (see Waltz 1983; Rypstra and Tirey 1991). The insect swarms were typically ephemeral (section 9.4.3), meaning each individual could not exploit them for long. Resource depletion was also unlikely if the insect swarms contained insects in superabundance, as often seemed true in our study area. The size of a foraging group had little impact on per capita food intake (section 10.9.1; Brown 1988b). But perhaps the main reason successful foragers did not disguise their success was that additional birds could track the insect swarm's movements more continuously, enabling a forager to locate the swarm more easily on subsequent returns. This also may be why cliff swallows called to attract other foragers once an insect swarm was discovered (section 9.6; Brown, Brown, and Shaffer 1991).

Taken together, the evidence was strong that cliff swallow colonies functioned as information centers in several ways. But how were foraging locations discovered initially if the birds often depended on others to find the food? Food sources at times petered out, as evinced by the return of unsuccessful foragers. Somebody had to search for new sources. Unfortunately the mechanisms of food discovery are still unclear, but they are the focus of our current studies.

9.6 FOOD CALLS

The previous section focused on presumably unintentional information transfer by the successful cliff swallows. They were simply observed by others, and their knowledge of food locations was parasitized by the unsuccessful individuals. There were contexts, however, in which cliff swallows actively signaled their discovery of food to conspecifics using special calls (Brown, Brown, and Shaffer 1991).

Food-finding calls were not used at the colony sites. The birds used them exclusively on the foraging grounds, where they appeared to aid foraging by local enhancement. Because the food calls were not colony based, we could not address how calling varied with colony size. We therefore present here only a summary of how cliff swallows actively communicated about food on the foraging grounds, and we refer interested readers to Brown, Brown, and Shaffer (1991), where our methods and results are presented in greater detail.

Cliff swallows have a rather limited vocal repertoire (Brown 1985b), but one of their vocalizations, termed the squeak call, was used exclusively to signal the presence of prey. The call was given by birds network foraging in loose flocks over fields and in canyons up to 4 km from their colony sites. On encountering an apparent insect swarm, birds would give several squeak calls in rapid succession. These calls obviously attracted other foragers, often leading them to converge into a dense and well-defined foraging group. These groups usually remained together at a relatively fixed site for several minutes, then either drifted away mostly intact or broke up entirely. A series of playback experiments using recorded squeak calls and the all-purpose "chur" call showed that the birds responded only to the squeak call. They converged on the source of the sound as soon as playbacks of the squeak call began, but the birds showed no recruitment to tape recordings of the chur call. We provided cliff swallows with food during poor foraging conditions (cool and cloudy weather) by flushing insects out of grass, and the natural per capita incidence of squeak calling increased up to fourfold, illustrating that the calls were in fact associated with food and were not simply generalized aggregation signals.

Squeak calls were not commonly used by foraging cliff swallows. They seemed to be restricted to bad weather when insects were relatively scarce and the birds fed close to the ground in ill-defined groups. Calling varied especially markedly with wind speed, declining significantly as wind speed increased, probably because high winds disrupted insect swarms so prey densities were not high enough to warrant calling. Squeak calls were primarily given when temperature was $\leq 17°C$, solar radiation was ≤ 500 w/m^2 (cloudy skies), and wind speed was ≤ 26 km/h. We did not hear any squeak calls given in hot, sunny weather. Consequently, squeak calling was most common in the early part of the nesting season when poor weather was more frequent. The latest in the season that squeak calls were ever heard was 2 July. The percentage of total days per season with weather conditions appropriate for squeak calling varied from 14.3% to 36.5% (although these data did not include the cold season of 1992, in which the percentage was likely much higher), with a mean of 24.5% of days each season (Brown, Brown, and Shaffer 1991).

Although we never heard a squeak call given naturally by a cliff swallow at a colony site, the birds did respond slightly to squeak call playbacks at colonies during the appropriate weather conditions. When we broadcast the recorded calls 25–100 m from the colonies, birds recruited to the tape recorder even though they typically did not feed that close to the colony sites. The birds probably did not use squeak calls at the colonies simply because they did not forage there. There is evidence, however, that cliff swallows in some areas may rarely use other kinds of calls on their return to a colony that alert the other colony residents that food has been found on the foraging grounds (Stoddard 1988; Brown, Brown, and Shaffer 1991). Such calls have not been heard in the Nebraska study area.

The cliff swallow's use of a call to signal discovery of a food source, thus recruiting conspecifics, is at first glance puzzling. Calling actively shares information with, and thus benefits, birds that are unlikely to be closely related to the caller. We (Brown, Brown, and Shaffer 1991) evaluated several possible explanations for active information sharing, and the most likely seems to be that calling enhances the caller's own foraging efficiency. Because the insect swarms the birds feed on vary in space and time (section 9.4.3), no individual can exploit a given swarm for very long, especially if it must travel back to the colony to feed its nestlings. Alerting other birds increases the number of foragers in the vicinity of the swarm, raising the odds that the insects' subsequent movements will be tracked by at least some members of the group so the caller can more easily find the swarm on returning from the colony. Even if the other birds do not also call, the original caller could benefit via local enhancement simply by watching nearby group members as some of them track the prey. If calling has little cost owing to the ephemeral nature and high density of the insect swarms, callers might benefit by ensuring effective tracking of a swarm even early in the season before nestlings have hatched and when travel back to the colony is unnecessary. The birds' use of squeak calls during bad weather when foraging was poor was not surprising, but we do not know why they apparently did not use the calls during better foraging conditions. We still do not fully understand the contexts that do and do not lead to calling (Brown, Brown, and Shaffer 1991).

9.7 SOCIAL FORAGING TO AVOID PREDATORS?

Information transfer both at the colony site (section 9.5) and among birds on the foraging grounds via food calls (section 9.6) and local enhancement (section 9.2) led to cliff swallows' foraging in groups. We interpreted group foraging as a consequence of information's being made available on the whereabouts of food. Perhaps, however, cliff swallows foraged in groups primarily to avoid predators, presumably the reason they preened and gathered mud in flocks (section 8.4). If so, individuals might only have informed others where to find a group to forage with to increase overall vigilance or reduce individual alertness (section 8.4). We examined whether social foraging was an antipredator response by measuring the mean and variance in prey capture success of solitary and group foragers and of birds foraging in different spatial positions within a flock (Brown 1988b).

9.7.1 Methods for Scoring Prey Captures and Foraging Group Size and Position

We observed cliff swallows foraging in fields up to 1 km from colony sites. As in the observations on spatiotemporal variability (section 9.4), we collected data only on relatively warm, sunny days when birds fed in well-defined groups high enough above the ground (5–15 m) that we would not lose sight of them among the irregularities of the terrain. A foraging swallow was selected and followed visually for as long as possible, usually between 45 seconds and 5 minutes. During this time we counted prey captures per unit time as a measure of foraging success. The birds' characteristic twists and turns when pursuing an insect made scoring prey capture attempts relatively easy (see Brown 1988b for further details).

We used prey capture *attempts* per unit time as a relative measure of foraging success, because we could not always be certain a swallow caught an insect it pursued, although in most cases it did (Brown 1988b). Total observation time for each bird was recorded with a stopwatch, and foraging success was expressed per foraging bout, defined as a period of up to four hours on a single day during which temperature and sky cover were virtually unchanged. Insect activity and absolute abundance varied in response to weather (Johnson 1969; Drake and Farrow 1988), and therefore all comparisons among foragers were done within each foraging bout when absolute prey capture rates were comparable (Brown 1988b).

For each forager observed, we scored the size of its foraging group. Because groups tended to be so distinct, a sufficient operational definition of a group was usually all the birds within our sight along a given directional radius (e.g., 90°) from our vantage point. Immediately after each observation of an individual forager, we determined the size of the group it fed in. If a forager under observation left its group or changed to a group of a different size, or if other birds joined or departed from the group and thereby changed that group's size by more than approximately 5%, the observation was terminated. We also observed cliff swallows feeding solitarily, and we recorded whether they subsequently joined a group while they were still foraging or remained solitary until their foraging ceased.

For cliff swallows foraging in groups of more than fifteen, we recorded whether a focal individual foraged on the edge or in the center of the flock. "Edge" birds maintained a position on the group's boundary in which no other individuals surrounded them on at least one side. "Center" birds were surrounded on all sides by others. Even though they were moving continuously, birds remained remarkably consistent in their positions as they foraged, and thus virtually all birds could be assigned reliably to either center or edge classes. An observation was terminated if a focal forager changed its relative position within the flock.

9.7.2 Foraging Success in relation to Group Size and Spatial Position

Prey capture attempts per minute were significantly greater for cliff swallows within a group than for birds foraging solitarily (table 9.4a; Wilcoxon matched-pairs signed-rank test, $p < .001$). This analysis combined all groups (two or more birds) regardless of size and used only foraging bouts with at least ten observations of both solitary and group foragers. These results (table 9.4a) are consistent with group foraging's being both antipredator behavior and a response to food information. Perhaps reduced individual vigilance requirements in groups allowed social foragers to devote more time to feeding and thus enhanced their prey capture rate.

The behavior of solitary foragers (table 9.4b) is instructive on this point, however. Although solitary foragers averaged fewer prey capture

Table 9.4 Prey Capture Attempt Rates for Cliff Swallows That Foraged Solitarily Compared with Group-Foraging Birds

Action	Prey Capture Attempts per Minute per Bird per Foraging Bout				Mean Within-Bout SD	Number of Bouts	Number of Birds Observed
	Mean[a]	SD[a]	SE[a]	Range[a]			
(a) Solitary forager	3.6	1.6	0.3	1.3–7.4	2.2	29	504
Group forager	5.9	7.9	1.4	2.2–10.1	2.0	29	1,615
(b) Solitary forager that joins group	3.4	1.3	0.4	1.8–7.7	3.1	8	70
Solitary forager that does not join group	7.1	2.0	0.7	4.2–10.7	7.8	8	57
Group forager active at same time	6.4	1.8	0.6	4.8–10.1	2.3	8	254

Source: Brown 1988b.
[a] Calculated on bout means.

attempts than social foragers in general (table 9.4a), some exceeded the expected success of group foragers active at the same time. Those that foraged successfully did not join a group, whereas those that did poorly (within the time of observation) joined a group (table 9.4b). These data were collected during eight foraging bouts in which we had at least fifteen foragers for which we knew whether they remained solitary or joined groups. Foragers that remained solitary had significantly higher prey capture attempt rates than did solitary foragers that later joined groups, and they also had significantly higher prey capture attempt rates than did birds foraging in groups that were active simultaneously (table 9.4b; Wilcoxon matched-pairs signed-rank tests, $p = .012$ for each). Solitary foragers that joined groups had significantly lower average success while they were solitary than did birds foraging in groups that were active simultaneously (table 9.4b; Wilcoxon matched-pairs signed-rank test, $p = .012$). Solitary foraging appeared to be risky; success could either exceed, equal, or fall below that expected in a group.

The behavior of these solitary foragers is consistent with group foraging's resulting from information transfer (Clark and Mangel 1984, 1986) rather than predator avoidance. When solitary foragers were relatively unsuccessful, they apparently looked to improve their success by joining a group where their expectation of success was higher. If they were relatively successful while solitary, they had no need to join a group and did not. This result follows from the patchy food sources these birds exploit; groups may on average be more efficient in locating and subsequently

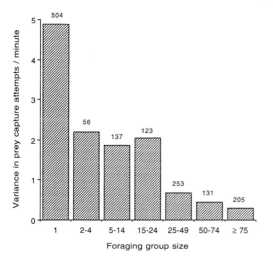

Fig. 9.5 Variance in prey capture attempts per minute per bird in relation to cliff swallow foraging group size. Numbers above bars are total number of individuals observed for each group size class. Variance decreased significantly with increasing group size ($r_s = -.96$, $p < .001$, $N = 7$ group size classes). From Brown (1988b).

tracking insect swarms, and thus individual success in groups is greater. That solitary foragers that were successful stayed solitary, however, argues against group foraging as primarily an antipredator response.

The observed pattern of variance in foraging success in relation to group size (fig. 9.5) illustrated in another way that solitary foraging could yield either a very high foraging success or a very low one, probably depending on whether the solitary forager happened to encounter an insect swarm. In this analysis (fig. 9.5), variances for each group size were calculated by averaging the respective variances of each group size class among twenty-four foraging bouts in which at least four group size classes were represented. These bouts included some that had no solitary foragers and were thus not used in table 9.4. The variance for solitary foragers in figure 9.5 is from the same birds as in table 9.4a.

Variance in success declined markedly with increasing group size (fig. 9.5), implying that individuals in the largest groups probably encountered food at the most consistent rate. Less variable prey encounter rates in larger groups, and in groups relative to solitaries, are predicted by various models of information transfer in patchy environments (e.g., Caraco 1981; Clark and Mangel 1984, 1986; Caraco et al. 1995) but do not necessarily follow from models in which animals group solely to avoid predation (e.g., Caraco and Pulliam 1984). We address further the effect of group size on foraging success in section 10.9, where we consider possible costs of social foraging.

Finally, data on foraging success of cliff swallows in different spatial positions within flocks suggested that these birds did not use social foraging primarily to avoid predators. If predators routinely attack groups, individuals closer to the edge of a group (and thus closer to the potential approach of a predator) should be more vigilant than those closer to the center (section 8.4; Milinski 1977; Jennings and Evans 1980; Inglis and Lazarus 1981; Lipetz and Bekoff 1982; Underwood 1982; Alados 1985; Petit and Bildstein 1987; Colagross and Cockburn 1993; Hoogland 1995), leaving them less time for foraging and reducing foraging success. Individuals on the edge of the group might also continually try to move toward the center, to increase their foraging success by decreasing the amount of time spent watching for predators. Cliff swallows on the edges of loafing flocks preened less than birds closer to the center, presumably because they spent more time scanning for predators (section 8.4).

For center birds in foraging flocks, the mean number of prey capture attempts per minute per bird per foraging bout was 4.5 (SD = 2.3, SE = 0.4, N = 33 bouts, 641 birds observed; calculations based on bout means). For edge birds, the mean was also 4.5 (SD = 1.8, SE = 0.3, N = 33 bouts, 808 birds observed). There was no significant difference in prey capture attempt rates between birds foraging at the center and on the edges of flocks (Wilcoxon matched-pairs signed-rank test, $p = .75$). This result suggests that cliff swallows feeding at the edges of groups were no more vigilant than ones nearer the center. Of 1,356 foraging birds observed to quantify prey capture attempt rates, only 156 (11.5%) changed their positions within the flock during the time they were observed. Individuals that moved from edge to center accounted for 4.6% of the total birds observed, and those that moved from center to edge constituted 6.9% of the total. Cliff swallows were thus likely to maintain their relative positions within a flock while foraging, contrary to the expectation if group foraging was primarily to avoid predators. Furthermore, we know of no predators in southwestern Nebraska that attack adult cliff swallows in foraging flocks (section 8.2). In sum, the data presented in this section favor social foraging as a direct response to food distribution and information transfer rather than primarily as a way to avoid predators.

This chapter has described the nature of the cliff swallow's food resources and how information transfer seems to occur. In the next chapter we ask how social foraging and its consequences vary in cliff swallow colonies of different sizes and to what extent the potential advantages of information transfer represent a benefit of coloniality.

9.8 SUMMARY

One potential advantage of living in a colony is the opportunity to obtain information from other residents on the whereabouts of food, in which case the colony serves as an information center. Cliff swallows usually foraged in groups of up to 2,000 birds and at relatively high altitudes, except during cool and cloudy weather, when they fed just above the ground or water surface and more often solitarily. Once eggs were laid and especially after nestlings hatched, the birds often commuted between the colony and nearby fields where they foraged, and then information transfer often occurred.

Ring-collar samples from nestlings revealed that cliff swallows fed on 84 insect families from 10 orders. Most of the commonly taken taxa were known to swarm or otherwise occur in aggregations, and thus the swallows' food sources were often patchy in distribution. The birds' foraging locations varied considerably across days and within a day, but neither between-day nor within-day variability varied significantly with colony size. A spatiotemporally variable food distribution was probably created and maintained by the insects' own tendency to swarm at different times of the day, by mass emergences of aquatic insects, by windbreaks, and by localized convection currents that concentrated insects and transported them aloft.

Cliff swallows that had been unsuccessful on their previous foraging trip returned to the colony, monitored other colony residents, and on their next trip followed birds that had just returned with food. Unsuccessful birds were unlikely to be followed. Birds leaving a colony together traveled to foraging sites and subsequently fed together. The birds apparently determined an individual's success by observing whether it carried insects back to its nest in its mouth and throat to feed its nestlings. Individuals seemed to monitor the success of neighbors and were more likely to follow their close neighbors to food than more distant neighbors. No birds were consistent followers or leaders, and thus all colony residents seemed to benefit from unintentional information sharing. Cliff swallows in very large colonies also gained information on food locations by joining streams of constantly departing foragers that led from the colony to foraging sites. Birds did not try to disguise their foraging success in any obvious way, perhaps because there was no cost to sharing a food source with conspecifics.

Cliff swallows at times gave a specific call, the squeak call, on discovering an insect swarm. The call recruited nearby birds so that a foraging group formed at the food source. Squeak calls were used only in cool and cloudy weather when foraging was presumably poor. Calling probably entailed little cost and may have benefited the caller by ensuring that the insect swarm's movements would continue to be tracked and thus remain obvious. Squeak calls were not used at colony sites.

Social foraging in cliff swallows does not seem to be used primarily to avoid predators. Although group foragers had higher average success, some solitary foragers exceeded the expected group success and remained solitary. Solitary foragers that did poorly joined groups. Perhaps as a result of information transfer in groups, variance in foraging success declined with group size and was highest for solitary foragers. Foraging success of birds feeding on the edges and in the center of groups did not differ, suggesting no increased vigilance for edge foragers. Cliff swallows generally maintained their spatial position within a flock while foraging, reflecting no attempts by edge birds to escape potential predators.

10 Social Foraging 2

Effects of Colony Size

> The forage range of the ... 800 ... adults belonging in this [cliff swallow] colony was for the most part within a radius of two miles. ... Some individuals foraged over the creek, others over dry grass pastures or fruit orchards, and often a considerable percentage of the whole population gave attention to a single one of these forage areas.
> Tracy I. Storer (1927)

The goal of this chapter is to examine how social foraging and information transfer in cliff swallows (chapter 9) affect an individual's foraging efficiency, and how that efficiency varies with colony size. Only if information transfer increases the food intake of birds living in colonies can social foraging be a strong enough advantage to cause or maintain coloniality. Little is known in general about how foraging-related costs and benefits vary with colony size or how they influence the evolution of colonial nesting in birds (section 9.1). In this chapter we use both direct observations of the birds' behavior and more indirect measures such as individual body mass to evaluate the net effect of social foraging on cliff swallows in different-sized colonies.

One difficulty in analyzing foraging success among birds at different colonies is that cliff swallows at different sites fed in different locations and at times may have experienced both qualitative and quantitative differences in local prey availability. This meant that between-colony measures of feeding efficiency, such as the number of food deliveries per unit time or the average distance traveled to find food, could have been affected in part by site-specific ecological characteristics. For example, birds occupying colonies on the edge of a lake had to go farther to feed because the water surface reduced the foraging habitat near the colony (section 9.4.3). Climatic differences between years (section 3.2.2), which could affect insect abundance (Bryant 1975), suggested caution in comparing foraging efficiency at colonies active in different years. We are cognizant of these pitfalls (which were anticipated by Hoogland and Sherman 1976), and therefore the years and sites used are specified where appropriate in the analyses in this chapter. Where possible, we present measures of foraging efficiency by year, to allow evaluation of seasonal effects. We also present

276 • *Social Foraging 2*

some within-colony analyses for single sites that varied in size, which reduced the problems associated with between-site comparisons.

10.1 DEPARTURE FREQUENCIES AND WAITING INTERVALS

A large colony might provide more foraging information to an individual than a small colony simply because the greater number of conspecifics ensures that birds will often be leaving the colony to forage. An unsuccessful forager might therefore more readily locate a departing successful forager in a large colony at any given time.

10.1.1 Departure Frequencies at Colonies

As a measure of the information potentially available to a forager, we recorded how frequently cliff swallows departed from colonies of different sizes. We timed with stopwatches the intervals between all departures and arrivals of cliff swallows on days when virtually all birds within a colony were feeding nestlings. We made observations in 1983–84 at colonies ranging up to 345 nests. These data yielded a continuous record of all departures at each colony, because the time of each was recorded to the nearest second. We then calculated the total number of seconds per hour when at least one bird left each colony. We used only hours for which we had a continuous time record for all 3,600 seconds and during which the birds were undisturbed by people or predators. At a 2,000 nest colony departures were so frequent that it was physically impossible to time the intervals between them, and there we simply recorded the cumulative seconds when at least one bird left.

As expected, the number of cliff swallows departing each hour to forage increased significantly with colony size (fig. 10.1). The pattern was a predictable outcome of increasing the number of potential departing birds in

Table 10.1 Period Cliff Swallows Waited after Arriving at Nest before Departing on Next Foraging Trip

Date	Colony Size	Waiting Interval (seconds)[a]		N
		Mean	SE	
4 July 1983	165	1.46	0.21	182
5 July 1983	160	1.33	0.21	139
6 July 1983	114	4.84	0.52	136
8 July 1983	75	4.44	0.53	196

Source: Brown 1988a.
Note: N = number of intervals.
[a] Intervals on 4–5 July (before colony size diminished) were significantly shorter than intervals on 6–8 July (after colony size diminished; Wilcoxon rank sum test, $p < .001$).

a fixed time period. Although the data came from two seasons, there was no reason to expect confounding seasonal or between-colony effects, and thus the results are probably robust. This analysis (fig. 10.1) supports the assumption that unsuccessful cliff swallows in large colonies had more opportunities to locate appropriate individuals to follow than did birds living in small colonies. The results also illustrate quantitatively, for the 2,000 nest colony, the stream foraging described in section 9.5.2, in which departing birds were constantly leaving the colony. More frequent departures represent a unique attribute of large colonies that is impossible to achieve in small colonies.

10.1.2 Individuals' Waiting Intervals

As another measure of how efficiently a forager might locate an appropriate bird to follow in colonies of different sizes, we timed the interval between a given bird's arrival at its nest and its subsequent departure on its next foraging trip. Our procedure was to randomly select a returning cliff swallow and time with a stopwatch the interval between its feeding its nestlings—or for a bird returning without food, its arriving at the nest—and its next departure for the foraging grounds. We observed only birds feeding relatively large nestlings (at least 10 days old), which partially or entirely blocked the entrance and thus kept the parents clinging to the nest's exterior in full view (section 9.5.1). Timing began when the parent arrived at the nest or actually fed the nestlings and ended when it flew away. We observed both successful and unsuccessful individuals.

We conducted the bulk of these observations at a single colony (8305) during one season as it diminished in size through fledging of nestlings. This approach minimized between-site effects on foraging behavior and eliminated seasonal (yearly) effects. It was also a useful analysis because birds nesting late in the season should have fewer opportunities to glean information from other colony residents, as earlier nesting individuals *within* the colony finish rearing their broods and depart (Emlen and Demong 1975). At this 165 nest colony, virtually no nestling cliff swallows had fledged before 5 July 1983. Between 5 and 8 July, a rash of fledgings occurred. As large numbers of juveniles and their parents left, the colony was reduced from 165 to 75 active nests over a four-day period.

We recorded waiting intervals for nest owners in a sample of forty-six focal nests. We thus collected data, in some cases, from the same individuals at the same site both before and after the colony decreased in size. Intervals between arrival at the nest and departure on the next foraging trip increased about three times when the colony size decreased (table 10.1). Assuming these within-colony results applied across colonies, they suggest that cliff swallows spent longer clinging to their nests after each foraging trip in smaller colonies.

That these waiting intervals in fact reflected the birds' attempts to locate appropriate foraging associates was suggested by a comparison of waiting intervals for individuals that were successful (returning with food) versus unsuccessful (returning without food) at a 2,200 nest colony (8905). At this site, where relatively few birds returned without food (section 9.5.2), the mean interval between an individual's arrival and its subsequent departure was 1.56 ($\pm$ 0.25) seconds for successful foragers ($N = 160$) versus 9.64 ($\pm$ 2.51) seconds for unsuccessful foragers ($N = 22$); the difference was highly significant (Wilcoxon rank sum test, $p < .001$). It seems likely that the increased waiting time for birds returning without food was related to their apparent lack of foraging success.

The difficulty with comparing different colonies between years was illustrated by data on waiting intervals collected at two colonies of 13 and 2,200 nests (8931 and 8905) in 1989. The mean waiting interval at the 13 nest colony was 0.50 ($\pm$ 0.084) seconds ($N = 194$) and at the 2,200 nest colony 2.84 ($\pm$ 0.51) seconds ($N = 182$). Surprisingly, the mean waiting interval at the 13 nest colony in 1989 was substantially *lower* than that at the 75 nest colony in 1983 (table 10.1), and the mean waiting interval at the 2,200 nest colony in 1989 was *higher* than that at the 165 nest colony in 1983. These results, which were contrary to expectation, are hard to interpret because of the possible seasonal and site-related differences among colonies. The cleanest and most interpretable analysis is the one restricted to a single colony site (8305) and year (table 10.1).

Waiting intervals for cliff swallows at their nests were much lower for the 13 nest colony than for the 2,200 nest colony in 1989, despite the birds' engaging in stream foraging at the latter site. This result was also surprising because it suggested that birds in very small colonies spent essentially no time looking for neighbors to follow. One possibility is that residents of very small colonies simply did not bother to look for neighbors because they had so few that their time might have been best spent searching for food. If so, some colony size threshold may exist below which too few residents are present for any information transfer at nests to occur.

Another possibility is that residents of very small colonies watched for successful foragers not *at* their nests, but by circling above the colony site. This was suggested by observations at another 13 nest colony (8329). At this culvert site, we often saw departing cliff swallows circle repeatedly above the colony before flying to a foraging site. Circling would last up to several minutes, usually until another individual or a small group of birds left the colony in a direct, straight flight toward a foraging site. The circling individual(s) would then immediately follow the straight-flying individual or group. Sometimes up to three cliff swallows, all leaving their nests at different times, would assemble in "circling patterns" above the colony and wait until another bird or group of birds departed from the colony in direct flight.

To quantify this behavior, we recorded whether individuals left their nests as a group (defined as two or more birds departing within five seconds of each other; Brown 1986) or whether they left individually, and then whether they made at least one 360° circle of the colony before flying to a foraging site. This colony was in open, treeless terrain, so we could observe with binoculars all foraging by colony residents in the surrounding fields and all circling above the colony. We watched for similar circling at larger colonies in equally open terrain the same year (1983). To avoid possible differences in interpretation of what constituted "circling," only one person (CRB) watched for this behavior in the different colonies.

Solitary cliff swallows that left their nests were more likely to circle the 13 nest colony than were departing groups of birds (total solitaries departing = 328, solitaries that circled = 234 (71.3%); total groups departing = 169, groups that circled = 23 (13.6%); $\chi^2 = 148.8$, df = 1, $p < .001$). Departing groups were likely to go straight toward a foraging site. These observations suggested that the more often an individual was able to depart as part of a group (that is, follow a simultaneously foraging individual), the less foraging time it wasted. In contrast, at larger colonies (≥ 100 nests) where departures were more frequent (fig. 10.1), we saw no cliff swallows circle in apparent attempts to locate successful foragers. Circling above the colony may be the most effective way to locate potential foraging associates when colony size is small and, though not observed at this colony, may also enable unsuccessful individuals to use

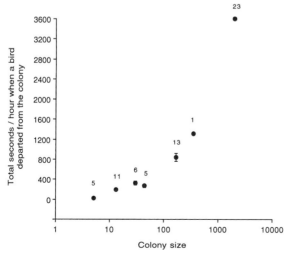

Fig. 10.1 Total seconds per hour when at least one cliff swallow departed from the colony in relation to colony size. Total hours each colony was observed are shown above dots. Seconds per hour with departing birds increased significantly with colony size ($r_s = .96$, $p = .001$, $N = 7$ colonies).

local enhancement to locate foraging sites that might be visible from the colony.

We conclude from the data presented in this section that the increased numbers of foragers active at any given time in larger colonies represented more potential sources of information for an unsuccessful colony resident.

10.2 FOOD DELIVERY RATES

With potential information on food sources more quickly obtained in large colonies (section 10.1), we predicted that the rate at which parent cliff swallows delivered food to their nestlings should increase with colony size. Birds in large colonies, which did not waste time circling above the colony, should have had more total time available for foraging. We examined this prediction by counting the number of food deliveries by parents in colonies of different sizes.

10.2.1 Counting Food Deliveries

At selected focal nests, we counted the total number of food deliveries made by both parents per unit time as a measure of their foraging success. By scoring only visits on which nestlings were fed, we minimized the chances of ever scoring a visit by a nonparent, because color-marked cliff swallows were never seen feeding nestlings at nests other than their own. We could watch up to forty-five focal nests at a time and count all food deliveries without missing any visits. For each nest observed, we recorded the number of nestlings present and their age, as determined from hatching dates known from nest checks (section 2.2).

These data were collected on the same days and during the same periods (usually 08:00 to 12:00 MDT) at the different colonies by separate teams of observers. This helped to minimize differences between sites owing to seasonal or circadian influences on either prey availability or the birds' foraging times (see Earle 1986). Watches were made for continuous one-hour periods. When the same nest was watched for two or more hours on a single day, we averaged the counts taken in the different periods and used a single mean value for that nest for the day. This meant each nest was represented equally in the analyses, even though some nests had been watched longer on certain days. In examining food deliveries in relation to nestling age, we averaged the daily means for all nests with nestlings of each age. For analyses of overall food delivery rates per colony, we averaged the mean food delivery rates for each nestling age between 10 and 17 days (section 10.2.2), which gave a single mean value per colony for the 10–17 day age period. All ages for which we had a mean food delivery rate thus were represented equally. Data were analyzed separately for each brood size.

10.2.2 Food Deliveries in relation to Nestling Age and Colony Size

Nestling age and brood size affect food delivery rates in most birds with altricial young (e.g., Moreau 1947; Royama 1966; Skutch 1976; Bryant 1975; Hails and Bryant 1979; O'Connor 1984; Jones 1987a). Very young nestlings require less food than older ones, and larger brood sizes place greater demands on the parents that must provision them.

Food deliveries in cliff swallows varied with nestling age (fig. 10.2). Parents delivered more and more food from the time of hatching until the nestlings were about 10 days old, at which point the number of food deliveries remained relatively stable until the nestlings were about 17 days old, whereupon they declined. This pattern of food delivery, not surprisingly, matches that of nestling body mass gain with age (Stoner 1945). The food delivery profiles shown (fig. 10.2) are typical of all those obtained, and the general pattern seemed similar among colonies of different sizes. We present data for broods of three and four because those were the most common brood sizes among the nests we observed and among nests in general (section 11.2). We observed too few broods of one, two, or five to analyze them with respect to nestling age or colony size, and therefore those brood sizes are not considered further in this section.

Because food deliveries seemed to stabilize between ages 10 and 17 days at each colony (fig. 10.2; Brown 1988a), we confined our comparisons of food delivery rates in different colonies to nests with nestlings of those ages. Average food deliveries for colonies of different sizes in each of three years are presented, by brood size, in table 10.2. The small number of colonies observed each year (a consequence of making observations at all sites concurrently; section 10.2.1) precluded rigorous within-year statistical analyses. However, the trend in 1983 and 1985 was clearly for food deliveries to increase with colony size for both brood sizes (table 10.2). No such trend was apparent in 1989, for which interpretation is difficult given data for only two colonies.

When we considered all data across years (and included data for two additional sites—8429, 8605), food delivery rates increased, though not significantly, with colony size for both brood sizes (fig. 10.3). The pattern seemed to be that food delivery rates peaked at intermediate-sized colonies. There are potential seasonal problems in comparing measures of foraging efficiency across years, but if these results (fig. 10.3) are real, they suggest costs of reduced information transfer in small colonies and competition for food or resource depression (see sections 10.8, 10.9) in large colonies. Based both on the within-year (table 10.2) and the between-year (fig. 10.3) analyses, the safest conclusion is probably that food delivery rates tended to increase with colony size but may have declined at the largest sites. Additional data for different-sized colonies within the same

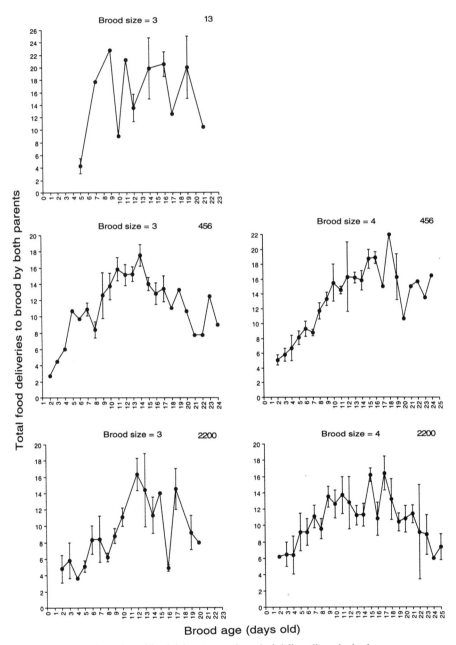

Fig. 10.2 Total number of food deliveries to a brood of cliff swallows by both parents per hour in relation to brood age (in days) for brood sizes of three and four nestlings. Mean shown is an average of each nest's mean food deliveries per hour on each day. Colony size is given in the upper right corner of each graph.

Table 10.2 Food Deliveries per Hour by Both Parents to Broods of Three and Four Nestlings

Colony Site	Colony Size	Brood Size = 3			Brood Size = 4		
		Mean	SE	N^a	Mean	SE	N^a
8305L[b]	2	3.4	0.9	2	—	—	—
8329	13	14.9	2.7	7	15.2	2.0	6
8305	165	17.8	1.8	7	18.4	2.8	7
8539	42	8.4	1.3	2	9.2	0.7	3
8540	85	9.0	0.7	8	10.5	1.0	8
8505	456	14.7	0.5	8	17.2	0.9	8
8931	13	16.1	2.1	6	13.4	2.2	3
8905	2,200	12.4	1.4	7	13.1	0.8	8

[a] Total means for different ages (from 10 to 17 days) on which grand mean was based.
[b] Two late-nesting pairs at site after all others had departed.

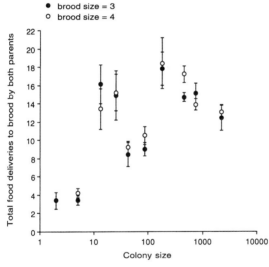

Fig. 10.3 Total number of food deliveries to a brood by both parents per hour in relation to cliff swallow colony size for broods of three and four nestlings. Total food deliveries per hour did not vary significantly with colony size (brood size three: $r_s = .44$, $p = .20$, $N = 10$ colonies; brood size four: $r_s = .40$, $p = .29$, $N = 9$ colonies).

year are needed, but these results on food delivery rates are broadly consistent with the hypothesis that information transfer and foraging efficiency generally increase with colony size.

10.3 AMOUNT OF FOOD DELIVERED

Food delivery rates per se cannot be conclusive as an index of foraging efficiency without our knowing how much food parents bring back (Royama 1966; Hoogland and Sherman 1976), because infrequently arriving parents may bring back more food than parents that arrive frequently. Net foraging efficiency is reflected both in how often birds return to feed their young and in how much food they collect per trip.

We measured the mass of food boluses delivered to nestlings as an index of how much food parents delivered. Boluses were collected by ring collaring nestlings (section 9.3). Wet mass of boluses was recorded to the nearest 0.01 g using an electronic toploading balance. Ring collaring was too time consuming to allow us to collect food boluses from a definitive range of colony sizes within a single year. We were thus forced to compare data collected from a total of eight colonies ranging from 10 to 2,200 nests during four seasons (1983, 1984, 1987, 1988). We ring collared only broods of three and four nestlings that were the same ages (10–17 days) as the broods for which food deliveries were scored (section 10.2). Because there was no significant difference in bolus mass delivered to broods of three versus four nestlings (all colonies combined, Wilcoxon rank sum test, $p = .48$), we disregarded brood size.

Bolus mass increased significantly with colony size (fig. 10.4). The pattern was dramatic, with parent cliff swallows in the largest colonies returning with 0.4–0.5 g more food per trip, on average, than parents in the smallest colonies. We would be more comfortable with these results if they all came from a single year. However, the overall trend was so strong (fig. 10.4) that we feel justified in concluding that the amount of food delivered per foraging trip increased with colony size.

Because both food delivery rate (section 10.2.2) and amount of food per delivery (fig. 10.4) varied with colony size, we examined the average net amount of food delivered to nestlings in different-sized colonies. We did this by fitting least-squares regression equations to the data on food

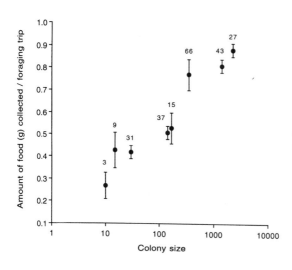

Fig. 10.4 Amount of food (bolus mass, g) collected by parent cliff swallows per foraging trip in relation to colony size. Numbers of boluses sampled are shown above dots. Amount of food brought back per trip increased significantly with colony size ($r_s = .98$, $p < .001$, $N = 8$ colonies).

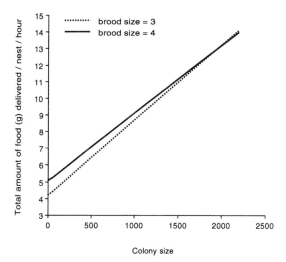

Fig. 10.5 Estimated total amount of food (g) delivered to a cliff swallow nest per hour in relation to colony size for broods of three and four nestlings. Estimates were based on regression equations derived from data in figures 10.3 and 10.4.

delivery rates (fig. 10.3) and amount of food (fig. 10.4), enabling us to calculate the average amount of food delivered to a nest per hour for the different brood sizes (fig. 10.5). We assumed a linear relation between food deliveries/amount of food and colony size, an assumption that perhaps was questionable for food deliveries but seemed reasonable for amount of food. Our calculations showed a striking net gain in food delivered per hour with colony size (fig. 10.5). For example, colonies of about 2,200 nests, the largest for which we had data on food deliveries, showed an increase of 8–10 g of food delivered per nest per hour relative to small colonies of about 50 nests. Results for the different brood sizes were similar. Cliff swallows living in the largest colonies brought back substantially more total food per hour than did birds at the smaller sites, despite apparently greater foraging-related costs in larger colonies (sections 10.8, 10.9). These analyses (figs. 10.4, 10.5) support the hypothesis that cliff swallows transferred information and foraged more efficiently as colony size increased.

Variation in food delivery with colony size could also reflect differences in the spatiotemporal distribution of food around colonies. Barta (1992), based on Horn (1968), argued that birds exploiting more unpredictably distributed food sources, in the absence of information transfer, should require longer search times and consequently would return to their nest with food less often (and see Schoener 1971). If small colonies routinely formed near more spatiotemporally variable food sources, the reduced foraging efficiency we observed in small colonies might have nothing to do with patterns of information transfer. However, there were no differences in spatiotemporal variability in foraging locations among colonies of different sizes (section 9.4.3), suggesting that resource distributions alone did not cause these results (figs. 10.3, 10.4).

10.4 NESTLING BODY MASS

Greater total amounts of food delivered by parent cliff swallows in larger colonies (fig. 10.5) presumably should translate into faster growth rates and greater body mass for nestlings in large colonies, all else being equal. In this section we examine differences in nestling body mass among colonies of different sizes. Other studies also have used nestling body mass as an index of parents' foraging efficiency (Hoogland and Sherman 1976; Snapp 1976) or local food availability (Nettleship 1972; Bryant 1978a; Moss et al. 1993). Body mass is a useful measure of net foraging efficiency for birds occupying colonies with potentially different resource bases, because it is expressed in the same currency (g/bird) for all sites.

Rather than calculate nestling growth curves for broods within each colony (since collecting data would have been time consuming, greatly restricting the number of colonies compared), we used a single measure of nestling body mass, recorded at a standardized time for each nest. Published nestling growth curves for cliff swallows (Stoner 1945) show a rapid rise in body mass to about 10 days, then it stabilizes and declines slightly before fledging. Therefore we selected 10 days as an age that might best reflect parents' foraging efficiency during the period of maximum nestling weight gain. Hatching times were known from nest checks (section 2.2), enabling us to return to each nest and record the nestlings' body mass at day 10 (section 2.3.1). Body masses of all nestlings within each nest were averaged to give a mean body mass per nest. Nestling body masses used in analyses of foraging efficiency were recorded at the same time as those used in assessing the effects of ectoparasites and in many cases were the same ones (section 4.7).

10.4.1 Nestling Body Mass in relation to Brood Size and Year

Before analyzing the effect of colony size, we evaluated how brood size and year potentially influenced body mass of nestling cliff swallows. Brood size perhaps has the most obvious effect (section 4.7), assuming limited parental resources. The usual pattern is for nestlings in larger broods to weigh less, per capita, at a given age than nestlings in smaller broods (e.g., Perrins 1965; Royama 1966; Nur 1984a, 1987a). Yearly effects on nestling body mass can indicate seasonal differences in resource availability (Bryant 1975).

We initially combined all mean nestling body masses per nest from all colonies for nonfumigated and for fumigated colony sites. Mean nestling body mass per nest varied significantly with brood size for fumigated nests

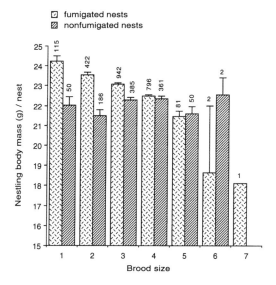

Fig. 10.6 Nestling cliff swallow body mass (g) at 10 days of age per nest for broods of different sizes in fumigated and nonfumigated colonies. Numbers of nests sampled are shown above error bars. Mean nestling body mass varied significantly with brood size for fumigated nests (ANOVA, $p < .001$) but not for nonfumigated nests (ANOVA, $p = .33$).

(fig. 10.6). The pattern was the expected one in which per capita body mass declined in larger broods. For nonfumigated nests, however, brood size had no significant effect on nestling body mass (fig. 10.6). Why nonfumigated nests showed this pattern was unclear. One possibility is that the smaller brood sizes in part represented nests where ectoparasitism was severe enough to have killed part of the brood before day 10 (section 4.7) and body mass of the surviving nestlings reflected this high level of ectoparasitism. The larger brood sizes among nonfumigated nests may have represented nests where ectoparasites were not numerous enough to have killed nestlings. Based on these results (fig. 10.6), we analyzed the effect of colony size on fumigated nests separately by brood size and combined nonfumigated nests of all brood sizes.

Nestling body mass also varied significantly with year for both fumigated and nonfumigated nests (two-way ANOVA with brood size, $p < .001$ for fumigated and nonfumigated, but no significant interaction between year and brood size for either). For nonfumigated nests we performed Scheffe's tests to determine which years differed significantly; those that showed no significant difference were combined into sets (to increase our sample size of colonies), and analyses were performed on each set separately (fig. 10.7). Among fumigated nests, 1984 and 1985 were not significantly different from each other (Scheffe's test) and were combined for analysis. We did not attempt to analyze the data for fumigated nests in the remaining years (1986–89) because we had three fumigated colonies or fewer in each year, an insufficient sample for any within-year analysis. The yearly differences in nestling body mass are

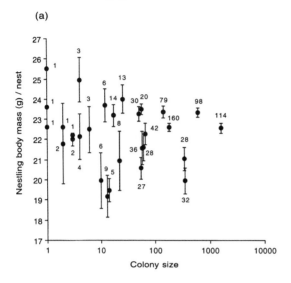

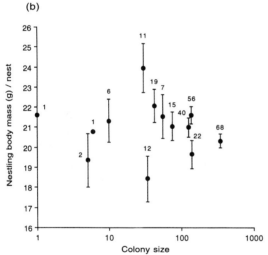

Fig. 10.7 Nestling cliff swallow body mass (g) at 10 days of age per nest in relation to colony size in 1982–83/1985/1987–89 (a) and 1984/1986 (b) in nonfumigated colonies. All brood sizes were combined. Numbers by dots are sample sizes (number of nests) for each colony. Nestling body mass did not vary significantly with colony size (a: $r_s = -.18$, $p = .34$, $N = 29$ colonies; b: $r_s = -.09$, $p = .76$, $N = 13$ colonies).

consistent with our premise that resource availability may have varied among years and possibly confounded between-year measures of foraging efficiency.

10.4.2 Nestling Body Mass in relation to Colony Size

Mean nestling body mass per nest was averaged for each colony. Among nonfumigated nests of all brood sizes combined, body mass did not vary significantly with colony size in either set of years (fig. 10.7). There was a suggestion that mean nestling body mass may have declined slightly with colony size in some years, although the correlation coefficients were not significant. These results seemed to suggest that, on balance, nestling cliff swallows grew at roughly similar rates and achieved similar mean body mass in all colonies regardless of size.

However, analyses of nestling body mass in nonfumigated nests may provide little insight into parents' foraging efficiency per se because of increased ectoparasitism and its concomitant effects in larger colonies (section 4.7). Nestling body mass presumably reflects both parents' foraging efficiency and the level of ectoparasitism. Thus, fumigating nests provides an opportunity to study how parents' foraging efficiency affected nestling body mass with ectoparasitism (at least by swallow bugs) removed (Brown 1988a).

Among fumigated nests in 1984–85, two patterns emerged depending on brood size (fig. 10.8). Mean nestling body mass per nest for broods of three nestlings increased significantly with colony size, while body mass for broods of four decreased significantly with increasing colony size (fig. 10.8). Body mass for broods of one or two nestlings increased, but not quite significantly, with colony size. (These brood sizes were combined because body mass in broods of one was not significantly different from that in broods of two; fig. 10.6 with Scheffe's test.) We did not have enough data from fumigated broods of five or more nestlings in 1984–85 ($N = 29$) for a meaningful between-colony analysis.

With ectoparasites removed, the results for brood sizes one or two and three were consistent with increased parental foraging efficiency in larger cliff swallow colonies. The average increase in body mass from the smallest to largest colonies was about 2 g for broods of three, and perhaps slightly more for the smaller broods (fig. 10.8a,b). Strangely, however, nestling body mass in broods of four declined by a similar amount over the same range of colony sizes and in many cases in the same sites (fig. 10.8c). One possible interpretation is that foraging-related costs (sections 10.8, 10.9) increased the difficulty of raising broods of four nestlings in larger colonies where these costs were highest. But why these costs would not also be reflected to some degree in smaller broods is not clear. Another possibility is that parents invested less in larger broods in the larger colonies where the odds were higher that at least one of the nestlings in the brood was unrelated to them (e.g., section 6.3.2; Moller and Birkhead 1993b; but see Whittingham, Taylor, and Robertson 1992).

The nestling body mass analyses for fumigated nests suggested increased foraging efficiency in larger colonies for parents of the small and average-sized broods, but this is a net benefit of coloniality only in colonies without ectoparasites. In natural colonies exposed to ectoparasitism, there was no apparent advantage for the birds in larger colonies

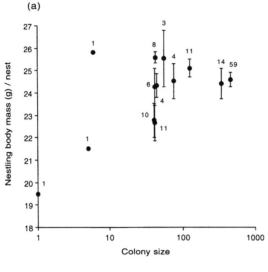

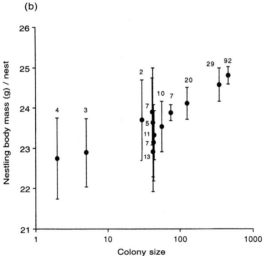

Fig. 10.8 Nestling cliff swallow body mass (g) at 10 days of age per nest in relation to colony size for broods of one and two (*a*), three (*b*), and four nestlings (*c*) in fumigated colonies in 1984–85. Numbers by dots are sample sizes (number of nests) for each colony. Nestling body mass increased almost significantly with colony size for broods of one and two ($r_s = .49$, $p = .089$, $N = 13$ colonies), increased significantly with colony size for broods of three ($r_s = .73$, $p = .004$, $N = 13$ colonies), and decreased significantly with colony size for broods of four ($r_s = -.71$, $p = .005$, $N = 14$ colonies).

(fig. 10.7). Any foraging advantages in larger colonies apparently are negated by the increased costs of ectoparasitism. The net result is little difference in nestling growth rates among those nests with surviving young (for further discussion see section 10.10).

10.5   ADULT BODY MASS

The previous section used nestling body mass as an indirect index of parents' foraging efficiency. Body mass of the adult foragers themselves, how-

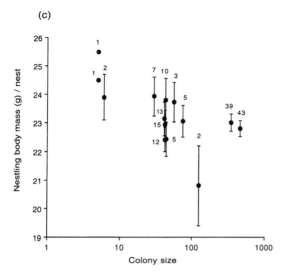

ever, is perhaps a more direct reflection of how efficiently birds feed. For example, weight is directly correlated with a bird's fat reserves (Owen and Cook 1977; Clench and Leberman 1978; Jones 1987b) and is associated with its foraging ability (Weimerskirch, Chastel, and Ackermann 1995). Body mass of adults may also reflect the parents' overall physical condition and subsequent expectation of survival (Haramis et al. 1986; section 12.5.3) or annual reproductive success (Gibbons 1989). Our goal in this section is to examine adult cliff swallow body mass in different-sized colonies as a measure of foraging efficiency and the potential importance of information transfer.

10.5.1 Methods for Recording and Analyzing Adult Body Mass

Adult cliff swallows were captured at colonies and weighed to the nearest 0.5 g (section 2.3.2). We rotated among active colonies in the study area, catching and weighing as many birds as possible during each capture session. Capture effort varied among sites for a variety of unavoidable reasons (section 12.2.1). Sample sizes (number of birds caught) tended to be larger at the larger colonies, although the percentage of the total colony residents caught was usually higher in the smaller colonies. Birds were classified into three time periods—early, mid, and late—corresponding to their colony's stage of nesting (section 2.3.2). Analyses in this section focus on cliff swallows captured "early" (during nest building and egg laying) and "late" (during feeding of nestlings).

Many individuals were caught and weighed repeatedly, both within the same day and within the same season. For any bird caught more than once on a single day, we averaged the separate weighings and used a single daily mean body mass. Within a day there was no systematic bias in masses recorded from birds caught more than once. For example, in 1989 we caught 351 birds twice within a single day, ranging from 11 to 459 minutes apart. Of these, 132 (37.6%) increased in mass between captures, 124 (35.3%) decreased, and 95 (27.1%) were unchanged. The average change in mass between successive weighings of birds within a single day was +0.09 g, indicating that weighings of the same individual on the same day were consistent.

Body masses of adult cliff swallows were recorded at various times of the day, usually between 08:30 and 17:30 MDT. As in other species (Van Balen 1967; Clark 1979), time of day had a moderate effect on body mass, with cliff swallows weighing less in the early morning and increasing in mass in mid- to late afternoon. This pattern was more pronounced early in the season, before the birds were feeding nestlings and when they foraged relatively more in the afternoons. For example, at one colony (9005) on 22 May 1990, we caught and weighed 591 adult cliff swallows between 08:18 and 17:02 MDT. We calculated the average body mass of birds captured during thirty-minute intervals throughout that day. Mean body mass was 23.50 ($\pm$ 0.17) g between 08:00 and 09:30 (N = 96), 24.49 ($\pm$ 0.10) g between 09:30 and 14:30 (N = 263), and 25.11 ($\pm$ 0.11) g between 14:30 and 17:30 (N = 232). At another colony (9205) later in the season (27 June 1992), we caught and weighed 492 adults between 08:23 and 17:33. There was little if any variation in average body mass per thirty-minute interval before 14:30 (21.89 $\pm$ 0.08 g, N = 265) and only a slight increase after 14:30 (22.60 $\pm$ 0.07 g, N = 227). Similar patterns were obvious on other days at other colonies.

To ensure that body masses recorded at all times of the day were represented in the data for each colony, we sampled each colony approximately equally in mornings and afternoons at each time of the year. Unless bad weather forced us to abandon netting, each single capture session at a colony lasted at least three hours, spanning either an entire morning or an entire afternoon and often the entire day. Our sampling method of rotating among colonies at different times of the day, and the large total number of birds caught and weighed at most sites, diminished any confounding effect of capture time on adult body mass. Incorporating time of day into the following analyses would have made them unnecessarily complex with little additional insight, and therefore we pooled all body mass data irrespective of capture time.

For any bird caught on two or more days during the same seasonal time period (early or late), we averaged its daily mean body masses for that period and used a single early or late mean for that individual in all analyses. Most cliff swallows lost mass as the season progressed (section 10.5.3), although within a given time period individuals sometimes gained mass from day to day. For example, in 1983–89, among 209 individuals caught on more than one day during the early period, 46 (22.0%) gained mass, 140 (67.0%) lost mass, and 23 (11.0%) were unchanged during that period; the greatest gain was 5.0 g and the greatest loss was 6.0 g. Among 1,396 individuals caught on more than one day during the late time period, 411 (29.4%) gained mass, 818 (58.6%) lost mass, and 167 (12.0%) were unchanged; the greatest gain was 6.0 g and the greatest loss was 6.5 g.

10.5.2 Adult Body Mass in relation to Colony Size

We examined how adult cliff swallow body mass varied with colony size both early and late in the season. Early-season patterns are important because they may reflect preferential settlement of certain birds in different-sized colonies (section 13.3.3; Brown 1988a). If heavier birds, for instance, settled in larger colonies at the start of the nesting season, a positive relation between body mass and colony size later in the season might reflect primarily the earlier settlement patterns and not necessarily enhanced foraging efficiency. In these analyses we combined fumigated and nonfumigated colonies, because nest fumigation had little effect on the ectoparasites (mainly fleas and chewing lice) that commonly travel and feed on the adult birds.

Analysis of how adult cliff swallow body mass varied with colony size was complicated by seasonal and intersexual differences in average body mass. Year had a significant effect on adult body mass when birds from all colonies were combined, for both early and late in the season (fig. 10.9). Therefore we analyzed how body mass varied with colony size separately for each year. We also analyzed patterns for each sex separately, because body mass differed significantly with sex in each period (fig. 10.9; analyses for 1984–85 combine the sexes because we did not begin sexing cliff swallows until 1986; section 2.4).

We had five years (1988–92) in which we sampled enough colonies ($N > 6$) to permit rigorous analysis of the effect of colony size on adult cliff swallow body mass early in the season (fig. 10.10). Among males, body mass declined, though not significantly, with colony size in four of the five years (table 10.3; early-season data in 1992 may not be interpretable owing to the severe weather in May of that year; section 10.6). The pattern was strongest in 1988 and 1991, in which males in the smallest colonies averaged 2–4 g heavier than males in the largest colonies early in the year (fig. 10.10). The remaining years showed no clear trend. Body mass of adult females (fig. 10.10) also declined with colony size in 1988 and 1991, while increasing with colony size in 1989 and 1990; the correlations were not statistically significant (table 10.3). On balance, these analyses suggest no clear relation between adult cliff swallow body mass

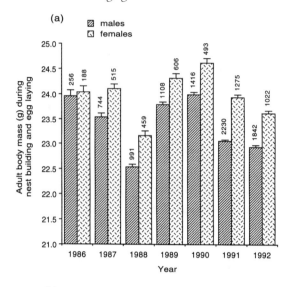
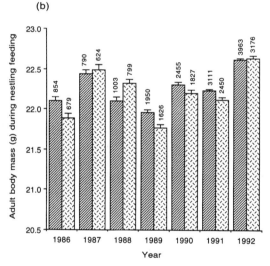

Fig. 10.9 Adult cliff swallow body mass (g) for males and females in each year for all colonies combined, early in the season (*a*) and late in the season (*b*). Sample sizes (number of birds) are shown above error bars. Early in the season, females had significantly greater body mass than males (Wilcoxon rank sum tests, $p < .001$ for each) in each year except 1986. Late in the season, male and female body mass differed significantly in each year except 1987 and 1992 (Wilcoxon rank sum tests, $p \leq .007$ for each; $p \geq .62$ for 1987, 1992). Year had a significant effect on body mass for each sex, both early and late in the season (ANOVA, $p < .001$ for each).

and colony size during nest building and egg laying (also see section 13.3.3). If anything, birds (especially males) in small colonies weighed slightly more early in the season.

The period of nestling feeding, when parents are constantly gathering food, is presumably the time of the year when adult body mass is most likely to reflect the birds' foraging efficiency. Late-season body mass may also reflect the physical condition of the birds as they prepare to migrate south (section 12.5.3). We predicted, based on the birds' use of

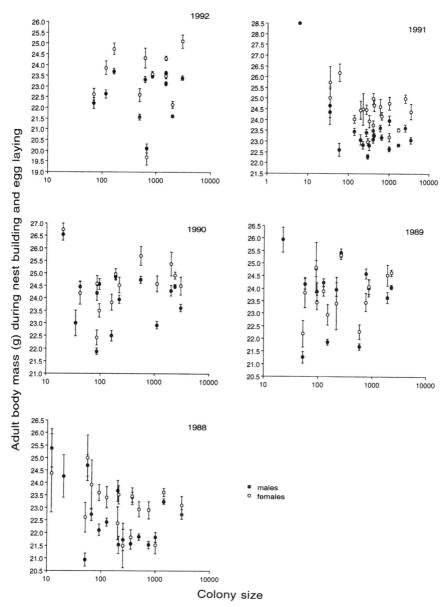

Fig. 10.10 Adult cliff swallow body mass (g) for males and females in relation to colony size early in the season, 1988–92. Correlation coefficients are shown in table 10.3.

Table 10.3 Spearman Correlation Coefficients between Mean Adult Cliff Swallow Body Mass per Colony and Colony Size

	Early					
	Male			Female		
Year	r_s	p	N^a	r_s	p	N^a
1992	.15	.65	11	.08	.82	11
1991	−.23	.33	20	−.32	.18	19
1990	−.05	.66	14	.22	.48	13
1989	−.16	.58	14	.35	.24	13
1988	−.30	.24	17	−.40	.12	16
1987						
1986						
1985						
1984						

Note: See also figures 10.10 and 10.11.
[a] Number of colonies.

information from conspecifics to find food (chapter 9) and the increased food delivery rates to nestlings in larger colonies (section 10.3), that adult cliff swallow body mass should have increased with colony size. Surprisingly, in only three years (1984, 1987, and 1992) did adult body mass increase with colony size late in the season (fig. 10.11). In three other years (1986, 1989, 1990) it declined with colony size. The remaining years (1985, 1988, and 1991) showed no clear trend. Males and females exhibited virtually the same pattern in each year, and most of the correlation coefficients were not significant (table 10.3).

This yearly variation (fig. 10.11) indicates that the birds annually have different expectations of foraging efficiency in different-sized colonies. In some years adult cliff swallows probably improved their body mass and overall condition by nesting in larger colonies (probably through information sharing), whereas in other years adult body mass was reduced in larger colonies (see sections 10.8, 10.9) or was unaffected by colony size. These results illustrate the value of long-term studies and reveal how misleading data from only one or two years might be. We were misled in our earlier report (Brown 1988a), in which we used data for only one year (1984), and consequently may have overstated the effect of colony size on adult cliff swallow body mass.

There was no obvious cause or correlate of the yearly differences (fig. 10.11) in adult body mass variation with colony size. Annual temperature and rainfall (section 3.2.2) did not explain these differences. Yearly (and perhaps site-related) variation in resource availability presumably was one factor contributing to these patterns, and perhaps yearly

			Late		
Male			Female		
r_s	p	N^a	r_s	p	N^a
.23	.33	21	.21	.37	21
−.10	.62	26	.03	.87	25
−.65	.004	18	−.35	.16	17
−.22	.33	21	−.19	.45	19
.04	.59	13	−.14	.63	13
.29	.34	13	.21	.56	10
−.29	.39	11	−.58	.06	11
Sexes Combined					
.18	.63	9			
.21	.56	10			

differences in costly ectoparasite loads of the adult birds (in relation to colony size) were also responsible (see section 4.8; Brown, Brown, and Rannala 1995). We could not identify any other specific ecological influences that might have led to the differences depicted in figure 10.11. How differences in adult body mass affected individuals' probability of surviving to the next breeding season is addressed in section 12.5.3

10.5.3 Reductions in Adult Body Mass in relation to Colony Size

Another potential body mass-related index of foraging efficiency is how much mass birds lose during the season. Mass reduction by breeding birds presumably reflects (at least in part) energy stress associated with reproduction (Ricklefs 1974; Nur 1984b; Jones 1987b,c, 1988; Bryant 1988a; Moreno 1989; Weimerskirch, Chastel, and Ackermann 1995; cf. Freed 1981; Norberg 1981; Ricklefs and Hussell 1984). Cliff swallows followed the typical passerine pattern and tended to lose mass as the season progressed, but presumably the birds that fed most successfully and were energetically less stressed lost the least mass. Reduction in mean body mass over the season represents a relative measure of foraging efficiency among different colonies, irrespective of seasonal differences in means (e.g., fig. 10.9), and consequently data from different years may be pooled for analysis.

We used the difference between early and late mean body mass as our measure of mass reduction over the season. Simply subtracting colony-wide early and late mean masses was not possible, however, because different subsets of birds could have been caught at the different times of the

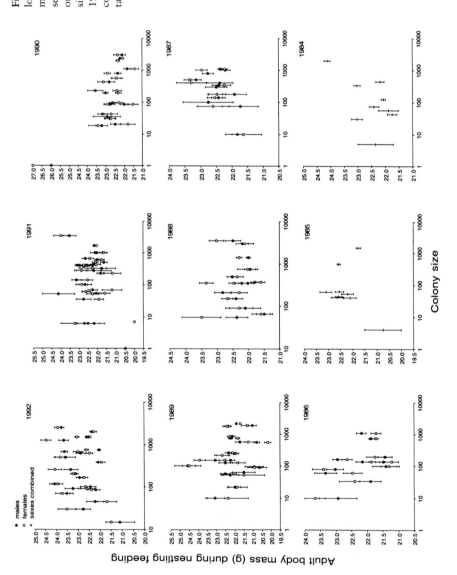

Fig. 10.11 Adult cliff swallow body mass (g) for males, females, and both sexes combined (1984–85 only) in relation to colony size late in the season, 1984–92. Correlation coefficients are shown in table 10.3.

Adult Body Mass • 299

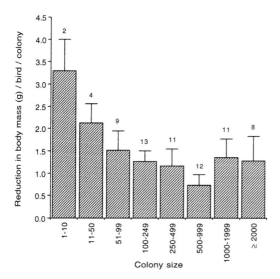

Fig. 10.12 Reduction in body mass (g) for cliff swallows caught both early and late in the season in relation to colony size. Sample sizes (total colonies) are shown above the error bars. Mean mass reduction per colony declined almost significantly with colony size ($r_s = -.64$, $p = .086$, $N = 8$ size classes).

year. Therefore in this analysis we used only individuals that were caught both early and late in the season at a colony. For each individual we calculated its mass reduction, and we used only colonies with three or more individuals for which we had both early and late data. Average body mass reduction was known for individuals in seventy colonies ranging from 5 to 3,500 nests.

The average body mass reduction per colony declined, almost significantly, with colony size for all colonies and years combined (fig. 10.12). This meant that cliff swallows in larger colonies lost less body mass over the course of a season than did birds in smaller colonies. Mass reductions were greatest in colonies of 50 nests and fewer (fig. 10.12). These results are consistent with the hypothesis that feeding efficiency is enhanced in larger colonies. Either cliff swallows in large colonies can find enough food to stave off mass loss, or birds in smaller colonies lose mass specifically to save energy expended in flight (Freed 1981; Norberg 1981; G. Jones 1987c; I. Jones 1994) and may consequently improve their feeding efficiency. Of course, if individuals in small colonies begin the season heavier than birds in larger colonies (fig. 10.10), they may be able to afford to lose more mass at less net cost. The implication in any case is that there are differences among colony sizes in an individual's expected foraging efficiency as reflected in adult body mass.

The observed pattern of mass reduction in relation to colony size (fig. 10.12) suggests a steady decline in mass loss to colony sizes of 500–999 nests, and then perhaps an increase in larger colonies. This sort of pattern suggesting an intermediate "optimal" colony size should not be discounted, although in practice distinguishing intermediate optima from noise in a single analysis is difficult. In this case there perhaps is reason to predict that foraging is most efficient at an intermediate colony size. There appear to be foraging-related costs in the largest colonies (sections 10.8, 10.9), which presumably depress foraging success there. Opportunities for information transfer increase with colony size, at least up to the point where foragers depart constantly (sections 9.5.2, 10.1.1). If information transfer levels off at colonies of 500–999 nests (which, from our observations of the birds' behavior, seems possible), the foraging costs in the largest colonies could create an optimal colony size at 500–999 nests, all else being equal. Other analyses also suggested an intermediate optimal colony size with respect to foraging: seemingly unusually long waiting intervals in a 2,200 nest colony (section 10.1.2); food delivery rates that peaked at intermediate-sized colonies (fig. 10.3), and mean adult body mass (fig. 10.11) that seemed to increase, then decrease slightly, with increasing colony size in some years (e.g., 1985, 1987, 1989, 1992). These analyses, and that in figure 10.12, do not permit confident determination of exactly where the optimal colony size may be, but they suggest it is somewhere between 100 and 1,000 nests.

The interpretation that seasonal mass loss is an index of foraging efficiency is complicated by the costly effects of ectoparasites (such as fleas and lice) on adult cliff swallows (section 4.8; Brown, Brown, and Rannala 1995). Some reduction in mass may be a direct result of being parasitized (e.g., Senar et al. 1994). If so, the reduction in body mass in the largest colonies (fig. 10.12) is perhaps a consequence of increased ectoparasitism of the adult birds there, whereas the reduction in body mass in the smallest colonies may be attributed to less foraging information at those sites. An apparent intermediate optimal colony size (fig. 10.12) is consistent with such a scenario.

10.6 STARVATION DURING COLD WEATHER

Information on the whereabouts of food might be especially important during times of prey scarcity. The aerial insects cliff swallows feed on are inactive during cold weather, so late spring cold spells are times of food scarcity for these birds. If the cold weather lasts several days, widespread cliff swallow mortality may occur (e.g., Audubon 1831; Kimball 1889; Bent 1942; R. Stewart 1972; Henny, Blus, and Stafford 1982; DuBowy and Moore 1985; Krapu 1986; Littrell 1992). Hoogland and Sherman (1976) used a naturally occurring cold spell to evaluate the importance of information sharing in bank swallow colonies of different sizes, and in this section we investigate how cliff swallow foraging efficiency may have varied with colony size during cold weather.

Unlike the situation in Hoogland and Sherman's (1976) study of bank swallows in Michigan, there were no severe cold spells in our study area during the period (June and July) when most cliff swallows were feeding nestlings. There were, however, two periods of severely cold weather in May that affected adult survival and that consequently offered some insight into these birds' foraging efficiency in colonies of different sizes during energetically stressful conditions.

Cliff swallows routinely experienced one and sometimes two consecutive days of cold weather during late April, May, and the first half of June in southwestern Nebraska. On such days when the daily high temperature was <13°C, the birds foraged in loose groups low over water or in canyons (section 9.2.1), often using food calls to signal discovery of prey (section 9.6). These one- or two-day cold spells were relatively frequent but did not seem to affect the birds in obvious ways. However, in 1988 and again in 1992 cold spells four days long occurred and were severe enough to cause widespread mortality among adult cliff swallows, likely through starvation.

In 1988 the cold and rainy period persisted from 19 through 22 May. During this time, daytime temperatures were mostly between 5°C and 9°C. The maximum hourly temperature readings from Arthur, Nebraska, north of the study area (section 3.2.2), were 16.1°, 8.5°, 10.2°, and 10.4°(C) on these four days. High temperatures from about 18°C to 23°C preceded and followed this cold spell. Immediately before the cold weather (on 10–18 May) we caught and weighed adults at eight colonies of different sizes (to profile the birds choosing different colonies; section 13.3.3). We also determined the colony sizes at these sites before the cold spell, using a combination of nest checks (section 2.1) and visual estimates of the number of birds present (some residents had not yet laid eggs). Immediately after the cold weather (on 23–25 May) we again caught and weighed adults, estimated colony sizes, and looked for dead birds.

Cliff swallows living in small colonies were clearly hit hardest by the cold weather in 1988 (table 10.4). We found dead adults at all four of the colonies smaller than 100 nests, but none at colonies of 100 nests or larger. These birds were found below nests or in some cases hanging out of nest entrances. All were emaciated, suggesting that they had starved to death. There was no way to standardize our searching for dead birds at different sites, but that no dead birds were found at the larger colonies, despite the greater number of birds that could have died, suggests that mortality was in fact heaviest in small colonies. This conclusion was also supported by the reduction in colony sizes during the cold weather. The small colonies declined in size by up to 80%, whereas the large colonies remained unchanged. If anything, birds may have moved into the largest colonies during the cold weather (table 10.4).

Table 10.4 Cliff Swallow Colony Sizes (Number of Nests) Immediately Before and After Spells of Cold Weather

Colony Site	Colony Size Before	Colony Size After	Dead Birds Found	Change in Colony Size (%)
1988				
8831	15	5	2	−66.7
8827	25	5	2	−80.0
8832	35	13	6	−62.9
8824	55	45	1	−18.2
8830	100	110	0	+10.0
8805	700	725	0	+3.6
8810	1,000	1,000	0	0.0
8845	2,000	2,400	0	+20.0
1992				
92MDO	5	5	0	0.0
9260	6	11	0	+83.3
9232	30	15	2	−50.0
9203	60	45	4	−25.0
9266	60	50	8	−16.7
9256	90	60	11	−33.3
9221	100	85	12	−15.0
9241	165	165	0	0.0
9223	400	380	2	−5.0
92OG	—	850[a]	6	—
9202	—	1,300[a]	7	—
9205	1,500	1,500	0	0.0
9210	1,800	2,000	0	+11.1
9216	2,000	1,500	6	−25.0
9245	3,000	1,250	90	−58.3

[a] Final 1992 colony size; colony size before the cold spell was unknown.

These census results suggested that adult cliff swallows were more likely to starve in the smaller colonies. We interpret this to be because they had less information on food and foraged less efficiently when conditions were harsh. Data on adult body mass reduction during the cold weather supported this conclusion. The average reductions in body mass during the cold spell were 3.04 g for males and 3.79 g for females in the four smallest colonies, versus 1.37 g and 1.53 g in the four largest colonies. These average reductions were based on the mean body masses for all birds sampled before and immediately after the cold weather. Unlike the situation in section 10.5.3, they are not based on data for the same individuals, because too few birds were caught during both time periods for meaningful analysis. Nevertheless, the results clearly suggested that cliff swallows in the smaller colonies lost more mass during the cold weather than did birds living in the larger colonies.

The most severe cold spell in the twelve years of the study occurred from 25 through 28 May 1992. Maximum hourly temperatures were 6.4°, 11.7°, 5.5°, and 12.2° (C) on these days, and temperatures were primarily between 1°C and 3°C on 25 and 27 May. Light snow fell on 27 May, and a low of −1.4°C was recorded on 26 May. Cliff swallows suffered greater mortality in the 1992 cold spell, probably because it had been preceded by ten days of below normal temperatures (highs from 12°C to 17°C) and perhaps also because it occurred relatively late in the season after many birds had laid eggs. The generally colder weather in 1992, which caused birds often to be away from their colonies foraging, prevented us from getting body mass data from a range of colony sizes before and after the cold spell.

On 29 May, after the cold spell broke, we visited fifteen colony sites to look for dead birds and estimate colony sizes. We searched for weather-caused fatalities only at sites where standing water or dry ground underneath the nests made finding any carcasses likely. Colonies over flowing water were not surveyed because bodies there would have been washed away. The number of dead cliff swallows found at each site and each colony's size before and after the cold weather are shown in table 10.4. The pattern of mortality was less striking than that seen in 1988 and was somewhat different. No mortality was evident in the two smallest colonies (<10 nests), whereas all colonies of 30–100 nests had fatalities (table 10.4). Mortality occurred sporadically among the larger colonies. Surprisingly, there was appreciable mortality in the two largest colonies, especially the single largest one. At this site (9245), we discovered large numbers of dead cliff swallows and saw scores of additional birds that were obviously weak and flying poorly and probably also soon succumbed. There was a major colony size reduction at 9245 during the cold spell (table 10.4).

Among the dead birds found in 1992, neither sex predominated among the casualties we dissected (47.8% males, $N = 23$; many birds were so emaciated that their gonads could not be identified). There was no indication among banded casualties that younger or older age classes predominated. Owing to the extremely warm temperatures in early May 1992 (section 3.2.2), many cliff swallows had begun laying eggs earlier than usual, and a few had small nestlings by the time of the cold spell. All nestling cliff swallows throughout the study area that had hatched before 25 May apparently starved, but eggs were unaffected.

The observed patterns of weather-caused mortality during these two cold spells led to contradictory conclusions. Cliff swallows in large colonies fared markedly better during the 1988 cold weather than birds in small colonies, but in 1992 far more deaths apparently occurred in the largest colony than in all others. Later in the summer in 1992 we discovered many dead birds, probably dating from the cold spell, inside nests at two other very large colonies (>2,000 nests), but no additional mortality was found in any smaller colonies. The differing mortality patterns for the two cold spells are consistent with the seasonal differences in how foraging efficiency apparently varied with colony size (section 10.5.2). Seasonal differences in prey availability and the prey's sensitivity to cold weather perhaps accounted for these different results. During a cold spell in California, Littrell (1992) also found cliff swallow mortality at some colony sites but apparently not at others (colony sizes were not stated), and mortality varied substantially between house martin colonies during a cold spell in Germany (Lohrl 1971). We conclude simply that during cold spells cliff swallows benefited from increased information transfer in larger colonies in some years, but apparently not in other years. Living in the largest colonies may have been costly in some years during cold weather when prey was scarce.

10.7 DIFFERENCES IN LOCAL FOOD RESOURCES AMONG COLONY SITES

The presumed foraging-related advantages of larger colonies described in the previous sections of this chapter are consistent with increased opportunities to transfer information as colony size increases. But foraging efficiency of cliff swallows in large colonies could also be enhanced if sites harboring large colonies were near particularly abundant or densely concentrated food resources (as in some spiders; Uetz, Kane, and Stratton 1982; Smith 1983, 1985; Rypstra 1983, 1985). If local food resources were sufficiently better near large colonies, the colony size effects described earlier could occur without any foraging enhancement via information transfer (Brown 1988a).

Direct sampling of food resource availability among colony sites was not practical or likely to yield meaningful results, owing to the extreme diversity of insects cliff swallows preyed on (section 9.3). Thus we used an indirect experimental approach to determine whether differences in foraging success among colonies could be attributed in part to differences in local food availability. The results (Brown 1988a) are summarized here but not presented in detail, because we have nothing new to add.

In this experiment we reduced the sizes of large cliff swallow colonies, thus removing many potential foraging associates (and sources of information) for the individuals that remained. If large colonies formed near abundant food resources and the colony size effects documented earlier merely reflected the locations of colonies of different sizes, enhanced foraging efficiency should still have occurred even when the number of birds at the large colony sites was reduced. If, however, enhanced foraging efficiency was attributable to efficient information transfer among the many residents in large colonies, when the large colonies were reduced we should have observed no differences in foraging success between the birds remaining at the formerly "large" colonies and the birds living in small, unaltered colonies of similar size. This type of experiment has been used in studies of spider coloniality (Rypstra 1985; Smith 1985; Spiller 1992) but not to our knowledge in other studies of colonial birds.

In 1985 large cliff swallow colonies were reduced by removing nests (see Brown 1988a for methodology). Colonies were not reduced until all laying there had ceased and the birds had thus "committed" themselves to a colony of that size at that site. Removing nests was sufficient to effectively reduce the number of adult cliff swallows present at the colonies (many birds moved to other colonies and renested), and no adult birds were killed or removed (Brown 1988a).

We examined body mass of both nestling and adult cliff swallows in experimental and control colonies that all had been fumigated to remove the confounding effects of ectoparasites. There were no significant differences among colony sizes in either nestling or adult body mass. Birds in the formerly large sites appeared to forage no more successfully, as measured this way, than individuals at the smaller colonies (Brown 1988a). These results supported the hypothesis that foraging efficiency in larger colonies was enhanced by information transfer and not merely affected by site-specific differences in local food availability.

We still believe the design of this experiment is sound and that this approach is the single best way, in principle, to address differences in local food availability among cliff swallow colony sites. However, for two reasons we are no longer as confident about the experiment's conclusions. One concern is that now we know there were appreciable yearly differences in how adult body mass varied with colony size (fig. 10.11). The colony-reduction experiment was done in one year only (1985). For any single year it is not immediately clear what sort of pattern should be "expected" from the experimental treatment. Another concern is the range of colony sizes over which the experiment was done. Among the large colonies experimentally reduced, the largest originally was only 120 nests (reduced to 63 nests; Brown 1988a). Thus even our largest site was smaller than the mean colony size for the population (section 3.5.5). For ethical reasons we obviously could not reduce a truly large colony (e.g., >1,000 nests) to 40–50 nests. Doing several such reductions in at least two seasons would probably be necessary to achieve the best results.

Despite these problems, however, there was no indication that the three larger sites were in better foraging areas than the four small colonies (Brown 1988a). In the absence of any better data or a more definitive and less destructive approach at present, we still agree with the earlier conclusion that enhanced foraging efficiency in larger cliff swallow colonies results primarily from information transfer and not merely from differences among sites in local food availability. Further study is needed, though, because the problem has not been definitively addressed in any colonial bird to date (see Cairns 1989).

10.8 FORAGING COSTS: INCREASED TRAVEL DISTANCES AND SEARCH AREAS

The previous sections examined potential foraging advantages associated with increasing cliff swallow colony size. Coloniality also presents some potential foraging-related costs. Increasing numbers of conspecifics with similar food requirements may deplete localized resources (Wittenberger and Hunt 1985), forcing colony residents to go farther to find food or to invest more time and energy in searching a given foraging arena. A colony resident's reproductive success consequently may be reduced whenever these costs prevent it from finding enough resources to sustain all its offspring or impair its prospects for surviving to the next year. In this section we examine to what extent travel distances and search areas increase for cliff swallows breeding in large colonies and evaluate how potentially costly these effects may be. In the following section (10.9) we address more directly to what degree resource depletion reduces cliff swallow foraging success and reproductive success.

Although Wittenberger and Hunt (1985) suggested that localized competition for food represents a net energy cost of coloniality for most animals, there are few direct data for colonial species to evaluate this claim. Indirect evidence from a variety of species suggests some competition for food in large colonies and social groups (Gaston, Chapdelaine, and Noble 1983; Furness and Birkhead 1984; Watts 1985; Wittenberger and Hunt 1985; Hunt, Eppley, and Schneider 1986; Janson and Van Schaik 1988; Seibt and Wickler 1988; Harris and Wanless 1990; Danchin 1992b; Roberts and Hatch 1993; Wiklund and Andersson 1994). Average travel distance while foraging is one of the more direct indexes of resource depletion and competition for food around colonies because, for a variety of reasons, individuals should feed as close to their nests as possible (Andersson 1978, 1981). Longer travel distances presumably mean food is insufficient close to the colony (Hamilton and Watt 1970; Orians 1971; Furness and Birkhead 1984; Wittenberger and Hunt 1985). For birds, only Erwin (1978) has examined specifically how colony size might affect travel distances of foragers, finding that tern species nesting in large colonies traveled farther offshore to feed than did species nesting in small colonies. However, Erwin attributed this pattern to interspecific differences in the fish preferred by the various tern species and not necessarily to intraspecific resource depletion. Among primates, Van Schaik, Van

Noordwijk, De Boer, and Den Tonkelaar (1983) and Janson (1988) found that daily travel distance increased with group size in long-tailed macaques and brown capuchins, probably as a result of competition for food within the larger groups.

10.8.1 Measuring Average Travel Distance and Foraging Areas

Average travel distances and foraging areas were calculated for the same colonies in which spatiotemporal variability in foraging locations was studied (section 9.4.2). Field methods of observing foraging cliff swallows at colonies are described in section 9.4.2 and in Brown, Brown, and Ives (1992). Average travel distance and foraging area were based on the positions and counts of foragers surrounding a colony at ten-minute intervals on given days while the birds in the colony were feeding nestlings. Because birds were commuting to and from the colony and foraging sites at relatively short intervals (e.g., section 10.2.2), the individuals in the foraging groups were constantly changing, and thus each scan of the foraging grounds was treated independently. As described in section 9.4.2, the daily average number of birds seen at ten-minute intervals at each point (x_i, y_i) was denoted b_i. The b_is from each day were then combined to give B_i, the number of birds seen per ten-minute scan at point (x_i, y_i) averaged across all days of observation. Because the b_is represented the average number of birds seen, the calculation of B_i weighted all days equally despite variations in the numbers of scans taken on different days.

In calculating travel distances and foraging areas, we assumed that the number of trips birds made to a particular foraging location was proportional to the total number of birds observed (e.g., section 9.4.1). If $B_s = \sum_{i=1}^{N} B_i$, then (B_i/B_s) is the fraction of trips made to location (x_i, y_i) over the period of observation, and the average one-way distance traveled from the colony to foraging locations, D_{col}, is given by

$$(1) \qquad D_{col} = \sum_{i=1}^{N} \sqrt{x_i^2 + y_i^2}\, (B_i/B_s).$$

In calculating the size of a colony's foraging area, we first determined the center of the foraging arena (Brown, Brown, and Ives 1992). A foraging arena consisted of the nearby pastures and fields over which colony residents foraged. We defined the center of the foraging arena as that point at which the average distance to observed foraging locations was minimized. The average distance from any point (x, y) to the foraging locations, denoted $D(x, y)$, is given by

$$(2) \qquad D(x, y) = \sum_{i=1}^{N} \sqrt{(x - x_i)^2 + (y - y_i)^2}\, (B_i/B_s).$$

308 • Social Foraging 2

Let (x_Δ, y_Δ) be that point that minimizes $D(x, y)$, and denote $D(x_\Delta, y_\Delta)$ as D_{cen}. Because there is no explicit formula for (x_Δ, y_Δ), it must be calculated implicitly from equation (2) using numerical methods.

Calculating the area of the foraging arena required defining the boundaries of the arena. Because there were no abrupt geographical boundaries of the foraging arenas, the boundaries were determined from the observed foraging patterns. The boundary could be defined as the smallest ellipse, or the smallest polygon, that circumscribes all foraging observations. However, this definition can give misleading results because it does not distinguish distant foraging locations that are used rarely from those locations that are used more frequently (Brown, Brown, and Ives 1992).

Instead of using only the most distant foraging locations to determine the boundary of the foraging arena, we derived a method for finding the boundary that used all observations of foraging. We defined the boundary of the foraging arena as that line along which every point is on average $2\,D_{cen}$ from all of the foraging locations. That is, the boundary is the line of points (x_b, y_b) such that

$$(3) \quad 2\,D_{cen} = \sum_{i=1}^{N} \sqrt{(x_b - x_i)^2 + (y_b - y_i)^2}\,(B_i/B_s).$$

This definition has the advantage that the average radius of the foraging arena will vary roughly proportionally with the average distance from the center of the foraging arena to the foraging locations. Distant foraging locations that are rarely used and therefore contribute little to the average distance from the arena center may lie outside the boundary. See Brown, Brown, and Ives (1992) for further details on our method of defining foraging arenas.

After determining the boundary of a foraging arena, we directly calculated the area of the arena. If $R(\theta)$ is the radius of the foraging arena at any angle θ from the calculated center of the arena (eq. 2), then in polar coordinates the area of the arena, A, is given by

$$(4) \quad A = 1/2 \int_0^{2\pi} R^2(\theta)\,d\theta.$$

10.8.2 Average Travel Distance in relation to Colony Size

We calculated average one-way travel distances for sixteen colonies ranging from 10 to 3,000 nests (fig. 10.13). These data came from colonies active in different years, 1985–91, and thus seasonal effects on food resources potentially influenced them. Two of these colonies were on cliff faces along Lake McConaughy, which restricted the birds' foraging space (section 9.4.3) and may have led to longer travel distances at those sites for that reason alone. All other colonies were in areas with broadly suit-

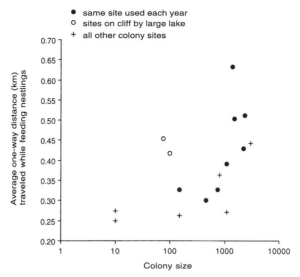

Fig. 10.13 Average one-way distance per colony (km) traveled while cliff swallows were feeding nestlings in relation to colony size for a single site occupied each year, colonies on a cliff adjacent to a large lake, and all other colonies. Travel distance increased significantly with colony size for all sites combined ($r_s = .61$, $p = .012$, $N = 16$ colonies), for all sites except cliffs ($r_s = .83$, $p < .001$, $N = 14$ colonies), and for the colonies at the single site only ($r_s = .79$, $p = .020$, $N = 8$ colonies).

able foraging habitat on all sides. Data for eight of the sixteen colonies (fig. 10.13) came from one single site (site 05) that was active each year of the study, varying in size each year (two temporally separate colonies formed at this site in 1991, one much later in the year than the other).

Average travel distance increased significantly with colony size for all sites combined, for all sites exclusive of the cliffs, and for the single site active each year (fig. 10.13). The strength of this pattern suggested that it was real and not an artifact of seasonal differences in resource use. Cliff swallows in the larger colonies traveled on average about 0.2 km (or 66.7%) farther each way per foraging trip than did birds in the smallest colonies (fig. 10.13). Interestingly, the colony with the longest average travel distance was active in 1988, the extremely hot summer (section 3.2.2). Our subjective impression, and that of others (A. Joern, pers. comm.), was that insect numbers in general were reduced that year, perhaps as a direct result of the heat.

These differences in average travel distance among colony sizes (fig. 10.13) were magnified because cliff swallows in larger colonies delivered food to their nestlings more often and thus traveled these distances more frequently than did parents in small colonies (section 10.2.2). For instance, if each parent in a 2,200 nest colony delivered food approxi-

310 • *Social Foraging 2*

mately 7 times per hour (from fig. 10.3) and traveled a round-trip distance of about 0.86 km per trip (from fig. 10.13), that bird would commute about 6 km each hour, compared with only about 0.85 km each hour for a parent in a 10 nest colony that delivered food 1.7 times per hour and traveled a round-trip distance of 0.50 km per trip. This suggests that cliff swallows in larger colonies experienced greater energy costs of commuting than did birds in small colonies. Nevertheless, birds in larger colonies were able to deliver more total food (section 10.3). More efficient foraging owing to information transfer may have enabled residents of large colonies to compensate for their longer travel distances.

10.8.3 Foraging Area in relation to Colony Size

The total area over which a colony's residents foraged provided another index of potential resource depletion. Although foraging area was determined in part by average travel distance (section 10.8.1), a measure of foraging area was especially useful for sites where the foraging arena was not centered on the colony site. For example, as shown in figure 9.2, the foraging areas of some colonies were centered on or near the colony site, but in other instances the birds fed principally in positions not central to the colony. If foraging habitat near a colony was unsuitable and contained no prey because of local topographic features (e.g., a lake), colony residents automatically were forced to fly farther to reach suitable foraging grounds. In these cases travel distance per se might not accurately reflect any resource depletion or competition for food. Our definition of the area of a colony's foraging arena (section 10.8.1) therefore may have provided a better index of potential resource depletion for some sites. A colony's total foraging area should increase with colony size if resource depletion forces the birds to range farther once they reach preferred foraging habitat.

Foraging areas were determined for the same sixteen sites for which we calculated average travel distances (fig. 10.14). Foraging area increased significantly with colony size for all sites exclusive of cliff sites; this increase was especially marked for the colonies located at the single site in different years. The correlation was not significant (barely) when the two cliff sites were included, illustrating that cliff swallows occupying the cliff colonies foraged over a wider total area than might be expected for colonies of that size (perhaps as a consequence of the nearby lake). There was surprising variation among colonies in total foraging area, and the largest colony (3,000 nests) had a foraging area that was the fifth smallest among the fourteen noncliff colonies (fig. 10.14). The result for this largest colony suggests that increased foraging areas in large colonies are not automatic, and some colonies may not necessarily experience resource depletion relative to smaller sites. However, the results (fig. 10.14) are

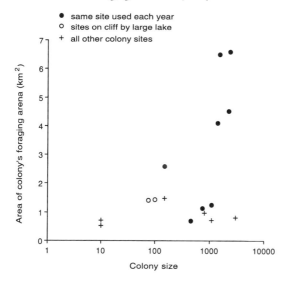

Fig. 10.14 Total area of a cliff swallow colony's foraging arena (km²) in relation to colony size. Total area increased with colony size, though not quite significantly, for all sites ($r_s = .47$, $p = .064$, $N = 16$ colonies); total area increased significantly with colony size for all sites except cliffs ($r_s = .57$, $p = .032$, $N = 14$ colonies) and for the colonies at the single site only ($r_s = .83$, $p = .010$, $N = 8$ colonies).

broadly consistent with the notion of increased search areas in larger colonies. They also support the conclusion, based on average travel distance (fig. 10.13), that cliff swallows generally traveled and ranged farther afield to find food as colony size increased.

Increased foraging areas in larger colonies also could result from greater spatiotemporal variability in prey location. If dense concentrations of prey are more unpredictable around some sites, larger total search areas may be required to find the prey (Schoener 1971; Barta 1992), independent of any resource depletion. However, the increased foraging areas of large cliff swallow colonies probably did not result from higher spatiotemporal variability in prey abundance, because spatiotemporal variability did not change with colony size (section 9.4.3). If anything, there was less variability within a day at larger colonies (fig. 9.3a). This result reaffirms the conclusion that increased foraging areas in larger colonies more likely reflected resource depletion. Although the cause(s) may still be problematic, increased travel distances and search areas in larger colonies represented a cost of cliff swallow coloniality.

10.9 FORAGING COSTS: PREY CAPTURE RATES AND NESTLING STARVATION

Data on travel distances and foraging areas reported in the previous section provided indirect evidence for resource depletion as a cost of cliff swallow coloniality. In this section we evaluate more directly to what extent coloniality may depress an individual's food harvest rate. We do this

312 • Social Foraging 2

by examining both observed prey capture success in foraging groups of different sizes and the incidence of nestling starvation in colonies of different sizes.

10.9.1 Prey Capture Rates in relation to Foraging Group Size

Cliff swallows usually fed in groups that ranged from 2 to over 2,000 birds (section 9.2.1). Mean foraging group size tended to increase with colony size (Brown 1988b), which was not surprising since larger colonies contained more potential foragers at any given time. At any single colony, however, foraging groups varied considerably in size even within a single day (section 9.4.3). This enabled us to measure an individual forager's success as a function of group size, revealing whether large groups depleted local insect swarms.

We observed cliff swallows at foraging sites near colonies and recorded foraging success and group size as described in section 9.7. For each foraging bout, we established arbitrary group size classes to which each observation was assigned based on its foraging group size. The mean number of prey capture attempts per minute was determined for each group size class. Only foraging bouts with at least four group size classes were used in this analysis, and solitary foragers were not included (see section 9.7 for a discussion of solitary foragers' success). Because the maximum number of size classes for any bout was only six, we did not test the resulting Spearman rank correlation coefficients for statistical significance.

Correlations between foraging success and group size are presented in table 10.5 for 24 separate foraging bouts. We also present the overall average prey capture rate for all birds within each bout, as a measure of relative prey abundance at the different times (bouts) these data were collected. In 16 of 24 (67%) foraging bouts there was a positive correlation between foraging group size and an individual's average prey capture rate (table 10.5). This preponderance of positive correlations was not significant (binomial test, $p = .152$). There was thus no evidence that average prey capture rate was routinely depressed in larger foraging groups, and if anything, larger group sizes may have more often increased individual foraging success (perhaps through food calls or local enhancement).

Potential resource depletion with increasing group size might be most likely on days when prey is less abundant, and bouts on those days might more likely result in negative correlation coefficients. We examined whether relationships existed between the sign and the magnitude of the correlation coefficient and the overall mean group foraging success per bout. Overall individual foraging success during bouts with positive correlations (3.96 ± 0.36 prey captures/min/bout, $N = 16$ bouts) was not significantly different from overall individual success during bouts with negative correlations (3.43 ± 0.40 prey captures/min/bout, $N = 8$ bouts;

Table 10.5 Spearman Rank Correlation Coefficients for Prey Capture Attempts per Minute versus Cliff Swallow Foraging Group Size, and Mean Foraging Success

Foraging Bout	r_s	Mean Prey Capture Attempts per Minute per Bird[a]	Total Number of Birds Observed
1	.99	6.5	32
2	.35	4.0	29
3	−.89	5.1	33
4	−.80	4.9	45
5	−.83	4.0	119
6	.54	5.1	60
7	.70	5.1	60
8	−.80	3.5	32
9	−.90	2.7	65
10	−.20	2.2	55
11	−.60	2.7	67
12	.40	5.5	48
13	.90	6.7	62
14	−.06	2.3	85
15	.71	2.5	68
16	.30	2.5	76
17	.94	2.8	87
18	.71	2.3	124
19	.77	2.2	75
20	.83	3.1	64
21	.56	3.7	106
22	.92	3.5	69
23	.77	3.4	148
24	.75	4.4	41

Source: Brown 1988b.
[a] For all foragers in all groups.

Wilcoxon rank sum test, $p = .39$). The magnitude of the correlation coefficient did not vary significantly with overall mean foraging success for bouts with either positive ($r_s = .03$, $p = .59$, $N = 16$ bouts) or negative ($r_s = .64$, $p = .09$, $N = 8$ bouts) correlations. These results thus suggest that any effect of group size on foraging success is not necessarily related to overall abundance of prey during foraging periods.

10.9.2 Nestling Starvation Rates in relation to Colony Size

Another measure of potential resource depletion and competition for food among birds is how frequently nestlings starve. The cause of nestling mortality in most species is generally difficult to determine, except during un-

314 • Social Foraging 2

usually harsh conditions (e.g., cold weather) when starvation is relatively obvious (Hoogland and Sherman 1976; section 10.6). Under normal conditions, starvation is a safe interpretation only when part of a brood is lost. In these instances resources are presumably insufficient to sustain all the chicks, and typically the youngest or weakest succumb (e.g., Ricklefs 1965; Bryant 1978a,b; Lamey and Mock 1991). We examined the incidence of partial brood loss (terminology of Mock 1994) in cliff swallow colonies of different sizes as an index of nestling starvation rates.

Our sample of nests for this analysis included all those in which a known number of eggs hatched and at least one nestling remained alive ten days later (section 2.3.1). Any nest in which the number of nestlings at day 10 was less than that at hatching was scored as having partial brood loss. We did not include nests in which total brood mortality occurred. Some of these nests may have in fact represented cases where the entire brood starved, but whole brood starvation could often not be distinguished from predation. Predators such as bull snakes (section 8.2.2) always took an entire brood upon gaining entry to a nest. Only fumigated nests were used for this analysis. This removed confounding effects of ectoparasites, which also often killed nestlings (section 4.7), and therefore we were more likely to measure any effect of food shortage. However, we could not rule out the possibility that some partial brood loss was caused by intruding conspecifics that occasionally killed chicks in neighboring nests (section 5.6.1), and this should be kept in mind in evaluating the results in this section. We collected partial brood loss data from twenty-nine colonies distributed among all years from 1984 to 1989, and all were combined for analysis, so the usual caveat about potentially confounding seasonal effects applies.

Partial brood loss was strongly related to brood size. Among all nests combined, 20.4% of nests with original brood size two ($N = 465$ nests) suffered partial brood loss, as did 26.8% of those with original brood size three ($N = 1,506$); 42.8% of those with original brood size four ($N = 1,879$); and 66.0% of those with original brood size five ($N = 268$; $\chi^2 = 248.1$, df $= 3$, $p < .001$). This pattern is consistent with that expected if partial brood loss is a consequence of insufficient food.

The percentage of nests with partial brood loss increased with colony size for each brood size (fig. 10.15). The correlation approached statistical significance for broods of four and five. The pattern seemed strongest for broods of five (fig. 10.15), which is perhaps not surprising given the additional food resources necessary to sustain broods that large. Although large colonies generally seemed to have more partial brood loss than small ones, variation among small colonies was great, especially those with fewer than 100 nests (fig. 10.15).

These results were consistent with the hypothesis that increased nestling starvation rates reflected greater resource depletion and competition

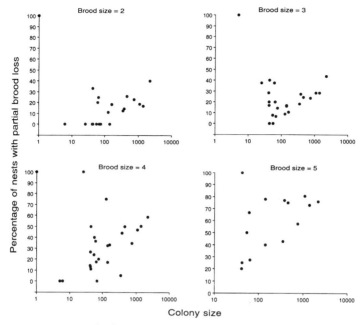

Fig. 10.15 Percentage of cliff swallow nests per colony that exhibited partial brood loss in relation to colony size, by brood size. Percentage of nests with partial brood loss increased, almost significantly, with colony size only for broods of four and five nestlings (brood size two: $r_s = .34$, $p = .11$, $N = 23$ colonies; three: $r_s = .03$, $p = .63$, $N = 24$; four: $r_s = .35$, $p = .071$, $N = 27$; five: $r_s = .45$, $p = .092$, $N = 15$).

for food in large colonies. The variation in apparent starvation rates among smaller colonies may have indicated site-specific differences in overall resources and the effectiveness of information transfer. Increased rates of partial brood loss in larger colonies represented another cost of cliff swallow coloniality. Apparently the advantages associated with information sharing were not great enough to enable many cliff swallows in larger colonies to raise broods of four and five nestlings without partial brood loss. Also note that these results (fig. 10.15) were used only as an index of potential nestling starvation, coming from fumigated nests. Cliff swallows in large colonies under natural conditions had to deal with both ectoparasites and the potential effects of food depletion.

10.10 BENEFITS OF INFORMATION TRANSFER

This chapter reports a variety of measures of foraging efficiency in cliff swallow colonies of different sizes. These included both behavioral measures (e.g., waiting intervals, food delivery rates, travel distances), which tended to be relatively direct, and physiological measures (e.g., body mass, starvation rates), which were more indirect. Different indexes of foraging success did not always yield consistent conclusions with respect to colony size variation. This could have been because of seasonal variation in food availability that either truly changed the expected payoffs in different-sized colonies between years or obscured any general pattern through, for example, unequal sampling on our part from year to year. Site-specific variation in food availability also may have muddled general patterns, although we tentatively concluded (section 10.7) that there were no obvious differences between large and small colonies in local food availability. Finally, some of the more indirect measures of foraging success (e.g., nestling body mass) might not have accurately reflected cliff swallows' foraging efficiency. For example, confounding variables, such as ectoparasitism (e.g., Moller 1994), may have influenced the birds' behavior such that parents in different-sized colonies allocated unequal proportions of their time or energy to foraging. Birds in large colonies may have responded to greater expected ectoparasitism by working harder to find food or investing proportionally less in self-maintenance and more in provisioning nestlings. The potential foraging-related advantages of coloniality are clearly complex and challenging to address empirically, and thus it is not surprising that we do not understand their effects on the evolution of avian coloniality (Wittenberger and Hunt 1985; Richner and Heeb 1995).

Despite these problems, however, several general patterns emerged. The behavioral measures of foraging efficiency revealed that cliff swallows in larger colonies harvested more food per unit time (sections 10.2, 10.3), likely because they had more opportunities to gain information and wasted less time (section 10.1), and this benefit persisted despite travel distances and search areas that also increased with colony size (section 10.8). That cliff swallows foraged more efficiently in larger colonies at most times of the year (but perhaps not during the 1992 cold spell; section 10.6) was relatively clear. But it was not clear how this increased foraging efficiency affected the fitness of birds in different-sized colonies.

How foraging behavior is directly translated into individual fitness is a general problem afflicting foraging theory. Most studies make the reasonable assumption that an individual's foraging decisions are eventually reflected in its probability of survival and its reproductive success, although there are few relevant empirical data from natural populations on this point (Morse and Fritz 1987; Nur 1987a). In cliff swallows, foraging behavior seems to most directly affect body mass of nestlings and adults. Body mass, which presumably reflects overall physical condition, in turn influences an individual's probability of survival (section 12.5) and thus its realized lifetime reproductive success. However, using body mass—even in fumigated colonies—as a direct index of foraging efficiency and condition is confounded by the apparent costs of storing fat and thus of maintaining high body mass (Witter and Cuthill 1993) and by the effects on adult birds of ectoparasites such as fleas and lice (Senar et al. 1994; Brown, Brown, and Rannala 1995) that were unaffected by nest fumigation.

Nestling body mass increased with colony size for broods of one to three nestlings in fumigated nests but declined with colony size for broods of four (section 10.4.2). Colony size had no net effect on nestling body mass in nonfumigated colonies under natural conditions, however, with nestlings from colonies of all sizes weighing about the same. Given the severe effects of ectoparasites in large colonies (section 4.7), enhanced foraging efficiency represents enough of a benefit to compensate birds in large colonies for the cost of ectoparasitism, at least in terms of mean nestling body mass. The results suggest no *net* advantage for birds in larger colonies, but this does not mean that they receive no foraging-related benefits. Information transfer and resulting increases in foraging success presumably allow cliff swallows to live in larger colonies and suffer no net cost relative to lightly ectoparasitized birds in small colonies.

Were there any net foraging-related advantages of cliff swallow coloniality? The data on adult body mass, taken largely from nonfumigated colonies exposed to natural levels of ectoparasites, suggested there might be net benefits in some years and at intermediate colony sizes. In years when adults weighed more in large colonies at the end of the season (section 10.5.2) or lost less mass in certain colony sizes (section 10.5.3), there probably was a net benefit of coloniality (see section 12.5.3 on survival-related advantages of larger body mass). Yet it seems doubtful that this effect alone was strong enough or frequent enough to compensate the birds fully for the substantial costs of coloniality they incurred (section 14.1.1). We are thus left with the conclusion that cliff swallow foraging efficiency increased with colony size, but how far this translated into a net benefit of coloniality was unclear. The principal foraging-related advantage may simply be that birds can nest in larger colonies at no net cost. Perhaps this result is a key to understanding the colony size variation seen in cliff swallows (section 3.5.5). But before addressing colony size variation (chapters 13, 14), we will more rigorously investigate whether there might be net fitness differences among birds in different-sized colonies, using data on annual reproductive success (section 11.3) and annual survivorship (section 12.4.2).

10.11 THE GEOMETRICAL MODEL AS AN EXPLANATION FOR COLONIALITY

Before we leave the possible food-related advantages of cliff swallow coloniality, Horn's (1968) "geometrical" model of the evolution of coloniality warrants some consideration. Wittenberger and Dollinger (1984) specifically suggested cliff swallows as one species in which the model might explain colony formation. We (Brown, Brown, and Ives 1992) have recently addressed Horn's model in detail, and here we summarize only the critical points.

Horn (1968) showed that when food is spatiotemporally variable within a given foraging arena, individuals may minimize their travel distances from the nest site to food by locating their nest in the geometric center of the foraging arena. Under such conditions, dispersed nesting requires a 31% increase in average distance traveled to find food. A colony forms as each individual attempts to locate its nest in the center of the arena, and coloniality may result even in the absence of any direct benefits of conspecific association (such as predator avoidance or information sharing). Although this is a popular explanation for the evolution of coloniality (see most recently Barta 1992) and often cited as such (Brown, Brown, and Ives 1992), there exist virtually no data for any species to evaluate either the model's assumptions or its predictions.

Cliff swallows fed on spatiotemporally variable prey (section 9.4), a necessary prerequisite for Horn's hypothesis to apply, but they did not meet the other assumptions inherent in the geometrical model. For example, the birds did not use externally bounded foraging arenas where topographic features or other habitat-related factors constrained foraging to a given area. Consequently the advantage of using a centrally located nest site disappears (Waser and Wiley 1980; Wittenberger 1981; Wittenberger and Hunt 1985). If food is found over a wide, continuous area, the distances of foraging trips are limited only by how far the birds can fly, and any one nest location is as good as another. Without an external boundary, birds may form smaller separate colonies or nest solitarily at regular intervals throughout the habitat. In all cases, cliff swallow colonies were in broadly uniform foraging habitat with other identical habitat that appeared suitable for foraging abutting the observed foraging arena on one or more sides (Brown, Brown, and Ives 1992). Cliff swallow foraging arenas seemed to be bounded only by how far the birds were willing to fly to find food.

Horn's model also implicitly assumes that individuals can choose to nest either dispersed uniformly throughout the foraging arena or together at a central site. If nesting sites are restricted within the observed foraging arena, however, individuals are constrained to nest in particular places that may or may not be the site that minimizes travel distance. Cliff swallows did not meet this assumption. Although cliff swallow coloniality did not seem to be a direct result of limited nesting sites when the study area as a whole was considered (section 7.4), in all cases potential nesting sites were restricted when we considered their availability only within a given foraging arena. Colonies generally had fewer than three alternative nesting sites available within their foraging arena (Brown, Brown, and Ives 1992).

At most sites cliff swallows fed centrally with respect to the colony (e.g., fig. 9.2), but this most likely followed from central-place foraging considerations (Smith 1968; Hamilton and Watt 1970; Orians 1971; Covich 1976). Individuals, whether colonial or solitary, that regularly return to a fixed nest site and exploit spatiotemporally variable prey should center their searching on the nest, minimizing the energy costs of foraging. Cliff swallows seemed to forage centrally with respect to their nest site rather than nest centrally with respect to their foraging sites (Brown, Brown, and Ives 1992). For these reasons, a potential reduction of travel distances sensu the geometrical model cannot represent an important foraging-related cause of coloniality in cliff swallows.

10.12 SUMMARY

Knowing how the costs and benefits of social foraging vary with colony size is essential in evaluating the importance of information transfer in the evolution of coloniality. As colony size increased, cliff swallows had more foraging information available to them and spent less time waiting to follow successful foragers. The frequency with which birds delivered food to their nestlings peaked at intermediate colony sizes. The amount of food delivered per trip increased with colony size; parents in the largest colonies delivered 0.4–0.5 g more food per trip, on average, than parents in the smallest colonies.

We used nestling body mass as an index of parents' foraging success and measured body mass for all birds at 10 days of age during the time of maximum weight gain. Among nonfumigated nests, nestling body mass did not vary significantly with colony size. When the effects of ectoparasites were removed by fumigating nests, nestling body mass increased with colony size for broods of one to three nestlings and decreased with colony size for broods of four.

We used adult body mass as another index of foraging efficiency. Adult body mass, especially among males, tended to decline with colony size early in the season when birds were building nests and laying eggs. There was no evidence that heavier birds consistently settled in larger colonies at the start of the season. While birds were feeding nestlings later in the season, adult body mass increased with colony size in three years, decreased with colony size in three years, and showed no relationship in three years. Residents of small colonies lost the most mass during the season, on average, and seasonal mass reductions were least for birds in colonies of intermediate size. Several kinds of data suggested that cliff swallows may have foraged most efficiently at medium-sized colonies of between 100 and 1,000 nests.

During a 1988 cold spell, mortality was highest in small colonies, and there was no evidence for any mortality at large colonies. Birds in small colonies also lost over twice as much body mass on average during the cold spell as did birds in large colonies. During a 1992 cold spell, however, cliff swallows living in the single largest colony suffered the highest mortality, illustrating how the potential benefits of social foraging did not consistently apply across years. An experiment in which we reduced the size of some cliff swallow colonies suggested no major differences in local resource availability among colony sites and suggested that enhanced foraging efficiency at larger colonies most likely stemmed from information transfer.

A foraging-related cost of coloniality was an increase in mean travel distance with colony size. This translated into greater energy costs of commuting for cliff swallows in large colonies. Total area over which a colony's residents fed increased with colony size, probably reflecting resource depletion in the larger colonies. Foraging group size did not affect observed rates of prey capture, but partial brood loss (an index of nestling starvation) increased with colony size, especially for broods of four and five.

The different measures of foraging efficiency often yielded contradictory conclusions and differed among years. Cliff swallows' foraging efficiency, by all behavioral measures, increased with colony size, but how this efficiency affected the fitness of birds in different-sized colonies was unclear. The principal advantage of information transfer may be that it improves the birds' foraging efficiency enough to compensate them for the increased cost of ectoparasitism in larger colonies and allows them to occupy a range of colony sizes. Cliff swallow coloniality does not seem to be a direct result of birds' nesting in the center of their foraging arena to minimize their travel distances (sensu Horn 1968).

11 Reproductive Success

> Where many cliff swallows nest together, the reproductive cycle proceeds as if one organism, rather than many, were laying its eggs, hatching them, and fledging its young.
> George Miksch Sutton (1986)

The costs and benefits of coloniality in cliff swallows described in the previous chapters ultimately expressed themselves in the birds' reproductive success. Each cost or benefit presumably influenced the number of offspring produced annually and over a lifetime by individuals in different-sized colonies. In some cases we could directly measure a potential cost or benefit in terms of its effect on annual reproductive success (e.g., sections 4.7, 5.6.3), but generally these costs or benefits could not be expressed in explicit net increments of reproductive success. For example, increased travel distance in larger cliff swallow colonies was a clear cost of coloniality (section 10.8), but how this cost per se influenced the number of young raised was unknown. In this chapter we focus on patterns of annual reproductive success, with the goal of using these patterns, in section 14.2, to better understand the evolution of cliff swallow coloniality.

11.1 BACKGROUND

Central to any understanding of why animals form colonies is knowing how individual reproductive success varies with colony size. Four basic patterns have been observed (Brown, Stutchbury, and Walsh 1990): per capita reproductive success may consistently increase, consistently decrease, remain constant with colony size, or it may peak at intermediate colony sizes. In the first two cases, the implications of the patterns are relatively clear. A consistent increase in reproductive success as colonies get larger means that the birds gain important net benefits (e.g., predator avoidance, enhanced food finding) from nesting together, and these benefits have probably led to coloniality. A consistent decrease in reproductive success as colonies get larger means that colony residents receive no net benefits from living together and are probably forced into colonies by limited nesting habitat. When reproductive success does not change with colony size, however, it is not obvious what sort of advantages (if any) individuals gain from colonial nesting or why they form colonies at all. Patterns in which reproductive success peaks at an intermediate colony size are easily understood, conceptually, in terms of optimal group size (e.g., Pulliam and Caraco 1984; Giraldeau 1988; Packer, Scheel, and Pusey 1990; Higashi and Yamamura 1993; Rannala and Brown 1994) but are often difficult to detect empirically (Brown, Stutchbury, and Walsh 1990).

Surprisingly few good data are available to evaluate how individual reproductive success varies with colony size in general. Most field studies have measured reproductive success in six colonies or fewer, which generally represents an inadequate range of colony sizes. A large sample of colonies of different sizes is necessary to detect any optimal peak in reproductive success at intermediate sizes. With these caveats in mind, the available evidence suggests that the two most common patterns are an increase in reproductive success with colony size or nest density (e.g., Darling 1938; Fisher and Vevers 1944; Orians 1961; Ahlen and Andersson 1970; Robertson 1973; Morris and Hunter 1976; Birkhead 1977; Veen 1977; Nelson 1978; Haas 1985; Robinson 1985; Robertson 1986; Burger and Gochfeld 1990, 1991; Wiklund and Andersson 1994) and no relation between reproductive success and colony size or nest density (e.g., Smith 1943; Patterson 1965; Dexheimer and Southern 1974; Hoogland and Sherman 1976; Snapp 1976; Knopf 1979; Smith 1982; Haas 1985; Emms and Verbeek 1989; Snyder, Beissinger, and Chandler 1989; Reville 1991; Holloway 1993; Jehl 1994; Lessells, Avery, and Krebs 1994; Post 1994; Wolff 1994). Other studies have shown a decrease in reproductive success with colony size (Burger 1974b; Pienkowski and Evans 1982a,b; Hunt, Eppley, and Schneider 1986; Van Vessem and Draulans 1986; Shields and Crook 1987; Wickler and Seibt 1993) or nest density (Bengtson 1972; Davis and Dunn 1976; Siegel-Causey and Hunt 1986; Kilpi 1989; Reville 1991). Highest success at an intermediate colony size (Veen 1977; Wiklund and Andersson 1980; Wiklund 1982; Gotmark and Andersson 1984; Spiller 1992) or nest density (Newton and Campbell 1975; Parsons 1976; Scolaro 1990) may occur in a relatively few species. The most obvious conclusion based on these studies is that there is no single pattern in how individual reproductive success varies with colony size across species.

Reproductive success is often measured as the number of offspring reared in a given year. Such annual indexes are relatively easy to obtain, because they do not require following the histories of known individuals for multiple years. All the work reviewed above used annual measures of reproductive success, representing a cross-sectional view of the population in a given season. But lifetime reproductive success is a more realistic expression of individual fitness (Birkhead 1985; Clutton-Brock

1988a; D. Brown 1988; Partridge 1989; Barrowclough and Rockwell 1993; section 14.2), taking into account both yearly and age-specific variation in annual reproductive success and measures of individual survivorship. Survival rates generally heavily influence the lifetime reproductive success of small birds such as cliff swallows (section 12.1). For example, if survival rates vary consistently with colony size, so could lifetime reproductive success, even when annual reproductive success does not. Virtually nothing is known for any species about how lifetime reproductive success changes with colony size (section 1.3.3). However, we defer consideration of this issue in cliff swallows (section 14.2) until we present information on survival rates (chapter 12). In this chapter we describe patterns of annual reproductive success, focusing especially on how it varies with colony size. We also examine other correlates of reproductive success that may explain reproductive variation within and between cliff swallow colonies. Our objective is to use the observed patterns to understand more fully the costs and benefits of coloniality described in earlier chapters.

11.2 VARIATION IN CLUTCH SIZE

The first step in assessing the effect of colony size on cliff swallows' reproductive success is determining how clutch size varies with colony size. Clutch size, a measure of fecundity among individuals, obviously directly affects how many offspring an individual may eventually rear. Clutch size also may reflect the condition, age, and overall phenotypic quality of an individual (at least of females), because in many birds clutch size tends to increase with age (Klomp 1970; Perrins and Moss 1974; Coulson and Horobin 1976; Brown 1978b; De Steven 1978; Finney and Cooke 1978; Ryder 1980; Nisbet, Winchell, and Heise 1984; Saether 1990; Sjoberg 1994) and energy reserves (Ryder 1970; Jones and Ward 1976; Murton and Westwood 1977; Fogden and Fogden 1979; Drent and Daan 1980; Houston, Jones, and Sibly 1983; Murphy and Haukioja 1986; Martin 1987). Therefore patterns of clutch size variation may provide clues to what subsets of individuals occupy particular nests or colonies (Coulson and Porter 1985). For example, if clutch size varies consistently with colony size, it could mean that individuals sort among colonies based on their age or condition. It could also mean that subsequent variation in reproductive success among colonies could be attributable to the initial settlement patterns and not necessarily to any colony size effects per se. Little information is available in general on how clutch size varies with breeding colony size in birds.

Cliff swallows in our study area exhibited relatively little variation in clutch size (fig. 11.1). Among all nests from all colonies for which we knew clutch size ($N = 8,094$), the range was one to fifteen eggs. Approximately 80% of all clutches were three or four eggs, with relatively few

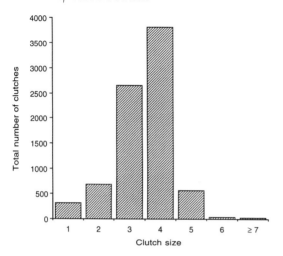

Fig. 11.1 Frequency distribution of cliff swallow clutches of various sizes, combined across all years and colonies.

clutches of other sizes (fig. 11.1). Probably all clutches greater than six eggs ($N = 27$) represented instances in which a nest had been heavily brood parasitized (section 6.3; Brown and Brown 1989). This rather limited variation in clutch size within the population should be kept in mind in evaluating the importance of clutch size differences reported later in this section.

Unless noted, the analyses in section 11.2 use clutch size data only from nonfumigated cliff swallow colonies. With one exception (nest size in section 11.2.3), a similar pattern was seen for fumigated colonies, suggesting little impact of ectoparasites on the clutch sizes produced by cliff swallows. Although conceivably the birds could adjust their output of eggs in response to ectoparasite loads (which the birds apparently can predict; Brown and Brown 1991), we detected no evidence of clutch size adjustment to parasites. For example, among all years mean clutch size per colony for nonfumigated sites ($N = 52$ colonies) was 3.41 ($\pm$ 0.06) eggs, versus 3.49 ($\pm$ 0.05) eggs for fumigated sites ($N = 27$); the difference was not significant (Wilcoxon rank sum test, $p = .21$). We may not have detected an effect of ectoparasites anyway, because often our fumigation did not begin at a site until the birds were about to lay or had already begun laying. We probably could have legitimately combined nonfumigated and fumigated colonies for clutch size analyses, although we did not because our sample sizes for nonfumigated sites were relatively large. Clutch size was defined as the maximum number of eggs appearing in a nest, which unavoidably included brood parasitic eggs in some cases (section 6.3).

11.2.1 Clutch Size in relation to Year and Colony Size

Before examining how clutch size of cliff swallows varied with colony size, we looked for seasonal differences in the number of eggs laid. We

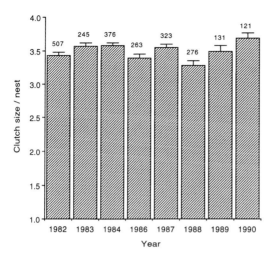

Fig. 11.2 Cliff swallow clutch size per nest in relation to year. Data from all colonies each year were combined. Sample sizes (number of nests) are shown above error bars. Clutch size did not vary significantly among years (Kruskal-Wallis ANOVA, $p = .10$).

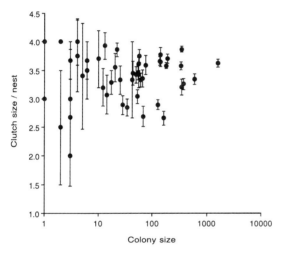

Fig. 11.3 Clutch size per nest in relation to cliff swallow colony size, 1982–90, nonfumigated sites only. Clutch size did not vary significantly with colony size ($r_s = -.09$, $p = .50$, $N = 52$ colonies). Sample size (number of nests) for each colony is the same as in figure 6.9.

knew clutch sizes for 2,242 nonfumigated nests, 1982–90 (except 1985). The nests from all colonies were pooled to examine potential yearly effects (fig. 11.2). There was no significant variation among years in average clutch size. Thus there was no evidence that resource availability, as reflected in females' abilities to produce eggs, varied appreciably between years within the study area over the eight years. These results (fig. 11.2) also allowed us to disregard yearly effects in subsequent analyses of clutch size.

Average clutch size per nest was known for fifty-two colonies ranging from 1 to 1,600 nests. Clutch size did not vary significantly with colony size (fig. 11.3). There appeared to be slightly greater between-colony variation in average clutch size among the colonies of fewer than 10 nests, although whether this pattern was real or simply reflected the unavoidably smaller sample sizes in the small colonies was unclear. If clutch size reflected phenotypic attributes of females or food availability among sites, these data suggest little sorting of individuals among colonies (but see other evidence for sorting in section 13.3) and support earlier claims (section 10.7; Brown 1988a) that local food resources do not differ greatly among colonies, at least early in the season when females are making eggs. Any subsequent variation in reproductive success among colonies (section 11.3.2) is therefore not attributable simply to birds' laying different numbers of eggs in colonies of different sizes.

11.2.2 Clutch Size in relation to Date

Time of year affects clutch size in a variety of species (e.g., Klomp 1970; Perrins and Moss 1974; Murton and Westwood 1977; Stutchbury and Robertson 1988; Steeger and Ydenberg 1993; Rowe, Ludwig, and Schluter 1994; Sjoberg 1994), usually leading to a decline in the number of eggs produced as the breeding season progresses. This pattern presumably reflects both decreased resources for making eggs late in the season and generally later breeding by younger birds. The date a clutch was initiated (first egg was laid) was used as our measure of time of year in these analyses. We assessed the effect of date on clutch size for birds in the study area as a whole and within each colony. To examine the broad seasonal pattern across colonies, we pooled nests from all colonies (because colony size had no effect on clutch size; section 11.2.1), lumped nests into arbitrary five-day periods, and determined the average clutch size among all nests for each period. For the within-colony analysis, we calculated Spearman rank correlation coefficients for clutch size as a function of date at each colony (see section 4.4).

Laying by cliff swallows in our sample of nonfumigated nests spanned 11 May to 28 July. (Laying in fumigated nests was detected from 4 May to 23 July.) When all nests were considered, clutch size declined significantly with date (fig. 11.4). There was a reduction of about 1.0 egg in average clutch size between the time of peak egg laying (about 25 May to 5 June) and the time when most laying ceased (after 15 July). Since yearling females laid slightly smaller clutches than older females (section 11.2.4), this seasonal decline in clutch size (fig. 11.4) perhaps could be attributed in part to later breeding by yearlings (Stutchbury and Robertson 1988). There was little evidence, however, based on capture times of known aged birds, that yearling females consistently arrived later than older females (section 3.5.3). This suggested that the seasonal decrease in clutch size could probably not be explained solely by settlement dates of yearlings and may also have resulted from diminishing food resources later in the summer (Bryant 1975; Hussell and Quinney 1987; Sydeman

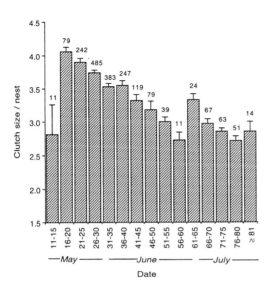

Fig. 11.4 Clutch size per cliff swallow nest in relation to date egg laying was initiated in nonfumigated colonies (01 = 1 May). Nests from all years and colonies were combined into arbitrary five-day periods, and sample sizes (number of nests) for each period are shown above error bars. Clutch size declined significantly with date interval ($r_s = -.62$, $p = .013$, $N = 15$ intervals).

et al. 1991), combined with the birds' gradual loss of body mass and energy reserves as the season progressed (section 10.5.3).

Reproduction within each colony occurred during a shorter span than the range of dates represented in figure 11.4. For example, the high degree of breeding synchrony in cliff swallow colonies (section 8.6.1) meant that all clutches at most sites were initiated over a period of twenty-one days or less. Therefore we examined whether date affected clutch size over these shorter periods within colonies. A series of correlation analyses for thirty-two colonies ranging from 5 to 1,600 nests showed that date also strongly affected clutch size within a single colony (table 11.1). In thirty of thirty-two colonies, clutch size declined with date, a significant preponderance of negative correlations (binomial test, $p < .001$). The correlation coefficients were relatively large for many of the sites, and twelve were significant or nearly so (table 11.1). Clutch size declined with date in colonies of all sizes. This same result was seen in separate analyses for both old and new nests. Declining seasonal resource availability seems less likely for these within-colony results, given the shorter times involved. Smaller clutches by later-settling birds within a colony may instead reflect competition for nest sites (section 5.2), with younger, less experienced females (which lay fewer eggs; section 11.2.4) taking longer to secure sites and initiate nesting.

11.2.3 Clutch Size in relation to a Nest's Spatial Position, Age, and Size

Cliff swallows often competed for nests based on the nests' spatial positions and ages (section 5.2.3), and there were differences among these

Table 11.1 Spearman Rank Correlation Coefficients between Clutch Size and Date Clutch Was Initiated within Each Cliff Swallow Colony

Colony Site	Colony Size	r_s	p	N
8429	5	−.67	.22	5
8709	10	.21	.28	28
8228	12	−.65	.16	6
8327	13	−.27	.46	10
8825	14	−.26	.37	14
8324	17	−.07	.77	17
8831	20	−.15	.58	16
8931	22	−.58	.008	20
8329	25	−.51	.019	21
8207	50	−.02	.93	25
8934	52	−.28	.060	47
8832	54	−.19	.25	36
8226	56	−.34	.023	44
8434	56	−.10	.69	17
8932	57	−.53	<.001	54
8203	60	−.13	.32	56
8824	66	−.30	.013	66
8841	68	−.23	.27	25
8421	75	−.10	.77	12
8405	125	−.01	.94	87
8641	137	−.05	.58	137
8630	140	−.16	.36	34
8732	140	−.51	<.001	126
8642	163	−.11	.50	42
8305	180	−.44	<.001	187
9030	190	.02	.87	58
8730	340	−.34	<.001	144
8430	345	−.34	<.001	223
8206	350	−.39	.003	55
8830	375	−.13	.29	71
8202	600	−.13	.17	120
8201	1,600	−.37	<.001	96

Note: N = number of nests.

nests in expectation of ectoparasitism (section 4.5.1). We examined how clutch size varied with a nest's spatial position within the colony and with nest age, both in preparation for later analyses of how reproductive success was affected by these variables (sections 11.5, 11.6) and to gain insight into the possible costs associated with occupying different nests within a colony. For example, reduced clutch sizes for birds in certain classes of nests may mean either that certain subsets of individuals (e.g., young ones) are forced into these nests or that competition for them is so energetically stressful that the birds able to occupy them must reduce egg production.

As in section 4.5, we used a nest's linear distance from the colony's center, and its nearest neighbor distance, as measures of spatial position within the colony. We also analyzed clutch size in the center and edge nests at each colony site (section 4.5). Neither a nest's distance from the colony's center nor its nearest neighbor distance had a significant effect on clutch size (table 11.2). A series of within-colony correlations showed no significant preponderance of positive or negative coefficients for either distance from the center (19 of 30 negative; binomial test, $p = .20$) or nearest neighbor distance (23 of 38 negative; binomial test, $p = .34$; table 11.2). The number of eggs females laid thus did not seem to vary consistently with where they settled in the colony, at least as measured by distance from the center and nearest neighbor distance.

In comparing center and edge nests, we used only colonies where we had clutch size data for both classes of nests (e.g., section 4.5.1). The number of nests in each category per colony varied from 13 to 42, depending on the geometry of the sites. Considering first all nests irrespective of nest age, cliff swallows occupying center nests averaged 0.24 more eggs laid per clutch per colony than did birds in edge nests; this difference approached statistical significance (fig. 11.5). This result suggests both that center nests are occupied earlier than edge nests (when birds laid more eggs), a conclusion in line with the observations of nest site competition (section 5.2.3), and that competition for center nests is not costly enough to reduce clutch sizes among their occupants relative to edge birds. This center-edge comparison (fig. 11.5) was the only evidence that spatial position within a colony had any effect on cliff swallow clutch sizes (and see section 11.8).

Clutch size seemed not to vary much with nest age. Among nests from all colonies, birds occupying old nests ($N = 1,056$) laid a mean 3.52 (± 0.03) eggs, versus 3.41 (± 0.03) eggs for birds in new nests ($N = 751$). The difference was significant (Wilcoxon rank sum test, $p = .005$) although so slight it is probably meaningless. We also examined clutch size in old and new nests both in the center and on the edge of colonies (fig. 11.5). Old nests showed the same center-edge difference as all nests combined: center old nests averaged 0.20 more eggs per clutch per colony than edge old nests. Among new nests, however, center nests averaged 0.29 fewer eggs per clutch per colony than edge nests, although the difference was not significant (fig. 11.5).

Finally, we evaluated how nest size potentially affected clutch size. Nest size (section 2.8.2) is a relative measure of the cost of building a nest. Larger nests are presumably more costly to build and thus might affect the energy reserves a female could use for egg production. We combined all nonfumigated nests from all colonies and found a significant reduction in clutch size as nest size increased (fig. 11.6a). This supports the assumption

Table 11.2 Spearman Rank Correlation Coefficients between Clutch Size and a Nest's Distance from the Colony's Center, and Nest's Nearest Neighbor Distance, within Each Cliff Swallow Colony

Colony Site	Colony Size	Distance from Center			Nearest Neighbor Distance		
		r_s	p	N	r_s	p	N
8227	4	—	—	—	−.83	.17	4
8429	5	—	—	—	−.53	.36	5
8827	6	—	—	—	.06	.91	6
8628	10	—	—	—	−.44	.20	10
8709	10	−.41	.031	28	.54	.003	28
8228	12	—	—	—	.13	.73	9
8327	13	−.43	.21	10	.67	.008	14
8825	14	—	—	—	.44	.12	14
8324	17	−.26	.32	17	−.06	.81	17
8831	20	—	—	—	−.03	.90	20
8931	22	.38	.16	15	−.28	.23	21
8329	25	.51	.017	21	.12	.60	21
8627	34	.23	.19	34	.11	.52	34
9032	42	−.40	.60	4	−.12	.83	6
8431	43	−.38	.25	11	−.27	.42	11
8207	50	−.11	.53	33	−.24	.17	33
8934	52	.07	.61	51	−.01	.92	51
8832	54	.04	.76	48	−.02	.91	51
8226	56	.06	.68	48	−.09	.54	48
8434	56	−.21	.37	20	−.22	.30	23
8932	57	−.33	.015	54	.09	.52	57
8203	60	—	—	—	.34	.031	40
8824	66	.01	.96	66	.05	.70	66
8841	68	−.01	.99	11	−.09	.64	30
8421	75	−.50	.039	17	−.63	.007	17
8405	125	−.12	.27	87	−.16	.13	87
8641	137	−.05	.59	137	.01	.92	136
8630	140	−.11	.54	35	−.01	.56	35
8732	140	.21	.010	144	.02	.77	145
8642	163	−.21	.15	47	−.09	.54	46
8305	180	−.01	.93	186	−.05	.49	186
9030	190	.04	.64	115	.01	.89	109
8730	340	.05	.53	147	−.17	.036	147
8430	345	−.13	.057	217	.08	.22	223
8206	350	−.26	.061	53	−.03	.81	53
8830	375	−.05	.66	75	−.01	.91	75
8202	600	−.01	.89	139	−.17	.044	145
8201	1,600	.16	.070	125	.16	.079	125

Note: Missing data are cases where we could not calculate nests' distances from center owing to colony size or geometry. N = number of nests.

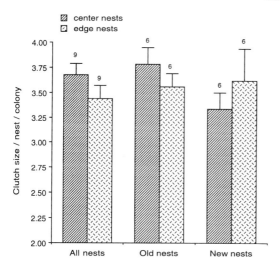

Fig. 11.5 Average clutch size per nest per colony for center and edge nests of different nest ages in nonfumigated colonies. Values shown are means of colony averages, and the numbers of colonies compared are shown above error bars. The difference in clutch size between center and edge nests approached significance for all nests combined (Wilcoxon matched-pairs signed-rank test, $p = .09$), but center and edge nests did not differ significantly among old or new nests separately (Wilcoxon matched-pairs signed-rank tests, $p \geq .12$ for each).

that nest building is costly and thus any benefits gained from sharing walls with neighbors (section 5.3.2) could be important in allowing a female to produce more eggs. Of course, some of the birds occupying the nests in this analysis (fig. 11.6a) had not constructed these nests themselves; that is, they had used existing old nests. The best analysis would be to use only new nests—nests built primarily by the birds currently occupying them—in evaluating the effect of nest size on clutch size. Unfortunately, we did not have sufficient new nests over the same range of nest sizes to permit an analysis exclusively of new nests. Nevertheless, the results (fig. 11.6a) indicate that, for whatever reasons, clutch size was reduced for cliff swallows using larger nonfumigated nests.

Surprisingly, we found the opposite pattern for fumigated nests (fig. 11.6b). Clutch size increased significantly with nest size for fumigated nests, meaning that birds in fumigated nests paid no apparent cost to construct or occupy larger nests. We can offer no reason why ectoparasitism should have caused these opposite patterns and in particular why fumigated nests exhibited an increase in eggs laid with nest size. Hole-nesting birds are reported to sometimes lay more eggs as cavity size increases (Nilsson 1975; Karlsson and Nilsson 1977; reviewed in Rendell and Robertson 1993). Cliff swallows might respond in similar ways, but why only birds in fumigated colonies would do so is not clear. A similar inconsistency in how nest size affected clutch size was reported for the barn swallow in Denmark (Moller 1982), in which positive or negative correlations were seen during different times of the year. No relation between nest size and clutch size was found for barn swallows in Michigan (Good-

332 • *Reproductive Success*

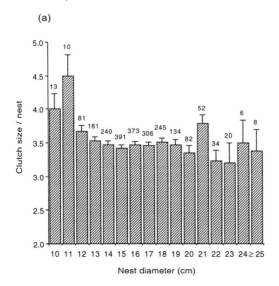

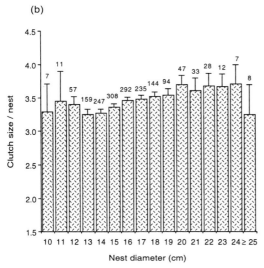

Fig. 11.6 Clutch size per nest in relation to cliff swallow nest size as measured by its diameter (cm) in nonfumigated (*a*) and fumigated colonies (*b*). Data from all colonies and years were combined. Sample sizes (number of nests) are shown above error bars. Clutch size declined significantly with nest size among nonfumigated colonies ($r_s = -.63$, $p = .009$, $N = 16$ nest sizes) but increased significantly with nest size among fumigated colonies ($r_s = .58$, $p = .019$, $N = 16$ nest sizes).

man 1982) or for eastern phoebes (another mud-nest building species) in Ontario (Conrad and Robertson 1993).

11.2.4 Clutch Size in relation to Female Age and Body Mass

Thus far we have assumed that clutch size reflects, to some degree, a female cliff swallow's age and energy reserves. Here we directly evaluate what effect age and body mass (the latter presumably a measure of individual condition) had on cliff swallow clutch size.

For our analysis of how age affected clutch size, we used data from 122 females banded as nestlings and subsequently found breeding in the study area, for which we knew clutch size. We used birds from both nonfumigated and fumigated sites in this analysis to increase our sample size, and we pooled colonies (legitimate because clutch size was unaffected by fumigation and did not vary with colony size; section 11.2.1). Clutch size was 3.34 ($\pm$ 0.09) eggs for females aged one year ($N = 67$); 3.67 ($\pm$ 0.16) for those aged two years ($N = 33$); 3.69 ($\pm$ 0.17) for those aged three years ($N = 13$); and 3.44 ($\pm$ 0.34) for females aged four years or more ($N = 9$). These differences were significant (Kruskal-Wallis ANOVA, $p = .020$). As might be expected based on the patterns in other species, yearling cliff swallows laid slightly smaller clutches than older birds (also see section 11.8).

In analyzing how body mass of females affected clutch size, we used all females that were captured and weighed during the "early" period (section 2.3.2), corresponding to the time of nest building and egg laying, for which we knew their subsequent clutch size. Thus these body masses reflected the birds' condition immediately before and during the times they were producing eggs. Our sample size was 180 birds distributed among seven years. We pooled these for a single analysis because yearly sample sizes were not sufficient to separate the data by year, although the trend in each year separately matched the overall results. Too few of these birds were of known age for us to analyze how body mass affected clutch size within age classes.

Body mass for female cliff swallows laying one-egg clutches ($N = 2$) was 22.50 g ($\pm$ 0.50); for two-egg clutches ($N = 8$), 24.19 g ($\pm$ 0.48); for three-egg clutches ($N = 50$), 24.65 g ($\pm$ 0.26); for four-egg clutches ($N = 102$), 24.05 g ($\pm$ 0.19); for five-egg clutches ($N = 16$), 24.66 g ($\pm$ 0.49); and for six-egg clutches ($N = 2$), 27.25 g ($\pm$ 0.75). Body mass varied almost significantly with clutch size (ANOVA, $p = .060$), and the mean body mass values increased significantly with clutch size ($r_s = .83$, $p = .042$, $N = 6$). These data generally support our assumption that clutch size reflected to some degree a female's energy reserves and overall condition. Heavy females—the ones with the greatest energy reserves—laid slightly larger clutches. Therefore any increase in females' foraging efficiency early in the season (e.g., through information sharing; sections 9.6, 10.6) may let them produce more eggs.

11.3 EFFECTS OF YEAR AND COLONY SIZE

How per capita annual reproductive success varies with colony size is central to determining the net fitness effect of colonial nesting. Our measure of annual reproductive success throughout this chapter is the number of nestling cliff swallows surviving in each nest to 10 days of age. All nests were checked for the presence of nestlings 10 days after hatching (section 2.3.1), unless the nest had lost all its eggs before hatching. Nests failing during incubation were scored as having zero young surviving and used in calculating the mean number of young surviving per nesting attempt. Conceivably some additional nestling mortality occurred after day 10 (when it was difficult to check cliff swallow nests without causing premature fledging) and went undetected, but survival to day 10 served as a useful relative measure of reproductive success among colonies or classes of nests. In the following analyses we used data from fumigated nests where possible to assess the net effect of ectoparasitism on cliff swallows' reproductive success. Fumigation (section 4.7) allowed us to remove ectoparasitism as an influence on the survival of nestlings and thus investigate other possible correlates of reproductive success.

11.3.1 Yearly Variation in Reproductive Success

Before evaluating the potential effects of colony size, we examined how cliff swallows' reproductive success varied among years. If, for example, seasonal patterns of resource availability affect the number of nestlings reared in each year, comparisons among colonies must be done for each year separately or only for combinations of years that do not differ significantly (e.g., sections 10.4, 10.5).

We combined all nests from all colonies and calculated average reproductive success in each year. Overall reproductive success varied significantly among years for both nonfumigated (fig. 11.7a) and fumigated nests (fig. 11.7b). Yearly variation was more pronounced among nonfumigated nests, suggesting seasonal variation in either absolute ectoparasite load or how resource availability interacted with ectoparasite load to affect reproductive success. Not all years differed significantly from each other, however. We performed Scheffe's tests to determine which of these yearly differences (fig. 11.7) were significant. Among nonfumigated nests, 1982 and 1990 were statistically indistinguishable, as was the set of 1984/1986/1987, and also the set of 1988/1989. Among fumigated nests, 1984, 1986, and 1987 did not differ significantly, nor did 1985, 1988, and 1989.

The yearly differences, for nonfumigated nests (fig. 11.7a), appeared to correlate generally with climate. Comparing these results with seasonal temperature (table 3.1) and rainfall (table 3.2) profiles shows that reproductive success tended to be highest (e.g., 1982 and 1983) in years when June temperatures were coolest (except 1990) and lowest (e.g., 1988) when June temperatures were warmest. Low rainfall during late June also seemed to be associated with higher reproductive success. Cooler temperatures may enhance cliff swallows' reproductive success by slowing development of swallow bug eggs (e.g., Myers 1928), provided of course

Effects of Year and Colony Size • 335

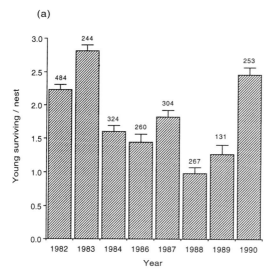

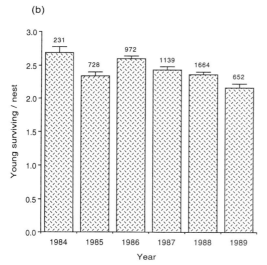

Fig. 11.7 Number of young surviving per cliff swallow nest in relation to year for nonfumigated (*a*) and fumigated colonies (*b*). Data from all colonies each year were combined. Sample sizes (number of nests) are shown above error bars. Number of young surviving varied significantly among years for both nonfumigated and fumigated sites (Kruskal-Wallis ANOVA, $p < .001$ for each).

that temperatures are not so cool as to reduce the abundance of flying insects (the birds' food). Low rainfall means little reduction in cliff swallow foraging time owing to rainy weather during feeding of nestlings, perhaps contributing to greater reproductive success. In contrast, hot years such as 1988 probably promote rapid development of swallow bug eggs and, if the heat is severe enough, reduce cliff swallows' food (e.g., section 10.8.2). There was not enough yearly variation in reproductive success among fumigated nests (fig. 11.7b) to suggest a clear temperature

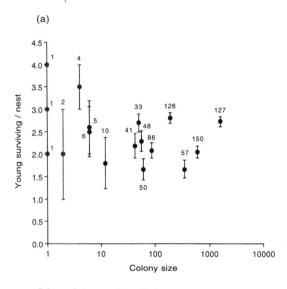

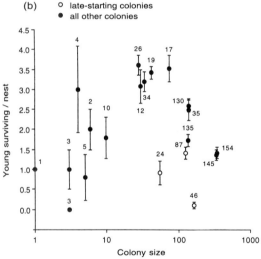

Fig. 11.8 Number of young surviving per nest in relation to cliff swallow colony size among nonfumigated sites in 1982 and 1990 (a), 1984–87 (b), and 1988–89 (c). Numbers by dots are sample sizes (number of nests) for each colony. In (a), number of young surviving decreased almost significantly with colony size ($r_s = -.44$, $p = .061$, $N = 19$ colonies); in (b), number of young surviving did not vary significantly with colony size ($r_s = .10$, $p = .68$, $N = 20$ colonies); and in (c), number of young surviving increased significantly with colony size exclusive of late sites ($r_s = .98$, $p < .001$, $N = 8$ colonies) and nearly so including late sites ($r_s = .55$, $p = .079$, $N = 11$ colonies).

or rainfall pattern, so the climatic differences among nonfumigated nests may have expressed themselves primarily through their effects on swallow bugs.

11.3.2 Reproductive Success in relation to Colony Size

We would have preferred to analyze the effect of colony size on reproductive success for each year separately, given these seasonal effects (fig. 11.7). Collecting data on reproductive success via nest checks was time consum-

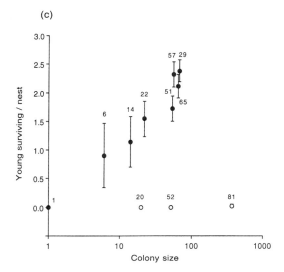

ing, however, and consequently we were unable to study enough colonies in each year to provide a sufficient sample size or colony size range. Thus we pooled all colonies that were active in the years when reproductive success did not differ significantly (section 11.3.1; figs. 11.8, 11.9). This gave us a greater, more definitive range of colony sizes for analysis of each year-set.

In our analyses of how colony size affected reproductive success, we distinguished late-starting colonies. These sites became active in late June and July of each year, with the first eggs typically not being laid until 5–10 July or later. We made this distinction because cliff swallows in late colonies experienced a very different climatic environment (section 3.2), with possibly large effects on both food availability and ectoparasitism. For instance, late colonies were often begun by birds that had failed en masse elsewhere and in the process introduced large numbers of swallow bugs to their late colony site (section 4.9.1). Cliff swallows occupying late colonies usually had substantially reduced reproductive success, probably owing in large part to high levels of ectoparasitism (sections 4.4, 4.9.2).

The variation in cliff swallow reproductive success with colony size differed among year-sets for nonfumigated sites (fig. 11.8). In 1982/1990, the number of young surviving per nest tended to decline, almost significantly, with colony size (fig. 11.8a). In 1984–87 the number of young surviving per nest among all sites exclusive of late colonies showed a peak at colony sizes of 30–80 nests, with reduced success in smaller and larger colonies (fig. 11.8b). In 1988–89, reproductive success at all nonlate colonies increased significantly with colony size (fig. 11.8c). Late colonies generally had poor reproductive success, especially in 1988–89, when virtually no young were reared at any of the three late-starting sites.

The most important conclusion from these results (fig. 11.8) is that cliff swallows' reproductive success in most years peaked at intermediate colony sizes. This pattern was strongest in 1984–87. The 1988–89 data also are consistent with intermediate-sized sites' being favored; we had no data for sites larger than 100 nests (except for one late colony) in those years, and thus the observed increase in reproductive success with colony size (fig. 11.8c) may have represented only the left half of a bell-shaped pattern. There was no strong evidence that intermediate colony sizes were the most successful in 1982/1990, and if anything, the smallest sites seemed to do slightly better in those years. The same pattern (fig. 11.8a) held for 1982 separately, and thus it probably was not attributable to our combining the 1982 and 1990 data.

We conclude that cliff swallow reproductive success under natural (i.e., nonfumigated) conditions peaked at intermediate-sized colonies in most years. These data (fig. 11.8) suggest that the "optimal" colony size is between 30 and 80 nests, substantially smaller than the average colony size of the population (394 nests; section 3.5.5). Increased success at intermediate sizes is consistent with the severe cost of ectoparasitism in larger colonies, coupled with less enhancement of foraging efficiency via information transfer in the smallest sites.

One caveat is in order, however. For a variety of logistical reasons, we were unable to estimate reproductive success in truly large cliff swallow colonies that were not fumigated. We had only two sites larger than 500 nests, both active in 1982. Possibly reproductive success could routinely be higher in the very large colonies not represented in our sample (fig. 11.8). If so, this might produce a bimodal pattern with respect to colony size. Although we think this is unlikely given the rather substantial costs of large colonies documented earlier, there is a need for more data on reproductive success in large nonfumigated colonies.

Surprisingly, reproductive success in fumigated nests also tended to peak at intermediate colony sizes (fig. 11.9). The pattern among fumigated nests was less striking than that seen in nonfumigated nests, perhaps owing to a smaller total number of colonies. We predicted that success should have increased consistently with cliff swallow colony size in the absence of ectoparasites and their effects in larger colonies, but this was clearly not the case: the colony sizes at which success seemed greatest among fumigated nests also were in the 30–80 nest range (fig. 11.9). Late-starting fumigated colonies had reduced success too, implying that increased ectoparasitism late in the year was not the sole cause of the seasonal reduction. The similarity in patterns for nonfumigated (fig. 11.8) and fumigated colonies (fig. 11.9) strengthens our conclusion that reproductive success indeed peaked at intermediate colony sizes, especially since several very

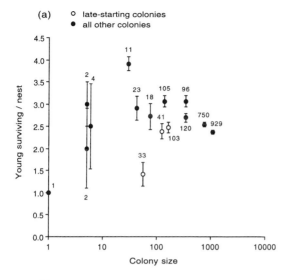

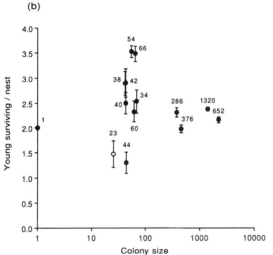

Fig. 11.9 Number of young surviving per nest in relation to cliff swallow colony size among fumigated sites in 1984, 1986–87 (*a*) and 1985, 1988–89 (*b*). Numbers by dots are sample sizes (number of nests) for each colony. In neither (*a*) nor (*b*) did the number of young surviving vary significantly with colony size (*a*: $r_s = .15$, $p = .58$, $N = 15$ colonies; *b*: $r_s = .00$, $p = .99$, $N = 14$).

large colonies (>1,000 nests) were included in the fumigated sample. We further address the implications of these results in section 14.2 when we consider lifetime reproductive success and the evolution of coloniality in cliff swallows.

11.4 EFFECTS OF DATE

Time of year can have a strong effect on reproductive success, since it reflects both changing food availability over the summer and population increases of ectoparasites, especially swallow bugs (section 4.4). Clutch sizes of cliff swallows decreased as the nesting season advanced (section 11.2.2), meaning that the number of young raised per nest inevitably decreased as well. In this section we examine seasonal patterns of reproductive success and evaluate whether any relation to time of year was independent of the seasonal decline in average clutch size.

As in our analysis of clutch size (section 11.2.2), we pooled all nests from all years into five-day periods based on when egg laying started in each nest. We initially analyzed each year separately, but the results were virtually identical and are therefore combined here. In both nonfumigated and fumigated nests, reproductive success declined significantly as the season progressed (fig. 11.10). The reduction in reproductive success was more pronounced among nonfumigated nests, and 25 June seemed to represent a threshold date after which success declined markedly (fig. 11.10a). The most successful nonfumigated nests were those initiated before 1 June. The decline among fumigated nests was more gradual but steady (fig. 11.10b).

These data, collected from across colonies, illustrated a clear effect of date when the season as a whole was considered. How much of the seasonal reduction in reproductive success was attributable to diminishing clutch sizes? Among nonfumigated nests, mean clutch size decreased by about 33%, on average, from the beginning to the end of the breeding season (fig. 11.4). Over approximately the same time span, the mean number of young surviving decreased by about 92% (fig. 11.10a). This means that the seasonal reduction in reproductive success cannot be explained solely by patterns in clutch size. Increased ectoparasitism by swallow bugs later in the summer is the most likely single factor accounting for the reduced success.

However, the results for fumigated nests (fig. 11.10b) illustrate that other factors also may account for some of the seasonal reduction in reproductive success. The mean number of young surviving in fumigated nests declined by about 43% over the season, slightly more than might be expected based solely on clutch size. Declining food supplies during the season and poor parental ability of younger, less experienced late breeders may also contribute to this seasonal pattern. The nonfumigated/fumigated comparison (fig. 11.10), however, suggests that most of the effects of date are caused by ectoparasitism, underscoring the seriousness of this cost of coloniality.

These seasonal patterns in reproductive success (fig. 11.10) span a longer time than any single colony site was active. As emphasized in section 11.2.2, it is also important to examine what effect date may have within the narrower time frame when each colony is active. We did this with a series of within-colony correlations between the number of young

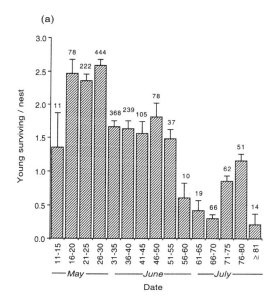

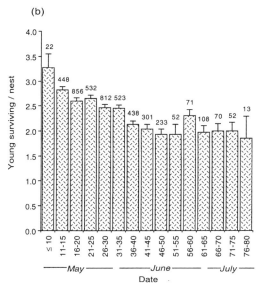

Fig. 11.10 Number of young surviving per cliff swallow nest in relation to date egg laying was initiated in nonfumigated (a) and fumigated colonies (b). Nests from all years and colonies were combined into arbitrary five-day periods (01 = 1 May), and sample sizes (number of nests) for each period are shown above error bars. Number of young surviving declined significantly with date for nonfumigated ($r_s = -.78$, $p < .001$, $N = 15$ periods) and fumigated sites ($r_s = -.87$, $p < .001$, $N = 15$).

surviving and the date egg laying began, per nest, for nonfumigated (table 11.3) and fumigated sites (table 11.4). Among nonfumigated nests, reproductive success declined with date in 26 of 30 colonies (a significant preponderance of negative correlations, binomial test, $p < .001$; table 11.3). Among fumigated nests, reproductive success declined with date in 19 of 24 colonies (also a significant preponderance of negative

Table 11.3 Spearman Rank Correlation Coefficients between Number of Young Surviving per Nest and Date Egg Laying Began within Each Nonfumigated Cliff Swallow Colony

Colony Site	Colony Size	r_s	p	N
8429	5	−.34	.57	5
8709	10	−.02	.92	26
8228	12	−.10	.85	6
8327	13	−.37	.29	10
8825	14	−.02	.94	14
8324	17	−.01	.70	17
8931	22	−.34	.15	20
8329	25	−.20	.37	21
9032	42	−.14	.73	8
8207	50	.11	.59	25
8832	54	−.39	.023	34
8226	56	−.04	.79	43
8932	57	−.28	.040	54
8203	60	.19	.26	38
8824	66	−.28	.022	65
8841	68	.41	.056	22
8421	75	−.26	.41	12
8405	125	−.30	.005	87
8641	137	−.57	<.001	135
8630	140	−.26	.13	34
8732	140	−.16	.09	116
8642	163	−.07	.66	41
8305	180	−.35	<.001	186
9030	190	.05	.69	57
8730	340	−.25	.002	142
8430	345	−.53	<.001	154
8206	350	−.07	.63	54
8830	375	−.20	.09	70
8202	600	−.07	.45	119
8201	1,600	−.55	<.001	95

Note: N = number of nests.

correlations, binomial test, $p = .006$; table 11.4). These analyses indicate that cliff swallows reared fewer young as the summer progressed even within the time a single colony site was active (also see section 11.2.2). Seasonal reductions in reproductive success thus applied to virtually all colonies regardless of when in the summer they were active and applied to colonies of all sizes. Presumably the same factors already discussed (e.g., ectoparasitism, declining food) contributed to the within-colony patterns. We conclude that anything that delays a cliff swallow's onset of egg laying, in terms of both its absolute laying date within the season and its laying date relative to the other birds within its own colony, is likely to be costly.

One way a late-breeding cliff swallow might partly compensate for its

Table 11.4 Spearman Rank Correlation Coefficients between Number of Young Surviving per Nest and Date Egg Laying Began within Each Fumigated Cliff Swallow Colony

Colony Site	Colony Size	r_s	p	N
8426	6	.83	.17	4
8842	25	−.86	.006	8
8531	41	−.30	.08	35
8537	42	−.38	.25	11
8539	42	−.39	.39	7
8431	43	−.29	.64	5
8524	44	−.13	.36	52
8434	56	.41	.059	22
8841	68	−.24	.19	32
8421	75	.05	.87	14
8525	81	−.25	.051	61
8530	90	−.24	.36	17
8538	120	−.45	.022	25
8405	125	.33	.035	41
8630	140	−.17	.09	102
8642	163	−.36	.001	73
8730	340	−.17	.033	154
8430	345	−.32	.001	96
8830	375	.00	.99	284
8505	456	−.09	.08	368
8605	750	−.21	<.001	686
8705	1,100	−.26	<.001	676
8805	1,400	−.08	.010	1,136
8905	2,200	−.37	<.001	621

Note: N = number of nests.

delayed start is by laying eggs before its nest is complete. This may improve an individual's temporal position relative to other colony residents, but laying in an incomplete nest can be disadvantageous for two reasons. Completing the nest requires that the nest owners leave more often to gather mud, and hence the clutch may be more vulnerable to neighboring cliff swallows that may toss out one or more eggs (section 5.6) or try to usurp the nest. Completing the nest also may reduce the daily time available for incubation, perhaps lengthening the incubation period and negating some of the temporal benefit of laying in an incomplete nest.

Cliff swallows often laid eggs in nests that were incomplete. We classified each nest by its completeness at the time laying began, using the categories such as those described in section 2.2 (see fig. 2.2). For this analysis we assigned each nest a completeness score as follows: 1, nest half complete or less when laying began; 2, nest two-thirds complete; 3, nest three-fourths complete; 4, no neck; 5, neck incomplete; 6, hole or crack in side; 7, mostly (but not fully) complete. We restricted this comparison only to

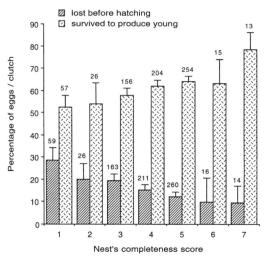

Fig. 11.11 Percentage of eggs per clutch lost before hatching or surviving to produce nestlings alive at day 10 in relation to a cliff swallow nest's completeness at the time egg laying began (see text for scores; 1 = least complete, 7 = most complete). All nests from all colonies were combined, and sample sizes are shown above error bars. Percentage of the clutch lost decreased significantly with increasing completeness ($r_s = -.98$, $p < .001$, $N = 7$ classes); percentage of the clutch surviving to produce nestlings increased significantly with completeness ($r_s = .96$, $p < .001$, $N = 7$).

nests that were incomplete to some degree when first occupied, because this presumably confined our comparison to the same general subset of birds (ones that had to build or refurbish nests). Cliff swallows occupying fully complete (old) nests tended to nest earlier and may have more often been older (section 11.8). These date- and age-related differences, therefore, might have confounded our analysis of the effect of nest completeness had we included all nests.

There was a disadvantage to laying in incomplete nests (fig. 11.11). The percentage of the eggs in a clutch lost before hatching decreased significantly with increasing nest completeness at the time of laying, and the percentage that eventually resulted in surviving offspring increased significantly (fig. 11.11; we used percentages to control for any clutch size or other systematic differences among nests of different degrees of completeness). Length of the incubation period, however, did not vary significantly with nest completeness ($r_s = -.43$, $p = .34$, $N = 7$ classes of completeness), although the trend was for eggs in less complete nests to require longer incubation times. Losses of eggs (fig. 11.11) were probably often caused by neighboring cliff swallows (section 5.6.1), and perhaps also by wandering birds trying to find a nest, which mainly tried to usurp incomplete nests (section 5.2.3). Laying in incomplete nests was thus risky, suggesting that the temporal advantage of earlier nesting must be substantial.

11.5 EFFECTS OF NEST SPATIAL POSITION

Another potential influence on reproductive success is a nest's spatial position relative to other nests within a colony. Extreme edge nests suffered higher rates of predation than more central ones (section 8.7), and birds living near the center have more neighbors and thus presumably more opportunities to glean information on food locations.

Our measures of spatial position were a nest's linear distance from the center of the colony, its nearest neighbor distance, and, for some, whether they were edge or center nests (sections 4.5, 11.2.3). Because colonies differed in how closely nests were clustered and the total space over which nests were spread, measures of distance from center and nearest neighbor distance were not directly comparable between sites. Thus these analyses could be done only as a series of within-colony correlations for each site (section 4.5).

Reproductive success declined as a nest's distance from the colony's center increased, for both nonfumigated (table 11.5) and fumigated sites (table 11.6). Among nonfumigated nests, the correlation coefficient was negative in 24 of 35 colonies, a significant preponderance of negative correlations (binomial test, $p = .040$; table 11.5). Among fumigated nests, the correlation coefficient was negative in 17 of 24 colonies, a marginally significant preponderance (binomial test, $p = .064$; table 11.6). These analyses indicate an advantage in most sites to being closer to the colony's center. This result probably was not due to clutch size variation, because there was no systematic variation with distance from the center (section 11.2.3). This result probably is also not explainable by differing levels of ectoparasitism within a colony, because fumigated nests exhibited the same pattern and there was no effect of distance from the center on incidence of swallow bug parasitism (section 4.5.3). Most likely, a combination of predation on edge nests (section 8.7), and perhaps enhanced foraging efficiency of center nesters, accounted for decreasing success toward the colony's edge (and see section 11.8). The birds' preference and competition for center nests (section 5.2.3) thus is easily understood.

A slightly different pattern emerged for nearest neighbor distance. Cliff swallow reproductive success in nonfumigated nests showed no consistent relation to nearest neighbor distance, declining with distance in 22 colonies and increasing in 22 (table 11.7). This indicated that occupying the more spatially isolated nests had no obvious advantage or disadvantage in terms of expected reproductive success. When ectoparasites were removed, however, reproductive success declined with increasing nearest

Table 11.5 Spearman Rank Correlation Coefficients between Number of Young Surviving per Nest and Nest's Distance from Colony's Center within Each Nonfumigated Cliff Swallow Colony

Colony Site	Colony Size	r_s	p	N
8227	4	−.78	.23	4
8709	10	−.25	.21	26
8327	13	−.24	.50	10
8324	17	−.29	.26	17
8931	22	−.13	.66	15
8329	25	.47	.030	21
8432	30	−.63	.039	11
8627	34	−.23	.19	34
9032	42	−.06	.75	32
8431	43	−.13	.59	19
8207	50	−.12	.51	33
9131	53	−.25	.38	14
8832	54	.01	.96	45
8226	56	.28	.055	48
8434	56	.25	.28	20
8932	57	−.21	.13	54
8824	66	−.27	.029	65
9124	68	−.10	.57	34
8841	68	−.55	.12	9
8421	75	.21	.42	17
9041	86	−.05	.69	76
8405	125	−.04	.74	87
8641	137	−.13	.13	135
8630	140	−.30	.08	35
8732	140	.06	.49	130
9130	140	−.02	.86	75
8642	163	−.05	.74	46
8305	180	.06	.43	185
9030	190	.11	.20	126
8730	340	−.24	.004	145
8430	345	−.33	<.001	149
8206	350	−.39	.005	52
8830	375	.05	.71	73
8202	600	.20	.021	138
8201	1,600	.06	.52	125

Note: N = number of nests.

neighbor distance in 17 of 22 fumigated colonies (table 11.8), a significant preponderance of negative correlations (binomial test, $p = .016$). This effect was not strong, with only two of the correlations significant or nearly so. The pattern nevertheless was consistent with there being advantages to clustering nests, such as sharing walls (section 5.3) and increased information sharing among close neighbors. We do not know why this result (table 11.8) was not seen in nests with natural levels of ectoparasites

Table 11.6 Spearman Rank Correlation Coefficients between Number of Young Surviving per Nest and Nest's Distance from the Colony's Center within Each Fumigated Cliff Swallow Colony

Colony Site	Colony Size	r_s	p	N
8842	25	.25	.26	22
8432	30	−.61	.14	7
8531	41	.32	.060	35
8537	42	.10	.55	39
8539	42	−.17	.27	42
8431	43	.28	.23	20
8524	44	.04	.75	52
8434	56	−.45	.011	31
8841	68	−.47	.005	34
8421	75	−.41	.09	18
8525	81	−.29	.026	61
8530	90	−.17	.19	58
8538	120	−.23	.067	62
8405	125	−.16	.33	41
8630	140	.14	.16	105
8642	163	.14	.15	103
8730	340	−.04	.54	209
8430	345	−.24	.018	96
8830	375	−.25	<.001	281
8505	456	−.29	<.001	348
8605	750	−.10	.006	762
8705	1,100	−.12	<.001	839
8805	1,400	−.06	.035	1,249
8905	2,200	−.11	.006	607

Note: N = number of nests.

(table 11.7), unless within-colony transmission of swallow bugs (section 4.9.2) was enhanced sufficiently in clustered nests to negate the possible nest building or foraging advantages of having close neighbors.

Finally, we examined how reproductive success varied between center and edge nests, which represented the two extremes in spatial positioning within a colony. As in other analyses of center and edge effects (sections 4.5.3, 11.2.3), we used only colonies where we had both center and edge nests for direct comparison. To avoid problems with different colonies' having different overall mean reproductive success, we analyzed these data with matched-pairs tests, which directly compared center and edge nests within each colony. Total number of center or edge nests in a colony ranged from 13 to 42.

Among nonfumigated nests, center nests averaged 0.12 more young reared per nest per colony than edge nests (fig. 11.12a), a nonsignificant difference and one that could be attributed to initial differences in clutch size of center versus edge birds (section 11.2.3; fig. 11.5). Among fumi-

Table 11.7 Spearman Rank Correlation Coefficients between Number of Young Surviving per Nest and Nest's Nearest Neighbor Distance, and Nest Size, within Each Nonfumigated Cliff Swallow Colony

Colony Site	Colony Size	Nearest Neighbor Distance			Nest Size		
		r_s	p	N	r_s	p	N
8227	4	−.82	.18	4	−.82	.18	4
8526	4	.00	.99	4	−.60	.40	4
8429	5	−.18	.78	5	.82	.18	4
8827	6	−.30	.56	6	−.65	.16	6
9132	6	.06	.94	4	−.50	.50	4
9160	6	.56	.24	6	.62	.19	6
8628	10	−.58	.075	10	.08	.82	10
8709	10	.45	.022	26	.19	.35	26
8228	12	−.18	.64	9	−.03	.93	9
8327	13	.50	.070	14	.18	.55	13
8825	14	.60	.024	14	.37	.19	14
8324	17	.02	.94	17	−.22	.39	17
8931	22	−.17	.46	21	−.09	.68	21
8329	25	.03	.89	21	−.25	.29	20
8432	30	−.77	.004	12	−.33	.30	12
8627	34	.08	.66	34	−.20	.26	34
9032	42	−.07	.67	40	−.10	.52	41
8431	43	−.07	.78	19	−.06	.80	19
8207	50	.06	.75	31	.06	.75	31
9131	53	.22	.35	21	.18	.46	20
8832	54	−.16	.29	48	−.11	.47	49
8226	56	−.19	.21	48	−.19	.20	48
8434	56	.26	.22	23	−.14	.52	24
8932	57	.12	.37	57	.05	.70	56
8203	60	−.13	.46	33	−.08	.69	26
8824	66	−.03	.82	65	.10	.42	65
8841	68	.23	.26	27	.03	.87	29
9124	68	.11	.55	34	−.09	.59	34
8421	75	.04	.87	17	−.04	.88	17
9041	86	.00	.99	86	.09	.41	86
8405	125	−.08	.45	87	−.08	.49	87
8641	137	−.10	.26	134	.10	.24	135
9130	140	−.06	.61	75	−.11	.34	78
8630	140	−.09	.59	35	−.24	.18	34
8732	140	−.10	.25	130	.03	.74	130
8642	163	.18	.24	45	−.03	.83	43
8305	180	.06	.42	185	−.13	.074	185
9030	190	−.09	.34	120	−.10	.27	118
8730	340	−.12	.15	145	.07	.42	145
8430	345	−.09	.28	154	−.18	.031	149
8206	350	.09	.52	52	.48	<.001	52
8830	375	−.07	.55	73	.02	.89	75
8202	600	.20	.016	144	.02	.77	150
8201	1,600	.11	.22	125	.00	.97	115

Note: N = number of nests.

Table 11.8 Spearman Rank Correlation Coefficients between Number of Young Surviving per Nest and Nest's Nearest Neighbor Distance, and Nest Size, within Each Fumigated Cliff Swallow Colony

Colony Site	Colony Size	Nearest Neighbor Distance			Nest Size		
		r_s	p	N	r_s	p	N
8426	6	−.50	.50	4	.63	.37	4
8842	25	.25	.26	23	.18	.43	23
8432	30	−.29	.39	11	.08	.81	11
8531	41	−.17	.30	38	−.21	.20	38
8537	42	.22	.18	40	.27	.087	40
8539	42	−.22	.17	42	—	—	—
8431	43	.13	.55	23	−.21	.34	23
8524	44	−.26	.063	52	−.14	.32	52
8434	56	−.11	.54	33	.39	.023	33
8841	68	.21	.23	34	.08	.64	34
8421	75	−.04	.89	18	−.43	.072	18
8525	81	−.06	.67	61	.03	.84	61
8530	90	.11	.41	58	−.12	.38	58
8538	120	−.15	.24	62	−.01	.98	62
8405	125	−.25	.11	41	−.05	.77	41
8630	140	−.14	.16	105	−.04	.71	105
8642	163	−.06	.52	103	−.10	.33	103
8730	340	−.07	.34	209	.06	.40	208
8430	345	−.03	.76	96	−.09	.38	93
8830	375	−.09	.13	276	−.08	.21	268
8505	456	−.06	.29	357	—	—	—
8905	2,200	−.16	.002	417	.09	.14	257

Note: N = number of nests.

gated nests, those in the center averaged 0.53 more young than edge nests (fig. 11.12b), a significant difference. This result for fumigated nests (fig. 11.12b) is consistent with edge nests' more often falling victim to predators (section 8.7). The increased predation rates on the edge of a colony apparently translated into about half an offspring reduction in reproductive success on average. The finding of no difference in reproductive success between center and edge for nests exposed to ectoparasites (fig. 11.12a) suggests that any predator-related advantages of being in the center are negated by the presence of ectoparasites. This implies that ectoparasites are more numerous in center nests than in edge nests and that they have greater effects on birds in center nests, although there was no strong evidence for increased parasite density in the centers of colonies (section 4.5.3). Regardless of what equalizing factors existed, under natural conditions cliff swallows occupying the center and edge nests exhibited roughly the same reproductive success. This comparison (fig. 11.12a) also suggests that the apparent decline in reproductive success with increasing

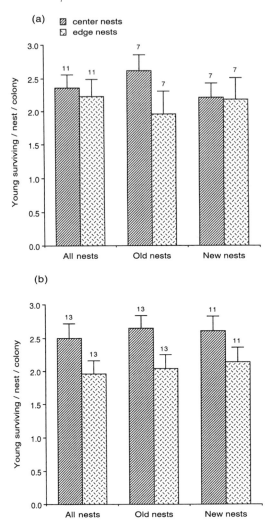

Fig. 11.12 Average number of young surviving per nest per colony for center and edge nests of different ages in nonfumigated (*a*) and fumigated colonies (*b*). Values shown are means of colony averages, and the numbers of colonies compared are shown above error bars. In (*a*), number of young surviving in center and edge nests did not differ significantly either for all nests combined or for new nests (Wilcoxon matched-pairs signed-rank tests, $p \geq .67$ for each), but center nests differed significantly from edge nests among old nests ($p = .042$). In (*b*), number of young surviving in center nests was significantly greater than in edge nests for all nests combined, old nests, and new nests (Wilcoxon matched-pairs signed-rank tests, $p \leq .020$ for each).

distance from the center in nonfumigated colonies (table 11.5) is probably not a strong effect. A nest's spatial position within a colony clearly influences its expectation of success, but there are apparent trade-offs that complicate the issue and obviously prevent clean interpretations.

11.6 EFFECTS OF NEST AGE AND SIZE

Nest age may affect reproductive success in cliff swallows for several reasons. Old, existing nests should theoretically harbor more swallow bugs, because the bugs can overwinter in them, and building new nests is one way cliff swallows may avoid parasites in larger colonies (section 4.10.3). On the other hand, using a new nest is costly in that additional time and energy must be invested in constructing it (Withers 1977; Gauthier and Thomas 1993), and consequently cliff swallows using new nests usually begin egg laying later than birds occupying old nests. Nest size may affect reproductive success in similar ways: building a large nest presumably is energetically costly and thus might reduce the birds' clutch sizes and the resources available for parental care. In addition, owing to its enclosed nature, the interior of a cliff swallow nest is usually several degrees warmer than the ambient temperature (Brown and Brown 1995). This allows the birds to leave their eggs and nestlings unattended more often than might be expected during cool weather when foraging conditions are poor. We do not know whether these thermal advantages vary with nest size (as reported for penduline tits by Hoi, Schleicher, and Valera 1994), but if they do, nest size might affect reproductive success for that reason.

11.6.1 Nest Age

In assessing the effect of nest age on reproductive success, we used only colony sites where we had data for both old and new nests within the same colony. As in the analysis of center and edge nests (section 11.5), we analyzed old versus new nests with matched-pairs tests, which controlled for differences among colonies in overall mean reproductive success. Among nonfumigated nests, old nests averaged 1.66 ($\pm$ 0.19) young reared per nest per colony, versus 1.62 ($\pm$ 0.17) young per nest per colony for new nests ($N = 25$ colonies). The difference was not significant (Wilcoxon matched-pairs signed-rank test, $p = .84$). Among fumigated nests, old nests averaged 2.37 ($\pm$ 0.14) young reared per nest per colony, versus 2.34 ($\pm$ 0.11) young per nest per colony for new nests ($N = 17$ colonies). This difference also was not significant (Wilcoxon matched-pairs signed-rank test, $p = .43$).

These comparisons show striking uniformity in reproductive success among old and new nests within a colony. The presumed costs of ectoparasitism in old nests and of increased construction in new ones appeared to balance each other among the nests chosen. This does not imply, however, that the birds selected nests indiscriminately with regard to age. They clearly avoided old, heavily infested nests (section 4.10.1), and the old nests they chose apparently gave them an expectation of success similar to that of cliff swallows occupying new nests.

We also examined whether a relationship existed between nest age and spatial position in effects on reproductive success. Again, we used only colony sites where we had both old and new nests represented among center and edge nests within the same colony, and we used matched-pairs tests for our analysis. Among nonfumigated nests, old nests in the center averaged 0.65 more young reared per nest per colony than old nests on the edge, a significant difference (fig. 11.12a). New nests, however, exhibited essentially identical reproductive success whether they were at the center or edge of the colony (fig. 11.12a). Among fumigated nests, center nests of both ages had approximately 0.60 more surviving young per nest per colony than edge nests of the same ages (fig. 11.12b), a significant difference in each case.

The advantage of center nests relative to edge nests of either age in fumigated colonies (fig. 11.12b) reflects their greater success in general when ectoparasites were removed (sections 11.5, 8.7). In the presence of ectoparasites (fig. 11.12a), only old nests showed a significant effect of nest spatial position. Why old nests, but not new ones, in the center would be more successful than edge nests of the same age under natural conditions is not clear.

11.6.2 Nest Size

Size of a cliff swallow nest presumably is an index of the relative amount of effort expended in constructing it (section 11.2.3; see Withers 1977). That females occupying larger (nonfumigated) nests laid slightly fewer eggs (fig. 11.6a) is consistent with nest building's being somewhat costly. The costs of building nests could be reflected in reproductive success if birds with larger nests were forced to forage more often and left their nests unattended more frequently during incubation (leading to egg loss to other cliff swallows; section 5.6), or if these individuals had reduced energy reserves to invest in feeding nestlings. Other disadvantages of large nests are that they offer more surface area and thus more nooks and crannies for swallow bugs to reside in and are perhaps more likely to be encountered by dispersing bugs (section 4.5.2). On the other hand, the thermal advantages of the enclosed nest (Brown and Brown 1995) might vary in ways that could compensate for (or magnify) these disadvantages of large nests.

We elected not to combine nests of various sizes from different colonies (in contrast to our clutch size analysis; fig. 11.6), given the variation in reproductive success among different colonies and years (e.g., fig. 11.8). There were also occasional difficulties in accurately measuring a nest's diameter in colonies where nests were tightly clustered (e.g., fig. 3.6b). Using a series of within-colony correlation analyses removed any biases introduced by either differences in mean reproductive success among different colonies and years or measurement difficulties in certain colonies.

There was no evidence that nest size affected reproductive success within a colony for either nonfumigated (table 11.7) or fumigated sites (table 11.8). Among nonfumigated nests, reproductive success declined with nest size in 25 of 44 colonies; the slight preponderance of negative correlations was not significant (binomial test, $p = .45$; table 11.7). Among fumigated nests, reproductive success declined with nest size in 11 colonies and increased in 9, showing no consistent effect of nest size (table 11.8). For cliff swallow nests exposed to ectoparasites (table 11.7), there was thus little evidence that larger nests were disadvantageous. Nest size probably did not substantially affect ectoparasite load (section 4.5.2). The results for both fumigated and nonfumigated nests suggested there was little reproductive cost to building or occupying larger nests. A similar result was found for barn swallows (Goodman 1982). The cliff swallow nest's thermal advantages did not seem to vary enough with nest size to directly affect reproductive success.

11.7 VARIANCE IN REPRODUCTIVE SUCCESS

This chapter thus far has focused entirely on mean reproductive success of cliff swallows. Means are most commonly used in analyses of reproductive success (e.g., Clutton-Brock 1988b, Newton 1989a). However, variance (or certainty) in reproductive success may also be important, because variation can affect an animal's expected probability of success (Gillespie 1974, 1977; Caraco 1981; Rubenstein 1982; Real and Caraco 1986; Stephens and Krebs 1986; Caraco et al. 1995). Variation in reproductive success among colonies has not been explicitly addressed in studies of colonial birds, yet it might influence an individual's choice of colony (Brown, Stutchbury, and Walsh 1990). In this section we examine how colony size affects variance in cliff swallow reproductive success.

As our measure of variance, we used the coefficient of variation (section 4.6). This statistic allowed us to compare colonies with different means and combine all colonies and years into a single analysis. We calculated the coefficient of variation in number of young surviving for each colony, except for colony sizes of one, which had no variation and thus could not be included. Any colony larger than one that had no variation among nests in number of young surviving was assigned zero in this analysis, including those in which all nests were unsuccessful in producing any young.

Nonfumigated colonies in general had higher coefficients of variation in number of young surviving than fumigated sites. The mean coefficient of variation per colony for nonfumigated sites ($N = 49$) was 94.80 (± 8.04), versus 60.67 (± 5.25) for fumigated sites ($N = 27$). This means that some of the within-colony variance in reproductive success is probably directly attributable to ectoparasites. That swallow bugs in particular would cause variation in the birds' reproductive success is not surprising, given the sometimes patchy distribution of bugs within a colony and their tendency to aggregate in certain nests (section 4.9.2).

We compared variance in reproductive success with colony size for 49

354 • *Reproductive Success*

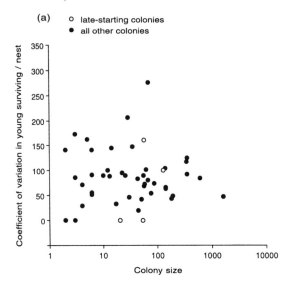

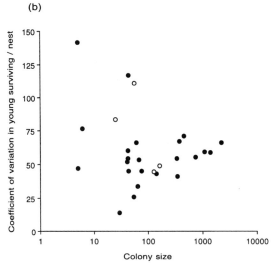

Fig. 11.13 Coefficient of variation in the number of young surviving per nest in relation to cliff swallow colony size among nonfumigated (*a*) and fumigated sites (*b*). Coefficient of variation did not vary significantly with colony size for either nonfumigated ($r_s = .10$, $p = .49$, $N = 49$ colonies) or fumigated colonies ($r_s = -.05$, $p = .80$, $N = 27$ colonies). Sample size (number of nests) for each colony in (*a*) is the same as in figure 11.8; (*b*) is the same as in figure 11.9.

nonfumigated colonies and 27 fumigated ones (fig. 11.13). Among nonfumigated colonies, there was no statistically significant relation between the coefficient of variation and colony size (fig. 11.13a). Colonies with the greatest variance in reproductive success occurred in the same size range (30–80 nests) where mean reproductive success was greatest, but there were also many colonies of those sizes with low variance. Late-starting colonies exhibited no clear pattern relative to the other sites or to colony size. We conclude that within-colony variation in reproductive success was generally unaffected by colony size, although we were surprised that the larger nonfumigated sites exhibited such low variation (fig. 11.13a). Within-colony variation in swallow bug parasitism was much higher in larger colonies (section 4.6), and thus we expected this patchy bug distribution to cause greater variance in reproductive success in large colonies, primarily through the swallow bugs' severe effects on late nests (sections 4.4, 4.9.2).

Within-colony variance among fumigated colonies also showed no relation to colony size (fig. 11.13b). If anything, smaller colonies showed greater variation in reproductive success, a not surprising result in the absence of ectoparasites. Birds in small colonies probably foraged less efficiently, without benefit of information transfer, and thus differences among individuals in their inherent ability to find food may have accounted in part for higher variation in reproductive success in the smaller fumigated colonies (fig. 11.13b).

Among both nonfumigated and fumigated sites, colonies of 100 or fewer nests exhibited substantial *between-colony* variation in their coefficients of variation (fig. 11.13). This meant that reproductive success varied little among cliff swallows in some of these colonies, while in others it varied greatly. There appeared to be less between-colony variation among colonies of more than 100 nests. We are not quite sure how to interpret this pattern, but it suggests that an individual's expectation of reproductive success, relative to other residents of the same colony, can perhaps be more reliably predicted among a given set of large colonies than among small colonies.

In preliminary analyses of reproductive success for this chapter, we examined how other factors—date, spatial position, nest age and size—affected the coefficient of variation in the number of young surviving. We found no obvious or consistent effects on variance in reproductive success among these factors.

11.8 AGE-SPECIFIC DIFFERENCES IN REPRODUCTIVE SUCCESS

The within-colony variation in reproductive success reported in the previous section could be generated in part by inherent differences in reproductive abilities among the individuals occupying a colony. Age is likely to account for some of these differences, because reproductive success is often influenced by parental experience (Kluijver 1951; Coulson and White 1958; Perrins and Moss 1974; Harvey et al. 1979; Ryder 1980; Nelson 1988; Languy and Vansteenwegen 1989; Saether 1990; Weimerskirch 1990). Yearling cliff swallows, for example, laid fewer eggs than two- or three-year-olds, and light females laid fewer eggs than heavy females (section 11.2.4). These differences in clutch size could themselves lead to differences in eventual reproductive success among the birds occupying a given colony. Age and experience may also affect other aspects of reproductive success, such as where in the colony an individual settles, how adequately it provisions its young, and how much interference from neighbors it experiences. Our goal in this section is to document age-specific differences in, and correlates of, reproductive success in cliff swallows.

Age-related data were available for all birds originally banded as nestlings and subsequently found breeding in a known nest where information on reproductive success was collected. These birds were residents of eight colonies in 1986–89, ranging from 80 to 2,200 nests, and all were fumigated except the 80 nest site. We did not have a sufficient sample of known-age individuals to examine each colony or year separately, so we pooled all colonies for analysis. Birds in each of the two main age classes (one year old and two years old or more) were represented in each colony, and sample sizes for each age class per colony varied from 2 to 63 individuals. Total sample sizes are given in table 11.9. For various reasons, not

Table 11.9 Comparisons among Cliff Swallows Aged One and Two or More Years

Variable	Age 1			Age $\geq$ 2			p^a
	Mean	SE	N	Mean	SE	N	
Clutch size	3.48	0.07	149	3.69	0.07	124	.037
Number of young surviving	2.44	0.10	136	2.71	0.09	120	.091
Relative nest position	0.672	0.022	155	0.576	0.029	126	.029
First egg date (01 = 1 May)	31.48	1.04	126	25.14	0.91	98	<.001
Laying synchrony	1.59	0.08	126	0.55	0.11	97	<.001
Nest diameter (cm)	15.63	0.26	57	15.74	0.25	65	.69
Nearest neighbor distance (cm)	23.58	4.50	62	18.12	4.11	75	.012
Days to build nest	9.83	1.04	18	10.09	1.06	11	.68
	%		N	%		N	p^a
Active nests that were new	67.8		115	24.8		109	<.001
Nests with brood parasitism via laying	7.4		136	4.7		107	.39
Nests with brood parasitism via transfer	5.1		136	9.3		107	.20
Nests with egg tossing	7.4		136	14.0		107	.090

Note: All colonies were combined for analysis; N = number of nests.

[a] Based on Wilcoxon rank sum tests for the first eight variables, χ^2 tests for the rest.

all of the information shown there was always known for each bird in the analysis.

Each known-age individual of each sex was treated independently, even when we did not know the age of its mate. We assumed that the age of one member of a pair was a reliable indicator of the age of the other member. This seemed justified because we had little evidence for widespread mixed-age pairing. In 29 instances in which the age of both members of a pair was known, 13 pairings (44.8%) were between yearlings, 10 (34.5%) were between birds older than one year, and 6 (20.7%) were between a yearling and an older bird. Among all mixed-age pairings ($N = 11$, using exact ages for those two years old or older), the average age of both males and females was 2.18 years, indicating no systematic bias for either sex to pair with younger or older birds. In purple martins, Morton and Derrickson (1990) also found that ages of mates tended to match. In estimating relative reproductive success in this section, we assume no differences among age classes in the probability that some young would result from extrapair copulation, although this is perhaps a problematic assumption (Morton, Forman, and Braun 1990).

We first examined reproductive success as measured by the number of young surviving per nest for parents of different ages. Reproductive success for yearlings ($N = 136$) was 2.44 ($\pm$ 0.12) young; for two-year-olds ($N = 78$), 2.68 ($\pm$ 0.13); for three-year-olds ($N = 26$), 2.85 ($\pm$ 0.23); and for four-year-olds and older birds ($N = 16$), 2.62 ($\pm$ 0.40). Reproductive success varied almost significantly with age class (Kruskal-Wallis ANOVA, $p = .07$). The analysis includes the individuals whose nests failed and that consequently had zero reproductive success. This pattern paralleled the age-specific variation in clutch size (section 11.2.4), in which yearlings did slightly less well than birds two years old or older.

Initial differences in the clutch sizes of different age classes were probably not the sole determinant of age-specific differences in reproductive success. We examined a number of factors that might vary among cliff swallows of different ages (table 11.9). In this analysis we combined age classes 3 and ≥4 with age class 2, because we did not have a large enough sample of older ages for detailed analysis. Table 11.9 shows mean values for all yearlings and birds two years old and older, combined from all colonies. As a measure of the spatial positions of these birds' nests, we calculated each nest's *relative position* within the colony by taking the maximum radius from the colony's center to the outermost nest and dividing a nest's distance from the center by this maximum radius. This gave a relative nest position that was comparable among colonies, regardless of each site's specific geometry. For example, a nest near the center would have a relative position near zero, whereas a nest near the colony's edge would have a relative position approaching one.

358 • *Reproductive Success*

Yearlings and birds two years old or older differed significantly in several ways that might influence their reproductive success (table 11.9). Yearlings settled closer to the colony's edge, on average, than older birds, and initiated egg laying later. Consequently, their nesting attempts were less synchronized (section 8.6.2) with those of the rest of the colony. Yearlings' nearest neighbor distances were greater than those of older birds (table 11.9), meaning that yearlings settled in more spatially isolated nests. Perhaps as a consequence of settling later, closer to the edges, and in more isolated situations, yearling cliff swallows were more likely to build new nests than reuse old ones. Yearlings also experienced more parasitism by fleas (section 4.6).

There were essentially no differences among age classes in nest size, time taken to build nests, or likelihood of suffering brood parasitism through either laying or transfer (table 11.9). Yearlings seemed slightly (though not quite significantly) less likely than older birds to have conspecifics toss eggs out of their nests (table 11.9). This could have reflected the yearlings' slightly later breeding times and more isolated spatial positions, since trespassing among neighbors (and thus egg tossing) declined seasonally and was less common among birds in widely spaced nests (section 5.5.4).

Of these differences between yearlings and older cliff swallows, the one most likely to affect reproductive success was the date on which egg laying began. Yearlings initiated egg laying over six days later, on average, than birds two years old and older (table 11.9). Because reproductive success declined with date (section 11.4), any delay in initiating nesting was costly. We were surprised that yearling cliff swallows nested later than older birds, because there was no consistent evidence that yearlings (especially females) arrived in the study area later (section 3.5.3). Perhaps yearling birds, being less experienced, required more time to assess colony sites (section 13.2.5) and establish ownership of a nest site within a colony.

11.9 SUMMARY

Knowing how per capita reproductive success varies with colony size is the key to understanding the net costs or benefits of animal coloniality. This chapter focuses exclusively on annual reproductive success in cliff swallows as measured by the number of young per nest surviving to day 10.

There was relatively little variation in clutch size among cliff swallows, with 80% of birds laying three or four eggs. It did not vary among years or with colony size. Clutch size declined as the breeding season progressed, both over the entire summer and within the time a single colony was active, and was largely unaffected by a nest's position relative to the colony's center, nearest neighbor distance, or nest age. Clutch size declined with increasing nest size in nonfumigated colonies but increased with nest size in fumigated colonies. Yearling females laid fewer eggs than birds two years old and older, and heavier females laid more eggs than light ones.

Reproductive success varied among years, with cliff swallows reproducing best in cooler, drier years and worst in hot, wet years. Reproductive success seemed to peak in most years at intermediate colony sizes of 30–80 nests and declined in smaller and larger colonies. This result for nonfumigated sites was consistent with the severe cost of ectoparasitism in larger colonies, coupled with reduced foraging efficiency via information transfer in the smallest sites.

Reproductive success across colonies and within a single colony declined as the season progressed in both nonfumigated and fumigated sites, more so than could have been accounted for by the seasonal reduction in clutch sizes. Ectoparasitism was a likely cause of this pattern for the nonfumigated colonies. Some cliff swallows compensated for late nesting by starting to lay before the nest was complete, but this was costly because eggs were more likely to be lost and fewer young survived in nests that were less complete when laying began. Reproductive success in general was slightly higher for cliff swallows nesting closer to the centers of colonies. Nearest neighbor distance did not affect reproductive success among nonfumigated colonies, but success declined with increasing distance among fumigated colonies, possibly reflecting advantages of sharing nest walls in the absence of ectoparasite transmission. Cliff swallows' reproductive success did not vary with nest age or nest size.

Within-colony variance in reproductive success did not vary significantly with colony size. Smaller colonies showed greater between-site variability in within-colony variance than did larger colonies, indicating that the likelihood of success may have been harder to predict in smaller colonies. Yearling cliff swallows had lower reproductive success than older birds. Yearlings settled closer to the colony's edge, initiated egg laying later, were less closely synchronized in time with the other colony residents, occupied more spatially isolated nests, and were more likely to build new nests than older birds within the same colony. Yearlings and older birds showed no significant differences in nest size, time taken to build a nest, or likelihood either of having their nests brood parasitized or of losing an egg to a conspecific.

12 Survivorship

> There are probably no birds whose past history would be more interesting than that of the [cliff] swallows.
> H. D. Minot (1877)

12.1 BACKGROUND

The two principal components of an animal's lifetime reproductive success are its life span and its annual reproductive success. Annual success, as reflected in the number of young surviving to fledge, is commonly measured in natural populations, and we used it as one way to assess the net effect of the different costs and benefits of cliff swallow coloniality (chapter 11). However, for at least two reasons annual fledging success alone may not accurately reflect the potential fitnesses of different individuals. If the costs or benefits of coloniality affect nestlings after they fledge (for example, by influencing their body mass and condition), young from different colonies may experience different levels of postfledging mortality. In such cases the most relevant measure of reproductive success is juvenile survivorship to the following spring and recruitment into the next year's breeding population. Also, some birds may live longer than others and consequently produce more young during their lifetimes. If these differences in survivorship consistently vary among sites, lifetime reproductive success may also vary, perhaps even when annual measures of nesting success are relatively constant.

Survivorship and average life span in natural populations have historically been difficult to estimate, especially among relatively long-lived birds and mammals. Several long-term studies have recently begun to yield information on typical life span of individuals and its potential effect on fitness (e.g., Clutton-Brock, Guinness, and Albon 1982; Woolfenden and Fitzpatrick 1984; Koenig and Mumme 1987; Clutton-Brock 1988b; Robinson 1988; Newton 1989a; Hoogland 1995). Among small birds such as cliff swallows, life span accounts for much of the variation in lifetime reproductive success between individuals (Bryant 1988b, 1989; Fitzpatrick and Woolfenden 1988; McCleery and Perrins 1988; Smith 1988; Van

Noordwijk and Van Balen 1988; Bunzel and Druke 1989; Dhondt 1989; Gustafsson 1989; Hotker 1989; Payne 1989; Sternberg 1989). However, surprisingly few data exist for any species to evaluate average life span or annual survival among birds living in different-sized colonies; Vehrencamp, Koford, and Bowen's (1988) study of anis in communal breeding groups of different sizes comes the closest. Wiklund and Andersson (1994) recently studied how colony size affected survival of fieldfares over a portion of the breeding season but did not address the birds' annual survivorship. Various other studies have examined survivorship and demography in colonial species per se but by focusing only on individuals within a single colony (section 1.3.3).

Collecting data on annual survivorship among birds in colonies of different sizes is challenging for several reasons. Most colonial species are not sedentary and thus do not remain in a year-round territory or home range where their presence can be continually monitored (unlike, for example, cooperative breeders; Woolfenden and Fitzpatrick 1984; Koenig and Mumme 1987). Colonial species tend to occupy breeding areas for relatively short periods, often migrating long distances during the nonbreeding season. Consequently both mortality and juvenile dispersal may be high, requiring that large numbers of individuals be marked initially to ensure that some will be encountered later. If most juveniles disperse long distances from a study area, there may be no way to ascertain the age of any of the breeders that are present.

Furthermore, the many individuals one must monitor if studying several colonies of different sizes inevitably means that some marked birds will simply be "missed" in a given season. This is perhaps one reason that most demographic studies, even of solitary and sedentary species, report a mean production of fewer than two offspring per pair (Grant 1990). The implication is that most of these populations are declining in size, but in many cases it may be that temporary and permanent emigrants were not detected by the observers and consequently did not figure into estimates of lifetime reproductive success. Finally, following the longest-lived individuals to the completion of their lives presents logistic (especially financial) challenges, and most published studies to date have included animals that had not yet died when the results were reported (Newton 1989b).

This chapter focuses on patterns of annual survivorship among cliff swallows. We use new statistical methods (section 12.2.2) to estimate birds' survival probabilities as a function of both breeding and natal colony size, and we also evaluate the associated effects of other variables treated in earlier chapters. Annual survival probabilities are estimated for birds originally marked as nestlings, to evaluate how the potential costs and benefits of coloniality directly affect juvenile (first-year) survivorship and parental reproductive success. In contrast to many migratory passer-

362 • Survivorship

ines, cliff swallows in southwestern Nebraska exhibit moderate natal philopatry, making it possible to infer relative degrees of juvenile survivorship among classes or subsets of birds. We also estimate survival probabilities for adults in subsequent seasons, to examine how the potential costs and benefits of coloniality affect breeders over the long term. We use these estimates of survival to better understand both the costs and benefits of coloniality and how lifetime reproductive success (section 14.2) varies among cliff swallow colonies of different sizes. We focus on annual survival probability, rather than life span, because the former is a more appropriate life-history parameter (Krementz, Sauer, and Nichols 1989).

12.2 METHODS FOR ESTIMATING SURVIVORSHIP

Our approach in this chapter is to estimate the probability of annual survival through recapture of previously banded individuals. For each banded bird we constructed a capture history recording whether the bird was known to be alive in each year subsequent to its banding and, if so, whether we encountered it that year. Any information concerning a bird's presence in a given year was used in constructing its capture history. This was usually based on mist netting or capture of birds in nests at different colonies (section 2.3), but finding a bird dead near a colony or having it recovered outside the study area also yielded information on its history.

12.2.1 Field Sampling for Mark-Recapture

Before 1988, all banding and subsequent recapture of cliff swallows in the study area was done at a relatively few colony sites during color-marking (e.g., section 5.5), for parentage studies (e.g., section 6.2.5; Brown and Brown 1988b), or to assess mass-related foraging advantages (section 10.5). All information on survivorship in the early years of our study therefore resulted from mark-recapture data initially generated for other purposes. In 1988 we began an extensive netting program specifically to examine survivorship and colony choice (chapter 13) through mark-recapture, and these efforts continue.

All accessible cliff swallow colonies within the primary study area (section 3.3) were sampled each year, 1988–92. Of the approximately one hundred colony sites in the primary study area, only eight that the birds used regularly (three years or more) were completely inaccessible to us for mark-recapture. Consequently the number of banded birds occupying unsampled sites and thus routinely escaping detection was relatively low. In addition, we netted less systematically at other selected colonies within the secondary study area and at two more distant colony sites, to search for long-distance movements by dispersing birds (sections 13.4, 13.5). The colonies sampled outside the primary study area were chosen largely for easy access, enabling us in most cases to catch a relatively large percentage of the colony residents in a short time.

We rotated among colonies, catching birds at intervals throughout the season (e.g., section 10.5.1). Capture effort varied among colony sites, owing to a variety of unavoidable factors. Chief among them was weather: developing bad weather or wind often shortened or canceled a sampling session at a given colony, and often these lost days simply could not be "made up" during the cliff swallow's relatively short breeding season. Weather also affected capture effort between years, given the substantial yearly variation in temperature and rainfall (section 3.2). Warm years tended to reduce overall capture effort, because birds were likely to overheat in nets on hot days, forcing us to curtail netting. The degree of between-colony nesting synchrony also affected yearly capture effort. We were able to sample each colony more intensively in years when nesting at different sites began at different times (e.g., 1991, 1992). In such cases we concentrated initially on the earlier colonies, then emphasized the later ones as the birds in the first colonies completed breeding. However, in years when nesting at all colonies began at about the same time (e.g., 1990), capture effort had to be proportionately reduced at each site. Yearly or site-specific differences in capture effort are unlikely to introduce serious bias, because the models we used to determine survivorship calculated probabilities of recapture in estimating the survival probabilities (section 12.2.2).

The numbers of cliff swallows of different ages marked each year are shown in table 12.1. The numbers of adults and juveniles banded annually best reflect relative sampling (i.e., netting) effort among years and the consequent opportunity to recatch previously marked birds. Most of the nestlings were marked in 1982–88, providing a large sample of known-age

Table 12.1 Number of Cliff Swallows Banded Each Year, and Percentage Recaptured in at Least One Subsequent Year

	Number of Birds Banded			Percentage Recaptured		
Year	Adults	Juveniles	Nestlings	Adults	Juveniles	Nestlings
1982	30	0	915	13.3	—	2.7
1983	418	0	637	13.4	—	9.7
1984	1,674	70	1,083	15.9	5.7	12.1
1985	937	31	1,512	26.5	16.1	17.5
1986	1,896	146	2,496	27.8	26.0	21.1
1987	2,162	216	2,022	28.8	27.8	21.3
1988	3,445	128	2,124	30.6	27.3	21.2
1989	4,521	598	240	29.9	28.4	15.8
1990	5,151	481	853	29.6	28.9	16.1
1991	7,808	1,685	890	19.7	21.5	16.6
1992	8,214	1,577	413	—	—	—

364 • *Survivorship*

birds during the later years of the study when our netting effort increased steadily. As a measure of the extent to which banded cliff swallows were recaptured in later years, we present in table 12.1 the percentages of marked birds that were recaptured in at least one year after banding. Recaptures among the 1982–84 cohorts were low owing to reduced sampling in those years. Recapture percentages for 1986–90 are highest because those cohorts were intensively sampled for several subsequent years. Among these well-sampled cohorts, 28–30% of cliff swallows banded as adults and juveniles, and about 21% of birds banded as nestlings, were subsequently recaught at least once (table 12.1). Cliff swallows exhibit a higher rate of return to the vicinity of their birthplace than that reported for many migratory songbirds, in most of which fewer than 10% of nestlings are ever encountered again (e.g., Lincoln 1934; Boyd and Thompson 1937; Stoner 1941; Farner 1945; Johnston 1961; Berndt and Sternberg 1968; Nolan 1978; Shields 1984b; Iverson 1988; Payne 1989; Weatherhead and Forbes 1994; cf. Rheinwald 1975).

12.2.2 Statistical Models

Unlike the case in nonmigratory or island-dwelling species, in which most surviving individuals within a local area can be annually detected by an investigator, the vast numbers of cliff swallows in our study area (section 3.5.1) made it impossible to census the population thoroughly for all marked birds each year. Consequently many birds were missed in a given year. For example, among the birds banded in 1988 and later recaptured (table 12.1), 999 were known to be alive in at least two subsequent seasons. Of those, however, only 254 (25.4%) were recaptured in *each* year they were known to be alive; the rest escaped detection in at least one year. Therefore estimation of survivorship must account for both the birds missed in a given year but known to be alive through later encounters and the fraction of banded birds in a cohort that are present but not detected in any subsequent year.

Models for estimating survival from mark-recapture data when a population is incompletely censused now exist and represent more realistic methods of assessing survival than ad hoc percentages based solely on the individuals actually recaught or resighted in a given year (Clobert et al. 1985, 1987; Nur and Clobert 1988; Lebreton et al. 1992; Lebreton, Pradel, and Clobert 1993). These models rely to some extent on the classical mark-recapture assumptions (e.g., Jolly 1965; Seber 1982) but are especially useful because they incorporate both the use of multiple recaptures of living animals (as opposed to band recoveries of dead individuals) and the calculation of recapture probabilities. The probability of recapture explicitly addresses instances when an animal is not seen in a given year. Based on observed capture histories, the models calculate the probability of recapture (a "nuisance" parameter; Lebreton et al. 1992) and use it in estimating survival probabilities. As long as a reasonable recapture effort is made, differences in probabilities of recapture can be dealt with statistically and modeled to examine biological influences on capture tendencies among animals. These models estimate parameters using maximum likelihood methods and yield a sampling variance for each estimate.

We used the program SURGE to estimate cliff swallow survival probabilities. SURGE (Clobert, Lebreton, and Allaine 1987; Pradel, Clobert, and Lebreton 1990; Lebreton et al. 1992; Pradel and Lebreton 1993) is a commercially available software package that computes maximum likelihood estimates of survival and recapture parameters on a logit scale and gives their asymptotic sampling variances and covariances (using an information matrix derived from recapture histories). This program calculates a standard error and 95% confidence interval for each parameter, providing the opportunity to compare parameters statistically. SURGE is particularly useful in selecting among different models, incorporating, for example, time- or age-dependence in survival or recapture probabilities to find the model that best fits the data. Further details on the statistical framework of SURGE are contained in Clobert, Lebreton, and Allaine (1987) and Lebreton et al. (1992), and examples in which this program has been applied to other data can be found in Clobert et al. (1987, 1988); Nur and Clobert (1988); Kanyamibwa et al. (1990); Pradel, Clobert, and Lebreton (1990); Blondel, Pradel, and Lebreton (1992); Lebreton et al. (1992); and Szep (1995).

In a preliminary series of analyses of recapture probabilities for birds banded both as adults and as nestlings or juveniles, we found that models incorporating time dependence in recapture probabilities provided a significantly better fit than models with constant recapture probabilities for virtually all data sets. Representative variations in yearly recapture probabilities are illustrated in table 12.2, in which annual probabilities of recapture were estimated for all adult cliff swallows in the 1982–89 cohorts (using a model in which survival was also time dependent; section 12.3.3). These results (table 12.2) indicate that a bird's probability of being recaught if alive increased consistently each year throughout the study, reflecting our increasing level of sampling effort (table 12.1). We therefore used models with time-dependent recapture probabilities in all analyses reported in this chapter. There was no evidence of age dependence in recapture probabilities, meaning in particular that yearling cliff swallows if present were no more or less likely to be recaptured than older birds. Consequently we did not routinely use models with age-dependent recapture probabilities.

To estimate survival probabilities, we began with a relatively simple

Table 12.2 Yearly Recapture Probabilities for All Adult Cliff Swallows in the 1982–89 Cohorts

Year	p	SE
1983[a]	—	—
1984	0.083	0.020
1985	0.097	0.012
1986	0.220	0.016
1987	0.204	0.012
1988	0.238	0.011
1989	0.301	0.010
1990	0.452	0.012
1991	0.487	0.016
1992	0.583	0.172

Note: $N = 15{,}083$. Probabilities were estimated using a model with time-dependent survival probabilities.

[a] Too few recoveries ($N = 2$) in 1983 to estimate recapture probability.

model and used likelihood ratio tests to compare it with increasingly complex, yet biologically relevant, models (Lebreton et al. 1992; see Kanyamibwa et al. 1990). For each analysis of survivorship among adults (which in most cases included birds of mixed or unknown age, especially for those first banded as adults), we began with a constant survival/time-dependent recapture model. In some analyses we then assessed whether a model that estimated survival separately for the next subsequent year versus all other years provided a significantly better fit; if so, such a model was used for parameter estimation. The extent to which more complex models could be used depended in part on the size of the data set, with small data sets generally permitting only the simpler models.

For each analysis of subsequent survivorship among birds initially banded as nestlings or juveniles, for which age was known, we began with a model of age-dependent survival/time-dependent recapture, using the two age classes of yearlings and all birds over one year old. This sort of model was biologically (and in almost all cases, statistically) appropriate, given the relatively high mortality during the birds' first year (section 12.3.1). In some cases we then tested and used other models with more complex age dependence, with additional age classes.

In the sections that follow, we specify in each analysis the model used to estimate survival probabilities. We use the notation of Clobert et al. (1985, 1988) and Clobert, Lebreton, and Allaine (1987), in which probabilities of survival and recapture are denoted s and p. For example, a model with age-dependent survival using two age classes and time-dependent recapture is denoted s_{a2}, p_t, whereas one with constant survival and time-dependent recapture is s, p_t.

12.2.3 Cohorts Used in Analyses

Given the frequency with which marked cliff swallows escaped recapture and detection in certain years, we examined how long (in years) a cohort had to be followed to achieve robust estimates of survival. To do this, we estimated survival for each cohort based on recapture data through 1990 and then reestimated those same parameters based on recapture data through 1991. For example, for all nestlings banded through 1989 we determined average survival probabilities for one-year-olds, two-year-olds, and birds three years old or more each year (model s_{a3t}, p_t). The cohorts from the early years of the study had thus been followed for multiple years, whereas the 1989 cohort had been followed for only one year in 1990 and two years in 1991.

For cohorts that had been followed for only one year (e.g., 1989 to 1990), survival estimates changed by an average +23.3% when an additional year of recapture was added (e.g., 1989 birds followed to 1991). For cohorts that had been followed for two years (e.g., 1988 to 1990), survival estimates changed by an average +7.0% when an additional year of recapture was added. In contrast, for cohorts that had been followed for three years or more (e.g., 1986 to 1990), survival estimates changed by an average 0.00% when an additional year of recapture was added. Thus, each yearly cliff swallow cohort had to be followed for three subsequent years before annual survival estimates ceased to vary with the addition of new data. This suggests that relatively few marked cliff swallows escaped detection for longer than three years, but it took this long to determine whether most birds had been/were present or not. (This three-year period applies specifically to our study and reflects both cliff swallow population size and our sampling effort, and it can be expected to vary for other species and study areas.) Similar analyses for adult cohorts also indicated that annual survival estimates did not stabilize until a cohort had been followed for three years or more.

Based on these results, we restricted all analyses in this chapter to cohorts from 1982–89 only. This gave three years of recapture for the 1989 cohort (1990, 1991, 1992) and more than three years for the earlier ones. It meant that all birds initially banded, or that otherwise would have begun being followed, in 1990–92 could not be used and are not included in these analyses. An unfortunate consequence of the three-year time lag between when a relevant cohort was marked and when it could be safely used in estimating survival was that we could include here recapture data for only two cohorts (1988, 1989) in years when we banded large numbers of birds (table 12.1). Another consequence is that the results reported in this chapter should be considered preliminary and subject to revision once additional cohorts come "on line" as our mark-recapture program continues.

12.2.4 Problems with Emigration

As in most bird population studies using mark-recapture (North 1988), we had no way to rigorously evaluate how many cliff swallows permanently emigrated out of the study area. Some surviving yearlings undoubtedly dispersed widely and hence could not figure in our calculation of survivorship. For example, one yearling from the study area was recovered in Edmonton, Alberta, about 1,700 km northwest of its birthplace. Some older adults also probably emigrated permanently. (Temporary emigrants that were absent one year but later returned were accounted for in the probability of recapture.) Absolute survival probabilities can likely be estimated only for completely sedentary species or species in which substantial numbers of marked individuals are encountered well beyond the study area, such as snow geese (Francis et al. 1992) or other waterfowl that are regularly hunted. That all field studies are done in a finite area and individuals generally cannot be recaptured or otherwise detected beyond that area is a recognized problem (e.g., Barrowclough 1978; Seber 1982; Payne 1990), but as yet no accepted statistical means exist to address it, especially among colonial species in which individuals are nonrandomly distributed (North 1988).

The survival probabilities we report in this chapter are relative ones and should not be taken to represent absolute measures of cliff swallow survivorship. We make the explicit assumption that long-distance dispersal out of the study area (both within and between years) is constant among the classes of individuals we consider. If this assumption holds, the survival probabilities we report are accurate relative estimates of survivorship. We note those cases in which the assumption of constant dispersal or emigration among classes is questionable (see section 14.2.2). Patterns of dispersal are addressed in sections 13.4 and 13.5.

12.3 EFFECTS OF AGE, SEX, AND YEAR

In this section we address how survivorship in our cliff swallow population was dependent on age, sex, and year. These analyses use all available banded individuals (through 1989; section 12.2.3) and are divided (in the case of nestlings) only by whether the natal colony was fumigated to remove ectoparasites. By including all birds in these analyses, we achieved large enough sample sizes to permit the use of more sophisticated models of survivorship than were possible in the other analyses based on fewer data. These results provide an overview of the demography of this population, useful in comparisons with other species.

12.3.1 Age-Specific Survivorship

The effect of age on cliff swallow survival probabilities could be addressed only for those birds initially marked as nestlings or juveniles, because

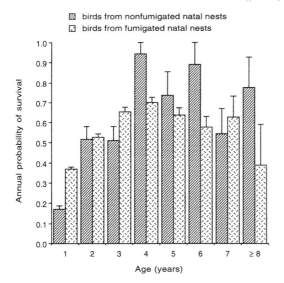

Fig. 12.1 Annual probability of survival for cliff swallows initially banded as nestlings from nonfumigated natal nests ($N = 3{,}321$ birds) and fumigated natal nests ($N = 7{,}708$ birds) in relation to age.

there are no reliable methods to determine the age of birds first caught as adults. Given the serious effect of ectoparasites on nestlings (section 4.7), we examined birds from nonfumigated and fumigated colonies separately (see section 12.6). Survivorship for one-year-olds (from fledging to the first breeding season) was lowest, gradually increasing and peaking at age four (fig. 12.1). Birds in the oldest age classes (seven years and older) seemed to show some reduced survivorship (senescence?), but the comparison is complicated by the large sampling errors associated with the survival estimates of the oldest age classes. The variance in annual survivorship estimates inevitably increases with age, because the sample size (birds of a cohort remaining alive) gets smaller each year. In this analysis (fig. 12.1) we used a model incorporating full age dependence (s_{a8}, p_t), which provided a significantly better fit than ones with constant survivorship (s, p_t) or fewer age classes (e.g., s_{a5}, p_t) for both nonfumigated and fumigated classes of birds (likelihood ratio tests, $p < .05$ for each).

Age-specific probabilities of survival differed depending on a bird's exposure to ectoparasites in the natal nest. Nestlings from fumigated nests were twice as likely to survive during their first year as birds from nonfumigated nests (fig. 12.1). This was a significant difference and one that reflected the long-term consequences (into a bird's first year) of being parasitized by swallow bugs (section 12.6). However, nestlings from nonfumigated nests had higher survival in age classes 4–6 and ≥ 8 years than did birds raised in parasite-free nests, and the difference was significant for four-year-olds (fig. 12.1). This may mean that swallow bugs cull the cliff swallow population so that the survivors of natal ectoparasitism tend to be the stronger birds that are apt to survive longer. A similar

370 • *Survivorship*

effect of parasites on purple martins was suggested by Davidar and Morton (1993). The markedly lower survivorship for first-year birds, compared with older age classes (fig. 12.1), is the basis for our using age-dependent models (usually s_{a2}, p_t) in all later analyses involving nestlings and juveniles.

12.3.2 Sex-Specific Survivorship

We found no evidence that overall survivorship of males and females in this population differed. Among birds initially caught as adults and sexed, beginning in 1986 (section 2.4), males (N = 6,523) had a 0.579 (± 0.006) average annual survival probability (model s, p_t) versus 0.574 (± 0.009) for females (N = 5,483). These nearly identical probabilities were not significantly different.

Similarly, there was no indication that males and females of certain age classes differed in survival probabilities. Because nestling and juvenile cliff swallows cannot be sexed, for this analysis we could use only birds at least one year old that were recaptured at breeding colonies and sexed. Those of known age and sex were placed into appropriate yearly cohorts, and their survival was estimated (fig. 12.2). Male and female survivorship did not differ significantly for any age class. Males, however, did show more of a decline in survivorship with age than did females.

These age- and sex-specific survival probabilities (fig. 12.2) are higher than the overall values for the population shown in figure 12.1. This was likely because the birds of known sex and age (fig. 12.2) were a nonrandom set of the population, defined by the fact that they had all returned to breed in the study area and were recaptured. However, the *relative* differences in survival probabilities between males and females should be ro-

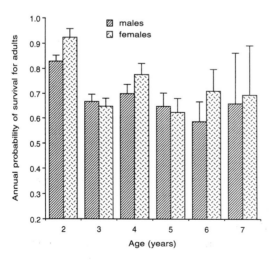

Fig. 12.2 Annual probability of survival for male (N = 1,233 birds) and female (N = 985 birds) cliff swallows initially banded as nestlings or juveniles (and subsequently recaptured at least once for sex determination) in relation to age.

bust. Given the similarity among the sexes in overall survivorship, we did not routinely separate the sexes in the later analyses in this chapter, many of which were based on smaller data sets where detection of significant sex-specific differences would have been even less likely.

12.3.3 Year-Dependent Survivorship

The probability of annual survival varied among years for our cliff swallow population. For all birds initially caught as adults, a time-dependent survivorship model (s_t, p_t) provided a significantly better fit than one with constant survival (s, p_t; likelihood ratio test, $p < .001$). Annual probability of adult survival during 1983–89 varied from 0.47 to 0.64 (fig. 12.3). Survival tended to vary with seasonal mean temperatures. Adults survived best (as measured to the next season) in 1986, 1987, and 1988, the years with the warmest June temperatures, daily highs averaging over 26.5°C (table 3.1). The years with the lowest adult survivorship (1984, 1989) were among the coolest. This pattern for adult survivorship was the opposite of that seen for annual reproductive success (section 11.3.1), in which the average number of young reared was lowest in warm years and highest in cool years. Warm years apparently enhanced adults' long-term survival prospects, perhaps through increased food availability and decreased incidence of weather-related mortality (e.g., section 10.6). The three years with the highest adult survivorship (1986–1988) differed significantly from the two years with the lowest (1984, 1989), with the remaining years (1983, 1985) not significantly different from any of the others (fig. 12.3).

First-year survivorship for birds initially banded as nestlings also differed among years (fig. 12.4). For cohorts from both nonfumigated and

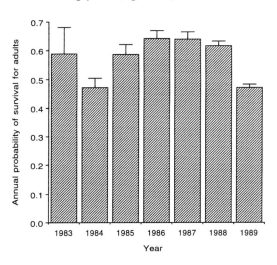

Fig. 12.3 Annual probability of survival for adult cliff swallows ($N = 15,083$ birds) in relation to year, for both sexes combined.

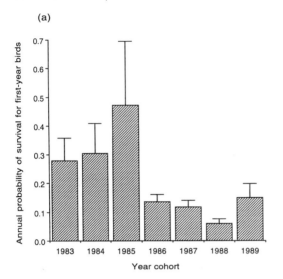
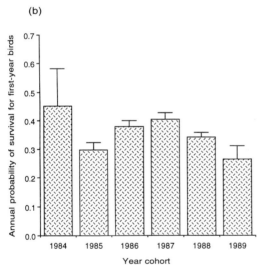

Fig. 12.4 Annual probability of first-year survival for cliff swallows reared in nonfumigated (a; N = 3,321 birds) and fumigated (b; N = 7,708 birds) nests in relation to the year they were hatched.

fumigated nests, a model with year-dependent survival (s_{a2t}, p_t) gave a significantly better fit than one without time dependence (s_{a2}, p_t; likelihood ratio tests, $p < .001$ for each). Among birds in nonfumigated nests exposed to swallow bugs, first-year survival (as measured to the next season) was lowest in the warm years of 1986–88 (fig. 12.4a). These were the same years in which adult survivorship was *highest* (fig. 12.3)! First-year survival was likely affected by ectoparasite load (section 12.6), and since warm years promoted swallow bug population growth, first-year survival reflected the consequences of being heavily parasitized in those years (fig. 12.4a). There was no such temperature-related pattern (or any other pattern) in the yearly differences among birds from fumigated nests (fig. 12.4b). These results parallel those for the number of young surviving to day 10 (section 11.3.1), indicating that the temperature/swallow bug interaction continued to express itself on nestlings after they fledged. In particular, first-year survival was extremely low following the hot 1988 season, with survivorship that year differing significantly from all other years (fig. 12.4a).

Although these analyses revealed that survival of adult and first-year cliff swallows varied among years, we were able to demonstrate these differences in large part because we used the entire data set that combined all birds from all colonies, band dates, and so on. Models with time-dependent survival were not statistically appropriate for other analyses in this chapter, which relied on smaller data sets. Where possible we investigated yearly differences, but readers should keep in mind that most, if not all, of the preliminary patterns in survivorship we report in this book may vary among years.

12.4 EFFECTS OF COLONY SIZE

Annual survivorship is one means through which the collective costs and benefits of coloniality can be measured (section 12.1). In cliff swallows, it seems reasonable that foraging efficiency, which may vary among colonies (chapter 10), influences an individual's survival from one year to the next, both by helping it avoid starvation during times of food scarcity (section 10.6) and by affecting its body mass and overall physical condition before fall migration (section 12.5). The risk of predation (chapter 8) obviously affects survival. The costly consequences of group living (e.g., chapters 4, 5) also may influence an individual's survival prospects, either directly (e.g., section 5.4) or indirectly through increased compensatory parental investment in offspring (that are heavily ectoparasitized, for instance). The first step in assessing these potential effects is to examine how cliff swallow survivorship varies with colony size.

12.4.1 Natal Colony Size

Nestling survivorship to fledging was used as our measure of reproductive success in chapter 11, in which we concluded that the highest number of young per nest were fledged in medium-sized colonies. Number of young fledged, however, cannot reflect any postfledging differences in parental abilities or inherent offspring quality. Recruitment into the next year's breeding population is perhaps the best measure of reproductive success, and to know this requires estimating survival probabilities of offspring after fledging.

We estimated first-year survival probabilities of cliff swallows banded as nestlings or juveniles as a function of natal colony size for both nonfumigated and fumigated sites. This reflected the birds' survivorship to their first breeding season (model s_{a2}, p_t). Survival probabilities among birds from nonfumigated natal colonies showed a clear peak at colonies of 100–249 nests (fig. 12.5a). Survivorship in colonies of that size was significantly greater than that in any other size class, except colonies of 1–10 nests (for which the estimate of survival was based on a relatively small sample and had a high standard error; fig. 12.5a). Thus we again found a

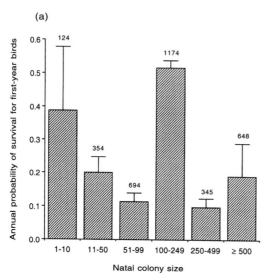

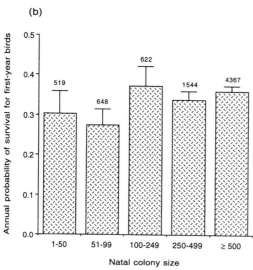

Fig. 12.5 Annual probability of first-year survival for cliff swallows reared in nonfumigated (*a*) and fumigated (*b*) nests in relation to natal colony size. Sample sizes (number of birds) are shown above error bars.

pattern in which birds in intermediate-sized colonies were the most successful (see section 11.3.2).

A different result held for fumigated colonies (fig. 12.5b). First-year survival probabilities for birds reared in the absence of ectoparasites showed no obvious relation to natal colony size. Survival was about the same for all birds, although, as among nonfumigated colonies, first-year survivorship was greatest (though not significantly so) for birds raised in colonies of 100–249 nests.

These results (fig. 12.5) emphasize especially the importance of ectoparasitism (chapter 4) in determining cliff swallow reproductive success. In the presence of swallow bugs, both the number of young fledged (section 11.3.2) and the survivorship of the nestlings that did manage to fledge were reduced in larger colonies. The enhanced survival probabilities for nestlings in the intermediate-sized colonies of 100–249 nests (fig. 12.5a) might reflect an "optimal" point where the colony is small enough that the costs of ectoparasitism are not sufficiently severe to seriously impair the nestlings' postfledging survival prospects (section 12.6), yet large enough that residents can exploit the various advantages of group living (e.g., information transfer, predator avoidance). The results on survivorship (fig. 12.5a), together with those on the number of young fledged (section 11.3.2), suggest that cliff swallow annual reproductive success under natural (nonfumigated) conditions peaks in colonies of about 50–200 nests; the exact peak within that range is not known with certainty and might fluctuate slightly in different seasons (see section 14.4.2). Unfortunately, sample sizes for the different natal colony size classes (fig. 12.5) were not sufficient to allow robust estimates of survival probabilities by year.

Did natal colony size affect survivorship of birds beyond their first year? Among nonfumigated colonies, there was no evidence for an effect of natal colony size on the birds' annual survival probabilities after their first breeding season (fig. 12.6). This suggests that the colony size effects (fig. 12.5a) were expressed largely through the extent of ectoparasitism nestlings experienced before and at the time of fledging (section 12.6). Once a bird survived to its first breeding season, conditions it experienced in its natal colony were probably irrelevant to its continued survival.

12.4.2 Breeding Colony Size

Once a cliff swallow returned to the study area to breed, its survival prospects were potentially influenced by the size of the colony it nested in. We assigned each breeding adult to one of eight colony size classes based on the site where it bred in a given year. We used a constant-survival model (s, p_t) because it was the only one statistically appropriate for the smaller colony sizes with smaller sample sizes. We used model s, p_t also because

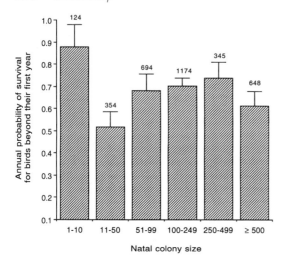

Fig. 12.6 Annual probability of survival beyond the first year for cliff swallows reared in nonfumigated nests in relation to natal colony size. Sample sizes (number of birds) are shown above error bars.

ages of adults were often unknown and a constant-survival model was less sensitive to potential differences in age structures of breeders among colonies (section 13.3.1). We separated nonfumigated and fumigated colonies, given the differences in parental provisioning that might be required for parasite-free versus naturally ectoparasitized broods.

Adult cliff swallows from nonfumigated and fumigated colonies showed the same pattern: annual survivorship increased for birds breeding in larger colonies (fig. 12.7). Annual probability of survival was lower in colonies of 1–50 nests than in all other colony sizes for both nonfumigated and fumigated sites, and the overall increase in survivorship with colony size was highly significant (fig. 12.7a) or approached significance (fig. 12.7b). Birds breeding in the larger nonfumigated colonies had about a 50% greater annual survival probability than those in the smallest colonies.

As with our analysis of natal colony size (section 12.4.1), we were unable to systematically investigate yearly differences in how adult survivorship varied with breeding colony size. Sample sizes through 1989 were not sufficient in most cases to permit use of a model with time-dependent survival (e.g., s_t, p_t), especially for the smaller colony size classes. Yearly differences in how breeding colony size affected adult survival seem possible, however, given the yearly variation in mean body mass among colonies (section 10.5.2) and the apparent effect of body mass on adult survival (section 12.5.3).

Among nonfumigated colonies, a model with time-dependent survival was appropriate only for colony size classes of 100–249 and 1,000–1,999 nests. For the other colony sizes, either sample sizes were not suffi-

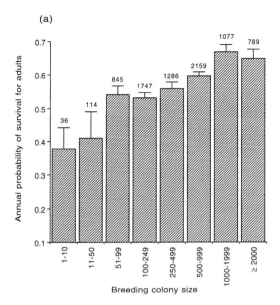

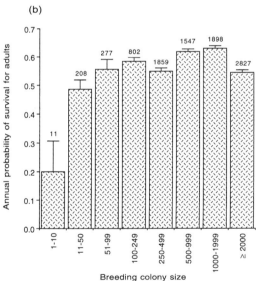

Fig. 12.7 Annual probability of survival for adult cliff swallows in relation to the size of the breeding colony they occupied. Nonfumigated colonies are shown in (a), fumigated ones in (b). Sample sizes (number of birds) are shown above error bars. Probability of survival increased significantly with breeding colony size for nonfumigated sites ($r_s = .95$, $p < .001$, $N = 8$ size classes) and approached significance for fumigated sites ($r_s = .57$, $p = .14$, $N = 8$ size classes).

cient to model survival as time dependent or there was not enough yearly variation in survivorship for a time-dependent model to provide a significantly better fit. In the 100–249 nest class, annual survival probabilities ranged from 0.319 (± 0.150) in 1984 to 0.794 (± 0.146) in 1987; in the 1,000–1,999 nest class, the range was from 0.261 (± 0.206) in 1984 to 0.834 (± 0.116) in 1987. In 1984 and 1989 (the cool years), annual probability of survival was greater for birds breeding in the 100–249 nest colonies than for ones in the 1,000–1,999 nest sites, significantly so in 1989. Thus, although larger colonies on average conferred an advantage in terms of increased survivorship for breeders (fig. 12.7), there was annual variation to the extent that in occasional years birds from smaller colonies fared better. We hope to explore these potential yearly differences in the near future as additional, larger cohorts become available for analysis.

Survival probabilities shown in figure 12.7 were estimated based on the size of colony each individual occupied in a single year. Birds captured in multiple years were included in multiple cohorts, based on their colony's size in each year. We also examined survivorship among birds known to occupy colonies of particular sizes in more than one year, to determine how longer-term history of colony use might influence survivorship. For this analysis we used only cliff swallows caught at breeding colonies in two consecutive years. We used the same colony size classes as in figure 12.7 and separated birds into those that used the same colony size in both years versus those that switched to larger or smaller colonies the second year. Annual survival probabilities (model s, p_t) were estimated from the second year on (because by definition all birds included in this analysis had survived from year 1 to year 2) as a function of colony size the first year (fig. 12.8). We combined nonfumigated and fumigated colonies, given the similarity among them in average survivorship in relation to colony size (fig. 12.7).

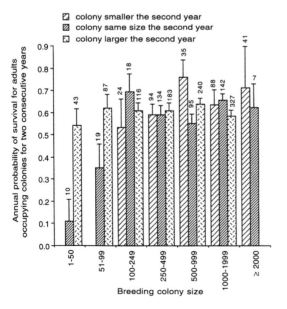

Fig. 12.8 Annual probability of subsequent survival for adult cliff swallows that occupied known colonies in two consecutive years in relation to the size of the breeding colony they occupied the first year. Sample sizes (number of birds) are shown above error bars.

As might be expected based on the earlier analysis (fig. 12.7), birds that remained in the same colony size in two consecutive years had greater expectation of survival as colony size increased (fig. 12.8; see section 13.4.2). However, switching colony sizes between years often had a major impact on subsequent survival. Birds occupying colonies of fewer than 100 nests the first year improved their survival probabilities substantially by switching to larger colonies the second year (fig. 12.8). This suggests that survivorship was indeed affected by colony size and that the earlier pattern (fig. 12.7) was not simply a result of preferential settlement in small colonies by inferior birds that were less likely to survive anyway. Most of the other differences among birds of different colony size histories were not significant (fig. 12.8). A more refined analysis, accounting for the exact sizes of colonies that birds switched to (beyond simply "larger" or "smaller"), might yield further insight into the consequences of using various colony sizes. Sample sizes through 1989 were too low for us to attempt a more detailed analysis at present.

The principal conclusion of this section is that adult cliff swallows improved their expectation of survivorship by breeding in large colonies, whether ectoparasites were present or not (fig. 12.7). This represents a potentially great benefit of coloniality through increased average life span and associated lifetime reproductive success (section 14.2). It is less clear, however, *how* this effect is achieved. Perhaps the most plausible explanation is that large colonies increase adults' foraging efficiency (chapter 10) and decrease energy costs (e.g., section 5.3.1) so that residents are in better physical condition at the end of the nesting season and thus more likely than birds from small colonies to survive the winter and the arduous round-trip between Nebraska and South America. The positive correlation between late-season body mass and adult survival probability (section 12.5.3) supports this conclusion. On the other hand, adult body mass late in the season did not vary with colony size in all years (section 10.5.2), suggesting that other unidentified colony size effects also may have contributed to the increased adult survivorship in larger colonies.

We believe that the increased survivorship of cliff swallows breeding in larger colonies is a direct result of colony size and not a demographic artifact. For example, one alternative possibility is that middle-aged and older birds might have settled preferentially in large colonies, with yearlings and younger birds in small colonies. Since survivorship varied with age (section 12.3.1), such an age-related colony settlement pattern might lead to the observed results (fig. 12.7). This is unlikely, however, because older birds tended to settle in smaller colonies (section 13.3.1). Thus we might expect, based on age alone, that birds in smaller colonies should have shown higher survival, which was not the case. Another possibility is that birds in small colonies were more likely to disperse out of the study area between years, leading to reduced apparent survivorship, but there was no reason to expect differential dispersal. If long-distance breeding dispersal was determined by the prior year's nesting success (as in other birds: Harvey et al. 1979; Part and Gustafsson 1989), there should have been no dispersal differences among residents of the smallest and largest colonies. This was because the number of young fledged per nest showed a bell-shaped pattern with birds in the very small and very large colonies doing about equally well (section 11.3.2). Birds in medium-sized colonies fledged the most young per nest, and thus if differential dispersal influenced the apparent survival probabilities, birds in colonies of 30–80 nests should have "survived" best, which was not the case.

12.5 EFFECTS OF BODY MASS

In earlier chapters we examined how body mass of adult and nestling cliff swallows varied with colony size and presumably reflected specific costs or benefits of coloniality (e.g., sections 4.7, 10.4, 10.5). We assumed that larger mass indicates greater energy reserves for an individual, thereby enabling it to better endure stressful periods (such as bad weather, weaning, or migration) or devote proportionately more resources to either self-maintenance (survival) or parental care (Martin 1987; Thompson and Flux 1988; Bryant 1991; Clutton-Brock 1991). In this section we investigate how annual survival varies with body mass.

12.5.1 Nestling Body Mass

Body mass of each nestling cliff swallow was recorded at 10 days of age (section 2.3.1). We separated nestlings from nonfumigated and fumigated colonies and assigned each bird to an arbitrary mass class based on its weight at day 10. The probability of annual survival was estimated (model s_{a2}, p_t) for each mass class, yielding a measure of survivorship for yearlings to their first breeding season. We combined birds from colonies of all sizes, because sample sizes were not sufficient to examine a range of masses within each colony size separately.

Body mass at 10 days did not seem to affect first-year survivorship for nestlings in nonfumigated nests, at least within the range of body masses observed (fig. 12.9a). This result was somewhat surprising, because we predicted that heavier nestlings should have had higher survival (as in other birds; e.g., Martin 1987; Krementz, Nichols, and Hines 1989; Tinbergen and Boerlijst 1990; Magrath 1991; Thompson, Flux, and Tetzlaff 1993). In contrast and as predicted, first-year survivorship for cliff swallows from fumigated nests increased significantly with 10 day body mass (fig. 12.9b). Among birds from fumigated natal nests, survival probabilities increased by over 50% across the observed range of nestling body masses.

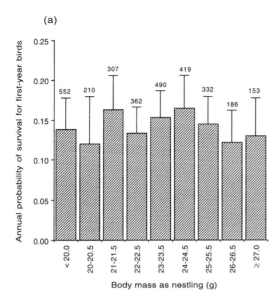

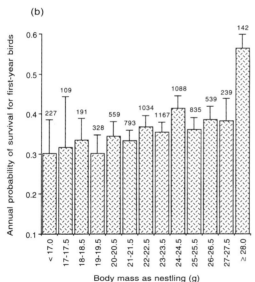

Fig. 12.9 Annual probability of first-year survival for cliff swallows reared in nonfumigated (*a*) and fumigated (*b*) nests in relation to their body mass (g) as nestlings at 10 days of age. Sample sizes (number of birds) are shown above error bars. Probability of survival did not vary significantly with body mass for birds from nonfumigated sites ($r_s = -.07$, $p = .86$, $N = 9$ classes) but increased significantly with body mass for birds from fumigated sites ($r_s = .90$, $p < .001$, $N = 13$ classes).

The results from fumigated colonies (fig. 12.9b) show that, in the absence of ectoparasites, the long-term survival prospects for nestlings were enhanced when nestlings achieved increased body mass. High nestling body mass for a given brood size was probably attributable to enhanced parental foraging efficiency (Snapp 1976; Hoogland and Sherman 1976; Brown 1988a; section 10.4). But why was the same result not seen with birds from nonfumigated colonies? Perhaps mass did not directly reflect fat reserves of birds exposed to parasites (Thompson, Flux, and Tetzlaff 1993), and thus no relationship between mass and survivorship would be expected. Why mass presumably would reflect fat reserves for nestlings from fumigated nests (fig. 12.9b) but not for those from nonfumigated nests is unclear, however.

A more likely reason that first-year survivorship did not vary with body mass in the presence of ectoparasites is that mortality in the first year was so high, relative to that of birds from fumigated colonies (section 12.3.1), that any effect of nestling body mass was swamped by the much larger effect of swallow bugs (sections 4.7, 12.6). When swallow bugs were removed, we observed not only an effect of body mass on survival but also a greater range in body masses (fig. 12.9), possibly because fumigation "allowed" more low-weight nestlings to fledge than would have been possible in the presence of ectoparasites.

The lack of a correlation between nestling body mass and first-year survival (fig. 12.9a) suggests that parents cannot enhance their nestlings' chances of postfledging survival by increasing their rate of provisioning beyond that necessary to compensate for ectoparasitism. This underscores the serious cost of ectoparasitism for these birds. The foraging advantages of larger colonies (chapter 10) may partly or fully compensate for the effects of swallow bugs, but these data (fig. 12.9a) imply that information transfer cannot represent a net benefit solely through *increased* nestling body mass (and resulting better first-year survival) whenever ectoparasites are also present (see section 10.10).

12.5.2 Juvenile Body Mass

Body mass taken when nestlings are 10 days old provides a relative measure of nestling condition among colonies but may not necessarily reflect relative differences among individuals at the time of fledging. This is because nestling cliff swallows remain in nests up to twenty-four days, giving them up to two additional weeks during which their relative weight may change before fledging. If low-weight nestlings at day 10 "make up" some of their deficiency before fledging (Shields and Crook 1987; Johnson, Eastman, and Kermott 1991; Johnson and Albrecht 1993), one might find no relation between 10 day body mass and subsequent survivorship (e.g., fig. 12.9a). Therefore we examined how nestling mass at day 10 correlated with mass at fledging, and how mass at fledging affected probability of survival. For these analyses we used birds caught as juveniles that had been out of the nest one day or more (see section 6.4.2) and that had earlier been weighed as nestlings at day 10.

Juveniles were caught and reweighed an average 17.35 ($\pm$ 0.43) days after their first weighing. This meant that their juvenile body masses were recorded about 3 days after fledging, on average. (We could not weigh

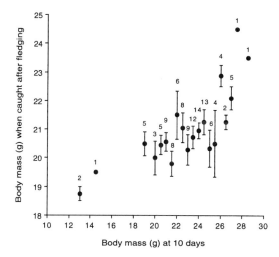

Fig. 12.10 Body mass (g) of juvenile cliff swallows when caught after fledging in relation to their body mass as nestlings at 10 days of age. Mean juvenile mass is shown for each 0.5 g increment in nestling mass. Numbers by dots are sample sizes (number of birds recaptured as juveniles) for each nestling mass class. Mean body mass as a juvenile increased significantly with body mass as a nestling ($r_s = .78$, $p < .001$, $N = 20$ mass classes).

older, unfledged nestlings near the time of fledging because often they would then fledge prematurely.) In these analyses we combined juveniles from nonfumigated and fumigated colonies, for two reasons: too few juveniles from nonfumigated colonies were caught to permit separate analyses; and combining gave us a greater range of nestling body masses to use in our correlation. Treating each individual as independent, a bird's nestling body mass and juvenile body mass were significantly positively correlated ($r_s = .37$, $p < .001$, $N = 119$ birds). Juveniles weighed an average 2.09 ($\pm$ 0.23) g less than they had at day 10. We also examined mean juvenile mass as a function of nestling mass (fig. 12.10). There was a significant positive correlation between mass at day 10 and mass soon after fledging. These results illustrate that differences among nestlings at day 10 were maintained to fledging and beyond. Underweight nestlings apparently did not make up the difference before fledging, and heavier ones did not lose their potential advantage.

Juveniles exhibited less total variation in body mass than did nestlings at 10 days. Survival probabilities were estimated (model s_{a2}, p_t) for all birds of different masses as measured to the nearest 0.5 g. Probability of survival in the first year after fledging increased significantly with juvenile mass (fig. 12.11), with a 100% increase in survivorship over the range of masses observed. This result is perhaps not surprising, given that differences in relative mass were maintained to fledging (fig. 12.10). The values of the survival probabilities for juveniles (fig. 12.11) are largely consistent with the survival probabilities of nestlings from fumigated colonies (fig. 12.9b), as might be expected.

Unfortunately, we do not know how body mass at fledging in the presence of ectoparasites affected first-year survival. Relatively few juveniles

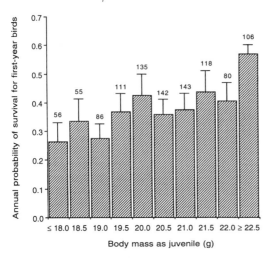

Fig. 12.11 Annual probability of first-year survival for cliff swallows in relation to their body mass (g) as juveniles soon after fledging. Sample sizes (number of birds initially caught as juveniles) are shown above error bars. Probability of survival increased significantly with juvenile body mass ($r_s = .87$, $p = .001$, $N = 10$ mass classes).

were recaptured at nonfumigated colonies. This was partly due to the logistics of sampling certain nonfumigated colonies late in the year when juveniles were likely to be caught, but much of it was also due to reduced nestling survival in nonfumigated colonies (section 4.7): there were relatively fewer juveniles to recapture in those colonies.

12.5.3 Adult Body Mass

Body mass of adult cliff swallows presumably reflects their foraging success and may vary with colony size at certain times of the year (section 10.5). Knowing how body mass affects an adult's long-term survival is critical in assessing the importance of potential foraging-related advantages of coloniality that influence body mass.

Adults were caught at various times during the nesting season (section 2.3.2), and their body masses at each of these times might influence their probability of survival. Body mass late in the season likely has the greatest effect on survivorship to the next season, however, reflecting the energy reserves with which an individual starts its fall migration. We used masses taken from adults during the "late" period of feeding nestlings (sections 2.3.2, 10.5). Because cliff swallows often left the study area soon after their young fledged (but see section 13.2.6), body mass taken while the birds were feeding nestlings probably reflected their condition at or near the start of migration.

Adults were weighed to the nearest 0.5 g. For those birds captured more than once during the late period, an average mass across days was used (section 10.5.1). In this analysis, all average mass measurements were rounded to the nearest 0.5 g, and survival probabilities were estimated for all the adults in each 0.5 g class. Given the similarity among the sexes in overall survivorship (section 12.3.2), we did not separate males and females in analyzing the effect of body mass. We also combined birds from nonfumigated and fumigated sites, given the apparent lack of any direct effect of nest fumigation on the ectoparasites found on adult cliff swallows (fleas, lice). We used a model with constant survivorship (s, p_t), because this was the only one statistically justified for most mass classes. This analysis pooled all colonies and years. Sample sizes were insufficient to allow time-dependent models or to provide enough mass variation within each colony size.

An adult's expectation of annual survivorship increased significantly with its late-season body mass (fig. 12.12). Probability of survival increased by about 14% over the observed range of body masses. This was a smaller effect of body mass than that seen for nestlings (fig. 12.9b) or juveniles (fig. 12.11), but it still seems to represent an advantage for adults that can attain higher mass at the end of the season. These results (fig. 12.12) suggest that breeders that allow their own energy reserves to deteriorate during the season (perhaps through increased provisioning of nestlings or greater ectoparasitism by fleas and lice; see Brown, Brown, and Rannala 1995) may reduce their likelihood of surviving to the next year. Combining years for this analysis obscured potential yearly trends, but it is possible that there was some yearly variation in this pattern, especially given the variation in adult body mass among years (section 10.5).

An increase in annual survival with late-season mass (fig. 12.12) is consistent with our premise (section 10.5) that increased body mass is ad-

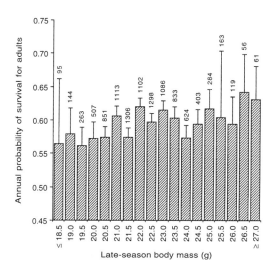

Fig. 12.12 Annual probability of survival for adult cliff swallows in relation to their body mass (g) late in the season during feeding of nestlings. Sample sizes (number of birds) are shown above error bars. Probability of survival increased significantly with late-season body mass ($r_s = .67$, $p = .002$, $N = 18$ mass classes).

vantageous for adults (but see Witter and Cuthill 1993). Patterns of late-season body mass variation among adults in different-sized colonies (section 10.5.2) are therefore likely to have an impact on average survivorship and expected lifetime reproductive success (section 14.2). Insofar as social foraging and information transfer increase an adult's body mass in larger colonies (section 10.5.3), foraging advantages represent a benefit of cliff swallow coloniality.

12.6 EFFECTS OF ECTOPARASITES

Earlier analyses (section 4.7) showed that ectoparasites, especially swallow bugs, in some colonies seriously reduced survivorship of nestling cliff swallows before 10 days of age and depressed the body mass and overall health of birds that survived. The ectoparasites' prefledging effects were substantial and represented the single greatest cause of nest failure in cliff swallows. But the effects of these nest-based ectoparasites may continue to be important even after nestlings fledge, if the birds fledge at lower body mass or experience greater physiological stress (Chapman and George 1991) when heavily parasitized as nestlings. In this section we evaluate the long-term effects of ectoparasites on first-year survivorship for birds reared in nests with differing numbers of ectoparasites. We recently addressed elsewhere the long-term effects of ectoparasites on adult birds (Brown, Brown, and Rannala 1995; see section 4.8) and refer interested readers to that paper for details on the survival-related costs of ectoparasitism for adult cliff swallows.

We analyzed the effects of ectoparasites on nestlings in two ways. One approach was to compare survivorship for birds raised in nonfumigated nests exposed to natural levels of ectoparasitism with that for birds from parasite-free (fumigated) natal nests (section 12.3.1). The other approach was to compare survivorship for birds from nonfumigated natal nests with different relative parasite loads, measured as the number of swallow bugs or fleas present on each surviving nestling's body at day 10 (section 2.6; Brown and Brown 1992). We emphasize that these counts provided only relative measures of ectoparasitism; nestlings without parasites were not necessarily from nests containing none, but their nests tended to have fewer than those of nestlings on whom, for example, we counted four parasites (section 2.6).

For all birds combined, probability of first-year survival for those raised in nonfumigated nests ($N = 3,321$ birds) was 0.171 ($\pm$ 0.018), versus 0.369 ($\pm$ 0.011) for those raised in fumigated nests ($N = 7,708$ birds). This significant difference in survival probabilities illustrates how the deleterious effects of parasites continue past the time of fledging. The presence of ectoparasites reduced annual survivorship for first-year birds by approximately one-half, relative to those from fumigated nests. This was about the same percentage decrease as seen during the prefledging stage (section 4.7). These long-term effects of parasites likely occurred through reduced body mass at fledging and other forms of physiological stress caused by exposure to large numbers of hematophagous swallow bugs (see Chapman and George 1991). As noted in section 12.3.1, however, the effects of ectoparasitism did not last beyond the first year.

Among nonfumigated nests from all colonies combined, swallow bug parasite load affected first-year survival (fig. 12.13a). Survivorship de-

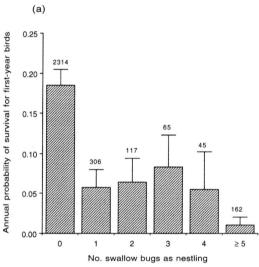

Fig. 12.13 Annual probability of cliff swallow survival during the first year (a) and beyond the first year (b) in relation to the number of swallow bugs counted on nestlings at 10 days of age. Sample sizes (number of birds) are shown above error bars. Probability of first-year survival declined almost significantly with extent of bug parasitism ($r_s = -.77$, $p = .072$, $N = 6$ classes); probability of survival beyond the first year did not vary significantly with extent of bug parasitism ($r_s = -.30$, $p = .62$, $N = 5$).

clined with rising bug parasitism; this negative correlation approached significance even with only six classes of parasitism (sample sizes were insufficient for finer classes). Virtually no nestlings that fledged from heavily infested nests (five or more bugs per bird) survived their first year, suggesting that the long-term cost of heavy natal ectoparasitism is high. This long-term cost may have been greater in larger cliff swallow colonies where nestlings were more heavily parasitized (section 4.3), although samples were too small to perform this analysis (fig. 12.13a) separately for each colony size. As in the comparison between nonfumigated and fumigated birds (above and section 12.3.1), there was little evidence for a consistent bug-related survival cost for birds beyond their first year (fig. 12.13b). Most bug-related mortality among fledged individuals probably occurs in the stressful first few weeks (or days) following fledging. At that time juveniles are learning to forage for themselves and are likely undergoing periods of reduced food intake before migration.

Parasitism by fleas did not seem to affect nestling survivorship to day 10 (section 4.7.1). We found a similar result when analyzing how flea load affected first-year survivorship (fig. 12.14). Among nonfumigated nests, probability of survival from fledging to the first breeding season did not vary consistently with natal flea load (fig. 12.14a). Overall probability of survival among this group of birds was low, however, probably owing to the overriding effects of swallow bugs. Fumigation, which seemed to affect fleas only partly (section 2.7), provided an opportunity to assess the potential effects of fleas on first-year survival in the absence of swallow bugs.

Probability of survival to the first breeding season did not vary significantly with natal flea load in the absence of swallow bugs (fig. 12.14b). If anything, birds with more fleas as nestlings seemed to have *increased* first-year survival. This result paralleled that found for the effects of fleas on nestling body mass (section 4.7.1; fig. 4.7): more fleas were associated with heavier nestlings. This was likely because heavier nestlings were larger and for that reason encountered more fleas in the nest. Heavier nestlings with more fleas probably survived better (fig. 12.14b) simply because of their weight (section 12.5.1). These results (fig. 12.14) support the conclusion (section 4.7) that flea parasitism does not represent an important cost to nestling (and thus first-year) cliff swallows.

The difference in first-year survival probabilities for birds from nonfumigated (0.171) versus fumigated (0.369) natal nests presumably reflects the long-term survival cost of swallow bug parasitism. However, these survival probabilities might also reflect permanent emigration out of the study area by yearlings (section 12.2.4): perhaps birds heavily parasitized as nestlings are more likely to emigrate. This possibility was suggested by our finding an effect of ectoparasitism on short-range yearling dispersal within the study area (Brown and Brown 1992). The more heavily ecto-

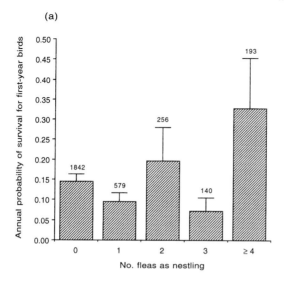
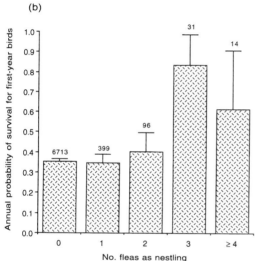

Fig. 12.14 Annual probability of first-year survival for cliff swallows reared in nonfumigated (*a*) and fumigated (*b*) nests in relation to the number of fleas counted on them at 10 days of age. Sample sizes (number of birds) are shown above error bars. Probability of survival did not vary significantly with extent of flea parasitism for birds from nonfumigated nests ($r_s = .30$, $p = .62$, $N = 5$ classes) but increased, almost significantly, with extent of flea parasitism for birds from fumigated nests ($r_s = .80$, $p = .10$, $N = 5$).

parasitized nestlings tended to move to nonnatal colonies to breed the next season, whereas lightly ectoparasitized nestlings returned to their natal colony. If high levels of ectoparasitism also make parasitized nestlings more likely to disperse long distances out of the study area, ectoparasites might "affect" only dispersal and not first-year survivorship.

Four kinds of evidence suggested that the survival probabilities reported here reflect real differences in relative survival rather than differential dispersal by parasitized versus nonparasitized birds. First, although nonparasitized nestlings were more likely to return to their natal colony (Brown and Brown 1992), there was no consistent difference in dispersal distances for first-year birds from fumigated versus nonfumigated colonies when *all colonies exclusive of the natal colony* were considered. For example, the mean linear distance between a bird's natal colony and its breeding colony the subsequent year (see section 13.5.1) was 6.26 ($\pm$ 0.95) km for birds dispersing from nonfumigated natal colonies (N = 109 birds), versus 4.12 ($\pm$ 0.53) km for birds dispersing from fumigated natal colonies (N = 242); the difference was not significant (Wilcoxon rank sum test, p = .46). Thus parasitized birds were not inherently more likely than nonparasitized birds to disperse over the distances we sampled within the study area.

Second, yearly recapture probabilities (sensu Clobert et al. 1985; Clobert, Lebreton, and Allaine 1987; Lebreton et al. 1992) were not consistently greater for birds from fumigated nests. If heavily parasitized nestlings were more likely to move long distances, they would be less likely to be recaptured their first year. We estimated age- and time-specific recapture probabilities for first-year birds from nonfumigated versus fumigated natal nests (model s_{a8}, p_{a2t}). This was done for the five years from 1985 to 1989, given that we began marking nestlings in fumigated colonies in 1984. The mean yearly recapture probability for first-year birds from nonfumigated colonies was 0.270, versus 0.346 for birds from fumigated colonies. In two of the years recapture probabilities were higher for birds from nonfumigated colonies, whereas in three years they were higher for birds from fumigated colonies. Thus there was little consistent difference in our encountering of first-year birds from nonfumigated versus fumigated colonies.

Third, because fleas seem to influence natal dispersal in cliff swallows (Brown and Brown 1992), one might expect birds with higher flea loads to be more likely to undertake long-range dispersal. That birds with higher flea loads showed *increased* apparent survival (fig. 12.14) suggests that fleas were not causing higher emigration by the more heavily parasitized nestlings. Although ectoparasites clearly influenced whether individuals returned to the natal colony, there was little evidence that they influenced long-range dispersal to a similar extent.

Fourth, the overall first-year survival probability estimated for cliff swallows from nonfumigated natal nests (0.171) was similar to that estimated for migratory bank swallows in Britain based on recoveries of dead birds outside the banding area (0.229; Mead 1979a). Thus the probabilities of survival reported in this section (figs. 12.13, 12.14) were unlikely to be artifacts created by dispersal differences among different classes of birds (see also section 14.2.2).

12.7 OTHER CORRELATES OF FIRST-YEAR SURVIVORSHIP

Survivorship of first-year cliff swallows was potentially affected by variables other than colony size, body mass, and ectoparasites. For instance, both date (section 11.4) and position in the colony (section 11.5) influenced prefledging reproductive success and might also affect first-year survivorship. In this section we evaluate whether survivorship to the first breeding season varied with hatching date, position of the natal nest within a colony, and natal nest age.

12.7.1 Hatching Date

Reproductive success of cliff swallows, as measured by the number of nestlings surviving to fledging, declined as the nesting season advanced (section 11.4). This pattern, which was strongest for birds in nonfumigated nests but also held for parasite-free nests (fig. 11.10), was attributable in part to the effects of ectoparasites but may also have reflected diminishing food availability over the summer (section 11.4). The postfledging survivorship of juveniles raised in fumigated nests is especially relevant to the issue of seasonal food availability, given that the confounding effects of swallow bugs were removed.

All birds banded as nestlings from all colonies were grouped into arbitrary three-day periods based on their hatching date. First-year survivorship was estimated for each period (model s_{a2}, p_t). The sample from fumigated nests was larger and probably for that reason spanned a greater range in dates. For both nonfumigated and fumigated nests, first-year survivorship declined significantly with hatching date (fig. 12.15). This meant that cliff swallows raised later in the summer had less chance of survival *after fledging* than did birds hatched earlier in the breeding season. This result (fig. 12.15a) is not surprising for birds from naturally infested nests, because the numbers (and thus the effects) of swallow bugs increased later in the season (sections 4.4, 12.6). However, the decline in first-year survivorship for birds from fumigated natal nests (fig. 12.15b) suggests that there are disadvantages besides ectoparasitism for later-reared birds.

First-year survivorship was reduced by about one-half over the observed range of hatching dates in the absence of ectoparasites (fig. 12.15b). Possible causes of the seasonal reduction include a general decline in food availability, fewer nests active late in the season that may be kleptoparasitized (section 6.4.2; and see Frumkin 1994), and less time to accumulate energy reserves after independence from parents and before migration (see Thompson and Flux 1988; Sullivan 1989; Magrath 1991; Wunderle 1991). Thus, as in other birds (reviewed in Martin 1987), there are inherent long-term disadvantages for cliff swallows hatched late in the year.

392 • *Survivorship*

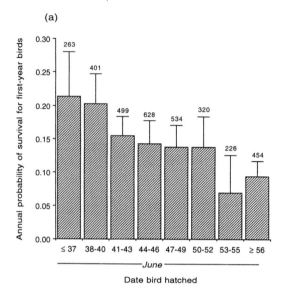

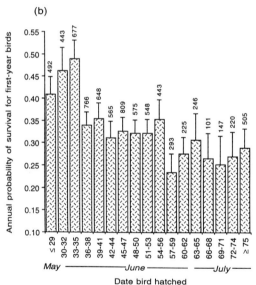

Fig. 12.15 Annual probability of first-year survival for cliff swallows reared in nonfumigated (*a*) and fumigated (*b*) nests in relation to the date they hatched (01 = 1 May). Sample sizes (number of birds) are shown above error bars. Probability of survival declined significantly with hatching date for birds from both nonfumigated ($r_s = -.97$, $p < .001$, $N = 8$ periods) and fumigated nests ($r_s = -.85$, $p < .001$, $N = 17$ periods).

These disadvantages are somewhat independent of ectoparasitism and compound the seasonal decline in number of young fledged per nest (section 11.4). First-year survivorship in the presence of ectoparasites (fig. 12.15a) was highest for birds hatched earliest in the year, suggesting that cliff swallows should nest as early as possible to minimize the parasites' effects (see section 4.4).

12.7.2 Position in the Colony

A cliff swallow nest's position in the colony influenced the number of young fledged (section 11.5), owing in part to both predation (section 8.7) and ectoparasitism (section 4.5.3) that could vary in different parts of the colony. A nest's spatial position also reflected the age of its owners, with younger birds occupying more peripheral nest sites within a colony (section 11.8). In this section we examine whether the natal nest's position in the colony affected the postfledging survivorship of birds raised in it. Predation at the nest was not relevant for birds after fledging, but position-related effects of ectoparasites (section 12.6) and parental age (section 11.8) might still influence first-year survivorship.

As a measure of a natal nest's spatial position, we used distance from the center nest (section 2.8.2) in this analysis. Because colonies varied in how closely nests were clustered (e.g., fig. 3.6), actual linear distances did not equally reflect, between colony sites, how peripheral a nest was relative to other nests in the same colony. Consequently, in earlier analyses of spatial position we analyzed each colony separately (e.g., sections 4.5, 11.5). This was not possible in studying survivorship, however, because prohibitively large sample sizes (number of birds banded) per colony site would have been necessary to yield useful estimates of survival probabilities. Therefore, for each nest with banded nestlings we determined the number of nests between it and the centermost nest in the colony, irrespective of the actual linear distance, similar to the methods used to determine "center" and "edge" nests within a colony (see section 2.8.2). Thus we could rank nests at different colonies on the same scale, reflecting their relative proximity to the center and enabling us to pool data across sites and achieve sufficient sample sizes for survivorship analyses. We estimated first-year survival probabilities (model s_{a2}, p_t) for arbitrary five- and ten-nest intervals, chosen to be large enough to yield estimates of survivorship with narrow confidence intervals and yet small enough to be biologically meaningful.

First-year survivorship declined with an increase in the distance of the natal nest from the colony's center for birds from fumigated nests but not for those from nonfumigated nests (fig. 12.16). Swallow bugs obviously can be ruled out as a cause of this pattern for fumigated nests, suggesting that birds raised in more peripheral nests may face reduced postfledging survival prospects for other reasons. The lower age of the parents at peripheral nests could be one possibility: perhaps younger parents provision their offspring less often than do older parents and consequently the offspring fledge at a lower body mass. The later laying times of these same parents (section 11.8) also could contribute to this pattern (fig. 12.16). (The relative importance of the various factors influencing first-year sur-

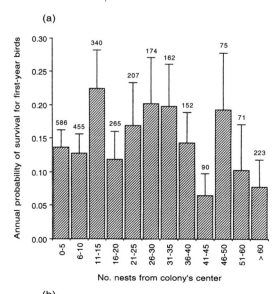
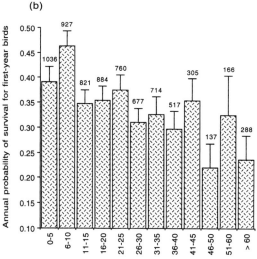

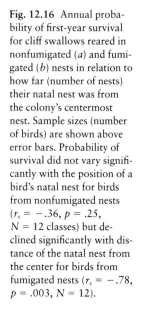

Fig. 12.16 Annual probability of first-year survival for cliff swallows reared in nonfumigated (*a*) and fumigated (*b*) nests in relation to how far (number of nests) their natal nest was from the colony's centermost nest. Sample sizes (number of birds) are shown above error bars. Probability of survival did not vary significantly with the position of a bird's natal nest for birds from nonfumigated nests ($r_s = -.36, p = .25, N = 12$ classes) but declined significantly with distance of the natal nest from the center for birds from fumigated nests ($r_s = -.78, p = .003, N = 12$).

vivorship could be addressed in principle through multivariate statistical tests, although no such tests exist at present for survival probabilities as estimated by SURGE. As we continue our mark-recapture program, we may eventually be able to tease apart some of these effects directly through appropriate sorting of the expanded data set.) The lack of a significant trend for nonfumigated nests (fig. 12.16a) may reflect the overwhelming long-term effects of swallow bugs that swamp the other influences on first-year survivorship (e.g., section 12.5.1). These results in general (fig. 12.16b) underscore the potential disadvantages associated with occupying edge nests and illustrate that not all birds in a given colony can expect similar reproductive success.

12.7.3 Natal Nest Age

Another potential influence on first-year survivorship was the age of the natal nest. Nest age was important because of possible differences in ectoparasite load (section 4.5.1) and in energy investment by parents. Parents constructing new nests, for example, might have reduced resources available for provisioning their young, leading to lower postfledging survival prospects for these offspring.

For all birds from nonfumigated colonies, the first-year survival probability for cliff swallows fledging from old nests ($N = 1,330$ birds) was 0.144 (± 0.016), versus 0.139 (± 0.024) for birds fledging from new nests ($N = 986$ birds). The difference was not significant. For all birds from fumigated colonies, the first-year survival probability for those fledging from old nests ($N = 3,624$ birds) was 0.346 (± 0.014), versus 0.322 (± 0.016) for those fledging from new nests ($N = 2,815$ birds). This difference also was not significant. There was thus no evidence that natal nest age affected postfledging survivorship of young in either the presence or the absence of ectoparasites. This result is consistent with our finding little difference among old and new nests in the number of young fledged (section 11.6.1). Use of old versus new nests seems to yield a roughly equivalent payoff in terms of reproductive success, despite the presumably different costs for each (ectoparasitism and nest building, respectively). Because first-year survivorship was so similar for old and new natal nests for the data set as a whole, we did not attempt finer-grained analyses of nest age and survival.

12.8 OTHER CORRELATES OF ADULT SURVIVORSHIP

In earlier sections we showed that annual survivorship for breeding adult cliff swallows was influenced by breeding colony size and late-season body mass. In this section we evaluate two other factors that might affect adult survival probabilities from year to year: the position of a breeder's nest within the colony and the nest's age.

Analyses in this section used only breeding adults for which we knew nest location within a colony. These individuals were primarily those we color-marked for the observations on trespassing (section 5.5) and brood parasitism (section 6.3), those used in the study of parentage (Brown and Brown 1988b), and birds caught in nests during nest checks. Consequently the sample sizes for these analyses were much smaller than in other comparisons that did not require a knowledge of the precise location within a colony where an individual lived (e.g., sections 12.4.2, 12.5.3).

12.8.1 Position in the Colony

Earlier we explored at length how a nest's spatial position in the colony affected reproductive success (section 11.5), finding that the most peripheral nests were more likely to fall victim to predators (section 8.7), but that owners of edge nests also fought less and suffered fewer intrusions by would-be nest usurpers (section 5.2.3). Predation risk and energy costs of fighting could be reflected in the breeding adults' annual probability of survival.

We did not have a sufficient sample of breeding adults with known nest locations to analyze survivorship as a function of distance from the colony's center, as we did for nestlings (section 12.7.2). Instead, we examined only birds from center and edge nests (see section 2.8.2), combined across colonies. Using these two extremes in nest position enhanced our chances of finding an effect of nest position on survivorship if one existed. Annual probability of survival (model s, p_t) was estimated for center- and edge-living adults. Probability of survival was that starting in the season in which a bird's nest position was known and estimated over all subsequent years. Nests from nonfumigated and fumigated colonies were combined to achieve adequate sample sizes.

Adult cliff swallows occupying center nests (N = 82 birds) had a subsequent annual survival probability of 0.638 ($\pm$ 0.041), versus 0.561 ($\pm$ 0.023) for adults occupying edge nests (N = 306 birds). The difference was not significant. Either a breeder's nest position had no effect on its subsequent survivorship, or the different costs and benefits associated with each position represented essentially equal trade-offs. The slightly lower survival probability for edge nesters may partly reflect yearlings' propensity to settle toward the colony's edge (section 11.8), because second-year survival in general was lower than that for older age classes (section 12.3.1). The only other study we are aware of that measured annual survival of breeders from the center versus the edge of a colony is that of Aebischer and Coulson (1990), in which center-nesting kittiwakes (like cliff swallows) had a slightly (but apparently not significantly) higher survivorship than birds on the edge. Aebischer and Coulson interpreted their result to mean birds that were inherently more fit consistently occupied the centers of colonies, because kittiwakes do not change colony positions in subsequent years.

12.8.2 Nest Age

Nest age is presumably most likely to affect a breeding adult cliff swallow's annual survival prospects through the differing energy expenditure required for constructing old versus new nests. Old nests often had to be refurbished only slightly by their occupants, whereas birds using new nests frequently built them entirely from scratch. Depending especially on how far mud must be carried from a source to the colony (section 5.3.3) and how much wall sharing is possible (section 5.3.2), new nest construction represents a sizable energy expenditure (see Withers 1977).

For adults in nonfumigated sites, annual probability of survival was 0.658 ($\pm$ 0.122) for birds using old nests ($N = 51$ birds) versus 0.624 ($\pm$ 0.063) for birds building new nests ($N = 40$ birds). The difference was not significant. For adults in fumigated sites, annual probability of survival was 0.547 ($\pm$ 0.034) for birds using old nests ($N = 592$ birds) versus 0.620 ($\pm$ 0.022) for birds building new nests ($N = 339$). This difference also was not significant.

These results show that adult survivorship did not vary with nest age. Either building a new nest is not as costly as it seems, or that cost can be recouped during the four to eight weeks after nest construction is finished and before the birds migrate. It is perhaps not surprising that nest age did not affect adult survivorship, given that other analyses found few consistent differences between old and new cliff swallow nests (e.g., sections 4.5.1, 11.6.1, 12.7.3).

12.9 SUMMARY

An important component of lifetime reproductive success is life span, although few data on survivorship exist for most colonial species. Analyses in this chapter estimate cliff swallow survival probabilities based on mark-recapture data from living individuals. Among cohorts followed for at least three years, about 30% of cliff swallows banded as adults and about 21% of birds banded as nestlings were recaught at least once in a later year. We used the software SURGE to estimate the probability of annual survival for different classes of birds; this method accounted for cases where a bird was known to be alive but escaped detection in a given year.

Annual survival was age specific, being lowest for one-year-olds and apparently peaking at age four. Males and females did not differ in annual survivorship. Probability of survival among breeding adults varied among years, with the highest survival occurring after warm summers. First-year survivorship was lowest for birds hatched in the warm summers, probably reflecting the long-term effects of increased swallow bug parasitism in the natal nest in hot years.

First-year survivorship of nestlings from nonfumigated nests peaked for those reared in colonies of 100–249 nests. Natal colony size apparently had little effect on first-year survivorship for nestlings from fumigated nests. Natal colony size had no effect beyond the first year for any birds. Annual survivorship of adult cliff swallows increased with breeding colony size, with birds in the largest colonies experiencing about a 50% greater survival probability than those in the smallest colonies. Birds that bred in a small colony one year and switched to a larger colony the next year improved their subsequent survival prospects relative to those of birds staying in small colonies for two consecutive years. Increased survivorship of breeders in larger colonies may be a major benefit of coloniality for cliff swallows.

Body mass at 10 days of age did not seem to influence first-year survival for birds reared in nonfumigated nests, but it varied directly with first-year survivorship for birds from fumigated natal nests. Annual survivorship of juveniles varied directly with their mass after fledging. Among breeding adults, probability of annual survival increased with an individual's body mass late in the season, suggesting that social foraging or other factors that lead to higher adult mass are important benefits of coloniality.

Nestlings from nonfumigated nests exposed to natural levels of ectoparasitism experienced about a 50% reduction in first-year survival probability relative to nestlings from parasite-free nests. Postfledging survivorship to the first breeding season declined with increasing swallow bug infestation in the natal nest, illustrating that the influence of bugs on nestlings continued after fledging. Parasitism by fleas did not seem to affect first-year survivorship in either the presence or the absence of swallow bugs. First-year survivorship declined with hatching date for birds reared in both nonfumigated and fumigated nests. This presumably reflected increasing ectoparasitism during the season, declining seasonal resource availability, or less time to amass energy resources before migration. First-year survivorship declined with the natal nest's distance from the center of the colony for birds from fumigated natal nests but not for birds from nonfumigated natal nests. Natal nest age had no apparent effect on first-year survivorship.

Annual survivorship of breeding adults did not vary between those occupying center and edge nests in the colony, or between those reusing old nests and those constructing new ones. Either there were no survival costs associated with particular positions in the colony and nest ages, or the different costs and benefits associated with each represented equal trade-offs.

13 Colony Choice

> It is pleasant to watch the establishment . . . of a colony of [cliff swallows]. Suddenly they appear—quite animated and enthusiastic, but undecided as yet; an impromptu debating society on the fly, with a good deal of sawing the air to accomplish, before final resolutions are passed.
> Elliot Coues (1878)

13.1 BACKGROUND

Data presented so far in this book show that breeding colony size influences many aspects of cliff swallow social behavior and demography. The choice of where to breed, and with how many other individuals, is therefore likely to be a key determinant of expected lifetime reproductive success (section 14.2). Patterns of colony choice potentially reveal the "best" colony size for a particular individual and consequently may yield insight into the evolution of group living (Brown, Stutchbury, and Walsh 1990). For example, if all individuals distribute themselves in colonies of similar sizes, that could mean that an optimal or near-optimal group size exists where the benefits of coloniality maximally exceed the costs. On the other hand, a completely random distribution of individuals among colonies of different sizes would suggest that no colony size is superior to another and might mean there are no net costs or benefits of breeding in different-sized groups.

No general theoretical framework has yet been developed to describe how animals should choose breeding colonies (Kharitonov and Siegel-Causey 1988; Brown, Stutchbury, and Walsh 1990). Most populations of colonial animals exhibit a wide range of colony sizes, but the factors generating this size variation are unknown in almost all cases (section 1.3.4). Colony size variation may result from individuals' distributing themselves in an "ideal free" way (sensu Fretwell and Lucas 1970) to reflect local resource abundance (Brown and Rannala 1995). Or they may distribute themselves such that each individual breeds in the colony size that is best for it, with the best colony sizes varying among different individuals (section 14.3.1). In this case each colony is presumably composed of individuals that experience their maximum expected reproductive success in the observed colony size. However, colony size variation may also result because incoming settlers have spatial, temporal, or cognitive constraints on how many sites they can visit and assess before choosing, at times preventing birds from selecting the "best" colony size. Further size variation may result when some individuals leave or others later arrive at a site, making it difficult for the earlier settlers to predict final colony size or the eventual relative spatial position of their nests (Kharitonov and Siegel-Causey 1988; Brown, Stutchbury, and Walsh 1990; Girard and Yesou 1991). In these cases an individual may not experience its maximum expected reproductive success in the observed colony size.

A satisfactory explanation of colony size variation will likely go a long way toward creating a general theory of breeding colony choice. Such a theory is beyond the scope of this book, but in this chapter we address some possible causes of colony size variation in cliff swallows by describing colony settlement patterns and the annual colony choices that different individuals make. Three hypotheses for colony size variation may be relevant to cliff swallows (Brown, Stutchbury, and Walsh 1990; section 1.3.4): size variation may reflect an ideal free distribution with regard to resources; phenotypic differences among individuals in what constitutes an optimal colony size; or intrinsic or extrinsic limitations on the birds' ability to sample, assess, and predict future colony sizes. Where possible, we interpret the data presented here in the context of these hypotheses. However, given our limited knowledge about how colonial birds choose nesting sites, we primarily describe the observed patterns of colony choice by cliff swallows, with the hope that more sophisticated theory about colony choice may later be developed. Cliff swallows are particularly suited for studies of colony size variation and the choices that lead to this variation, because colony size ranges from 2 to 3,700 nests and the birds also breed solitarily (section 3.5.5).

We begin by describing the birds' behavior during the early part of the breeding season when they are selecting where to live. We examine to what extent individuals sort themselves among colonies, based on phenotypic differences such as age and ectoparasite load, and address the between-year colony choices made by banded individuals, focusing on how consistent these choices are from year to year. We present data on natal dispersal and the patterns of colony choice by first-year birds and describe instances of colony switching within a season. Finally, we evaluate the reliability of the information on colony size at a site that is available to settlers at different times during the breeding season.

13.2 BEHAVIOR DURING COLONY SELECTION

On its return to the local breeding area in the spring, each cliff swallow must select a colony site at which to breed. Birds in our study area have many colony sites they can choose among (fig. 3.3). To what extent does an individual move among these sites before settling? Extensive movement between colonies means birds have the opportunity to assess different colony sites, based perhaps on the number of birds present or the quality of the available nest sites. For example, in the highly colonial quelea, birds followed early in the year visited at least four colonies before settling in one (Jaeger, Bruggers, and Erickson 1989). In contrast, if individuals settle at the first colony site they encounter and do not visit others, selection of an optimal colony may not be important, or birds may be incapable of evaluating the suitability of sites. In this section we describe cliff swallows' behavior early in the season while settling, focusing especially on how many colony sites they visited. We followed the birds during this time of year by placing radio transmitters on them and tracking their daily movements.

13.2.1 Radio Telemetry Methodology

After clipping the feathers, we used Skin Bond colostomy cement to attach transmitters to the bare skin of the birds' middorsal region. The transmitters were manufactured by Holohil Systems and weighed approximately 0.75 g, about 3% of the birds' body mass in early spring. With the exception of one bird that initially had difficulty flying, all radio-tagged birds seemed to adjust well to the transmitters, and their subsequent behavior appeared normal. Brigham (1989) found no difference in total foraging time between tagged and untagged birds in the closely related barn swallow; birds in Brigham's study carried transmitters that equaled 4.1 to 5.6% of their body mass. We assumed that the behavior of radio-tagged cliff swallows did not differ from that of untagged birds and specifically that tagged birds did not spend more time than normal foraging.

Once a bird was radio-tagged, we began monitoring its location at intervals throughout each day. All observations were done on cliff swallows that eventually settled at colonies along or near the Sutherland Canal and North Platte River just east of the Cedar Point Biological Station (fig. 3.3). A road ran parallel to the canal and river, enabling us to locate the birds' signals by driving from colony to colony and using a car-mounted antenna.

Our procedure was to begin searching for a radio-tagged bird at the last colony where it was known to have been present. If it was not there, we systematically searched adjacent colonies, gradually moving farther away, until we located it. We were interested primarily in the colonies a bird visited or passed near, and we did not attempt to triangulate exact positions of foraging cliff swallows away from colony sites. We could monitor up to six birds with transmitters at a time. Birds were generally followed for 1–3 hours each morning and each afternoon, and for 1–1.5

hours immediately preceding nightfall. We noted the sleeping location for each bird each day, the colonies it visited, and any other behavior that could be determined from its radio signals.

The biggest challenge in studying colony choice behavior in cliff swallows was identifying individuals that had not yet chosen a place to live. We could not know where birds had been in the days immediately before tagging and thus which colonies they might have already visited. We also could not reliably catch cliff swallows away from colony sites early in the season, which forced us to mist net the birds we radio-tagged while they were at colony sites. For these reasons we could not rule out the possibility that the birds had already chosen a site at the time we tagged them. However, we minimized this chance by catching birds for tagging only on the first or second day that cliff swallows had been seen at a site during daylight hours. We monitored colony sites closely and knew exactly the day (and in some cases the hour) when the birds first visited a site. In most cases tagged birds could have been at the colony where they were first captured for a maximum of 6–36 hours. Their subsequent behavior also indicated that most of the birds we studied had probably not already chosen to live at the capture site; 10 of 18 birds (55.5%) eventually settled at another colony, and all but 2 birds (90.0%) subsequently visited at least one other colony after tagging.

13.2.2 Profiles of Radio-Tagged Birds

Cliff swallows were difficult to catch early in the season. Except at late-starting colonies, the birds generally had to be present at a colony for at least four or five days before they would tolerate our presence and associated mist netting. Therefore netting early in the season when the birds had just arrived was inefficient and never yielded large numbers. At most colonies, through persistence (and luck) we managed to catch between 3 and 30 birds during this period. Consequently we had a limited number of previously banded birds to choose among for radio-tagging. Of the 18 birds tagged early in the season, 13 had nested in the study area in previous years; the remaining 5 consisted of 1 bird tagged as a yearling (born in the study area) and 4 previously unbanded birds whose histories were unknown. Among the 13 birds with breeding histories, 9 had lived in previous years at a colony other than the one at which they were tagged and 4 had lived at the same colony in previous years. We tagged 12 males and 6 females; the male bias was due to the presence of more males at colonies early in the season (section 13.2.5).

Transmitters remained on the birds from 2 to 23 days (table 13.1). In most cases the glue failed and the transmitter fell off before battery power was lost. Although there was substantial variation among birds in the time the transmitter remained attached, there did not appear to be any

Table 13.1 Summary of Data for Eighteen Cliff Swallows Radio-Tracked While Choosing Colony Sites

Bird Number	Sex	Date Tagged	Total Number of Days Followed[a]	Range along Valley (km)	Number of Colonies Known to Have Visited	Number of Colony Sites within Range	Chosen Colony Site/Size
1	Male	26 April 1992	9	3.25	1	9	9201/850
2	Male	30 April 1992	2	0.75	2	4	9256/70
3	Male	2 May 1992	23	3.75	2	10	9244/1,500
4	Male	4 May 1992	11	4.25	6	10	9205/1,500
5	Male	4 May 1992	13	2.00	2	6	9211/750
6[b]	Male	29 April 1993	10	>13.25	2	>14	9344/750
7	Male	2 May 1993	4	3.75	2	10	9344/750
8	Male	2 May 1993	5	2.25	2	5	9344/750
9	Male	7 May 1993	17	15.00	2	15	9344/750
10	Male	7 May 1993	7	3.75	1	4	9344/750
11	Male	7 May 1993	11	3.75	2	8	9344/750
12	Male	11 May 1993	5	6.25	3	12	9344/750
13	Female	2 May 1992	13	23.75	5	23	9223/380
14	Female	7 May 1993	9	10.25	5	9	9317/250
15	Female	15 May 1993	6	15.00	2	15	9344/750
16	Female	15 May 1993	7	10.25	3	12	9311/425
17	Female	18 May 1993	6	9.00	5	12	9319/425
18	Female	18 May 1993	23	14.50	6	17	9316/2,700

[a] Reflects total time transmitter remained functional and includes time after bird chose its colony.
[b] Bird left immediate study area for three days and then returned.

systematic biases among classes of birds in the length of time they could be followed. For example, we followed the 12 males a mean of 9.75 ($\pm$ 1.72) days, versus 10.67 ($\pm$ 2.69) days for the 6 females; the difference was not significant (Wilcoxon rank sum test, $p = .78$).

13.2.3 Number of Colonies Visited

Radio-tracking revealed that cliff swallows typically visited several colonies before settling (table 13.1). Because we were not monitoring these birds continually from dawn to dusk (which would have required automatic tracking stations), we could only estimate the *minimum* number of colonies they were known to visit or the minimum area over which they ranged. They could have moved elsewhere during times we were not following them, especially because visits to colony sites were often brief and easily missed. The cliff swallows we tagged confined their activities primarily to the North Platte River Valley east of Kingsley Dam (e.g., fig. 3.3), and accordingly we determined (to the nearest 0.25 km) the maximum linear distance or range along the valley that each bird moved during the time we followed it (table 13.1). Within this range we counted the total number of active or inactive colony sites the bird presumably visited or "knew about."

The behavior of the radio-tagged cliff swallows suggested that they moved among colonies and assessed sites on some basis before finally settling. (We assumed that a bird had settled at a colony after it slept in a nest there for four consecutive nights.) Birds were known to have visited up to six colonies, and most ranged over sections of the valley that included substantially more sites (table 13.1). Most visits to colonies occurred during the first one to four days after tagging, although one female was still visiting different colonies a week after we tagged her. Birds visited both active sites and ones that were ultimately not used that year.

Visits to colonies often consisted of going into nests. Some birds, especially females, entered up to five nests on a single (< 0.5 hour) visit to a colony. Other birds merely hovered in front of nests or flew in synchronized flights above the colony with up to several hundred other swallows (section 13.7.1). During midmorning on three consecutive days, a male (bird 4, table 13.1) made the same foray from his apparent colony of choice to four neighboring sites, visiting each one in turn and entering nests at each site. He spent about 1.5 hours each day on his tour before returning to the colony where he eventually nested.

Cliff swallows spent much time foraging early in the season and often left the colony sites for extended periods to feed, especially in the afternoons. Once they had apparently settled at a site (as determined from their sleeping there), their foraging tended to center on that colony. Foraging residents periodically passed over their colony site and seldom were more than about 2 km away. Before settling, however, cliff swallows ranged relatively far along the valley while foraging and did not center their activities on a given location.

Males and females differed in the number of colonies they visited and the area over which they ranged (table 13.1). The 12 males were known to have visited an average of 2.25 ($\pm$ 0.37) colonies, versus 4.33 ($\pm$ 0.61) for the 6 females. Males ranged an average of 5.17 ($\pm$ 1.27) km along the valley, versus 13.79 ($\pm$ 2.23) km for the females. These differences were significant (Wilcoxon rank sum tests, $p \leq .012$ for each). The greater range for females translated into more colony sites available to them (table 13.1). The greater movement tendency of females was also illustrated by the fact that none of the 6 females eventually settled at the colony where they were tagged, whereas 8 of the 12 males did.

We would like to have addressed how a bird's colony size in a previous year, or those available to it in early spring at the time of colony selection, influenced its movements and colony visits. Unfortunately, the 18 tagged birds—although an adequate sample by radio telemetry standards—were too few to permit us to address any effects of colony size. Limitations on our ability to catch a large number of birds with known histories early in the year (section 13.2.2) also prevented a rigorous study of colony size effects.

13.2.4 Sleeping Patterns of Early Settlers

Repeated sleeping at a colony seemed to mean that a cliff swallow had settled there, and therefore we determined sleeping locations for all radio-tagged birds each night as an indication of their interest in the colony sites they had visited during the day. Males and females differed in their sleeping patterns. Among the 12 males, 9 slept at the same colony each night during the time we followed them. Another slept at the same colony each night except one, when he slept in a tree. The remaining 2 slept in the same colony where they were initially caught for one and two days, then switched to the site where they eventually settled and slept there. Males thus seemed to choose a colony relatively soon after their arrival and generally slept there each night, despite making short forays to other sites during the day. Visits to other colonies, and movement range along the valley, declined with each passing day that a male had been settled. No bird of either sex was detected visiting another colony after it had been sleeping at the same site for four or more consecutive nights.

Females either were inherently more likely to sleep in trees or took longer to select a colony site. Of the 6 females we tagged, 4 slept in trees for at least two nights during the colony selection phase. The 2 not using trees moved to a colony different from the one where they were initially caught and slept there each subsequent night. One female slept in a tree two nights, visiting various sites during the day, before apparently choosing one where she slept each subsequent night. Another bird slept in a tree for the first seven nights after tagging (in the same tree on the first four nights), ranging widely during the day and visiting various sites until she confined her activities to one colony where she slept on all subsequent nights. The remaining 2 females slept at various colonies on one and three nights, respectively, but used trees on all other nights during the time they were followed. These 2 birds continued to use trees for sleeping even after they had confined most of their daylight activities to a single colony, where they presumably settled. Although the smaller number of tagged females dictates some caution in making conclusions, these observations on sleeping patterns suggest that females may not select colony sites as quickly as males. Birds that sleep at a colony may gain information on the site's ectoparasite load, especially because swallow bugs are active mostly at night and begin feeding on adults in early spring (section 4.2.1).

13.2.5 Insights into Colony Selection

We began section 13.2 by asking to what degree cliff swallows assess and really "choose" colonies. The radio-tracking observations revealed that these birds in fact moved among colony sites, visited nests at several, and (for unknown reasons) selected one of the sites they had visited. Although our data provided only a minimum estimate of the number of colonies the birds visited and the area they ranged over, it was clear that these birds did not simply settle at one site (perhaps the first they encountered) without ever visiting others. Presumably they at least had information on several neighboring colony sites, and they may have been relatively familiar with a much larger area and set of sites. For example, one male (bird 6, table 13.1) vanished from the immediate study area for three full days, only to reappear and eventually settle at a colony that he was not known to have visited before his absence. This bird left the North Platte River Valley entirely and presumably was visiting colonies (the next closest) at least 10 km away along the South Platte River.

Although the birds visited several colonies before settling, the radio-tracking revealed that they generally did not spend a long time looking at colony sites before presumably choosing one. All sporadic daylight forays to other sites occurred within four days of the time when birds began consistently sleeping at the same ("chosen") site; after that time, they ignored all other colony sites. The interval from the time of tagging until a bird first slept at its eventual colony of residence averaged only 1.50 ($\pm$ 0.42) days for males and 3.25 ($\pm$ 1.65) for females. Such a short span of time for assessing colonies perhaps represents a limitation on the birds' ability to make optimal choices of colony sites (Brown, Stutchbury, and Walsh 1990; see Abrahams 1986; Shapiro and Boulon 1987). Yet continued sampling of other colonies for a few days after apparently making a provisional colony choice may enable the birds to quickly resettle elsewhere should the colony size at their chosen site suddenly change (see section 13.7).

Radio-tracking also revealed that cliff swallows first chose a colony site and then found a nest there. In the first few days after tagging, birds repeatedly visited and entered different nests in a colony during daylight hours. Most also slept in a different nest each night for the first several nights they spent at their chosen colony. For example, one male (bird 1; table 13.1) that was never seen visiting another colony slept in different nests at his chosen colony on the first five nights we tracked him; another male (bird 4) slept in different nests at his chosen colony on the first three nights; and a female (bird 18) slept in different nests at her chosen colony on the first twelve nights! This last bird apparently chose the colony site well before she established ownership of a nest. Other birds behaved similarly.

In addition to choosing a colony and nest, females had to choose a mate. Radio-tagged females, on visiting a colony, went from nest to nest, presumably assessing both the nest sites and the males associated with them (see Emlen 1954). The need to assess males in addition to nest and colony sites may have caused females to visit more colonies and range over a wider area, as we observed (table 13.1; also see Dale et al. 1990).

By the time a female arrived and began looking at colonies and nests, most nests at active sites probably had male owners, since the earliest arrivals were predominantly males. We caught a total of 183 birds at nine colonies in April and early May 1992–93 when we were radio-tagging birds. Of those, 164 (89.6%) were males; some of the birds sexed as females might also have been males without obvious cloacal protuberances (section 2.4). This heavily skewed sex ratio suggests that colony sites may often be chosen, and settled, mostly by males, at least early in the season. Females therefore may have less choice in total colony sites, being restricted to ones males have already chosen. This constraint may be another factor causing females to move around more before accepting a site. (The 0.896 sex ratio reported above came from data collected earlier in the year than those used to explore sex ratios in section 3.5.2. By mid-May, when we could begin netting large numbers of cliff swallows, the sex ratio—though still male biased—had become more even; see sections 3.5.2, 3.5.3.)

The picture that emerges from the radio-tracking study is that cliff swallows probably assess colony sites before choosing, but most individuals make their choices relatively quickly. Within our study area, a colony-seeking bird may not have information on, or have visited, every colony site in the North Platte River Valley, but it likely has ranged over a sizable fraction of the valley by the time it settles, even given the short time when choices are being made. The greatest constraint on an individual's choice of colony may simply be time: there are many advantages to early nesting (section 11.4) and residency in a group (e.g., section 8.4), and thus birds should probably form a colony and begin nesting as soon as possible upon their return in the spring. This may not leave them enough time to visit large numbers of colony sites, depending on how long it takes to assess each one (but see Hovi and Ratti 1994). The outcome could be that birds sometimes end up in colonies that are not of optimal size or composition for those particular individuals (section 14.3.2; Brown, Stutchbury, and Walsh 1990).

13.2.6 Postbreeding Colony Assessments?

Cliff swallows also may gain some information on colonies late in the season after they finish breeding, as perhaps do bank swallows (Mead and Harrison 1979), purple martins (Brown and Bitterbaum 1980), tree swallows (Lombardo 1987), and various other birds (e.g., Danchin et al. 1991; Baker 1993; Halley and Harris 1993; Reed and Dobson 1993; Boulinier and Danchin 1994). In late June and July in our study area, as young fledged, large numbers of postbreeding adults and independent juveniles began appearing at various colonies. These birds entered nests, clung to the substrate, gathered mud for nests, and exhibited other behavior similar to that seen in early spring. These activities occurred both at colony sites that up to then had been inactive that year and at active sites with still-breeding residents. During this period the visitors inadvertently transmitted dispersing swallow bugs between colony sites (section 4.9.1). Most of the activity at colony sites late in the year occurred during the mornings and again at dusk, with the birds away foraging during the afternoons.

Spending time at colonies late in the season may inform cliff swallows about sites, and that information could be useful the following spring. For example, entering nests may enable the birds to predict future ectoparasite loads at a colony site (e.g., Brown and Brown 1991). As a comparison with the behavior of birds selecting colonies early in the spring, we radio-tagged three birds near the end of the season (8 July 1992) to track their movements. One bird, a female, vanished permanently from the study area the day after tagging, presumably having migrated. The other two, both males, remained in the study area with functioning transmitters for eight days and six days.

These birds ranged over a much larger section of the valley (15.0 km and 19.5 km) than did males early in the year and were observed to visit more colony sites than most early males (seven and four). Both slept in trees on all nights except one, when they slept in nests at colonies. They spent considerable time each day moving up and down the valley, although each bird seemed to focus its activities loosely on one colony (which for one bird was not the colony it had nested in; the other bird's history was unknown). Their behavior was consistent with the hypothesis that birds, freed from nesting tasks, moved widely among colonies late in the season, possibly to gain information for the following spring. These birds spent a relatively large fraction of their nesting season engaged in this late-season colony visitation and thus delayed the start of their migration south. They had undoubtedly been wandering among colonies for some time before being tagged, and one bird probably continued doing so after the eight-day tracking period, since his transmitter fell off while he was still in the study area. This rather substantial investment in time alone suggests that cliff swallows gain some benefit from postbreeding movements and visitation among colony sites. This issue could perhaps be resolved more directly by searching for banded birds in colonies they had visited and on which they had focused their postbreeding activities the previous year, although postbreeding cliff swallows were difficult to catch (for initial marking) at most sites.

13.3 SORTING OF BIRDS AMONG COLONIES

One outcome of the colony choices individuals make can be a nonrandom sorting of birds into different colonies. This is most likely to occur if certain subsets of the population do best in particular kinds or sizes of colonies. In this section we examine how cliff swallows sorted themselves among colonies of different sizes, based on age, ectoparasite load, condition, and body size.

13.3.1 Sorting by Age

We evaluated how age affected the distribution of cliff swallows among colonies by examining the age structures of colonies of different sizes. We knew ages only for birds originally banded as nestlings or juveniles and subsequently recaught as breeders. For each colony we calculated the percentages of the total known-age birds consisting of yearlings and birds two, three, and four or more years old. The number of known-age birds per site varied, principally owing to colony size and capture effort; the range was from 2 (for a solitarily nesting pair) to 680 birds per site. We had a total 3,150 birds of known age distributed among fifty-six colonies in 1989–92 (in this total any individual present in multiple years was counted for each year).

Although we had captures of known-age birds in years before 1989, the earlier data were not used in this analysis because we had not at that time conducted the study long enough to have known-age birds of the older age classes represented. The average percentage of birds four or

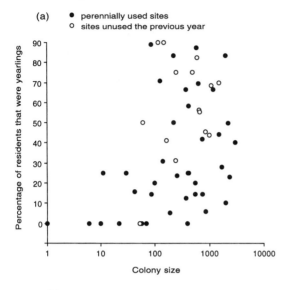
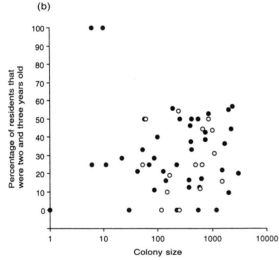

Fig. 13.1 Percentage of known-age cliff swallows in a colony that were yearlings (*a*), two to three years old (*b*), and four or more years old (*c*) in relation to colony size for perennially used sites and sites unused the previous year. For all sites combined, the percentage of yearlings increased significantly with colony size ($r_s = .39$, $p = .003$, $N = 56$ colonies), that of two- to three-year-olds did not vary with colony size ($r_s = .09$, $p = .53$, $N = 56$), and that of four-year-olds and older birds decreased significantly with colony size ($r_s = -.31$, $p = .020$, $N = 56$). None of the percentage age classes varied significantly with colony size when only the sites unused the previous year were considered ($p \geq .49$ for each, Spearman rank correlations).

more years old per site did not increase across the four years from 1989 to 1992, being 28.4% per site in both 1989 and 1992. This permitted us to combine data for these years without apparent yearly capture-effort biases. This analysis assumes that the overall age structure of our sample of known-age birds, all of local origin, was the same as that for those birds immigrating into the study area. Even if this assumption was unmet, however, our analysis still yielded a relative age distribution among colonies,

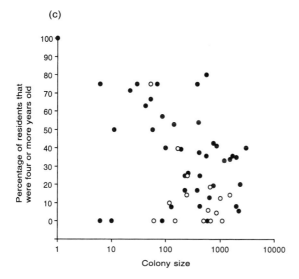

provided immigrants used the same age-related rules (if any) to select colonies and did not preferentially settle at certain sites solely because they were immigrants.

Cliff swallows chose colonies nonrandomly with respect to age (fig. 13.1). The percentage of residents that were yearlings increased significantly with colony size (fig. 13.1a). Most striking was an absence, in most cases, of yearlings in small colonies of fewer than 100 nests. The percentage of birds four years old and older declined significantly with colony size; the same small colonies that had few yearlings consisted mostly of the older individuals (fig. 13.1c). Only birds two and three years old, which were combined in this analysis for presentation purposes, seemed not to sort among colonies based on colony size. The percentage of two- and three-year-olds per colony did not vary significantly with colony size (fig. 13.1b).

History of a colony site also influenced how cliff swallows settled with respect to their ages. Of the fifty-six colonies for which we had data on the ages of the residents, fifteen had been unused the previous year. These fifteen tended generally to be more erratically used from year to year than the remaining forty-one, which were active (though with varying colony sizes) in most years of the study. Disregarding colony size, the mean percentage of yearlings per colony was 58.3 ($\pm$ 6.3) for the former fifteen sites, versus 30.2 ($\pm$ 4.4) for the latter forty-one. The complementary pattern for birds four years old or more was also evident (fig. 13.1c): the respective percentages were 14.7 ($\pm$ 5.2) and 37.8 ($\pm$ 4.1). Thus, in many cases most of the residents of the less regularly used colonies were yearlings, and the oldest birds in the population tended to avoid such sites (fig. 13.1).

The apparent preference of yearling cliff swallows for large colonies could be generated, in part, by age-specific settlement times. Yearlings might select colonies later than older birds, perhaps owing to their slightly later arrival on the breeding grounds (section 3.5.3) or their taking longer (because of inexperience) to assess the relevant features of colony sites. If so, a yearling might be more likely to choose a large colony simply because large colonies are more asynchronous than small ones (section 8.6.1). Thus an incoming yearling will encounter large colonies in which other birds are still settling at that time, and it will be less temporally isolated in a large colony than in a small one (and see section 14.4.2). Naive first-year birds might also be more attracted by larger colonies through conspecific aggregation processes (Burger 1988a; Stamps 1988; Podolsky and Kress 1989; Brown and Rannala 1995), perpetuating continued growth of the large colonies. Other reasons cliff swallows might sort among colony sizes based on age are discussed in section 14.3.1, but we can offer no hypothesis at present for why yearlings should prefer the less regularly used sites.

Although the age-related patterns discussed above (fig. 13.1) were striking, there was substantial variation among sites in age structure, especially among colonies of 100 to 1,000 nests. Some of this variation could reflect sampling error, in cases where a relatively small percentage of a colony consisted of banded birds of known age, but in other instances the results are probably real and may mean that at times birds of a given age choose a site based on cues other than colony size or regularity of use. In some cases birds may have initially settled in a colony of the predicted size, only to have that colony increase or decrease after they had initiated nesting, "stranding" them there (section 13.7).

Preferential settlement of older, presumably more experienced birds in smaller colonies may have accounted in part for the generally higher than expected average annual reproductive success in smaller colonies in some years (section 11.3.2). Nevertheless, some colonies in the 80–250 nest range, in which annual reproductive success often seemed to peak (section 11.3.2), consisted of mostly yearlings (fig. 13.1a), showing that their presence did not automatically depress mean reproductive success. The three colonies with the highest percentages of yearlings were each about 100 nests, a size several earlier analyses (e.g., sections 11.3.2, 12.4.1) pointed to as perhaps the most successful.

13.3.2 Sorting by Ectoparasite Load

One explanation for the increase in ectoparasite load with cliff swallow colony size (section 4.3) might be preferential settlement of already-parasitized birds in the large colonies. If birds distribute themselves among colonies based on their ectoparasite loads, sites chosen by the more heavily parasitized birds both would contain more total parasites at the time of initial settlement and might have additional parasites introduced throughout the time birds are selecting colonies. Unfortunately, there were logistical challenges in assessing ectoparasite loads of cliff swallows at the time they settled, primarily owing to difficulties in catching large numbers of birds upon their arrival (section 13.2.2) and to our not knowing their colony visitation histories before we caught them. There were also problems with knowing whether the ectoparasites on the birds had been on them at the time of settlement or were acquired after they settled at their chosen sites.

Despite these difficulties, we collected some data on parasite loads of cliff swallows soon after settlement, in an attempt to determine if and how ectoparasites might influence colony choice. Our method was to catch birds as soon as possible after they had arrived at colony sites. In most cases this was four to eight days; because we needed larger samples of birds for this analysis, we waited slightly longer after the birds had arrived to begin netting than in the radio-tracking study (section 13.2.2). All colonies studied were sampled during a five- to seven-day period in mid-May each year, 1988–90, with netting at each site usually confined to a half day during this time. All cliff swallows captured were placed in ectoparasite-sampling jars, and the parasites on them were collected and counted (section 2.6). Only fleas could be used in this analysis because too few swallow bugs were found on birds in jars (section 4.8).

Only a relatively few colonies could be sampled during this period; the short sampling time was to ensure that we would catch only newly arrived birds. We concentrated our efforts specifically at small and large colonies and ignored the medium-sized sites. In this analysis "small" colonies were those with 10–95 nests, and "large" colonies had 500–3,000 nests. In 1988 we sampled four small colonies and three large colonies; in 1989 and 1990 we sampled four of each. The distribution of fleas on male versus female cliff swallows at the time of colony settlement did not differ significantly in any of the three years (χ^2 tests, $p \geq .39$ for each), although parasite counts from birds slightly later in the season showed males with higher mean flea loads (section 4.8). Because there was no intersexual difference at the time of settlement, we combined males and females for this analysis.

The distributions of birds with respect to flea loads differed significantly between small and large colonies in each year (fig. 13.2). All heavily infested birds (those with nine fleas or more) were found in large colonies. These distributions (fig. 13.2) suggest that the bulk of the parasitized birds at or near the time of colony settlement occupied large colonies.

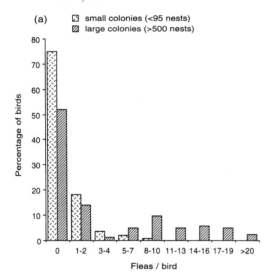

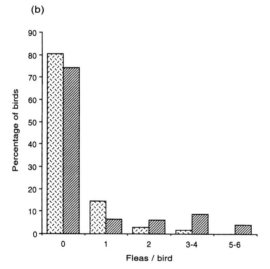

Fig. 13.2 Percentage distributions of cliff swallows carrying various numbers of fleas while settling at small colonies and large colonies early in the spring in 1988 (a), 1989 (b), and 1990 (c). The distributions for small and large colonies differed significantly each year (a: $\chi^2 = 45.2$, df = 8, $p < .001$; b: $\chi^2 = 9.50$, df = 4, $p = .049$; c: $\chi^2 = 30.3$, df = 11, $p = .001$). Sample sizes for small colonies were 132, 73, and 120 birds in (a) to (c), respectively; for large colonies they were 206, 214, and 308 birds.

The question whether these birds acquired their fleas before or after settling in their chosen colony is not easily answered; only if they acquired them before settling could ectoparasite load be construed as a potential predictor of colony size choice. The dispersal behavior of *Ceratophyllus* fleas (section 4.2.2; Bates 1962; Humphries 1969) suggests that the fleas we took off the birds' bodies had recently been picked up from other, probably unoccupied colony sites. Fleas cluster at nest entrances in unoccupied colonies early in the season, in apparent attempts to jump on pass-

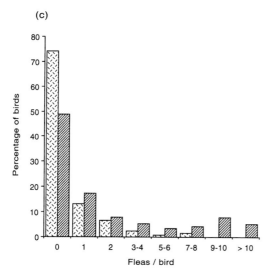

ing birds that may visit a site (section 4.3.2). In occupied colonies, we never saw fleas clustering at nests early in the season. At active sites the fleas are presumably inside the nests, beginning to reproduce, and consequently travel less on the adult cliff swallows. Since the colonies where we captured these birds (fig. 13.2) were active, the fleas we found were probably recently acquired on visits to other colonies.

Why would more heavily infested cliff swallows settle in large colonies? One possibility is that fleas per se had no effect on colony choice, but that settlers in large colonies visited more sites before choosing than did birds selecting small colonies and consequently were exposed to more fleas in the unoccupied sites they had previously visited. The other possibility is that parasite load reflected an individual's phenotypic "quality," and lower-quality birds settled in large colonies. This could be the case if parasitized birds (or ones otherwise inferior) require the foraging- or predator-related benefits of large colonies. Perhaps if a cliff swallow is already heavily parasitized by fleas, there is not much additional ectoparasite-related cost of living in a large colony. If so, individuals might use the advantages of large colonies to compensate fully or partly for their inferior condition. How an individual's phenotype may influence colony choice is explored further in section 14.3.1.

13.3.3 Sorting by Body Mass

Another potential measure of a bird's condition is its body mass (section 10.5). This may be especially true early in the season at the time of colony settlement, when cliff swallows have just completed their spring

416 • *Colony Choice*

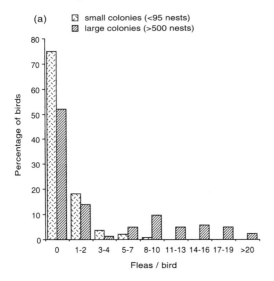

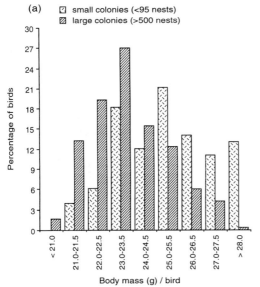

Fig. 13.3 Percentage distributions of cliff swallows of various body masses (g) while settling at small colonies and large colonies early in the spring in 1988 (*a*), 1989 (*b*), and 1990 (*c*). The distributions for small and large colonies differed significantly in (*a*) ($\chi^2 = 58.7$, df = 8, $p < .001$) but not in (*b*) ($\chi^2 = 7.23$, df = 8, $p = .51$) or (*c*) ($\chi^2 = 10.34$, df = 8, $p = .24$). Sample sizes for small colonies were 99, 73, and 119 birds in (*a*) to (*c*), respectively; for large colonies they were 233, 225, and 308 birds.

migration. Body mass often reflects fat reserves, and therefore mass may indicate the energy stress undergone by arriving migrants.

We recorded body masses of cliff swallows caught during the sampling at small and large colonies described in section 13.3.2. Individuals were weighed immediately before being placed in the parasite-sampling jars.

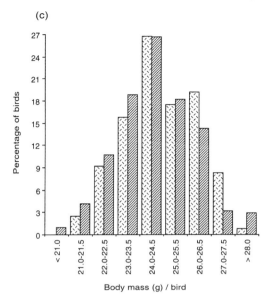

The body mass distributions for males versus females did not differ significantly in any of the three years (χ^2 tests, $p \geq .13$ for each), and thus we combined the sexes for this analysis.

The distribution of body masses for cliff swallows selecting small colonies differed significantly from that of birds choosing large colonies in 1988 (fig. 13.3a). In that year, small colonies included a markedly greater proportion of relatively heavy birds than did large colonies; 59.6% of the birds in small colonies ($N = 99$) weighed ≥ 25.0 g, versus only 23.2% of birds in large colonies ($N = 233$). The trend (though not significant) was the same for 1989, with lighter birds tending to avoid the smallest colonies (fig. 13.3b). Most of the relatively light birds (those ≤ 21.5 g) at the time of settlement were found in large colonies.

This pattern (fig. 13.3) parallels that reported earlier (section 10.5.2) in which cliff swallow body mass tended to decline with colony size early in the season but varied substantially between years. Because these results (fig. 13.3) came from a period just after the birds arrived, the lower 1988 body masses in the large colonies reflect the initial settlement patterns and not any sort of interaction (e.g., resource depression) among birds after they formed the large colonies. The results also are consistent with those reported in the previous section: cliff swallows in presumably worse condition (light ones) settled in large colonies (in some years). Although the reasons ectoparasitized or energetically stressed individuals should choose large colonies are not clear, these analyses (figs. 13.2, 13.3) suggest a nonrandom distribution of birds among colonies and support the hypothesis that different individuals have different optimal breeding colony sizes (see section 14.3.1).

13.3.4 Sorting by Body Size

Cliff swallows conceivably could sort among colonies based on overall body size. This might be especially likely if, as in many species, body size varies directly with competitive ability (e.g., Reiter, Panken, and LeBoeuf 1981; Clutton-Brock, Guinness, and Albon 1982; Petrie 1983; Rubenstein 1984; Robinson 1986; Hoogland 1995) and if certain individuals actively exclude others from particular colony sites (Brown, Stutchbury, and Walsh 1990). Body size differed among individuals occupying different colony sites in Atlantic puffins (Moen 1991) and yellow-rumped caciques (Robinson 1985, 1986).

Studying the relation between body size and colony settlement patterns in cliff swallows was easier than studying the effects of ectoparasites (section 13.3.2) or body mass (section 13.3.3), because skeletal size of breeding adults is unlikely to change during the nesting season. Thus birds could be caught at any time after settling rather than exclusively upon their arrival. This permitted us to collect data from a greater range of colony sizes than in the previous two analyses (figs. 13.2, 13.3). Skeletal body size was also unlikely to be directly affected by the costs and benefits of coloniality arising after birds settled.

We used the unflattened wing chord as our measure of body size. Wing chord is a common index of avian body size, although there is debate on how accurately it reflects skeletal dimensions (section 2.4). There were no significant differences in wing chord between yearlings and birds two or more years old (section 2.4), so birds of all ages were combined for this analysis. We used data collected throughout the 1989 season, as part of the survivorship studies (chapter 12), to examine the effect of body size on colony settlement patterns.

There was relatively little variation in mean wing chord among cliff swallows occupying colonies of different sizes (fig. 13.4). There was a range of only 2.1 mm among the means of the colonies studied. Nevertheless, there was a slight tendency for wing chords of birds in smaller colonies to be larger, and the trend for females was significant (fig. 13.4). This suggests that slightly larger birds tended to settle in the smaller colonies. Assuming that larger birds are competitively superior, this pattern also supports the general conclusion from the ectoparasite (fig. 13.2) and body mass data (fig. 13.3) that small colonies seemed to be preferred by the "better" birds (see section 14.3.1). In house martins, there is evidence that larger individuals are more successful breeders (Bryant and Westerterp 1982). We have no observational evidence at present that cliff swallows

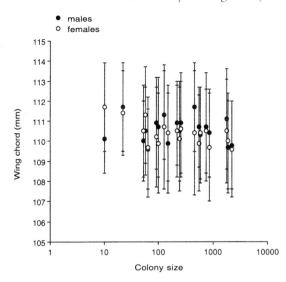

Fig. 13.4 Wing chord (mm) for male and female cliff swallows in relation to colony size in 1989. Mean wing chord declined significantly with increasing colony size for females ($r_s = -.51$, $p = .023$, $N = 20$ colonies) but did not vary with colony size for males ($r_s = -.08$, $p = .73$, $N = 20$ colonies).

settling in small colonies attempted to regulate colony size by excluding other would-be settlers (section 14.3.3).

13.4 HISTORIES OF BREEDING COLONY CHOICES BY INDIVIDUALS

Perhaps the best way to study colony choice is to follow known individuals and observe where they settle from year to year. Determining the colony choice histories of known individuals across years allows the birds themselves to reveal their preferences (if any) for particular colony sites or sizes. In this section we examine the between-year choices of banded cliff swallows, focusing on the breeding colony sites and sizes the birds chose in successive years. In particular, these colony histories allow us to determine whether a bird selects a colony independent of its past choice(s) or whether an individual prefers colonies that either do or do not differ in size from that used the previous year. This section addresses colony choices made by birds that have bred at least once ("breeding dispersal"); natal dispersal and the resulting settlement patterns of cliff swallows in their first year are examined in section 13.5.

The data on known individuals used in this section rely on our long-term mark-recapture program (chapter 12). Sampling at colonies was done as described in section 12.2. In each year an individual was captured, we knew the breeding colony site and the colony size it occupied. We caught all of these birds at sites that had been active for at least five days, and thus the birds captured there presumably were colony residents (with

420 • *Colony Choice*

a few exceptions; section 13.6). The only systematic error in designating an individual's breeding colony may have been for birds caught late in the year after postbreeding colony visitation had begun (section 13.2.6), although relatively few postbreeding birds were caught because cliff swallows became difficult to net at most sites late in the year.

At present we are continuing to compile colony choice histories for our sample of banded birds. As in the analyses of survivorship (chapter 12), the results in this section should be considered preliminary because the bulk of our banding has been done since 1988, meaning we have had relatively few years to follow most of our birds. There was also the problem that some individuals escaped detection in a given year (section 12.2.2). Consequently, for these analyses we had relatively few individuals followed for continuous periods (three or more years) in which colony choice was known in each year. Our sample of birds with multiyear histories is increasing each year, but in the meantime our analyses here use primarily birds for which colony choices were known in two consecutive years only.

13.4.1 Site Fidelity and Dispersal Distances

One choice each individual must make is whether to return to the same colony site where it bred the previous year or move to another site. For cliff swallows, this choice is probably not independent of colony size (in either the previous or the current year), but in this section we disregard the effect of colony size and examine only fidelity to breeding colony sites (philopatry) between years. We investigate the distance between successively used colony sites as an index of the spatial scale over which the birds make their choices of where to live. In all the analyses in this section, we consider only birds that were recaught in the study area in consecutive years. We do not try to estimate overall site fidelity for all cliff swallows, because if a bird was not observed in successive years it was impossible to know whether it dispersed out of the study area, died, or was present but simply missed detection.

The percentages of cliff swallows (among those recaptured) that returned to their colony site of the previous year are shown in table 13.2, grouped by the number of consecutive years in which the birds were recaptured and irrespective of an individual's age. Some individuals counted more than once in this tabulation; for example, if a bird's colony was known in years 1 and 2 and also in years 4 to 6, it counted in both the two- and three-year histories. On average, about 59% of the cliff swallows encountered in the second year returned to breed at the same colony site they had occupied the previous year. This percentage was unaffected by the length of time (years) for which we had data on an individual (table 13.2).

For each recaptured bird that chose a different colony site the next year,

Table 13.2 Colony Site Fidelity for Cliff Swallows Recaptured in Subsequent Years

	Consecutive Years Bird Recaptured				
	2	3	4	5	6
Percentage using same site all years	57.6	58.4	60.3	59.6	63.2
Percentage using different site in one year or more	42.4	41.6	39.7	40.4	36.8
Number of birds	8,159	1,988	494	99	19

Note: $\chi^2 = 2.09$, df = 4, $p = .72$.

we determined the linear distance in km between its site the first year and that chosen the second year. We also determined the "site distance," in terms of how many other colony sites were closer to the first year's site than the one chosen the second year. This was done by counting all colony sites contained within a circle centered on the first year's site with radius equal to the linear distance between the two sites. For example, a colony chosen the second year that was the single closest colony site to the previous year's site was given a rating of 1; a colony site with five other sites closer to the previous year's was given a 6. In our analysis of site distances we used all colony sites regardless of whether they were active in a given year (because birds visited even inactive sites during their selection period; section 13.2.3). Expressing dispersal distances in terms of the number of colony sites is perhaps more useful than giving absolute distances for colonial species that are restricted to nesting sites distributed nonrandomly in the environment. Because dispersal tendencies of males and females differ in some species (e.g., Greenwood 1980; Greenwood and Harvey 1982), we separated the sexes in analyses of dispersal distances.

Linear distances dispersed by breeders in consecutive years showed no significant difference between male and female cliff swallows (fig. 13.5). Of the recaptured birds that switched colony sites between years, about 86% moved to a colony within 3.5 km of the one they had used the previous year (fig. 13.5). This distance agrees well with the observed home ranges of radio-tagged birds during the time of colony selection (table 13.1), especially males. Females ranged farther early in the year (section 13.2.3), but despite this movement they did not ultimately settle farther from the previous year's site than did males (fig. 13.5). There was a low frequency of long dispersal distances in both sexes, as in many species (e.g., Rheinwald 1975; Mead 1979b; Shields 1982; Chepko-Sade and Halpin 1987; Coulson and De Mevergnies 1992), which may have reflected in part an increasing difficulty in detecting dispersers farther from the initial marking point (Barrowclough 1978). Some adults undoubtedly emigrated permanently out of the study area between years, and we had no direct way to estimate how many did so.

When changing colony sites, the birds did not always move to the clos-

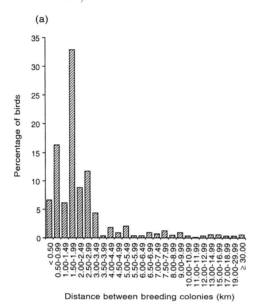

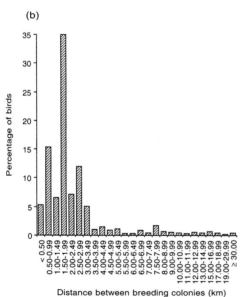

Fig. 13.5 Percentage distributions of cliff swallows moving various linear distances (km) between breeding colony sites in consecutive years, for males (a) and females (b). The distributions for males and females did not differ significantly ($G = 3.39$, df = 25, $p = .99$). Sample sizes were 2,075 males and 1,365 females.

est one available. Although about 30% of dispersing birds moved to either the closest or second closest site in the following year, about 40% moved to colonies that were six to nine sites away (fig. 13.6). About 88% of birds moved to sites within ten colonies of their previous year's site, which agreed broadly with the number of sites contained within each bird's

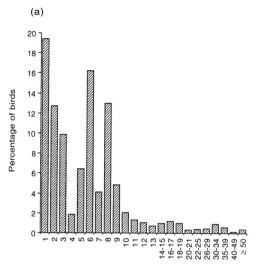

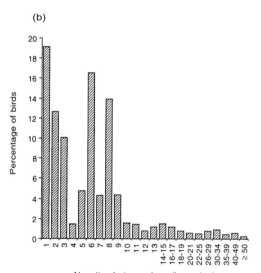

Fig. 13.6 Percentage distributions of cliff swallows moving various distances between breeding colony sites in consecutive years, as measured by the number of colony sites that were closer to the first year's site (see text), for males (*a*) and females (*b*). The distributions for males and females did not differ significantly ($G = 2.23$, df = 22, $p = .99$). Sample sizes were the same as in figure 13.5.

early-season home range (table 13.1). Dispersal distances as measured by the number of closer colony sites showed no intersexual difference (fig. 13.6). The linear and site distances (figs. 13.5, 13.6) suggest that many cliff swallows moved to colony sites they were relatively familiar with through early-season exploration.

About 12% of recaptured cliff swallows dispersed to colonies relatively far away (ten sites or more; fig. 13.6); the farthest we detected a bird moving was 66 km, representing a site distance of seventy-four colonies. Individuals dispersing long distances between years perhaps had information on more colony sites, assuming this information was retained and could apply between years, than did birds settling close to their previous year's breeding site. If lack of information on sites limits the birds' ability to make optimal choices of colonies, we would predict that cliff swallows dispersing long distances, both between and within (section 13.6) years, would make better choices of colony site and size. Unfortunately, we had no relevant data on reproductive success to test this prediction, although we are now approaching a relatively large sample of long-distance dispersers (over 400 birds dispersing more than ten colonies away) and eventually can compare their survivorship with that of short-distance dispersers and nondispersers (as in Clobert et al. 1988). However, more philopatric individuals could be compensated for their reduced information on other colony sites by greater familiarity with local food sources and the habits of local predators near their previous year's breeding site (e.g., Baker 1978; F. Buckley and P. Buckley 1980; Shields 1984b; Cuthbert 1988; Saitou 1991).

What caused some cliff swallows to return to the same colony they had used the previous year, whereas others dispersed varying distances? There is a large literature on the causes and correlates of breeding dispersal, but there are relatively few general patterns (reviewed in Greenwood and Harvey 1982; Chepko-Sade and Halpin 1987; and Part and Gustafsson 1989). The most pervasive result across species seems to be that dispersal is often correlated with reproductive success: if an individual is successful one year, it returns to the same site the next year, but if it is unsuccessful it disperses elsewhere. We examined correlates of breeding dispersal for cliff swallows whose nest location within a colony we knew and that were subsequently found in a breeding colony the next year (table 13.3). Nesting data from dispersers and nondispersers came from the same set of colonies and were thus directly comparable. The smaller sample size for dispersers probably reflects the fact that most of the colonies in this analysis were perennially active, and relatively few birds dispersed between years.

Two major results emerge from our comparison of dispersers and nondispersers. Dispersers had had significantly lower reproductive success (number of surviving young) the previous year, and had initiated laying significantly later, than nondispersers (table 13.3). There were trends for dispersers to have laid fewer eggs, to have built bigger nests, to have nested farther away from neighbors, and to have had more fleas on their nestlings than nondispersers during the previous year, although these differences were barely or not significant. Dispersers were not significantly younger than nondispersers. Thus, although younger cliff swallows seemed more

Table 13.3 Comparisons between Cliff Swallows That Returned to the Same Colony Site the Next Year and Those That Dispersed to Another Site

Variable	Mean	SE	N	Mean	SE	N	p^a
Age in first year	1.75	0.11	60	1.67	0.28	12	.63
Clutch size	3.59	0.046	224	3.26	0.14	46	.053
Number of young surviving	2.78	0.081	225	1.87	0.24	47	<.001
Nest diameter (cm)	15.28	0.18	90	16.54	0.59	24	.051
Nearest neighbor distance (cm)	28.60	12.28	103	97.16	38.39	25	.63
Fleas per nestling/nest	0.14	0.038	95	0.36	0.18	15	.69
First egg date (01 = 1 May)	25.72	0.76	202	30.86	1.86	35	.004
	%		N	%		N	p^a
Active nests that were new	39.9		193	36.4		33	.70
Nests with brood parasitism via laying	3.7		217	5.7		35	.57
Nests with brood parasitism via transfer	6.5		216	5.7		35	.86
Nests with egg tossing	7.9		216	14.3		35	.21

Note: Data refer to the birds' reproductive attempt during the first year. All colonies were combined for analysis. N = number of birds.

[a] Based on Wilcoxon tests for the first seven variables, χ^2 tests for the rest.

likely to change colony sizes between years (section 13.4.2), dispersal in this sample of birds (table 13.3) was probably unrelated to age per se.

These data (table 13.3) show that reproductive success the previous year influenced colony site fidelity in cliff swallows. Another major determinant of site fidelity and dispersal is whether a colony site is used at all in a given year (section 7.4). Returning birds that find their previous site unused have the option of recolonizing the site themselves or moving elsewhere. What determines their choice is unclear. Some colonies where many nests were successful were completely deserted the next year, so some successful birds chose to disperse elsewhere. Thus we are left with the conclusion that breeding dispersal in cliff swallows was determined in part by reproductive success the previous year and in part by whether the previous year's site was active during the current year. Undoubtedly, other unidentified factors also influenced whether birds reused the same site.

13.4.2 Choice of Colony Size

A second choice facing each cliff swallow on its return in the spring is what size colony to breed in. This choice must be made by all birds regardless of whether they ultimately select the same site as the previous year, because most colonies even at the same location vary in size between years (section 7.4.2). Changes in colony size at a site between years may have caused some residents to move the next year, accounting for some of the observed dispersal (section 13.4.1). In this section we examine the consistency of cliff swallows in choosing colony sizes between years and consider how choice of colony size may have been related to age and site fidelity.

We designated eight arbitrary colony size classes (section 2.10), which generally matched those used in earlier analyses. For each bird for which we knew its colony site in two consecutive years, we assigned its colony size in each year to the appropriate size class. Birds that had known colonies for three or more years had multiple, overlapping two-consecutive-year histories, and therefore some individuals were counted more than once in our analyses. The colony size classes used in the following analyses were: 1, 1–10 nests; 2, 11–50 nests; 3, 51–99 nests; 4, 100–249 nests; 5, 250–499 nests; 6, 500–999 nests; 7, 1,000–1,999 nests; 8, 2,000 nests or more.

Determining to what extent birds may prefer colonies of certain sizes requires knowing how they would distribute themselves if their choice was independent of colony size in the previous year. This is not a straightforward problem, because the observed distribution of birds is heavily influenced by the sampling effort directed at each colony. For example, imagine two colonies of 10 and 100 nests. If all residents of each are caught and banded the first year, we would be ten times more likely to encounter a bird from the 100 nest site the following year (all else being equal). If, however, all residents at the 10 nest site were caught but only 10% of the residents at the 100 nest site, we would have an equal probability of encountering a bird from either site. Capture effort during the second year also influences the probability of observing a bird in a given colony size. A site with 100% of the residents captured the second year would reveal any bird from an earlier year that is present, but as the percentage of the total residents at a site caught the second year declines, so do the odds of detecting a bird from an earlier year. Therefore the observed pattern of colony choice reflects colony size availability and capture effort in both the first year and the second.

We devised a method for calculating the expected distribution of two-year colony size histories assuming that birds moved among colonies independent of the previous year's colony size. As shown in table 13.4, the proportion of the total birds caught in year 1 at each colony is multiplied by the proportion of the total birds caught in year 2. This yields the expected proportion of birds that should be observed in each colony size history at year 2, assuming encounters are determined solely by capture effort at each site. All colonies in which we netted birds each year, 1982–92, were included in calculating the overall expected size histories. The same approach (table 13.4) could be extended for multiyear periods, to generate expected n-year size histories. Calculating n-year histories

Table 13.4 Method Used to Calculate Expected Proportion of Birds Exhibiting Each Colony Size History in Two Consecutive Years

Year 1		Year 2				Expected
Colony Size Class	Birds Caught at Site	Colony Size Class	Birds Caught at Site	Calculations		Proportion of Each Size Class
1	5	2	10	(.05) (.10)	=	.005, 1–2
3	10	4	90	(.05) (.90)	=	.045, 1–4
7	85			(.10) (.10)	=	.010, 3–2
				(.10) (.90)	=	.090, 3–4
				(.85) (.10)	=	.085, 7–2
				(.85) (.90)	=	.765, 7–4

beyond 2 becomes laborious, however, for as many colonies as we have. With our mark-recapture program still ongoing, we defer analyses of three-year or more size histories to future works.

We tabulated the observed and expected frequencies of two-year colony size histories for all birds over all years (fig. 13.7). This tabulation includes both the birds that remained at the same colony site in each of the years and those that dispersed elsewhere the second year. Males and females were combined, given their similarity in site fidelity and dispersal distances (section 13.4.1). The observed and expected distributions differed significantly (fig. 13.7). In particular, colony size histories in which birds remained in colonies of roughly similar size between years were overrepresented, and size histories with great changes in colony size were underrepresented, relative to frequencies expected if birds were choosing independent of the first year's size (fig. 13.7). In only 21.0% of observed two-year size histories did the colonies chosen differ by two or more size classes, and 37.6% had the same colony size class in each year ($N = 8,159$). Among the three-consecutive-year size histories ($N = 1,988$), 68.9% differed by no more than one size class among all three years.

Although our sample of birds with known colonies of residence for six or more consecutive years is still limited, the available long-duration size histories are presented in table 13.5. With occasional exceptions, these individuals tended to occupy colonies that differed little in size throughout much of their lives. Data like these (table 13.5) are especially useful in determining whether birds tend to prefer colonies of certain sizes, but they are not available for most colonial animals. We conclude that, overall, cliff swallows were likely to settle in a colony similar in size to the one they occupied the previous year (fig. 13.7).

Not all cliff swallows, however, occupied colonies of similar size each year. Some made rather drastic changes from year to year (fig. 13.7). Perhaps birds changing colony sites between years might be more likely also

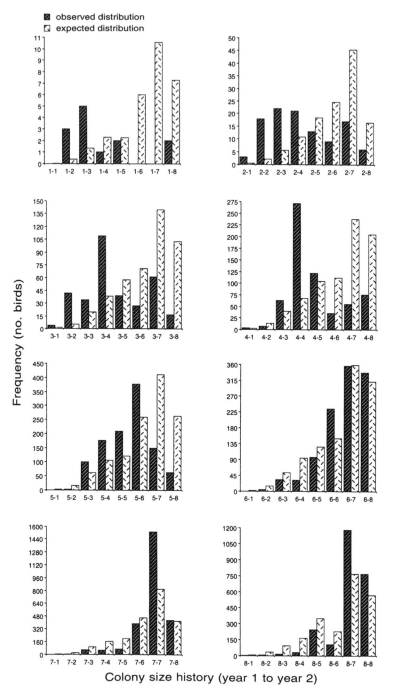

Fig. 13.7 Observed and expected frequency distributions of cliff swallow breeding colony size histories in two consecutive years. Expected distribution assumes choice of colony size is independent of breeding colony size the previous year. Colony size classes are defined in the text. Observed and expected distributions differed significantly ($G = 3522.1$, df = 63, $p < .001$).

Table 13.5 Colony Size Histories of Cliff Swallows for Which Breeding Colony Sizes Were Known in Six or More Consecutive Years

Colony Size History	Number of Birds
4-5-5-4-4-4	1
4-5-5-5-4-4	1
4-5-6-7-7-8	1
5-3-4-5-5-5	1
5-5-5-4-4-4	1
6-7-6-8-8-7	1
6-7-7-8-8-7	4
7-5-8-8-7-6	1
7-7-8-8-5-7	1
7-7-8-8-7-7	3
7-7-8-8-7-8	1
6-7-7-8-8-7-7	2

Note: Colony size classes are defined in the text.

to change colony sizes. We tested this by comparing the percentage distributions of the observed two-year size histories for birds remaining at the same site in both years with those for birds dispersing elsewhere the second year (fig. 13.8). The distributions differed significantly, with dispersing birds less likely to choose a colony of similar size the second year than nondispersers (fig. 13.8). Therefore cliff swallows exhibiting colony site fidelity between years were also more likely to exhibit colony size fidelity, possibly because colony size at a given site—though varying—was more likely to be similar between years than were sizes of any two different sites chosen at random.

Given the covariance between colony size and site, it was difficult to know with certainty whether the birds returning to the same colony site were faithful to the site itself or to the colony size. The cliff swallows that dispersed to other colony sites the next year, however, provided the opportunity to assess whether this subset of birds selected colony size independent of their previous year's colony size. To address this, we compared the percentage distribution of two-year size histories for the birds dispersing to another site the second year with that expected (as in table 13.4) if choice of colony size was independent of the previous year's size. These expected values were calculated only for birds moving between different sites and excluded the expected size histories for individuals using the same colony site in successive years. The distributions did not differ significantly ($G = 36.1$, df $= 63$, $p = .99$), meaning that, overall, the birds changing sites between years chose colony size independent of that of the previous year.

However, the expected distribution of colony size histories for birds

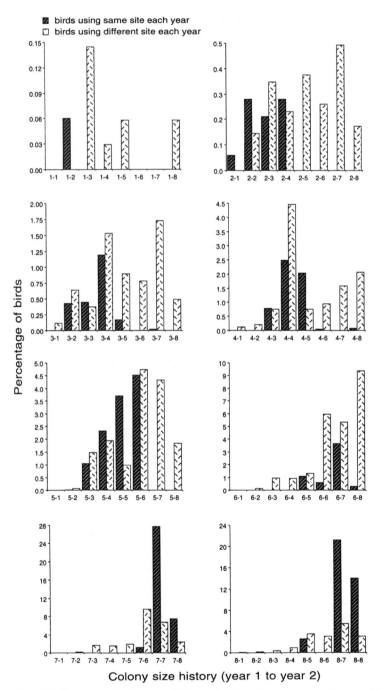

Fig. 13.8 Percentage distributions of observed cliff swallow breeding colony size histories in two consecutive years for birds occupying the same site in both years ($N = 4,697$ birds) and those using a different site each year ($N = 3,462$ birds). Colony size classes are defined in the text. The distributions differed significantly ($G = 195.9$, df $= 58$, $p < .001$).

changing sites predicted that 43.3% of birds should settle in colonies differing from that of the previous year by no more than one size class. We observed this pattern in 56.6%. This discrepancy was significant ($\chi^2 = 160.8$, df $= 1$, $p < .001$), suggesting a tendency for some cliff swallows changing breeding colonies between years to choose colonies similar in size to their site of the previous year.

Finally, we asked how age affected a cliff swallow's choice of colony size. The age distributions within colonies (section 13.3.1) were nonrandom and showed that small colonies contained more older birds than did large colonies. Could this have reflected birds' choosing colonies in different ways depending on their age? Specifically, the observed age distributions among colonies could result if individuals switched to small colonies as they got older. To explore this fascinating question in depth would have required a large sample of known-age birds whose colony choices were known in multiple consecutive years, but the data available indicated that age did influence cliff swallows' choice of colony size.

This analysis used only birds originally banded as nestlings or juveniles (hence of known age) and recaptured at breeding colonies in at least two consecutive years. Their two-year colony size histories were constructed as described above using the same colony size classes. The percentages of birds of given ages that chose a colony larger than, smaller than, and of the same size as their colony the previous year are shown in table 13.6. The percentage of cliff swallows remaining in colonies of the same size (identical size classes each year) remained largely constant across ages, but the percentage of birds choosing a smaller colony the next year increased with age (up to age class five to six years, which had a small sample size and may have been unreliable; table 13.6). There was a corresponding decrease with age in choosing a larger colony the next year.

These data (table 13.6) support the hypothesis that cliff swallows were more likely to switch to a smaller colony as they became older. Such a shift was presumably unrelated to reproductive success per se, because if anything, older birds on average were more successful (section 11.8) in a

Table 13.6 Percentage of Known-Age Cliff Swallows Choosing Colonies Larger Than, Smaller Than, and Same Size as Previous Year's Colony

Age (Year 1–Year 2)	Colony Size Change between Years (% of birds)			N
	Larger	Smaller	No Change	
1–2	27.3	20.8	51.9	428
2–3	18.9	30.6	50.5	366
3–4	8.7	34.4	56.8	241
4–5	4.8	40.5	54.8	126
5–6	8.2	29.5	62.2	61

Note: N = number of birds.

given year and consequently should have been less likely (for that reason) than younger birds to move to another site the next year. Thus these age-related colony size changes may be in part attributable to inherent advantages for older birds in moving to smaller colonies (section 14.3.1).

There was also a suggestion that younger birds experienced a greater change in colony size between years. Among birds aged one to two years that changed colony sizes ($N = 206$), 19.4% changed by two or more size classes. The comparable percentages for the other age classes were 11.0 for birds two to three years old ($N = 181$), 10.6 for birds three to four years old ($N = 104$), 14.0 for birds four to five years old ($N = 57$), and 13.0 for birds five to six years old ($N = 23$). Why cliff swallows would be more likely to undertake a more drastic change in colony size between their first and second breeding seasons is not obvious, unless they were still prospecting for the colony size that was best for them.

The principal conclusion from the colony choice data presented in this section is that many cliff swallows apparently preferred particular colony sites and sizes. The extent to which site-faithful birds chose colonies based on size independent of site could not be addressed in the analyses presented here, although finer-scale comparisons as our study continues can potentially separate the effects of site and size. Regardless of how the birds chose, however, the consequence was that many cliff swallows bred in a colony about the size of the one they used the previous year. These results are consistent with the hypothesis that different birds experience different colony size optima (Brown, Stutchbury, and Walsh 1990), a point we address in more detail in section 14.3.1

13.5 NATAL DISPERSAL AND COLONY CHOICE BY YEARLINGS

Like many species (Greenwood and Harvey 1982; Shields 1982; Chepko-Sade and Halpin 1987; Johnson and Gaines 1990), cliff swallows that survive their first winter must choose whether to return to their birthplace or disperse elsewhere to breed for the first time. The colony choices made by yearlings differ in one major way from those of older birds: yearlings have had no breeding experience at a given site. Their choice of a first breeding colony is influenced in part by characteristics of their natal colony (Brown and Brown 1992), but we understand relatively little about how yearlings decide which colony to occupy. Settlement patterns of first-year birds may help reveal whether certain phenotypes consistently sort among colonies of different sizes.

13.5.1 Patterns of Natal Dispersal

In many migratory passerines young birds disperse widely, and few are ever recaptured at or near their birthplace. Cliff swallows, however, exhibited remarkable natal philopatry in our study area, with over 20% of birds banded as nestlings or juveniles recaught as breeders in a later year (section 12.2.1). This enabled us to examine natal dispersal patterns with a relatively large data set and establish possible causes of natal dispersal (Brown and Brown 1992).

Through 1992, we compiled records on 2,090 cliff swallows originally banded as nestlings or recently fledged juveniles and recaptured the following year in a breeding colony; these data include the 1,580 birds in Brown and Brown (1992) plus 510 additional birds from the 1992 breeding season. Of the 2,090 birds, 1,539 (73.6%) returned to breed at their natal colony site. In this sample, there were significantly more males (58.3%) encountered at breeding colonies, perhaps meaning that first-year females were more likely to disperse long distances from the study area. However, this percentage of males agreed closely with the population's apparent sex ratio (section 3.5.2) and thus may have meant only that the (unknown) factors causing the skew toward males were also operating on first-year birds. Among those birds recaptured in the study area, in preliminary analyses we found no differences between males and females in natal site fidelity or dispersal distances, so the sexes were combined in the analyses that follow.

We examined the linear distance between a bird's natal colony and its first breeding colony for the 551 birds that dispersed their first year (fig. 13.9a). The results were similar to those for dispersal of breeders between years (section 13.4.1): most yearling cliff swallows that were recaptured settled within 3 km of their natal colony. These birds occupied sites they probably were familiar with through explorations as independent juveniles the previous summer (section 13.2.6). We also examined how many other colony sites were closer to a bird's natal colony than the breeding site chosen (see section 13.4.1). The distribution of distances as measured by nearest colonies was markedly bimodal (fig. 13.9b). Relatively large numbers of first-year cliff swallows settled at the colony site closest to their natal colony, and many (for unknown reasons) settled at sites seven to eight colonies away (fig. 13.9b). As with breeders between years (fig. 13.6), most birds chose colonies within ten sites of their natal colony.

These natal dispersal patterns (fig. 13.9) suggest a striking similarity between first-year birds and older age classes in where they settled relative to the previous year's site. Roughly the same fraction of first-year birds dispersed widely (ten or more colonies) within the confines of the study area, and these yearlings may have had information on more colony sites than did the more philopatric individuals (section 13.4.1). Why, then, did only some individuals disperse long distances?

Brown and Brown (1992) evaluated potential correlates of natal dispersal in cliff swallows and found that the extent of ectoparasitism a bird

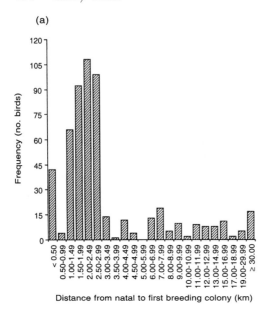

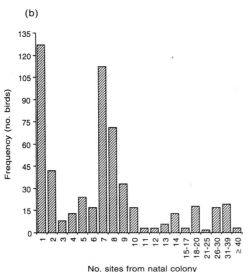

Fig. 13.9 Frequency distributions of first-year cliff swallows settling at breeding colonies of various linear distances (*a*), and distances as measured by number of closer colony sites (*b*), from the natal colony. Only birds dispersing to nonnatal sites are included.

experienced as a nestling seemed largely to determine whether it dispersed. Heavily parasitized birds moved to other sites to breed, whereas lightly parasitized individuals returned to the natal site. We suggested that cliff swallows use the extent of natal ectoparasitism as a rule of thumb to gauge the probable extent of infestation at a site the next year. Birds that survive a heavy parasite infestation as nestlings may be able to save time the following spring when choosing a breeding colony by automatically avoiding the natal site, which may not be active anyway. Further details on the differences between natal dispersers and nondispersers may be found in Brown and Brown (1992).

The relatively high percentage of cliff swallows encountered during their first breeding season that returned to the natal colony (73.6%) in our study area probably reflects the fact that most of the birds that survived to return came from relatively lightly parasitized nests and colonies. Those from heavily parasitized sites, which would have been unlikely to return to the natal colony, simply may not have survived their first winter, given the apparent long-term effects of ectoparasites, especially swallow bugs (section 12.6). The strong fidelity of first-year birds to colonies lightly infested the previous year, along with the birds' direct assessment of parasite load early in the season (section 4.10.1), works to ensure reuse of such sites (and abandonment of heavily infested colonies).

13.5.2 Choice of Colony Size

In addition to choosing a colony site, first-year cliff swallows also must choose a colony size. As in the older age classes of birds, yearlings' choice of site and size was not independent, especially since the same site varied less in size between years, on average, than any two sites selected at random (section 13.4.2). We evaluated to what extent a yearling's first breeding colony differed in size from its natal colony by constructing a colony size history for each nestling or juvenile recaptured in a breeding colony the following year. These size histories use the same colony size classes as those for breeding adults between years (section 13.4.2), with the natal colony size given first, followed by the next year's breeding colony size. We first determined the expected frequency of size histories if birds were choosing breeding colony sizes independent of the previous year's (natal) colony size (table 13.4). We used the number of nestlings and juveniles banded in each colony in year 1 and the total captures in each colony in year 2 to calculate the expected frequencies for each colony size combination (as in table 13.4).

The frequency distributions for the observed and expected colony size histories differed significantly (fig. 13.10). Most birds settled in colonies about the size of the natal site. Cases in which birds changed sizes by more than two size classes were grossly underrepresented relative to the expected frequencies in most cases (fig. 13.10). First-year cliff swallows clearly were not choosing breeding colony sizes independent of natal colony sizes. Overall, only 9.2% of first-year birds ($N = 2,090$) settled in a breeding colony that differed from the natal site by two or more size classes. First-year birds thus showed much greater size fidelity between

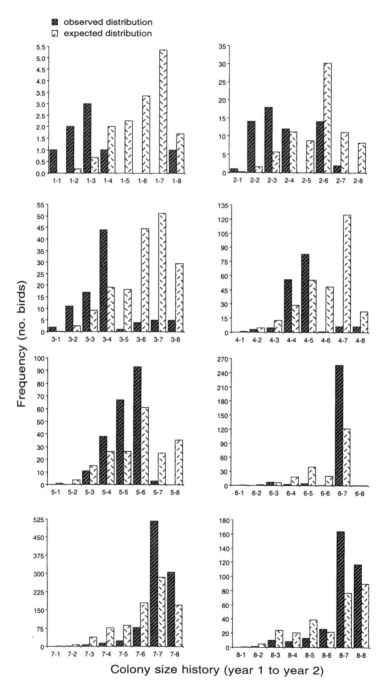

Fig. 13.10 Observed and expected frequency distributions of cliff swallow colony size histories for birds in their natal colonies and their first breeding colonies. Expected distribution assumes choice of breeding colony size is independent of natal colony size. Colony size classes are defined in the text. The observed and expected distributions differed significantly ($G = 1758.9$, df = 62, $p < .001$).

their first breeding sites and natal colony sites than did older breeders between years (section 13.4.2).

The extreme size fidelity of first-year cliff swallows could have resulted from their strong tendency to return to the natal site. Therefore we examined the colony size preferences for the 551 first-year birds that bred in nonnatal colonies. The percentage distribution of dispersers' size histories was compared with that expected if the birds moved among colonies independent of the natal colony's size (fig. 13.11). Unlike those for older breeders, which seemed more likely to choose colonies independent of their previous year's colony size (section 13.4.2 and fig. 13.8), dispersing first-year birds' size histories differed significantly from the expected distribution (fig. 13.11). In particular, it appeared that birds from smaller natal colonies (up to size class 5, 250–499 nests) preferentially settled in similar-sized colonies to breed when occupying nonnatal sites. Dispersing birds from natal colonies of size class 6 and larger (≥ 500 nests) seemed to be distributed in breeding colonies largely independent of natal colony size and did not exhibit the apparent size preference shown by individuals from smaller natal colonies (fig. 13.11).

The patterns of breeding colony choice for first-year birds suggest that cliff swallows evaluate and choose colony sizes relative to their natal sites. How and why they do this is not clear. There is no evidence at present that certain individuals have genetically based colony size preferences, although the issue has not been addressed. The apparent fidelity of first-year birds to breeding colonies of the same size as their natal sites, especially among birds that dispersed to other locations (fig. 13.11), is consistent with the notion that certain individuals are specialized (at least phenotypically) for certain-sized colonies (section 14.3.1). These natal dispersal patterns also indicate that settlement by first-year birds tends, overall, to mirror that of older age classes.

13.6 SWITCHING BETWEEN COLONIES WITHIN A SEASON

Another choice that some cliff swallows may face is whether to switch to another breeding colony within the same season after initially settling at a site. Birds should desert one site and move to another only if they can substantially improve their odds of success by switching (see Sasvari and Hegyi 1994). Moving to another colony is probably most likely a response to nest failure; in observations of color-marked birds (e.g., section 5.5), owners of nests that failed always disappeared immediately after the failure and were not seen in other portions of the same colony. Given the decrease in reproductive success for later nests within a colony (section 11.4), cliff swallows that lose their eggs or nestlings are unlikely to

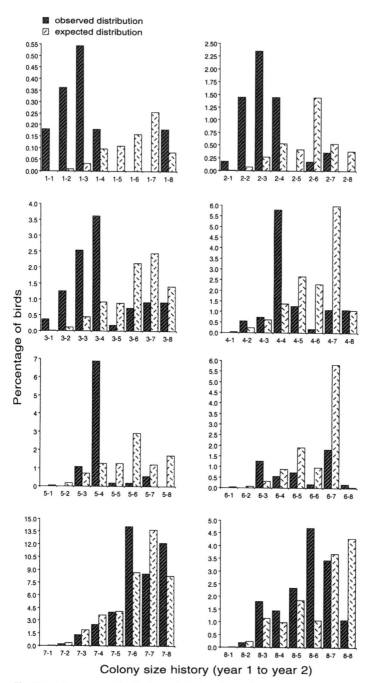

Fig. 13.11 Percentage distributions of cliff swallow colony size histories (natal colony to first breeding colony) for birds dispersing to nonnatal sites ($N = 551$ birds) and that expected if breeding colony size choice is independent of natal colony size (as in figure 13.10). Colony size classes are defined in the text. The distributions differed significantly ($G = 103.9$, df $= 62$, $p < .001$).

rear any offspring if they initiate a second clutch within the same colony. Their best odds of success would be to find another, later-starting colony where they could nest more synchronously with the rest of the birds there. Some individuals might also move between colonies because their chosen site suddenly increases or decreases to a nonoptimal size for them before they initiate nesting (Brown, Stutchbury, and Walsh 1990; section 13.7).

In the course of our mark-recapture program (section 12.2), we encountered cliff swallows that had moved between colonies within the same season. Of approximately 50,000 adults caught at breeding colonies from 1982 to 1992, only about 5% were caught at a second colony within the same year. Nevertheless, the movements of these relatively few birds that switched colonies provide another indication of the spatial scale over which cliff swallows range, and presumably assess colony sites, during a season.

For the 2,664 captures of birds at a second breeding colony within the same year, we determined the linear distance between the colonies at which they were caught and the number of other colony sites that were closer to the original site (as in sections 13.4.1, 13.5.1). Most cliff swallows switched to colonies within 3.5 km of the original capture site (fig. 13.12a). This distribution of distances suggested that the birds, when moving, for the most part were settling within the same general area over which they chose colonies between years (section 13.4.1). Switchers, however, seemed to avoid the colony sites closest to where they first settled, and most went five to nine colonies away (fig. 13.12b). Within-year movement distances for males and females were not significantly different in preliminary analyses, so the sexes were combined in figure 13.12. The spatial scale over which most birds made their within-season choices of second breeding sites was, overall, similar to the area over which individuals ranged early in the season when first selecting colonies (section 13.2.3). This suggests that many switchers moved to colonies they were somewhat familiar with from their early-season explorations.

Although most birds moved within a season to colonies relatively close to their initial site, a measurable fraction of cliff swallows moved long distances (fig. 13.12). Approximately 14% of the birds known to switch sites ($N = 2,664$) moved more than ten colonies from the original capture site, and 6% moved twenty or more colonies away. The farthest linear distance a bird was detected as having moved within a season was 64.2 km. These long-distance moves showed that some individuals ranged over very large areas within a season and presumably had access to information about many colony sites within that area (up to fifty-eight sites in the case of the longest movement).

The cliff swallows that switched colonies within a season did so at all

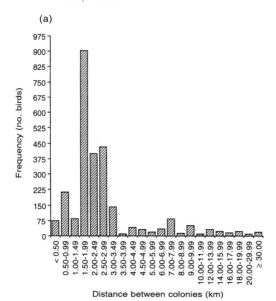

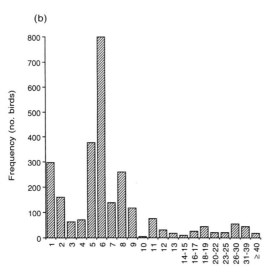

Fig. 13.12 Frequency distributions of cliff swallows switching within the same season to a second breeding colony at various linear distances (*a*), and distances as measured by the number of closer colony sites (*b*), from the original capture site.

times during the summer. We detected more switchers later in the season, but this may have been an artifact of our having caught more birds by then, so that we automatically were more likely to detect movements as our capture totals rose. Unfortunately, determining the absolute fraction of a colony's residents that move at a given time, and where they go, is a thorny statistical problem (B. Rannala and J. Hartigan, pers. comm.) and beyond the scope of this book. The earlier birds switching sites were probably ones whose first breeding attempts failed or that were seeking another breeding site for other reasons, but the individuals moving later in the season (after about 25 June) were likely postbreeders investigating other still-active colonies (section 13.2.6). The birds that moved within a season (fig. 13.12) presumably introduced ectoparasites to new colony sites (section 4.9.1).

If within-year switchers are primarily birds whose nests failed, we might expect to see these individuals moving to colonies of a different size than their first sites. Changing colony sizes might improve their odds of success, especially if the first site was not an appropriate size for them (section 14.3.1) and thus contributed to their reproductive failure. We examined the colony size histories for switchers, comparing the size of the first colony site with that of the second. Using the same colony size classes as in sections 13.4 and 13.5, we found that 36.4% of birds ($N = 2,664$) moved to a colony that differed by two or more size classes from their original site. There was nothing with which to compare this percentage directly, to gauge whether it was relatively high or low, although some size change would be expected as an automatic consequence of changing sites (section 13.4.2). We did not attempt to calculate expected size histories for randomly choosing switchers based on colony size and capture effort (sensu table 13.4), because of the statistical challenges associated with estimating within-year movement.

13.7 PREDICTABILITY OF COLONY SIZE

If we assume animals choose colonies at least in part based on colony size (section 13.1), an individual must be able to predict with some accuracy, at the time it settles, the number of other individuals that will eventually reside at the site. This is less of a problem for the later-settling animals because by then most individuals in the population have already settled (Brown, Stutchbury, and Walsh 1990; cf. Kharitonov and Siegel-Causey 1988), but earlier-settling individuals face the possibility that the colony they choose will increase or decrease markedly. Thus, how accurately can settlers predict the eventual size of the colony at the site they select?

We investigated this question for cliff swallows by recording the population increases at colony sites during the time of settlement. This yielded information on the predictability of a colony's size change early in the season. We determined the date at each site when the residents began building nests, as an index of when individuals committed themselves to nesting at that colony, and examined how much the colony's size changed after that time.

13.7.1 Measuring Colony Size

We used the number of cliff swallows sleeping at a site as a measure of colony size early in the year when birds were settling. This was appropriate because observations of the radio-tagged birds (section 13.2.4) showed that individuals (especially males) usually sleep at the colony where they eventually settle, as do other swallows (Brown 1980). Estimating population size at a site was aided by the birds' preroosting behavior early in the season. About half an hour before sunset, the cliff swallows at each site assembled above the colony in a loose flock that remained nearby as the birds seemingly continued to forage. Gradually the flock stayed directly above the colony site, and birds within it began a highly synchronized moving back and forth above the colony, with all individuals rapidly flying in unison and extremely close together. As the sun set and it became darker, the flock would eventually fly closer to the nests, making swift, synchronous flights under a bridge or through a culvert. Gradually birds would break out of the flights and enter nests, until finally all had either found a nest to sleep in or, in a few cases, vanished to sleep presumably in a tree or on a cliff face.

We estimated the numbers of cliff swallows gathered in these synchronized preroosting flights at various colonies each day or every second or third day. We began our observations when the first birds arrived at a site, usually in mid- or late April, and continued them until the residents ceased their synchronized flights. Birds at the earliest colonies exhibited these preroosting displays for several weeks, enabling us to chart colony size for up to three weeks after the first settlers arrived at some sites. For unknown reasons, birds at later-starting colonies engaged in preroosting flights for only a few days, and no synchronized flights were seen at colonies that started after late May. We could not accurately estimate daily colony sizes once arrivals ceased their synchronized displays. For each site we also determined the final colony size (section 2.1) and observed colonies daily to determine when the birds began collecting mud for their nests.

13.7.2 Population Changes at a Site

We collected data as described above for a total of twenty-one colonies in 1984, 1989, 1992, and 1993. These ranged in final colony size from 5 to 2,200 nests. Population sizes each day are shown for eight representative sites in figure 13.13. Also shown is the final colony size and the day when nest building began at each site. Several general points emerge. Colony sizes generally increased steadily each day during the time cliff swallows were choosing colonies. Colonies that showed a relatively steady increase across days (fig. 13.13a,b) were perhaps the most common among our sample of twenty-one, but there were others that showed some daily fluctuation during this time (fig. 13.13c,d). In a few cases the colony peaked

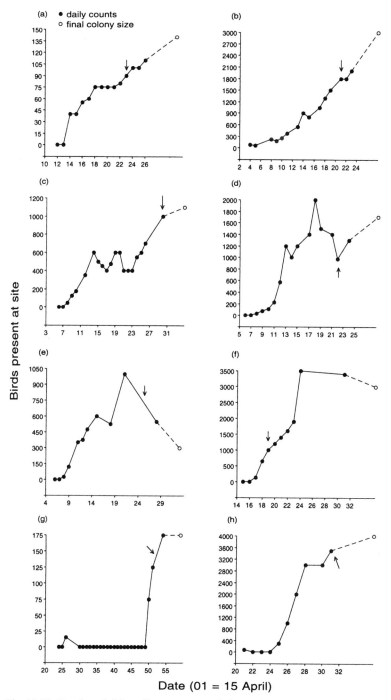

Fig. 13.13 Number of cliff swallows present each day early in the season while birds were choosing colonies, and final colony size (in total number of birds), at eight colony sites (*a–h*). Date nest building was first observed at the site is marked with an arrow. Time interval between the last daily count of birds at a site and when the site achieved final size is unknown; dotted line does not reflect actual number of days at any of the sites.

in size, then markedly decreased (fig. 13.13e). Some sites were colonized suddenly and by large numbers of birds, with 75–100% of the colony's total residents arriving during about a four-day period (fig. 13.13g,h).

Cliff swallows at most sites in our sample did not begin nest construction immediately upon their arrival at a site, typically waiting one to two weeks (fig. 13.13). Consequently, many of the daily size changes had already occurred before the birds apparently committed themselves to nesting at the site by building nests. There were exceptions; at one colony (fig. 13.13f), for example, birds began nest building relatively soon after arrival and the colony later tripled in size. In almost all cases, however, a colony continued to change in size at least slightly after the first birds began nest building (fig. 13.13).

Among the twenty-one colonies in our sample, sixteen continued to increase after nest building started, three decreased, and two did not change. The two colonies in which no further increase or decrease occurred were both small: 5 and 125 nests. Of the sixteen sites with population increase (e.g., fig. 13.13a), the mean percentage of the eventual colony residents that were present when nest building began was 56.8 ($\pm$ 5.9), ranging from 17.4 to 90.9. Thus, on average, only about half of a colony's eventual residents were present when the first birds committed themselves to nesting there. Of the three colonies that decreased (e.g., fig. 13.13e), the mean percentage change in size after mud gathering began was 51.5 ($\pm$ 9.5), meaning some colonies were reduced by half after the first birds committed themselves to nesting there.

These results suggest that cliff swallow colony sizes can be relatively unpredictable early in the season when birds choose colonies. The first individuals settling at a site can probably expect the colony size there generally to increase as additional birds arrive in the local area, but the extent or pattern of increase at a given site can probably not be accurately predicted. The short-term fluctuations in population size at some colonies (e.g., fig. 13.13c,d) indicate some daily movement in and out of sites during the colony selection period, in line with the radio telemetry observations (section 13.2.3) and the thesis that birds visit and assess colony sites. This apparent unpredictability in colony size, for perhaps all but the last settlers, may contribute to the birds' at times occupying colonies of apparently nonoptimal sizes (Brown, Stutchbury, and Walsh 1990). Unpredictability of colony size at the time of settlement may make it more difficult to detect any real patterns of colony size preference. Unfortunately, in the absence of comparable studies on other species, we do not know how unpredictable cliff swallow colony sizes really are in a relative sense.

One other way cliff swallows might predict colony size is by the number of old nests remaining from the previous summer (Shields et al. 1988). This may give a very gross prediction of whether a site tends to be large or small, because obviously more old nests exist at historically larger sites (depending on exposure and the quality of mud used to construct the nests). But existence of old nests is not necessarily informative, because many colony sites with existing nests are unused in some years (section 7.4), and other completely new sites can be colonized by large numbers of birds (section 7.6.2). If cliff swallows try to predict eventual colony size at a site based on old nests, their predictions are not likely to be very accurate.

The synchronized preroosting flights of cliff swallows (section 13.7.1) were perplexing, because there was no obvious benefit from them. The behavior was similar to the "epideictic" displays described for other colonial species by Wynne-Edwards (1962). Perhaps the cliff swallows' synchronized flights have an antipredator function and deter attacks by hawks into the dense mass of flying birds; we saw two sharp-shinned hawks attack cliff swallows near dusk early in the season (section 8.2.1).

Preroosting flights also may provide the birds some information on daily colony size at a site, as Wynne-Edwards suggested. Each individual in the group benefits from being able to assess colony size, presumably enabling each to decide whether to stay at the site. Massed flights at dusk may be the most effective way to convey and receive this information, especially early in the year when the colony residents spend much time foraging and often are not at the colony site during much of the day. Synchronized flights may further advertise colony size to incoming individuals that have not yet chosen a site. Accurate advertisement of colony size potentially benefits both current colony residents and birds seeking to join colonies (Rannala and Brown 1994), and no sort of altruism (sensu Wynne-Edwards 1962) need be invoked.

13.8 CONCLUSIONS ABOUT COLONY CHOICE

The principal conclusion emerging from the data presented in this chapter is that many cliff swallows did not choose colony sites or colony sizes independent of their previous year's choice. The birds clearly evaluated several sites early in the spring before settling, and their choices seemed to reveal preferences for particular sites (often the one used the previous year) and sizes. Many individuals used colonies of similar size from year to year, and switching to other sites was often a response to nest failure.

Among the hypotheses for how colonies are chosen and why sites vary in size (section 13.1), two were supported for cliff swallows. That individuals prefer different colony sizes based on their phenotypic characteristics (sensu Parker and Sutherland 1986; Houston and McNamara 1988) was suggested by the birds' occupying similar-sized colonies between years, their trying a new colony size when reproductively unsuccessful, and the nonrandom distributions of birds among sites based on age and apparent quality. We emphasize, however, that we do not yet know why birds with different phenotypic characteristics apparently prefer different colony sizes, although we have some ideas as starting points (section 14.3.1). It is clear, though, that there is no single optimal colony size that all birds occupy.

A second hypothesis, that settlers have constraints on the time (and energy) available for colony assessment and the ability to accurately predict final colony size at a site, was also supported for cliff swallows. The birds typically invested a few days in visiting colonies early in the season, but they were apparently constrained to choose a site relatively quickly and start nesting as soon as possible. Early choice was probably forced on them by the marked date-related reduction in reproductive success they could expect if they took too long to settle and began nesting after other cliff swallows at a site had initiated nests.

The relatively short time available for colony selection meant that most birds could probably visit only a subset of the available sites before making a decision. Colony choice represents the same kind of problem that animals face in searching for and selecting food patches or mates, and models developed for these activities (e.g., Janetos 1980; Janetos and Cole 1981; Bell 1991; Dale and Slagsvold 1994) could apply to colony-seeking individuals. Most likely, birds choosing colonies use relatively simple behavioral search strategies, such as those proposed by Janetos (1980) and Janetos and Cole (1981). Colony size at most sites changed daily during the settlement period, and thus the earlier-arriving birds probably could not accurately predict colony size for some time. Early settlers had the option of waiting to begin nesting until the colony size was more definitive, but the wait could be costly in terms of the seasonal decline in reproductive success. The birds may often have compromised by waiting to begin nest construction until the bulk of the daily size fluctuations had ceased but not until the colony had reached its final size.

The third hypothesis to explain colony size variation, that colony size at each site is matched to local resource capacity in an ideal free way (Brown and Rannala 1995), was not directly testable with our data on colony choice. A direct test would require quantifying resource abundance at different colony sites, which was difficult for the cliff swallow's insect food (the resource perhaps most likely to influence colony size in an ideal free way; section 10.7). However, several kinds of indirect evidence suggested that colony sizes were not ideal free distributions. First, extent of nesting substrate resources did not affect colony size (section 7.3.1). Second, a stochastic model of ideal free colony choice, which assumed that birds settled strictly in proportion to the habitat's overall resources at each colony site, generated explicit colony size predictions that did not match those observed for cliff swallows (Brown and Rannala 1995). Third, the apparent preferences birds showed for particular colony sites and sizes did not necessarily follow from a resource-based ideal free colony distribution.

13.9 SUMMARY

The way animals distribute themselves among colony sites and whether they prefer certain colony sizes may reveal whether optimal colony sizes exist or whether all colony sizes yield the same net cost or benefit. No general theory of colony choice is available at present, although several hypotheses for why colonies vary in size have been proposed.

Soon after their arrival in the spring, radio-tagged cliff swallows visited several colonies before settling and ranged over areas containing up to twenty-three colony sites with which they were presumably somewhat familiar. Individuals made brief forays to neighboring colonies for several days after apparently choosing a colony site. Birds appeared to choose a colony site before they established ownership of a nest site. Postbreeding birds also visited different colonies and may have gained site-related information to be used the following spring.

Cliff swallows sorted among colonies nonrandomly, with larger colonies containing a higher percentage of yearlings and small colonies a higher percentage of birds at least four years old. Birds carrying heavier infestations of fleas seemed to prefer larger colonies. There was a trend for heavier and slightly larger birds to settle in smaller colonies, perhaps indicating that the better-quality or competitively superior individuals preferred small colonies.

Among the banded adults recaptured the next year, about 59% returned to breed at the same site where they had nested the previous year. The recaptured birds not returning to the same site mostly settled at colonies within about 3.5 km of their previous sites. Some birds dispersed as far as 66 km, however, and long-distance dispersers presumably had information on more colony sites than did birds returning to the previous year's site. Reproductive failure the previous year was often associated with dispersal to a new site the next year. Most cliff swallows tended to remain in colonies of roughly similar size between years, perhaps partly through fidelity to a given site. Birds exhibited a shift in colony size preference with age; as an individual got older it was more likely to move to a smaller colony the next year.

Over 73% of the cliff swallows recaptured during their first year chose to breed at their natal colonies. Many of those dispersing to nonnatal sites moved to colonies within 3 km of their natal site, although a few yearlings moved long distances within the study area. First-year birds, including those dispersing to nonnatal sites, exhibited a strong preference for colonies about the size of their natal sites.

A relatively few cliff swallows switched breeding colonies within the same season, usually choosing colonies relatively close to their original sites, although some moved as far as 64 km between sites within a season. The spatial scale of the within-year movements showed that switchers chose second colony sites with which they were probably familiar. Daily censuses of population sizes at colony sites during the time of settlement in early spring showed that colony sizes tended to increase gradually during that time, but there were daily fluctuations, and some colonies later decreased. About 56% of a colony's eventual residents were present, on average, when the birds committed themselves to breeding there by starting nest construction. Early-settling birds, therefore, probably could not always accurately predict the eventual size of a colony at their chosen sites. The colony choice data in this chapter suggest that cliff swallows may have different optimal colony sizes depending on individual phenotypic characteristics.

14 *The Evolution of Coloniality*

> When you hang around cliff swallow colonies for any length of time, you can't help being caught up in them philosophically, biologically. The wondering, thinking, the romantic analysis of a cliff swallow colony is something everyone does when given a chance.
> John Janovy Jr. (1981)

We began this book by posing two questions: Why do cliff swallows live in colonies? And why do these colonies vary so much in size? Answers to both are necessary if we are to understand the evolution of colonial nesting in cliff swallows and many other species. In this chapter we evaluate how far we have come in answering these questions, drawing on material presented throughout the previous chapters, and highlight unanswered questions and directions for future research.

14.1   THE HISTORICAL CAUSES OF COLONIALITY

The question why birds breed in colonies can be asked in two ways. One is to focus on the historical cause(s) that originally led the birds presumably to shift from solitary to colonial nesting (Siegel-Causey and Kharitonov 1990, 1996). Alternatively, one can examine the observed patterns in reproductive success among birds in different-sized colonies and see how these at present reflect the net effect of the relevant costs and benefits of group living (section 14.2). This section addresses why cliff swallows originally might have become colonial and discusses some of the inherent limitations in trying to answer this sort of historical question, even when many of the relevant costs and benefits of coloniality have been identified. We also discuss Wagner's (1993) recent hypothesis that the opportunities for females within a group to pursue extrapair copulation represent a cause of avian coloniality.

14.1.1   The Traditional Cost-Benefit Approach

The predominant (if not exclusive) approach in field studies of colonial animals over the past two decades has been to focus on one or a few of the potential socioecological costs or benefits of group living, collect relevant data from a relatively small number of colonies, and suggest that coloniality historically arose in response to the potential benefit(s) that were observed to increase with colony size. Often, enhanced predator avoidance represented an easily measured advantage of coloniality (section 8.1), and many have concluded that in general animals form colonies primarily to avoid predators (Crook 1965; Lack 1968; Veen 1980; Burger 1981; Van Schaik, Van Noordwijk, Warsono, and Sutriono 1983; Wittenberger and Hunt 1985; Forbes 1989; Brown, Stutchbury, and Walsh 1990; Van Schaik and Horstermann 1994; Hoogland 1995; cf. Rodgers 1987; Clode 1993). A major problem in most studies is that not all the potential effects of group size have been addressed. This may have contributed in large part to the apparent conclusion that different species form colonies for different reasons (e.g., Bertram 1978; Krebs 1978; Coulson and Dixon 1979; Pulliam and Caraco 1984; Wittenberger and Hunt 1985; Siegel-Causey and Kharitonov 1990; Kopachena 1991; Clode 1993, 1994; Heeb and Richner 1994), whereas in reality the conclusion perhaps should be only that different factors have been studied for different species. To date, the only relatively complete studies addressing most of the potential socio-ecological costs and benefits of coloniality have been those of Hoogland and Sherman (1976), Hoogland (1981, 1995), and Moller (1987a)—too few to make any general statements about the evolution of animal coloniality.

There are three principal conclusions, which seem well supported by the empirical evidence, about why cliff swallows historically formed groups. One is that these birds were not forced into colonies by a shortage of suitable nesting substrates (section 7.6). There was little evidence that nesting sites have been limited, either at present or in the past. When presented with an abundance of artificial nesting substrates (bridges, buildings, and highway culverts), cliff swallows continue to aggregate at times in large colonies, despite the high costs inherent in using such sites.

Another conclusion is that cliff swallows do not form colonies solely as an ideal free response to the fraction of the habitat's total resources contained at each site (Brown and Rannala 1995). If resources necessary for reproduction (most likely food) were distributed patchily enough, the birds might aggregate in direct proportion to local resource abundance and coloniality could result (Siegel-Causey and Kharitonov 1990). Observed colony sizes at sites are too inconsistent for this to be true (Brown and Rannala 1995). This leads to the third conclusion that cliff swallow colonies form because of the direct benefits derived from associating with conspecifics.

Studies on other species have mostly attributed the direct social benefits of coloniality to either enhanced predator avoidance or increased foraging efficiency. These are not mutually exclusive, however (e.g., Krebs 1978;

Table 14.1 Summary of Costs and Benefits of Coloniality in Cliff Swallows and How Each Varies with Colony Size

Effect	Variation with Colony Size
Costs	
Ectoparasitism by swallow bugs	Increases
Ectoparasitism by fleas	Increases
Entombment by neighbors	Increases
Loss of eggs to conspecifics	Increases
Incidence of extrapair copulation	Increases
Incidence of conspecific brood parasitism	Increases
Efficiency at locating fledglings	Decreases
Incidence of kleptoparasitism	Increases
Within-colony breeding synchrony	Decreases[a]
Physical accessibility of nests to predators	Increases
Attraction of predators and risk of predation for adults	Increases
Travel distance to food and total foraging area	Increases
Incidence of nestling starvation	Increases
Benefits	
Time spent fighting for nests	Decreases
Days required to build nest	Decreases
Incidence of trespassing	Decreases
Opportunity for extrapair copulation	Increases
Opportunity for conspecific brood parasitism	Increases
Detection of predators	Increases
Group vigilance and time available for other activities	Increases
Predation risk during mobbing	Decreases
Efficiency at locating departing bird to forage with	Increases
Amount of food delivered to nestlings per hour	Increases
Body mass loss while breeding	Lowest at intermediate sizes
Fledging success and postfledging survivorship to first year	Highest at intermediate sizes
Annual survivorship of breeders	Increases

[a] Costly only for individuals breeding late.

Kopachena 1991; Siegel-Causey and Kharitonov 1996), and additional social factors may enhance the fitness of individuals breeding in colonies. We identified at least twenty-six effects of group size in this study of cliff swallows (table 14.1). All of these varied in a systematic way with colony size, usually consistently increasing or decreasing across the observed range of sizes but peaking at intermediate colony sizes in a few cases. These could be divided into thirteen that presumably reduced the fitness of birds living in colonies (relative to solitary nesters) and thus represented costs and thirteen that presumably increased the fitness of colonial nesters and represented benefits (table 14.1). In addition, there were effects of group size that did not vary in a systematic way or that changed among years (e.g., adult condition at the end of the breeding season; section 10.5.2) and therefore did not consistently represent either costs or benefits.

Given these many effects of colony size, determining whether a single one historically "caused" coloniality is probably impossible. There were clear differences in how much these various effects may have influenced fitness: for example, the relatively small risk of entombment by one's neighbor in a large colony (section 5.4) affected a trivial number of birds compared with those affected by increased infestations of swallow bugs in large colonies (section 4.7). Evaluating the relative importance of these factors (table 14.1) is further complicated because most of the major costs and benefits increased consistently across colony sizes for the range of sizes we studied. Only those that do not vary with colony size can probably be ruled out as historical influences on the evolution of coloniality, and there were few if any of this sort for cliff swallows. Furthermore, the observed costs and benefits were necessarily measured in different currencies in many cases (e.g., the risk of predation on adults versus the probability of being able to parasitize the parental care provided by one's neighbor), which complicated any sort of direct cost-benefit analysis.

The conclusion that there is no way to determine directly which benefit of coloniality was historically the single one that caused the birds to live in groups is perhaps less than satisfying, but it underscores the complex dynamics of this colonial system and most others. We predict that other species will show a similar array of group-size effects if studied to the same degree. Therefore previous conclusions on the historical causes of coloniality in other animals, based largely on study of only one or a few of the potential costs and benefits, are likely to be simplistic at best and wrong at worst. One also has to be especially careful not to confuse the incidental consequences of coloniality with those responsible for causing it in the first place (Siegel-Causey and Kharitonov 1990, 1996; Hoogland 1995).

With these caveats in mind, however, we suggest that coloniality in cliff swallows originally arose as a means of exploiting spatiotemporally variable insect prey. Cliff swallows forage in large groups even when not nesting (Brown 1988b), suggesting that social foraging is not a seasonal consequence of the birds' being forced together into high density in an area only while breeding. Colony formation, initially, may have simply allowed the birds to maximize the time spent in low-risk group foraging (section 9.7.2; Brown 1988b; and see Gillespie 1987; Rypstra 1989), eventually leading to more sophisticated mechanisms of information transfer such as food calls (section 9.6) and observing other colony residents' success (section 9.5.1). Nest clustering (section 5.3) probably followed to allow more efficient information transfer. Once nests were clustered, the other incidental benefits and all the costs of coloniality (table 14.1) would have been almost automatic consequences. As long as the cumulative effect of these costs remains less than that of the original plus the incidental benefits, coloniality persists. It could easily be that the original impetus for coloniality—increased foraging efficiency—is no longer sufficient, by itself, to compensate for all the costs of coloniality that developed once the birds grouped their nests (section 10.10). The suite of more incidental benefits (e.g., predator avoidance, energy savings in nest building) is perhaps what now keeps the net fitness effect of coloniality positive.

Comparisons with other swallows support this evolutionary scenario. Among the mud nest builders (all in the genera *Hirundo* and *Petrochelidon*), the cliff swallows are the most phylogenetically derived (Sheldon and Winkler 1993; Winkler and Sheldon 1993) and also the most colonial. The more primitive *Hirundo* species also build mud nests on the sides of cliffs but for the most part are not colonial (Turner and Rose 1989). If ecological conditions associated with mud nest building (such as a shortage of suitable nesting sites or high predation rates) led to coloniality in the cliff swallows, those same conditions should have promoted coloniality in the *Hirundo* group. That most *Hirundo* species remain solitary to semicolonial suggests that the pronounced coloniality in cliff swallows is a response to something unrelated to the type of nesting site per se. This brings us back to the most likely cause of cliff swallow coloniality being the need to forage socially and transfer information about ephemeral food supplies. This conclusion is supported in that the solitarily nesting *Hirundo* species typically do not feed in groups during the breeding season (Turner and Rose 1989; C. and M. Brown, pers. obs.). In *H. rustica*, for example, Hebblethwaite and Shields (1990) found no evidence of information transfer among foragers. An absence of social foraging in barn swallows is not surprising, given that they typically feed on larger, more evenly distributed single insects rather than on swarms (section 9.3).

Another obvious difference between the *Petrochelidon* cliff swallows and the *Hirundo* group is that the cliff swallows build largely enclosed, retort-shaped nests, whereas the nests of the *Hirundo* species are mostly open cups. Winkler and Sheldon (1993) suggested, based on a phylogenetic analysis, that coloniality arose in the *Petrochelidon* group in response to retort nesting, because enclosed nests reduced the incidence of extrapair copulation and thus allowed (for unspecified reasons) coloniality. Although cliff swallows certainly engage in extrapair copulation, often at incomplete nests (sections 5.5, 6.2.1), it seems more likely to us that retort nesting was instead a response to coloniality and the frequent nest usurpation attempts (section 5.2) and trespassing among neighbors (section 5.5). Enclosed nests help ameliorate these costs of coloniality by reducing the perimeter of defense for nest owners (section 5.2.3).

The suite of observed benefits of coloniality in cliff swallows (table 14.1) is substantially greater than that reported for other swallows in the previous studies of coloniality. In the weakly colonial North American barn swallow, no reproductive advantages that increased with colony size were observed by either Snapp (1976) or Shields and Crook (1987); only costs of coloniality were obvious. In the European barn swallow (Moller 1987a) and purple martin (Morton, Forman, and Braun 1990), the only reported benefits were an increased opportunity for some individuals to engage in extrapair copulation or brood parasitism. Because these applied to only a subset of colony residents, they probably cannot explain why all birds form colonies (section 1.3.2) and at best represent only incidental consequences of coloniality (section 14.1.2). In the highly colonial bank swallow, Hoogland and Sherman (1976) concluded that coloniality was primarily a response to predation, although their data on reproductive success showed no clear advantage to being in a larger colony. The greater number of both costs and benefits of coloniality we observed in cliff swallows is probably attributable to our more comprehensive study, a greater natural range in colony sizes over which these effects could be expressed, and coloniality's being a direct result of social benefits accruing to birds nesting together (e.g., social foraging, as opposed to nesting site limitation).

14.1.2 Extrapair Copulation and Colony Formation

The presumably greater opportunities for both sexes to seek extrapair copulation in larger colonies (section 6.2.3) provide potentially enormous genetic benefits to some individuals (Westneat, Sherman, and Morton 1990). This has led to suggestions that coloniality has resulted directly from birds actively grouping for the purpose of extrapair copulation (Morton, Forman, and Braun 1990). Wagner (1993) recently expanded this hypothesis, arguing that colonies may result when females actively pursue extrapair copulation. Males should cluster and form a colony because females will ignore any solitary male. Females' preference for clustered males occurs because settling solitarily will deprive a female of opportunities to copulate with additional males (see Moller 1992b). The process is posited to be similar to models of lek evolution in which females prefer to mate with males residing on larger leks (Alatalo et al. 1992; Lank and Smith 1992; Hoglund, Montgomerie, and Widemo 1993), or in some cases with males in larger colonies (Draulans 1988; Reville 1988).

There is little direct evidence to support Wagner's (1993) hypothesis, and he notes that an unequivocal test will be difficult because other factors surely influence the evolution of coloniality. It still seems unclear why males that are routinely cuckolded (as in yearling purple martins; Morton, Forman, and Braun 1990) should not abandon a colony altogether and take their chances as solitaries, especially in monogamous species where two parents may be important (or essential) for successful reproduction. If all the males in a colony are paired and mated males give no parental assistance to second females, some females may have to accept a solitary male if they want their offspring to receive any male parental care.

Nevertheless, Wagner's hypothesis may have some merit with cliff swallows, given the observed incidence of extrapair copulation attempts (section 6.2). In particular the hypothesis requires that females gain from pursuing extrapair copulation with multiple males. There are a number of potential benefits a female may receive (Westneat, Sherman, and Morton 1990; Wagner 1992b; Dunn, Robertson, et al. 1994), but in cliff swallows one obvious benefit is the opportunity to mate with certain males that may be less susceptible to ectoparasitism. Given the severe effects of ectoparasites on reproductive success (section 4.7), females within a colony that can discern male resistance should seek to copulate with the resistant males even when not paired with them. Although there is no direct evidence for male resistance in cliff swallows, in other animals resistance is acquired through exposure and subsequent immunological response to both hematophagous ectoparasites (Randolph 1979; Willadsen 1980; Chiera, Newson, and Cunningham 1985) and blood parasites (Davidar and Morton 1993; and see Clark and Swinehart 1966). Some males also may be genetically less susceptible to parasitism (Wakelin 1978; Wakelin and Blackwell 1988).

That cliff swallows seem able to assess the ectoparasite load of neighbors (Brown and Brown 1991) supports the idea that females might seek extrapair copulation on that basis. Also, some females seemed to be receptive to extrapair copulation attempts (section 6.2.5), implying that they received a gain of some sort from multiple mating. However, given the many other costs and benefits of cliff swallow coloniality (table 14.1) and the occurrence of extrapair copulation in the ecologically similar but largely solitary barn swallow (Moller 1985), it seems unlikely that the pursuit of extrapair copulation by females has been a primary cause of coloniality in cliff swallows. A more direct test of Wagner's (1993) hypothesis for cliff swallows must await results of DNA profile testing (section 6.2.5) that will estimate more precisely what fraction of offspring result from extrapair copulation.

We conclude by emphasizing that trying to determine the "reason" animals initially formed colonies is difficult if not impossible for most species, even when extensive information is available on the costs and benefits of coloniality. An alternative approach is to use phylogenetics to deduce a historical scenario for the evolution of coloniality (Siegel-Causey and Kharitonov 1990). However, this type of historical approach is useful only when phylogenetic relationships among groups of related colonial and noncolonial species are well known (Siegel-Causey and Kharitonov

456 • *The Evolution of Coloniality*

1996). In the absence of such information for cliff swallows at present, progress toward understanding the evolution of coloniality also can be made by examining ecological patterns in reproductive success for individuals in different-sized groups.

14.2 LIFETIME REPRODUCTIVE SUCCESS

The individual (or Darwinian) fitness of individuals breeding in colonies of different sizes represents the net effect of the many observed costs and benefits of cliff swallow coloniality (table 14.1) and thus may reveal whether certain colony sizes are favored over others. That a range of colony sizes occurs in nature is not, per se, evidence that they all yield equal or similar fitness, because the birds may have constraints on both the time available to sample colony sites before selecting one and the ability to predict the eventual colony size at a site (section 13.7). These limitations may result in some individuals' breeding at sites that may yield lower reproductive success than they would have achieved had they been able to find the "optimal" colony size (but presumably higher reproductive success than the alternative of not nesting at all).

Measuring Darwinian fitness in natural populations is a major challenge in most cases, and consequently it is usually estimated using lifetime reproductive success (LRS), the total number of surviving offspring an individual produces during its lifetime (Clutton-Brock 1988b; Grafen 1988; Newton 1989c; Partridge 1989; Hoogland 1995). There is disagreement, however, over how well LRS approximates fitness (Grafen 1988; Murray 1992), with Murray arguing that the proper measurement of fitness is the Malthusian parameter (m_{ij}) of population growth (Murray 1985, 1990; cf. Nur 1984c, 1987b). This caution should be kept in mind in this section, where we at times use LRS as an index of fitness. In analyzing reproductive differences among individuals in cliff swallows, LRS is especially important because these animals live for several years and exhibit differences in their probabilites of annual survival (chapter 12). For example, consistent differences in survivorship among birds in colonies of different sizes could result in differences in LRS even when annual reproductive success is similar among those colony sizes.

14.2.1 Estimation of LRS

LRS is usually determined by following a cohort of known individuals throughout their lives and counting the actual number of offspring each produces each year (Clutton-Brock 1988b; Newton 1989a; Hoogland 1995). We could not estimate LRS in this way because we had too few instances in which we followed a given cliff swallow throughout its life and measured its annual reproductive success each season. This was in large part due to the difficulty of detecting all birds that were present in the study area each year (section 12.2.2). Instead, we used the method of Vehrencamp, Koford, and Bowen (1988) to estimate mean LRS for birds consistently breeding in colonies of a given size. The values of mean LRS we present in the following analyses are intended only as relative estimates for cliff swallows occupying colonies of different sizes.

In the estimates of LRS that follow, we make the assumption that individuals remain in the same-sized colony their entire lives. This assumption held for some birds; for example, 37.6% of cliff swallows recaught in the study area used the same colony size class in the successive year (section 13.4.2). Other individuals changed colony sizes between years, however, and our estimates of LRS in this section do not apply to them. We are in the process of estimating LRS for birds that change colony sizes, but such estimates are complicated and must be presented in detail elsewhere. In the meantime, estimated LRS for individuals that hypothetically use the same-sized colony each year are useful in evaluating the relative costs and benefits of living in colonies of different sizes.

The mean LRS for birds of each colony size was calculated by multiplying the mean number of young surviving to fledging (section 11.3.2) times the average breeding life span for birds that entered the breeding population (that is, birds surviving to age one year). Breeding life span was calculated from the survival probabilities of adult breeders (section 12.4.2) using the method of Brownie et al. (1985, 208). For estimating LRS, mean breeding life span was that calculated from Brownie et al. plus 1, because survival probabilities of breeders (section 12.4.2) measured subsequent survivorship, and by definition all birds entering the breeding population began with a life span of at least one year. Mean life span of course would have been lower had we included the birds that died before reaching their first breeding season, but those individuals were irrelevant to consideration of how long birds breeding in particular colonies lived. In calculating mean life span for breeders, we used an estimated survival probability that was constant across ages (model s, p_t; section 12.2.2). This was appropriate because there was no consistent effect of age on adult survivorship beyond age one (section 12.3.1).

We combined colonies into size classes that were similar to those used in the earlier chapters, except for using a single class for all colonies of 500 nests or more. This wider size class was necessary because of the relatively few nonfumigated colonies of at least 500 nests for which we had data on the number of young fledged (section 11.3.2). All breeders using colonies of 500 nests or more were pooled, and a single probability of annual survival was calculated for that colony size class using model s, p_t (section 12.4.2). The mean number of young surviving to fledging was determined for each colony, and the means were averaged for all colonies in each size class. Only data from nonfumigated colonies were used in the analyses and estimates of LRS presented in this chapter.

We estimated the mean LRS for each colony size in another way by considering the fraction of birds surviving to fledging that also survived their first year and entered the breeding population. There were differences among colonies in the apparent survival probabilities for juveniles as measured to their first breeding season (section 12.4.1). The number of young surviving to breed is a more accurate measure of reproductive success than simply the number of young fledged, assuming we have reason to believe that the independent variable in question (sensu Grafen 1988)—in this case, colony size—affects first-year survivorship (as it seems to do; section 12.4.1). We multiplied the mean number of young surviving to fledging for each colony size by the probability of their surviving to breed, yielding a revised estimate of annual success, which was then multiplied by breeding life span as calculated above.

14.2.2 LRS in relation to Colony Size

Depending on the way it was defined, cliff swallow LRS showed different patterns with respect to colony size. If it was expressed solely in terms of the number of young surviving to fledging, mean LRS was markedly higher for cliff swallows occupying the largest colony size class (fig. 14.1a). Mean LRS was about the same for the other colony sizes, with a slight reduction at colonies of 250–499 nests (fig. 14.1a). When LRS also included survivorship of young to their first breeding season, there was a clear peak at intermediate-sized colonies of 100–249 nests, with the next most successful colonies being the smallest and largest ones (fig. 14.1b). We emphasize that these estimates assume that birds remain in the same-sized colony their entire lives, and as such they represent the expected LRS for birds pursuing each colony-size based "strategy."

Meaningful interpretation of these results (fig. 14.1) depends on knowing how close the estimates come to measuring actual LRS. The estimated mean number of young surviving to fledging during a bird's life span (fig. 14.1a) is probably relatively accurate and does not contain obvious biases: the number of offspring in a nest can be counted directly, and the only assumption we make is that the probability of adult survival (and hence breeding life span) can be inferred with the same accuracy among all breeding colony sizes (section 12.4.2). The disadvantage of this measure as an index of LRS and relative fitness (fig. 14.1a) is that it does not account for differences among colony sizes in the probability that those young that fledge will survive to breed the next year (section 12.4.1). Adding the probability of first-year survival of young gives, theoretically, a better estimate of LRS and real fitness (fig. 14.1b).

However, a potential problem with our measure of LRS that incorpo-

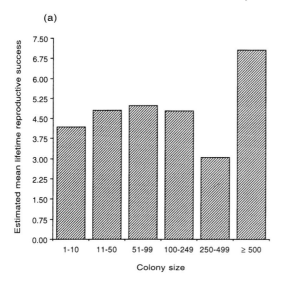
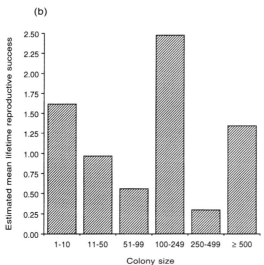

Fig. 14.1 Estimated mean lifetime reproductive success (LRS) of cliff swallows occupying colonies of different sizes, assuming a bird remains in the same-sized colony each year of its life. In (*a*), LRS is measured in terms of the number of offspring surviving to fledging; in (*b*) in terms of the number of offspring surviving to their first breeding season.

rates first-year survival is that natal dispersal biases among colonies may substantially affect the observed first-year survival probabilities. First-year cliff swallows tend to disperse away from natal colonies that were heavily infested with ectoparasites (Brown and Brown 1992). Since larger colonies tend to have more ectoparasites (section 4.3), we may be less likely to encounter birds raised in those heavily infested colonies in subsequent years. This could result in a lower apparent first-year survival probability for birds from larger natal colonies that is not directly comparable to that for birds from smaller and less heavily infested ones.

Although several kinds of evidence indicated that first-year survival probabilities for birds exposed to different levels of natal ectoparasitism were not systematically biased by dispersal differences (section 12.6), we are still left with a nagging doubt that our more "realistic" estimate of LRS (fig. 14.1b) is in fact unbiased. For example, it is heavily influenced (indeed, almost completely determined) by the probability of first-year survival (see fig. 12.5a); fledging success and breeding life span seem not to count for much. This estimate of LRS (fig. 14.1b) is also, on average, fewer than two offspring produced per breeding pair, implying either that the population is decreasing (it is not) or that some offspring are being "lost" to dispersal (Grant 1990). An average of fewer than two offspring per pair would not be a problem for our purposes if the estimates were accurate *relative* measures of LRS among colony sizes, but this would require no natal dispersal biases based on ectoparasite load or other factors. Both estimates of LRS (fig. 14.1) should be considered rather crude, but the one excluding first-year survivorship (fig. 14.1a) is potentially less biased.

14.2.3 Implications of Estimated LRS

What do the estimated mean LRS values suggest about the evolution of cliff swallow coloniality? Considering for the moment only those estimates that exclude first-year survivorship (fig. 14.1a), it appears that the largest colonies yield the greatest net benefit for birds that can consistently occupy them across years. Most of this advantage was realized through breeders' increased probability of annual survival, probably attributable to foraging and energy benefits (section 12.4.2), especially since annual reproductive success—as measured by the number of young surviving to fledging—seemed to decline in the larger colonies (section 11.3.2). The preferences many birds show for the larger colonies (section 13.4.2) are consistent with large colonies' being the most successful, on average. These estimates (fig. 14.1a) may explain why most cliff swallows in southwestern Nebraska breed in colonies of 500 nests or more each year (section 14.3.1).

If LRS excluding first-year survivorship (fig. 14.1a) accurately reflects fitness, however, there is no reason why small colonies should persist. Birds would do best by always settling in large colonies, and if forced into a smaller colony, they should switch to a larger colony the next year (see fig. 12.8). But some cliff swallows consistently occupied small colonies, and many actively switched to smaller colonies as they got older (section 13.4.2). Perhaps colony size preference has no heritable basis, and thus natural selection cannot favor birds choosing large colonies over those choosing small ones. It is also possible that the higher mean LRS for the largest colonies (fig. 14.1a) is incorrectly estimated owing to a small sample of large colonies for which there were data on annual reproductive success (section 11.3.2). This could mean that LRS might be more similar among all colony sizes, in which case the occurrence of small colonies would not be so unexpected.

Formation of small cliff swallow colonies is less perplexing if the estimates using first-year survivorship (fig. 14.1b) accurately reflect LRS and relative fitness. Birds consistently using the smallest and largest colonies did about equally well, along with birds in a single intermediate optimum of 100–249 nests. These estimates (fig. 14.1b) suggest that individuals should avoid consistent use of the three least successful colony size classes. It follows, therefore, that birds in colony size classes 1, 4, and 6–8 should exhibit greater colony size preference from year to year than should individuals in size classes 2, 3, and 5.

Determining the absolute or even relative percentages of birds in different colony size classes that change or do not change colonies is a difficult statistical problem, in large part because colony size, capture effort, and between-year survival probabilities all vary and all affect the observed percentages. Some birds are likely to show colony size "preferences" even if they choose colonies strictly at random, because they will end up in similar-sized colonies the next year. Colony size in particular will greatly affect how often individuals (regardless of how they choose) will be encountered at a given site. One way around this problem is to compare the observed colony size preferences *within each size class* with those that would be expected if birds chose colony sizes at random (section 13.4.2). The extent of agreement between observed and expected values within each size class is comparable among size classes.

Using the method of calculating expected frequencies of two-year colony size histories based on the chosen colony size's being independent of the previous year's colony size (table 13.4), we compared the observed and expected frequencies of birds' using the same colony size class in two consecutive years. In all but one class (the smallest), there were more birds using the same colony size the next year than would have been expected (see fig. 13.7). But how much did the observed and expected frequencies differ in each case?

For colony size class 1–10 nests, there were no instances (0.00%) of birds using the same colony size the next year, although none was expected, suggesting that the sample size (number of birds initially using this size class) was too small for meaningful analysis. For size class 11–50 nests, there was an 800% increase in observed over expected values (although again the small number of birds initially using this size class (18) bordered on being too small for meaningful analysis); for size class 51–99 nests, the increase was 70% ($N = 34$ birds); for size class 100–249 nests, 299% ($N = 271$); for size class 250–499 nests, 72% ($N = 208$); and for size class ≥ 500 nests, 30.4% ($N = 5,339$).

Thus these data suggest that cliff swallows using colonies of 100–249 nests were more likely to use the same-sized colony the next year than were birds in the other size classes (not counting size class 11–50 nests, which was based on a dubious sample size). This is consistent with the highest estimated mean LRS's being for birds perennially occupying colonies of 100–249 nests (fig. 14.1b), and it suggests that our measure of LRS incorporating first-year survivorship might be accurate after all. On the other hand, we did not find greater fidelity for colony size class ≥ 500 nests than for classes 51–99 and 250–499 nests, which would have been expected from the fitness estimates of figure 14.1.

The safest conclusion, based on the two estimates of mean LRS (fig. 14.1), is that size-faithful cliff swallows have about the same expectation of LRS in colonies of all sizes. Birds in the largest colonies perhaps experience slightly higher LRS, on average, than individuals in other colony sizes (fig. 14.1a). The peak in estimated LRS at intermediate colony sizes (fig. 14.1b) is consistent with intuition (and other analyses; sections 11.3.2, 12.4.1) that medium-sized colonies might represent the point where benefits are maximally greater than costs (see section 14.3.1); but unfortunately the potential natal dispersal biases among colonies taint the estimates of LRS using first-year survivorship.

In calculating the estimated values for LRS (fig. 14.1), we assumed that parents exclusively raise their own offspring. If, however, some of the young that survive to fledging resulted from extrapair fertilization (section 6.2) or a parasitic egg laid in the nest by a neighbor (section 6.3), actual LRS would be lower than that observed. This is not a problem in estimating *mean* LRS if all perpetrators of extrapair copulation and brood parasitism are residents of a colony and direct these activities exclusively at residents of the same colony or, assuming extrapair copulation and brood parasitism are perpetrated by nonresidents of colonies, if birds in each colony size are equally likely to suffer them. These conditions were probably met for brood parasitism, because all known parasites parasitized nests within the same colony (and many were in turn parasitized themselves), and there was no evidence of wandering birds that recruited to colonies of particular sizes to parasitize nests (section 6.3). However, there was evidence that extrapair copulation was often attempted by nonresident males (section 6.2.2), and the opportunities for engaging in extrapair copulation presumably increased with colony size (section 6.2.3). The result could be higher uncertainty of paternity for resident males in large colonies (also see Birkhead and Moller 1992).

If male cliff swallows are less likely to raise their own offspring in larger colonies, mean male LRS in those colonies will be lower than estimated (fig. 14.1). Depending on how frequently cuckoldry occurs and how many young within a brood are sired by nonresident males, the cost to resident males could be high enough to reduce their LRS to a level equal to or below that of males living in small colonies. If the threat of cuckoldry (from either nonresident or resident males) was severe enough, it could drive some males to settle in smaller colonies (section 14.1.2). Unfortunately, until we complete DNA profile testing to assign parentage (section 6.2.5), we cannot know to what extent cuckoldry affects LRS and estimated Darwinian fitness of cliff swallows in colonies of different sizes.

14.3 VARIATION IN COLONY SIZE

The observed variation in colony size for cliff swallows in southwestern Nebraska both served as an impetus for this study in the first place and also permitted us to measure the many group-size effects (table 14.1) among birds within a single population. Although cliff swallows show more variation than some species, colony size varies to some extent in almost all species studied so far (Brown, Stutchbury, and Walsh 1990). We contend that explaining colony size variation will in large measure explain the evolution of coloniality more generally. Thus, what progress have we made in understanding why cliff swallow colonies vary in size?

14.3.1 A Hypothesis of Phenotype-Based Colony Choice

Based on our analyses of colony size preferences of individuals in successive years and the sorting of birds among colonies of different sizes, we concluded in section 13.8 that phenotypic differences among cliff swallows could mean that individuals experience their maximum expected reproductive success in colonies of different sizes. Cliff swallow colonies consist of individuals that voluntarily live together, and the observed variation in colony size presumably represents a summation of individual-specific strategies in choosing where to live. If birds move freely among colonies and have complete information on what colonies are available and their eventual sizes, they should be able to select the single best colony for their needs.

Consider two individuals that differ in their inherent ability to find food. One might be younger and therefore less experienced as a forager and the other older and consequently more adept at food finding (see Burger 1988b; Wunderle 1991). Advantages associated with information transfer (section 9.5) in colonies might be less important for the experienced individual, and thus it would gain less of a foraging benefit from living in a large colony. Information transfer, on the other hand, could be critical in enabling the less experienced bird to find food efficiently enough to feed its offspring adequately. (Let us assume for now that enhanced

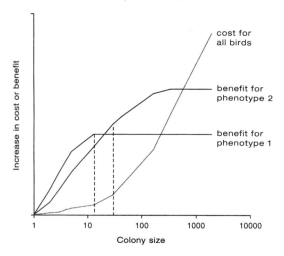

Fig. 14.2 Hypothetical costs and benefits of coloniality as a function of colony size for two individuals of different phenotypes. The cost is the same for all birds, but the benefits differ depending on phenotypic characteristics. For each phenotype, the point of maximum net positive difference between benefit and cost (*dashed line*) represents the optimal colony size.

foraging is the only benefit of coloniality.) How foraging-related advantages vary with colony size would differ for the two individuals (fig. 14.2). Let us also assume that the primary cost of coloniality (e.g., ectoparasitism) is fixed for all individuals and varies similarly with colony size for each (fig. 14.2). This might be likely if swallow bugs infested and dispersed among nests within each colony independent of their bird hosts' "quality" (or in this case foraging ability). The result is that the optimal colony size—where the greatest net positive difference between benefits and costs occurs—is different for the two types of individuals (fig. 14.2).

This discussion has assumed only one benefit (enhanced foraging) and one cost (e.g., ectoparasitism) of coloniality. However, the result (fig. 14.2) is generalizable to any combination of costs and benefits. If the sum total of all benefits and costs (e.g., table 14.1) yields a total benefit curve and a total cost curve that differ for certain classes of individuals, one can get different optimal colony sizes for each class of birds. The number of optima will depend on how many different phenotypes, with their associated cost and benefit expectations, exist in the population. Variation in colony size could thus reflect variation in bird phenotypes. If so, it is perhaps not surprising that mean LRS differed little among birds in different-sized colonies (section 14.2.2), although one might still see unequal expectations of fitness among individuals in different groups depending on their phenotypes (Parker and Sutherland 1986; Jones 1987d; Houston and McNamara 1988). Regardless of how LRS varies between birds in different-sized colonies, the important question is whether the individuals occupying each observed colony size are doing the best there that they can expect anywhere. If the answer is yes, colony size variation as described here should persist. Somewhat similar arguments for phenotype-based sorting among groups have been developed for stickleback schools by Ranta and Lindstrom (1990), Ranta, Lindstrom, and Peuhkuri (1992), Ranta, Rita, and Lindstrom (1993), and Ranta, Rita, and Peuhkuri (1995) and for ruff leks by Hoglund, Montgomerie, and Widemo (1993). Terborgh (1983) proposed a related (though not phenotype-based) model to explain interspecific variation in group size among primates, and Brown (1982) used the same logic to explore differing group sizes among cooperatively breeding birds.

A direct test of the phenotype-based hypothesis as an explanation of colony size variation has not been done for any species to date. (Ranta and Lindstrom's 1990 study of sticklebacks comes closest, although they emphasized sorting based on body size rather than group size.) The best test would be to follow known individuals for two or more years to determine their presumed optimal breeding colony size and their typical reproductive success within that colony, then manipulate colony size (e.g., section 10.7) and measure the same individuals' reproductive success when the colony size is changed. We do not have such data for cliff swallows at present. However, several of our observations are consistent with phenotype-based colony choice.

Perhaps the most suggestive evidence that different birds do best in different-sized colonies comes from the apparent sorting of cliff swallows among colony sizes based on age, parasite load, and body mass and body size (section 13.3). Older cliff swallows preferred smaller colonies (section 13.3.1), a not surprising result if individuals become more experienced at finding food, avoiding predators, or competing for nest sites or mates as they age. There may be little need for birds to pay the costs of living in larger colonies when they become older, if age per se improves their abilities to the extent that the foraging- or predator-related gain from residency in a large colony becomes minimal. Not all individuals switch to smaller colonies with age, however (table 13.6), so for some birds there may be less age-related enhancement of these abilities.

Similarly, birds that are good at avoiding or minimizing ectoparasites may gain disproportionately more from a small colony, where there are few existing parasites, than would birds that are already heavily parasitized. Depending on the density-dependent responses of the parasites, hosts that arrive at a large colony already parasitized may not suffer much increased ectoparasitism. In this case the net expected gain from settling in a large colony would differ among birds depending on their initial extent of ectoparasitism. The result could be heavily infested birds' preferentially selecting large colonies, as we observed (section 13.3.2). The same pattern could occur for competitively inferior individuals, as measured by body mass or size (sections 13.3.3, 13.3.4), in which the poorer-quality individuals gain disproportionately more from the net benefits of coloniality and therefore settle in larger colonies. Ranta, Rita, and Lindstrom (1993) make essentially the same argument, though reversed, for sticklebacks' foraging in groups versus solitarily, in which they suggest that low-ranking (= inferior?) individuals should choose to forage solitarily rather than in a group, based on increased competition for food within groups.

Phenotypic differences among individuals could often be subtle and may not always correlate well with age, parasite load, or body condition (which just happened to be variables we could measure). For example, age may improve some individuals' foraging abilities but not affect others', while simultaneously impairing (via physical deterioration) the ability of some birds to avoid predation. Consequently, some birds might shift to small colonies as they get older, whereas others might move to large colonies. Expectations may vary among years (see Sasvari and Hegyi 1994) or sites (Ranta, Rita, and Peuhkuri 1995), depending on annual levels of ectoparasitism, resource abundance, or overall population size, causing an individual's optimal colony size to change between years. If expectations do change markedly across years, empirically determining what colony sizes given individuals do best in may be hopeless. Changing optima across years would help to explain why cliff swallows seem to show colony size preferences in some years but not in others (section 13.4.2).

The proportions of cliff swallows occupying each colony size presumably indicate, roughly, the relative frequencies of each phenotype in the population. The number of *colonies* was surprisingly uniform across the size range (see section 3.5.5). Among the size classes used in the analyses of LRS (section 14.2.2), through 1993 there were 116 colonies of 1–10 nests; 134 of 11–50 nests; 88 of 51–99 nests; 106 of 100–249 nests; 115 of 250–499 nests; and 167 of ≥ 500 nests. However, the *percentage of the total birds* using each colony size was highly skewed toward the larger colonies. Through 1993 and combining all years, only 0.2% of all birds ($N = 570,664$) occupied colonies of 1–10 nests; 1.4% were in colonies of 11–50 nests; 2.2% in colonies of 51–99 nests; 5.6% in colonies of 100–249 nests; 14.0% in colonies of 250–499 nests; and 76.6% in colonies of ≥ 500 nests.

The high percentage of birds using colonies of 500 nests or more suggests that the individuals that gain most from large colonies (young or inexperienced ones, birds competitively inferior) predominate within the population and that "small colony phenotypes" are comparatively rare. These frequencies, in combination with the data showing sorting of individuals (section 13.3), may mean that small and medium-sized colonies are used only by the above-average individuals, whereas the large colonies are preferred by the average and below-average types, which presumably constitute the majority of most animal populations.

14.3.2 Constraints on Phenotype-Based Choice

The phenotype-based colony choice hypothesis as described in the previous section assumes an ideal free process. This idea endows birds with the ability to sample many (if not all) colony sites, accurately predict eventual colony size at each, and then settle where their expectation of success is greatest based on the social costs and benefits of coloniality. We discovered, however, that there are constraints on how many sites a cliff swallow can sample before settling (section 13.2.5) and how accurately colony size can be predicted from the number of initial settlers (at least by early-arriving birds; section 13.7). This means that individuals often must select colonies from a relatively small set of potentially available sites, and it seems likely that an individual at times cannot find the "perfect" colony size. Presumably each bird makes the best choice among the options available, but it probably has little control over whether its chosen colony subsequently increases or decreases. If these size changes come so late in the current nesting attempt that switching to another colony (section 13.6) is not a viable alternative, an individual may have to remain where its expectation of success is less than its optimum. This could contribute to observed patterns (section 11.7) in which mean annual reproductive success varied markedly among colonies of equivalent size: perhaps they initially began at different sizes and were settled by different kinds of individuals.

There has been theoretical debate over the stability of optimal group sizes (Sibly 1983; Pulliam and Caraco 1984; Giraldeau and Gillis 1985; Giraldeau 1988; Giraldeau and Caraco 1993; Higashi and Yamamura 1993; Ranta 1993; Rannala and Brown 1994). Most authors have assumed that intermediate-sized groups yield the highest fitness (even though there is relatively little empirical evidence for such an assumption in colonial birds; section 11.1). Given this assumption, it has been pointed out repeatedly that solitary individuals seeking to enter groups should join an existing group that is already at optimal size, thereby making it nonoptimal, rather than settling as a solitary. Individuals should continue to do this until the expected fitness of birds in the ever-increasing group falls below that of solitaries, at which point the next incoming settler will remain as a solitary. The net result may be groups routinely larger than the optimal size (Pulliam and Caraco 1984; Higashi and Yamamura 1993; Rannala and Brown 1994). This prediction is in accord with one estimate of lifetime reproductive success (fig. 14.1b) and the data on annual reproductive success (section 11.3.2) in cliff swallows, in that the most successful colony size (small to intermediate) was not the most commonly used (section 14.3.1).

However, whether optimal group size is smaller than, or the same size as, the most commonly observed depends on the pattern of payoffs expected from different-sized groups; that is, how the fitness curve varies with group size (Giraldeau and Gillis 1985). Previous models have implicitly assumed that all individuals have similar abilities and thus similar expected gains from each group size. If phenotypic differences create unequal payoffs among individuals, not all may gain from joining an existing optimal group size (which is perhaps optimal only to the "average" bird). Ranta (1993) notes that the highest fitness can be expected at different group sizes for different phenotypes (see section 14.3.1). Furthermore, the existing theory of optimal group size assumes that each individual operates independently. If groups of incoming settlers (perhaps of similar phenotype) coordinate their choice of colony, they can join existing groups en masse (or later fragment a group by leaving) and therefore perhaps come relatively close to achieving an optimal group size. This seems possible in cliff swallows, given that the birds arrive in the study area in groups (e.g., section 3.5.3) and move in and out of colonies in groups early in the season (section 13.7). The behavior of birds seeking to join a group from the outside represents a potential constraint on group members' realizing their phenotype-based optimum colony size, but at this time no clear predictions are possible as to how this constraint should affect cliff swallow colony choice.

14.3.3 Other Potential Causes of Colony Size Variation

Two other potential causes of colony size variation in cliff swallows are that colony size reflects an ideal free match to the fraction of the habitat's total resources at each site (Cairns 1989) and that birds regulate colony size by actively excluding others from certain sites (Robinson 1985, 1986; Ekman 1989). The only evidence that avian colony size is a direct response to local resource availability (i.e., food) comes from studies of common terns (P. Buckley and F. Buckley 1980b) and great blue herons (Gibbs et al. 1987; and see Fasola and Barbieri 1978; Gibbs 1991), in which much of the variance in colony size was explainable by the amount of foraging habitat near a colony. In these cases the birds were distributed in an approximately ideal free way with respect to total food resources.

Local food availability at colony sites could not be measured directly for cliff swallows, although indirect observations suggested no differences among sites supporting colonies of various sizes (section 10.7). A variety of analyses (section 13.8) suggested that cliff swallows probably did not distribute themselves among colonies of different sizes in an ideal free way with respect to environmental resources (see Brown and Rannala 1995). Furthermore, even if the birds initially chose colony sites based on local resources, they could then expect a variety of socioecological costs and benefits of coloniality (table 14.1) that would substantially alter any expected payoff based strictly on resource availability.

An intriguing possibility is that cliff swallows might regulate colony size by aggressively excluding later settlers (section 13.3.4). This could mean that certain colony sizes, such as small or medium ones (fig. 14.1b), are in fact best for some (or all) birds, but the competitively superior individuals settle first and keep the colony at the optimal size. This is consistent with larger, older birds' being present in the smaller colonies but inconsistent with our observations of cliff swallows in general. Although we have not intensively watched small colonies early in the year, in observing color-marked birds in the course of our studies on trespassing (section 5.5) and brood parasitism (section 6.3), we never saw any indication that cliff swallows in colonies of 125–1,100 nests tried to prevent other birds from settling in the colony. Birds with nests attacked new arrivals that tried to build near the entrances of their own (section 5.4) but ignored birds that settled anywhere else in the colony. Thus at present there is no clear evidence that colony size variation in cliff swallows represents a despotic distribution (sensu Fretwell and Lucas 1970 and Ekman 1989).

14.4 WITHIN-GROUP ASYMMETRIES: SPATIAL AND TEMPORAL EFFECTS

Our discussion thus far in this chapter has addressed the potential mean differences in Darwinian fitness of individuals between cliff swallow colonies of different sizes. A focus on mean differences between groups may yield clues to why colonies vary in size in a general sense, but group means obscure the variation in reproductive success within each colony. As reported in earlier chapters, virtually all of the documented effects of colony size in cliff swallows (table 14.1) varied to some extent for different birds within the same colony. Even if birds are selecting the colony size that is best for their own needs (section 14.3.1), not all individuals selecting a given colony are doing equally well. The within-colony asymmetries in the expected costs and benefits of group living may have an important influence on the evolution of coloniality, because the implication is that not all birds in a colony receive the same expected payoffs (Moller 1987a; Morton, Forman, and Braun 1990) or that not all birds formed colonies for the same reason (Krebs and Barnard 1980; Weatherhead 1983; Still, Monaghan, and Bignal 1987; Summers, Westlake, and Feare 1987; Rayor and Uetz 1990; Jenni 1993). The individual differences within cliff swallow colonies in large part are caused, proximately, by spatial and temporal effects on reproductive success.

14.4.1 Spatial Effects

Both in our analyses in this book and in those of others (e.g., Hoogland and Sherman 1976; Snapp 1976; Van Vessem and Draulans 1986; Moller

1987a; Shields and Crook 1987; Burger and Gochfeld 1990, 1991; Wiklund and Andersson 1994), colony size is often treated as if the numerical difference in number of nests is the chief distinction between groups of different sizes. What has generally not been appreciated is that usually colonies of different sizes also differ substantially in their internal spatial structure, and these spatial differences may be as important as size per se. For example, in most species an increase in colony size typically reduces average nearest neighbor distance, as nests become more closely packed. This occurred in cliff swallows (section 7.2.2). The increased proximity of individuals therefore exacerbates the costs of coloniality or enhances the benefits beyond what might be expected from a simple numerical change in colony size between hypothetical sites of similar spatial configuration. Besides creating potentially nonlinear cost and benefit curves across colonies, increased crowding (or other spatial effects) may also lead to fundamental asymmetries in the social environment experienced by individuals within the same colony. The most obvious example is the difference between nests in the center and on the edge of a colony.

Nests on the edge of a colony generally differ spatially in two major ways from those closer to the center. Edge nests, by definition, have fewer neighbors, because they have none on at least one side. They also may be slightly more isolated, meaning they are farther from the neighbors they do have. For example, a cliff swallow nest's linear distance from the colony's center was positively correlated with its nearest neighbor distance in twenty-eight of thirty-six colonies, a significant preponderance of positive correlations (binomial test, $p = .001$). The center-edge differences are often magnified in large colonies, where the central nests tend to be more densely clustered. Consequently, if spatial configuration influences reproductive success, we should see the greatest differences between center and edge nests in the larger colonies (section 5.1).

In earlier chapters we investigated a number of costs and benefits of coloniality that were potentially affected by nest position and colony spatial structure. The two likely to be most important were ectoparasitism and predator avoidance. Transmission of ectoparasites within cliff swallow colonies was probably enhanced when nests were more closely spaced (section 4.9.2). This led to our prediction that ectoparasitism should be highest in center nests and should decrease with a nest's distance from the colony's center. Center nests should be less likely to be preyed on, however, because predators such as bull snakes encounter edge nests first and often become satiated before reaching the center of the colony. There was strong evidence that edge nests were in fact more heavily preyed on (section 8.7), but surprisingly we found no evidence for increased levels of ectoparasitism the closer a nest was to the center (section 4.5.3). Unfortunately, our center-edge comparison of parasite load was confounded by edge nests' starting later in the year than the more central ones (section 4.5.3), and this temporal effect (section 14.4.2) may have obscured the likely differences in initial ectoparasite load between center and edge nests.

Annual reproductive success declined in most colonies with a nest's distance from the center of the colony (section 11.5; fig. 12.16b), although the effect was not strong. Nevertheless, the result was a slightly lower expectation of success on average for individuals that nested in the more peripheral positions within a colony. Given the relatively high probability of edge nests' being lost if predators gained access to a colony (section 8.7), we are faced with the problem of explaining why birds chose to occupy peripheral nests at all. For a colony to exist, some individuals must settle for the edge; otherwise, edge birds will successively peel off until either no colony is left or there is only a very small one with no clearly defined center and edge.

The cliff swallows that occupy edge nests in a given colony presumably do at least as well there as they would as solitaries or in more central positions in other colonies. Perhaps, as Weatherhead (1983) suggested for communal roosts, the peripheral nesters accept the higher risk of predation in exchange for the foraging-related benefits of being in a group. This is consistent with our observations that first-year birds often settle on the edge (section 11.8), and presumably these less experienced first-year birds are the ones to gain the most from the foraging advantages of a colony (section 14.3.1). A similar pattern was seen in a colonial spider, in which peripheral settlers experienced a higher risk of predation while increasing their foraging efficiency by being in the colony (Rayor and Uetz 1990).

Another potential spatial effect on reproductive success is colony geometry. For example, in some colonies nests are clustered in a honeycomb pattern (fig. 3.6b), whereas in others they are arranged in single or double rows (fig. 3.6a,d). Both types of nest spacing can occur in the same colony, especially on the more irregular cliff faces. In addition to the effects attributed to differing nest densities and nearest neighbor distances (above), visibility of neighbors is likely to be very different in these two types of spatial configurations. In the honeycomb pattern, more neighbors will be visible within a given radius from one's nest, potentially enhancing both the opportunities for trespassing (section 5.5) and its associated outcomes (such as brood parasitism; section 6.3) and the efficiency of information transfer (section 10.1). Why birds elect to nest in these different spatial configurations, and the configurations' effect on reproductive success independent of nest density per se, has not been established.

14.4.2 Temporal Effects

Although spatial effects within cliff swallow colonies were potentially great, the empirical evidence did not demonstrate large asymmetries in apparent fitness that were directly attributable to nest spacing (sections 14.4.1, 11.5). There was much stronger evidence for temporal effects on within-colony reproductive success (e.g., sections 11.4, 12.7.1); we found a marked seasonal decline in success, both over the nesting season as a whole and during the time each single colony was active. This effect was so severe that most nests in which egg laying was initiated after about 25 June were unsuccessful and birds resorted to laying in incomplete nests to compensate in part for a late start (section 11.4). A seasonal increase in the swallow bug population at each site (section 4.4), and the bugs' tendency to aggregate at late nests (section 4.9.2), were probably responsible for much of the seasonal reduction in reproductive success, but other factors were involved because the same pattern was seen among fumigated nests (section 11.4).

Given the substantial reduction in reproductive success for late nesters within a colony (some individuals had almost zero expectation of success), why did any birds initiate late nesting? In almost all colonies regardless of size, there were variable numbers of individuals that began egg laying well after the rest of the colony had finished (figs. 8.6, 8.7). These late nesters were in all likelihood yearling birds (or otherwise inferior ones) that either arrived on the breeding grounds late or took longer than normal to sequester the necessary resources for reproduction (e.g., nest site, energy for making eggs). Their options thus were either to join an existing colony that was already well into nesting or to breed solitarily. Solitary nesting was apparently not favored by young birds or those in poor condition (section 14.3.1). Given the average annual probability of survival for first-year cliff swallows (section 12.3.1), which was likely even lower for birds already in relatively poor condition (section 12.5.3), these individuals' best bet was probably to try to breed in the current year rather than wait until the next year. And assuming they made the choice to breed, perhaps they could make the best of a bad job by joining an existing colony (often a large one; section 8.6.1), even though their odds of success there generally were not good. Late nesting in large colonies should persist as long as late individuals do as well as (or better than) they can expect to do anywhere else (and assuming their condition leading to late nesting is phenotypically, not genetically, based).

Another sort of temporal effect is that of broader yearly influences on reproductive success and expected fitness. There were large differences between years in annual reproductive success (section 11.3.1), survivorship (section 12.3.3), and food availability (chapter 10). It is likely that many, if not most, of the costs and benefits of cliff swallow coloniality we studied (table 14.1) varied in importance between years, although in most cases we could not investigate these yearly effects rigorously. Climatic differences among years (section 3.2.2) probably caused much of the annual variation in reproductive success between sites and probably also altered within-colony expectations through climatic influences on populations of ectoparasites and predators. Consequently, center-edge or early-late dichotomies may be particularly pronounced in some years and virtually nil in others.

14.4.3 Old versus New Nests

A potential cause of within-colony asymmetry in reproductive success that is both spatial and temporal is nest age. Given that many cliff swallow nests remain intact from year to year, individuals returning in the spring have a choice of whether to reuse an old nest or build a new one (section 4.10.3). All else being equal, an animal behaving "optimally" should reuse an old nest, to save the time and energy costs of nest construction. However, most cliff swallow colonies included birds that used both old and new nests. The fitness expectations of using the two types of nest are not necessarily the same, at least theoretically. Building a new nest takes time (which varies depending on colony size; section 5.3), and hence usually delays the start of nest building so that the temporal effects discussed above may become important. New nests also must be built on substrate not occupied by existing nests, and thus new nests often are constructed on the edges of groups of old nests. This may lead to some of the spatial effects on reproductive success (section 14.4.1).

We predicted that new nests should have lower ectoparasite loads than old nests, on average, because new nests do not contain preexisting overwintering populations of ectoparasites and tend to be closer to the edges of colonies where parasites are less likely to reach (section 4.5). This seems intuitively reasonable, although an empirical test was confounded by the temporal effects that tended to concentrate dispersing swallow bugs in the later-starting new nests. Furthermore, the increase in time and energy required for nest building could itself reduce success, on average, for birds using new nests relative to those occupying old nests within the same colony.

Surprisingly, we found no significant overall differences in either annual reproductive success (section 11.6.1) or survivorship (section 12.8.2) of cliff swallows using old versus new nests. Either the increased time required for constructing new nests was not long enough to cause a date-related reduction in reproductive success, or other asymmetrical costs (increased ectoparasitism?) similarly reduced the success of birds using old nests. Because nest-building time was greatly reduced in large colonies owing to wall sharing (section 5.3.2) and increased efficiency at gathering mud (section 8.4.2), building a new nest in those colonies may not have represented much of a delay and thus not imposed much temporal cost. Perhaps for that reason alone, the proportion of birds using new nests increased with colony size (section 4.10.3) and resulted in no net reduction in fitness for most birds building new nests.

Cliff swallows clearly assessed old nests early in the spring and avoided those with heavy ectoparasite infestations (section 4.10.1); thus choice of nest age was probably not random. What causes individuals to use old versus new nests, and the inherent trade-offs of each type, is still not clear, but the evidence available at present suggests that nest age per se probably does not generate strong within-colony asymmetries in Darwinian fitness.

Overall, the potential spatial and temporal effects within cliff swallow colonies were not overwhelming, at least as measured by annual reproductive success and survivorship. There were clear differences in realized success of birds within the same colony, but in each case there was no reason to expect that the individuals were not doing as well by living in the observed colony as they could expect to do anywhere.

14.5 RELATEDNESS AND RECIPROCITY

We have assumed throughout this book that cliff swallow colonies consist largely of nonrelatives and that mean degree of relatedness among colony members is low (section 1.2). The extent to which animal groups are composed of relatives can be important in determining the stable group size (Giraldeau and Caraco 1993; Higashi and Yamamura 1993; Rannala and Brown 1994) and in interpreting social behavior (e.g., alarm calling and mobbing of predators; section 8.5) in terms of its effect on inclusive fitness (e.g., Hamilton 1964; Roeloffs and Riechert 1988). How valid is our assumption of minimal overall relatedness within cliff swallow colonies?

Given the patterns of colony choice by cliff swallows, in which both first-year birds (section 13.5.1) and older age classes (section 13.4.1) often returned to the same colony site the next season, parents and their offspring from a previous year were sometimes present in the same breeding colony. From a total of thirteen colonies ranging from 125 to 2,350 nests, we recorded fifty-six instances in which one parent and at least one offspring from a previous year bred at the same colony site and eight instances in which both parents and at least one offspring were present. However, the actual number of parent-offspring associations is relatively meaningless because of the many individuals that missed being caught in a given year (section 12.2.2); in addition, the observed number of associations is dependent on how many nests we had (relatively few) in which parents and offspring were both identified, the annual survival probabilities for adults and first-year birds, and age-specific dispersal patterns.

We also documented cases in which siblings settled together within the same colony. Through 1992 there were 153 instances in which two putative siblings from the same natal colony (putative siblings because of the prevalence of conspecific brood parasitism; section 6.3) were recaught the next year; in 132 cases they were both in the natal colony, and in an additional 5 cases both had dispersed to the same nonnatal colony. Thus, in 137 of 153 pairings (89.5%) siblings were present together in the same colony. In the remaining instances, only one sibling dispersed to a nonnatal colony or both dispersed to different nonnatal colonies. Among 21 cases in which three putative siblings were recaught their first year as breeders, in 14 all three birds remained in their natal colony and in 2 all three birds dispersed to the same nonnatal colony. In 2 of the remaining 5 cases, two of the three siblings remained in their natal colony.

These sorts of associations between siblings are expected, given the fidelity of first-year cliff swallows to their natal site (section 13.5.1). However, among the individuals recaptured during their first breeding season, their return to their natal site when they were known to have a surviving sibling (85.1%) was greater than the overall value seen for all birds encountered in the study area their first year (73.6%; section 13.5.1); the difference was significant ($\chi^2 = 11.03$, df $= 1$, $p < .001$). This suggests that siblings may associate in breeding colonies in subsequent years more often than might be expected of any two returning first-year birds chosen at random from the same natal colony.

The observed instances of parent-offspring and sibling associations in breeding colonies suggest that average relatedness within each cliff swallow colony is greater than zero. The strong effect of ectoparasites on natal dispersal (Brown and Brown 1992) might also suggest that average within-colony relatedness is higher in smaller colonies, because there is generally less ectoparasitism there and a greater fraction of the nestlings raised in a small colony would be more likely to return to that site, all else being equal. However, the documented cases of kin association may give a distorted picture of within-colony relatedness. They overlook, for instance, 1,721 cases in which a first-year breeder was not known to have a sibling present in any colony in the study area. Apparent absence of siblings could of course reflect our not catching birds that were actually present; nevertheless it still seems likely that many cliff swallows did not associate with siblings in breeding colonies. The seasonal dissolution and reforming of colonies and the annual movements of these birds between Nebraska and South America also seem to work against the maintenance of exclusive long-term kin associations. Even a small amount of migration (in a population-genetics sense) between colonies is enough to greatly reduce overall levels of relatedness within each colony (Hamilton 1975), and cliff swallows often moved between colonies (sections 13.4.1, 13.5.1).

A variety of methods exist to calculate the average relatedness of members within a group (e.g., Hamilton 1975; Seger 1977; Murray 1981, 1985; Pamilo and Crozier 1982; Pullum 1982; Coresh and Goldman

1988; Queller and Goodnight 1989), and these techniques could perhaps be used to estimate relatedness within cliff swallow colonies. Unfortunately, most of the currently available methods that are based on demography make several restrictive assumptions that are not met for cliff swallows. In particular they implicitly require a constant (equilibrium) population or subpopulation size from generation to generation in order to estimate rates of immigration and emigration. Cliff swallow colonies varied widely in size between years (generations), however (section 7.4.2), and at present we are not aware of any appropriate demographic methods to estimate relatedness that can be applied to our data. For the time being, the assumption of relatively low levels of relatedness within cliff swallow colonies still seems reasonable, although there may be greater kin association than we originally thought. The use of molecular tools to estimate degree of kin association within groups (e.g., Parker, Waite, and Decker 1995) would likely yield better information on relatedness than that obtainable with demographic methods, but applying the molecular techniques on the scale necessary in most of our colonies would be logistically prohibitive. Assuming low levels of relatedness, kin (indirect) selectionist interpretations of cliff swallow social behavior (e.g., Higashi and Yamamura 1993; Hoogland 1995) are not applicable.

A final selective process that might influence the evolution of coloniality in cliff swallows is reciprocity. Reciprocal interactions can either be of the classical sort, in which returning aid to a donor is costly to the original recipient (Trivers 1971; Axelrod and Hamilton 1981), or of the sort termed pseudoreciprocity, in which returning aid to a donor also benefits the original recipient (Connor 1986; Rothstein and Pierotti 1988). Although clear-cut evidence for reciprocity in birds in general is lacking (Koenig 1988), we interpreted calling to alert conspecifics to food as an example of pseudoreciprocity (Brown, Brown, and Shaffer 1991; section 9.6), and other forms of cliff swallow social behavior could have a reciprocal basis. For example, apparent cooperation among colony residents in not disguising their foraging success when returning to their nests and thereby informing neighbors of food locations (section 9.5.3) could theoretically represent reciprocity or pseudoreciprocity (see Newman and Caraco 1989; Richner and Heeb 1995; cf. Waltz 1983). Swallow colonies in which each individual's spatial location is fixed by its nest site seem to be good candidates for the evolution of spatially based cooperative interactions described by Nowak and May (1992), in which no memory of past encounters is required and both cooperators and noncooperators can persist indefinitely.

A rigorous exploration of the role of reciprocity in cliff swallow social evolution is beyond the scope of this book, but we did evaluate one major prerequisite. Most models of reciprocity require long-term associations between individuals, so that cooperators and cheaters can be identified based on repeated interactions (Trivers 1971; Axelrod and Hamilton 1981; Wilkinson 1985; Rothstein and Pierotti 1988). Perennial associations between individuals were relatively common in cliff swallows, occurring when the same birds returned to the same colony site in successive years (section 13.4.1). As our mark-recapture program continues, we are accumulating more instances where the same birds reside together in the same colony for three to five years or longer. It is probable that many, if not most, cliff swallows in our population maintain multiyear associations with unrelated individuals through use of a common colony site each year.

Being in the same colony does not guarantee that individuals interact, however, especially if the issue is information transfer occurring primarily among close neighbors. Therefore we examined where in the colony a bird settled each year, as a measure of the regularity with which it might encounter the same neighbors in successive years. If individuals return to the same part of a colony each year (as in snow geese; Cooke et al. 1983), they will likely interact repeatedly with some familiar neighbors (or perhaps kin).

For this analysis, we used only adults whose nest locations in the same colony were known in two or more years. All data came from two culvert sites, mostly in 1986–88, the years when we plugged nests at night and caught owners at their nests (section 2.3.2). Each bird's nest location each year was plotted on a schematic map of the colony, and the distance between locations was expressed in terms of the number of intervening nest diameters (as in Brown and Brown 1989, 1991). For example, a bird occupying a nest adjacent to the one it used the previous year had a distance of one nest diameter. At these sites nests remained largely intact from year to year, giving the birds the potential opportunity to reoccupy the same nest they had used the previous year.

The distribution of nest distances for 49 birds settling in the same colony in successive years (fig. 14.3) indicated a nonrandom settlement pattern. Since nest density on these culvert substrates was uniform throughout the colony, birds should have been equally likely to settle anywhere in the colony if they chose nest sites independent of the previous year's location. The observed distribution (fig. 14.3) differed significantly from the one expected if nest choice was random ($\chi^2 = 29.7$, df = 6, $p < .001$; this analysis was done similarly to that of figure 6.14, in which we concluded that kleptoparasites chose host nests independent of their natal nest). Cliff swallows generally tended to return to the same part of a colony if they reused the colony site the next year (fig. 14.3), although they seldom reused the same nest (only 3 observed cases). Among those birds moving large distances within the same colony the next year, many were originally on the periphery of the colony. Of 15 birds using nests

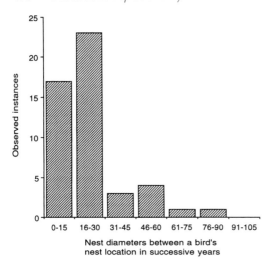

Fig. 14.3 Frequency distribution of distance between a cliff swallow's nest location within the same colony in successive years as measured by the number of intervening nest diameters. Nest density throughout the colony was uniform, predicting an equal number of frequencies (7.0) in each of the categories if birds chose nests independent of their location in the previous year.

classified as edge (section 2.8.2) in the first year, 13 (86.7%) moved away from the edge and closer to the center the next year. This is perhaps not surprising, given the slight disadvantage to using edge nests (section 14.4.1).

The spatial associations within colonies and the return of many individuals to the same colony site from year to year ensure that cliff swallows have the opportunity for reciprocally cooperative social interactions. Although we doubt that reciprocity has been a major force in the evolution of cliff swallow coloniality, advantages obtained through reciprocity or pseudoreciprocity could augment the other benefits of colonial nesting and promote certain kinds of behavior, such as active information transfer (Brown, Brown, and Shaffer 1991). Further study of individuals' annual spatial associations within colonies is needed.

14.6   CONCLUSIONS AND UNANSWERED QUESTIONS

We conclude with tentative answers to the two questions with which we started this chapter and the book. Cliff swallows live in colonies because of net social benefits that vary both with colony size and with an individual's phenotype. Colonies vary in size probably because birds seek to settle where their expectations are greatest, and phenotypic differences among individuals lead to some birds doing best in small colonies, others doing best in medium-sized colonies, and still others doing best in large colonies. Comparisons with other swallows suggest that cliff swallows did not initially form colonies because of limitations in suitable nesting sites or in response to high predation rates, but rather to aid social foraging. Once aggregation began, other benefits and the costs of coloniality were automatic, and the benefits may have further reduced nest spacing.

We believe that the suite of costs and benefits we explored in cliff swallows (table 14.1) apply generally to most colonial birds. We also believe that the observed colony size variation in many species can be explained by phenotype-based colony choice by individuals, with the distribution of colony sizes reflecting the nature of the particular cost-benefit curves and the relative frequency of the different phenotypes in the population. There are, however, spatial, temporal, and social constraints on an animal's ability to choose the best colony size for its needs, and not all individuals in a given year may be able to optimize their colony choice. In these cases one may measure a reduced reproductive success for animals in certain colonies, but that does not mean their colony choice is maladaptive. Phenotype-based colony choice also seems probable for those that are less successful than others within the same colony; the critical question is how well the less successful birds would do elsewhere, and the answer may be not as well.

Given the phenotype-based colony choice hypothesis, it follows that researchers interested in coloniality need to focus on individuals. We need to determine what colonies an individual uses during its lifetime and its actual lifetime reproductive success in those colonies. Measurement of mean differences in the socioecological costs and benefits between groups of different sizes, the predominant approach to date, will provide useful information on how average costs and benefits of coloniality vary with colony size, but this approach is unlikely to tell us definitively why colonies formed initially (perhaps an unanswerable question) or why different birds are affected differently by the known costs and benefits as colony size changes.

We have only begun to understand the complexities of cliff swallow social evolution and coloniality. Many unanswered questions remain. We need more and better data on actual lifetime reproductive success of known individuals, so that we can better estimate fitness for birds in different-sized colonies. Such data may allow us to determine conclusively whether there are a few "best" colony sizes that the birds jockey among (as the estimates from fig. 14.1b might suggest) or whether phenotype-based sorting among colonies yields roughly equal fitness for all individuals. We need to work out exactly how different phenotypes are favored in different-sized colonies, by following individuals' reproductive histories and manipulating colony sizes or settlement opportunities. Molecular techniques to assign exact parentage must be applied, because the potential biases in observed reproductive success owing to extrapair copulation and conspecific brood parasitism are enormous. Additional manipulations and field experiments could be designed, since our approach to date (with a few exceptions; sections 4.7, 8.3, 10.7) has largely avoided experimental alterations. Manipulations should be used with caution, because they may disrupt long-term patterns of behavior or demography that have taken years to detect, but field experiments—such as manipulating cliff swallow colony size, perhaps in an adjacent study area—may be the only way to address certain issues (sections 14.3.1, 10.7).

Some of the more important questions yet to be answered fully for cliff swallows include: How often, and in what pattern, are ectoparasites (especially swallow bugs) transmitted between colonies, and how does between-colony transmission affect average parasite load per colony (section 4.9.1)? Why do cliff swallows destroy their neighbors' eggs (section 5.6)? Which birds of each sex benefit from extrapair copulation, by how much, and for females, in what way (section 6.2.5)? Is the apparent variation among females in their propensity to brood parasitize other nests in the colony (or to be parasitized by others) real, and if so, what accounts for such variation within and between colonies (section 6.3.3)? How often do parent cliff swallows fail to recognize their own young after fledging, and what effect does this failure have on offspring survivorship (section 6.4.1)? What effect do local resource characteristics have on whether a colony site is used in a given year, and do they affect the colony's size if the site is used (section 7.6.2)? Which birds typically discover food sources, and are there differences in individuals' propensity to rely on foraging information available at the colony versus discovering food themselves (section 9.5.3)? Why are food calls used only in bad weather, and do Nebraska cliff swallows use the other kinds of food calls that have been noted elsewhere (section 9.6)? Why do the presumed benefits of information transfer (chapter 10) apparently vary so much between years and sites? Does annual reproductive success in the largest colonies continue to decline with colony size, as suggested by the existing data (section 11.3.2)? Does the variation in first-year survivorship with colony size (section 12.4.1) and ectoparasite load (section 12.6) reflect dispersal biases by birds from the natal nests that were most heavily infested with ectoparasites? To what degree do related individuals associate within the same colony (section 14.5)?

We end by emphasizing that there are many interesting and fertile issues yet to explore in studies of coloniality, both for animals in general and for cliff swallows in particular. The greatest progress will likely come from long-term studies of single populations in which the same individuals from different colonies are followed across years. Despite the many obstacles to long-term studies in general (e.g., Tinkle 1979; Nisbet 1989; Malmer and Enckell 1994), we cannot envision any other approach to studying coloniality in cliff swallows or other species.

14.7 SUMMARY

This chapter addresses two questions posed at the beginning of the book: Why do cliff swallows live in colonies? And why do colonies vary in size? Cliff swallows likely gain social benefits as a direct result of conspecific association and are not colonial because of nesting site limitation. Comparisons with other mud nest building swallows of the genus *Hirundo*, which mostly forage solitarily, suggest that cliff swallows initially formed breeding colonies to permit low-risk group foraging. The other benefits, and all the costs, of coloniality were automatic consequences.

We estimated mean lifetime reproductive success (LRS) for birds that occupied the same colony size for life. Estimated LRS when expressed in terms of the number of offspring produced to fledging was highest for birds using the largest colonies (≥ 500 nests) but was about the same for those birds using all the other colony sizes. The preference many birds showed for the larger colonies was consistent with large colonies' being the most successful, on average.

Estimated LRS expressed in terms of the number of young surviving to their first breeding season showed a peak for cliff swallows perennially using medium-sized colonies of 100–249 nests and was next highest for birds using the smallest (1–10 nests) and largest (≥ 500 nests) colonies. A measure of LRS incorporating survival of young birds to their first breeding season might have been biased by preferential, ectoparasite-caused dispersal of birds raised in the larger and more parasite-infested colonies. That LRS was highest for individuals in colonies of 100–249 nests was suggested by a greater between-year size preference by birds for colonies of that size. However, the LRS estimates did not allow us to conclude confidently that average fitness differed markedly among birds in different-sized colonies.

We hypothesize that colony size variation in cliff swallows reflects phenotype-based colony choice, with individuals selecting the colony sizes that are best for them. Experience and inherent differences in individual quality and ability alter the cost-benefit curves for each individual so that some birds can expect their highest reproductive success in small colonies, others in medium-sized colonies, and still others in large colonies. Sorting of birds among colony sizes based on age, parasite load, and body condition supports this hypothesis. Colony size variation is probably not caused by ideal free matching of population size to local resource availability or aggressive exclusion of birds from certain sites by earlier-arriving "despots."

There were within-colony asymmetries in reproductive success, caused largely by spatial positioning of nests and temporal effects on breeding success. Nests closer to a colony's edge were slightly less successful than more central ones, yet birds continued to nest on the edge instead of deserting the colony. Late-nesting cliff swallows were less successful than earlier birds within the same colony. Nesting on the edge of an existing colony, or later in the year within a colony, may have been a better option than solitary nesting for younger, later-arriving birds.

The movement of birds between colonies suggested that mean relatedness within colonies was too low for indirect selection to be important in the evolution of cliff swallow coloniality. The same individuals often associated from year to year within a colony, and birds returning to the same part of a colony the next year probably interacted repeatedly with some of the same neighbors. At present no potentially reciprocal social interactions other than food calling have been identified in cliff swallows.

Appendix

Common and Scientific Names of Species Mentioned in the Text

American elm	*Ulmus americana*
Boxelder	*Acer negundo*
Chokecherry	*Prunus virginiana*
Eastern cottonwood	*Populus deltoides*
Eastern red cedar	*Juniperus virginiana*
Green ash	*Fraxinus pennsylvanica*
Hackberry	*Celtis occidentalis*
Peachleaf willow	*Salix amygdaloides*
Russian olive	*Elaeagnus commutata*
Silver maple	*Acer saccharinum*
Bird flea	*Ceratophyllus celsus*
Swallow bug	*Oeciacus vicarius*
Honeybee	*Apis mellifera*
Three-spined stickleback	*Gasterosteus aculeatus*
Snapping turtle	*Chelydra serpentina*
Common garter snake	*Thamnophis sirtalis*
Rat snake	*Elaphe obsoleta*
Bull snake	*Pituophis catenifer*
White-tailed tropicbird	*Phaethon lepturus*
Great blue heron	*Ardea herodias*
Cattle egret	*Bubulcus ibis*
Snow goose	*Chen caerulescens*
Shelduck	*Tadorna tadorna*
Common eider	*Somateria mollissima*
Turkey vulture	*Cathartes aura*
Osprey	*Pandion haliaetus*
Sharp-shinned hawk	*Accipiter striatus*
Red-tailed hawk	*Buteo jamaicensis*
Swainson's hawk	*Buteo swainsoni*
American kestrel	*Falco sparverius*

Prairie falcon	*Falco mexicanus*
Ruff	*Philomachus pugnax*
Herring gull	*Larus argentatus*
Black-legged kittiwake	*Rissa tridactyla*
Common tern	*Sterna hirundo*
Common guillemot	*Uria aalge*
Razorbill	*Alca torda*
Atlantic puffin	*Fratercula arctica*
Groove-billed ani	*Crotophaga sulcirostris*
Barn owl	*Tyto alba*
Great horned owl	*Bubo virginianus*
Eastern phoebe	*Sayornis phoebe*
Say's phoebe	*Sayornis saya*
Purple martin	*Progne subis*
Tree swallow	*Tachycineta bicolor*
Bank swallow	*Riparia riparia*
Northern rough-winged swallow	*Stelgidopteryx serripennis*
House martin	*Delichon urbica*
Barn swallow	*Hirundo rustica*
Angolan cliff swallow	*Hirundo (Petrochelidon) rufigula*
Cave swallow	*Hirundo (Petrochelidon) fulva*
Cliff swallow	*Hirundo (Petrochelidon) pyrrhonota*
Fairy martin	*Hirundo (Petrochelidon) ariel*
Indian cliff swallow	*Hirundo (Petrochelidon) fluvicola*
Preuss's cliff swallow	*Hirundo (Petrochelidon) preussi*
South African cliff swallow	*Hirundo (Petrochelidon) spilodera*
Black-billed magpie	*Pica pica*
Penduline tit	*Remiz pendulinus*
House wren	*Troglodytes aedon*
Fieldfare	*Turdus pilaris*
Loggerhead shrike	*Lanius ludovicianus*
Common grackle	*Quiscalus quiscula*
Yellow-rumped cacique	*Cacicus cela*
Quelea	*Quelea quelea*
White-browed sparrow-weaver	*Plocepasser mahali*
House sparrow	*Passer domesticus*
Evening bat	*Nycticeius humeralis*
Raccoon	*Procyon lotor*
Long-tailed weasel	*Mustela frenata*
Badger	*Taxidea taxus*
House cat	*Felis catus*
Black-tailed prairie dog	*Cynomys ludovicianus*
Deer mouse	*Peromyscus maniculatus*
Bison	*Bison bison*
Brown capuchin	*Cebus apella*
Long-tailed macaque	*Macaca fascicularis*

References

Abrahams, M. V. 1986. Patch choice under perceptual constraints: A cause for departures from an ideal free distribution. *Behav. Ecol. Sociobiol.* 19:409–15.

Aebischer, N. J., and J. C. Coulson. 1990. Survival of the kittiwake in relation to sex, year, breeding experience and position in the colony. *J. Anim. Ecol.* 59:1063–71.

Ahlen, I., and A. Andersson. 1970. Breeding ecology of an eider population on Spitsbergen. *Ornis Scandinavica* 1:83–106.

Alados, C. L. 1985. An analysis of vigilance in the Spanish ibex (*Capra pyrenaica*). *Z. Tierpsychol.* 68:58–64.

Alatalo, R. V., J. Hoglund, A. Lundberg, and W. J. Sutherland. 1992. Evolution of black grouse leks: Female preferences benefit males in larger leks. *Behav. Ecol.* 3:53–59.

Alcock, J. 1973. The mating behaviour of *Empis barbatoides* Melander and *Empis poplitea* Loew (Diptera: Empididae). *J. Nat. Hist.* 7:411–20.

Alexander, R. D. 1971. The search for an evolutionary philosophy of man. *R. Soc. Victoria Proc.* 84:99–120.

———. 1974. The evolution of social behavior. *Ann. Rev. Ecol. Syst.* 5:325–83.

Allan, J. D., and A. S. Flecker. 1989. The mating biology of a mass-swarming mayfly. *Anim. Behav.* 37:361–71.

Allchin, D. 1992. Simulation and analysis of information-center foraging. *Behaviour* 122:288–305.

Allee, W. C. 1931. *Animal aggregations*. Chicago: University of Chicago Press.

———. 1938. *The social life of animals*. New York: Norton.

———. 1951. *Cooperation among animals*. New York: Schuman.

AOU. *See* American Ornithologists' Union.

American Ornithologists' Union (AOU). 1957. *Check-list of North American birds*. 5th ed. Baltimore: American Ornithologists' Union.

———. 1983. *Check-list of North American birds*. 6th ed. Lawrence, Kans.: American Ornithologists' Union.

Anderson, D. J., and P. J. Hodum. 1993. Predator behavior favors clumped nesting in an oceanic seabird. *Ecology* 74:2462–64.

Anderson, R. M. 1978. The regulation of host population growth by parasitic species. *Parasitology* 76:119–57.
Anderson, R. M., and R. M. May. 1978. Regulation and stability of host-parasite population interactions. I. Regulatory processes. *J. Anim. Ecol.* 47:219–47.
Anderson, R. M., P. J. Whitfield, and A. P. Dobson. 1978. Experimental studies of infection dynamics: Infection of the definitive host by the cercariae of *Transversotrema patialense*. *Parasitology* 77:189–200.
Andersson, M. 1978. Optimal foraging area: Size and allocation of search effort. *Theor. Pop. Biol.* 13:397–409.
———. 1981. Central place foraging in the whinchat, *Saxicola rubetra*. *Ecology* 62:538–44.
Andersson, M., and C. Wiklund. 1978. Clumping versus spacing out: Experiments on nest predation in fieldfares (*Turdus pilaris*). *Anim. Behav.* 26:1207–12.
Arnold, S. J., and R. J. Wassersug. 1978. Differential predation on metamorphic anurans by garter snakes (*Thamnophis*): Social behavior as a possible defense. *Ecology* 59:1014–22.
Arnold, W., and A. V. Lichtenstein. 1993. Ectoparasite loads decrease the fitness of alpine marmots (*Marmota marmota*) but are not a cost of sociality. *Behav. Ecol.* 4:36–39.
Audubon, J. J. 1831. *Ornithological biography*. Vol. 1. Edinburgh.
Aumann, G., and J. T. Emlen Jr. 1959. The distribution of cliff swallow nesting colonies in Wisconsin. *Passenger Pigeon* 21:95–100.
Axelrod, R., and W. D. Hamilton. 1981. The evolution of cooperation. *Science* 211:1390–96.
Baerg, W. J. 1944. Ticks and other parasites attacking northern cliff swallows. *Auk* 61:413–14.
Bailey, R. E. 1952. The incubation patch of passerine birds. *Condor* 54:121–36.
Baird, T. A., and T. D. Baird. 1992. Colony formation and some possible benefits and costs of gregarious living in the territorial sand tilefish, *Malacanthus plumieri*. *Bull. Marine Sci.* 50:56–65.
Baker, R. R. 1978. *The evolutionary ecology of animal migration*. London: Hodder and Staughton.
———. 1993. The function of post-fledging exploration: A pilot study of three species of passerines ringed in Britain. *Ornis Scandinavica* 24:71–79.
Barclay, R. M. R. 1988. Variation in the costs, benefits, and frequency of nest reuse by barn swallows (*Hirundo rustica*). *Auk* 105:53–60.
Barlow, J. C., E. E. Klaas, and J. L. Lenz. 1963. Sunning of bank swallows and cliff swallows. *Condor* 65:438–40.
Barnard, C. J. 1984. The evolution of food-scrounging strategies within and between species. In *Producers and scroungers: Strategies of exploitation and parasitism*, ed. C. J. Barnard, 95–126. London: Croom Helm.
Barnard, C. J., and J. M. Behnke, eds. 1990. *Parasitism and host behaviour*. London: Taylor and Francis.
Barnard, C. J., and R. M. Sibly. 1981. Producers and scroungers: A general model and its applications to captive flocks of house sparrows. *Anim. Behav.* 29:543–50.

Barrowclough, G. F. 1978. Sampling bias in dispersal studies based on finite area. *Bird-Banding* 49:333–41.
Barrowclough, G. F., and R. F. Rockwell. 1993. Variance of lifetime reproductive success: Estimation based on demographic data. *Amer. Nat.* 141:281–95.
Barta, Z. 1992. The effects of the patchiness and the nest location on mean flight distance: A model. *Ornis Hungarica* 2:37–44.
Barta, Z., and T. Szep. 1992. The role of information transfer under different food patterns: A simulation study. *Behav. Ecol.* 3:318–24.
——. 1995. Frequency-dependent selection on information-transfer strategies at breeding colonies: A simulation study. *Behav. Ecol.* 6:308–10.
Bates, J. K. 1962. Field studies on the behaviour of bird fleas. I. Behaviour of the adults of three species of bird fleas in the field. *Parasitology* 52:113–32.
Batschelet, E. 1965. *Statistical methods for the analysis of problems in animal orientation and certain biological rhythms*. Washington, D.C.: American Institute of Biological Sciences.
Bayer, R. D. 1982. How important are bird colonies as information centers? *Auk* 99:31–40.
Beal, F. E. L. 1907. Birds of California, in relation to the fruit industry. *U.S. Dept. Agri. Bull.* 30:1–100.
——. 1918. Food habits of the swallows, a family of valuable native birds. *U.S. Dept. Agri. Bull.* 619:1–28.
Beauchamp, G., and L. Lefebvre. 1988. Food finding in colonially nesting birds. *J. Theor. Biol.* 132:357–68.
Beecher, M. D. 1988. Kin recognition in birds. *Behav. Genetics* 18:465–82.
——. 1991. Successes and failures of parent-offspring recognition in animals. In *Kin recognition,* ed. P. G. Hepper, 94–124. Cambridge: Cambridge University Press.
Beecher, M. D., and I. M. Beecher. 1979. Sociobiology of bank swallows: Reproductive strategy of the male. *Science* 205:1282–85.
Beecher, M. D., I. M. Beecher, and S. Hahn. 1981. Parent-offspring recognition in bank swallows (*Riparia riparia*): II. Development and acoustic basis. *Anim. Behav.* 29:95–101.
Beecher, M. D., I. M. Beecher, and S. Lumpkin. 1981. Parent-offspring recognition in bank swallows (*Riparia riparia*): I. Natural history. *Anim. Behav.* 29:86–94.
Beecher, M. D., P. K. Stoddard, and P. Loesche. 1985. Recognition of parents' voices by young cliff swallows. *Auk* 102:600–605.
Behle, W. H. 1976. Systematic review, intergradation, and clinal variation in cliff swallows. *Auk* 93:66–77.
Beidleman, R. G. 1956. The 1859 overland journal of naturalist George Suckley. *Annals of Wyoming* 28:68–79.
Bell, W. J. 1991. *Searching behaviour: The behavioural ecology of finding resources*. London: Chapman and Hall.
Bengtson, S.-A. 1972. Reproduction and fluctuations in the size of duck populations at Lake Myvatn, Iceland. *Oikos* 23:35–58.
Bennett, G. F., and T. L. Whitworth. 1992. Host, nest, and ecological relationships of species of *Protocalliphora* (Diptera: Calliphoridae). *Can. J. Zool.* 70:51–61.

Bensch, S., and D. Hasselquist. 1994. Higher rate of nest loss among primary than secondary females: Infanticide in the great reed warbler? *Behav. Ecol. Sociobiol.* 35:309–17.

Bent, A. C. 1942. Life histories of North American flycatchers, larks, swallows, and their allies. *U.S. Natl. Mus. Bull.* 179.

Bentall, R. 1989. Streams. In *An atlas of the Sand Hills*, ed. A. Bleed and C. Flowerday, 93–114. Lincoln: Conservation and Survey Division, University of Nebraska.

Benton, A. H., and H. Tucker. 1968. Weather and purple martin mortality in western New York. *Kingbird* 18:71–75.

Berg, G. L., ed. 1981. *1981 farm chemicals handbook*. Willoughby, Ohio: Meister.

Berndt, R., and H. Sternberg. 1968. Terms, studies and experiments on the problems of bird dispersion. *Ibis* 110:256–69.

Bertram, B. C. R. 1978. Living in groups: Predators and prey. In *Behavioural ecology*, ed. J. R. Krebs and N. B. Davies, 64–96. Oxford: Blackwell.

———. 1979. Ostriches recognise their own eggs and discard others. *Nature* 279:233–34.

———. 1980. Vigilance and group size in ostriches. *Anim. Behav.* 28:278–86.

Bildstein, K. L. 1983. Why white-tailed deer flag their tails. *Amer. Nat.* 121:709–15.

Binford, G. J., and A. L. Rypstra. 1992. Foraging behavior of the communal spider, *Philoponella republicana* (Araneae: Uloboridae). *J. Insect Behav.* 5:321–35.

Birkhead, T. R. 1977. The effect of habitat and density on breeding success in the common guillemot (*Uria aalge*). *J. Anim. Ecol.* 46:751–64.

———. 1978. Behavioural adaptations to high density nesting in the common guillemot *Uria aalge*. *Anim. Behav.* 26:321–31.

———. 1985. Coloniality and social behaviour in the Atlantic alcidae. In *The Atlantic Alcidae*, ed. D. N. Nettleship and T. R. Birkhead, 355–82. London: Academic Press.

Birkhead, T. R., K. Clarkson, M. D. Reynolds, and W. D. Koenig. 1992. Copulation and mate guarding in the yellow-billed magpie *Pica nuttalli* and a comparison with the black-billed magpie *P. pica*. *Behaviour* 121:110–30.

Birkhead, T. R., and A. P. Moller. 1992. *Sperm competition in birds*. London: Academic Press.

Bitterbaum, E. J., and C. R. Brown. 1981. A martin house is not a home. *Nat. Hist.* 90 (5):64–69.

Bjorn, T. H., and K. E. Erikstad. 1994. Patterns of intraspecific nest parasitism in the High Arctic common eider (*Somateria mollissima borealis*). *Can. J. Zool.* 72:1027–34.

Black, F. L. 1966. Measles endemicity in insular populations: Critical community size and its evolutionary implication. *J. Theor. Biol.* 11:207–11.

Bleed, A. 1989. Introduction to plants and animals. In *An atlas of the Sand Hills*, ed. A. Bleed and C. Floweray, 123–26. Lincoln: Conservation and Survey Division, University of Nebraska.

Blondel, J., R. Pradel, and J. D. Lebreton. 1992. Low fecundity insular blue tits do not survive better as adults than high fecundity mainland ones. *J. Anim. Ecol.* 61:205–13.

Boness, D. J. 1990. Fostering behavior in Hawaiian monk seals: Is there a reproductive cost? *Behav. Ecol. Sociobiol.* 27:113–22.

Bonnot, P. 1921. Sparrow hawk captures swallow. *Condor* 23:136.

Borror, D. J., D. M. Delong, and C. A. Triplehorn. 1981. *An introduction to the study of insects.* Philadelphia: Saunders.

Boulinier, T., and E. Danchin. 1994. Information transfers on breeding patch quality and the evolution of coloniality. *J. Ornithol.* 135:192.

Boyd, A. W., and A. L. Thompson. 1937. Recoveries of marked swallows in the British Isles. *British Birds* 30:278–87.

Brigham, R. M. 1989. Effects of radio transmitters on the foraging behavior of barn swallows. *Wilson Bull.* 101:505–6.

Brockmann, H. J., and C. J. Barnard. 1979. Kleptoparasitism in birds. *Anim. Behav.* 27:487–514.

Brodskiy, A. K. 1973. The swarming behavior of mayflies (Ephemeroptera). *Entomol. Rev.* 52:33–39.

Brown, C. R. 1976. Minimum temperature for feeding by purple martins. *Wilson Bull.* 88:672–73.

———. 1978a. Post-fledging behavior of purple martins. *Wilson Bull.* 90:376–85.

———. 1978b. Clutch size and reproductive success of adult and subadult purple martins. *Southwestern Nat.* 23:597–604.

———. 1980. Sleeping behavior of purple martins. *Condor* 82:170–75.

———. 1984. Laying eggs in a neighbor's nest: Benefit and cost of colonial nesting in swallows. *Science* 224:518–19.

———. 1985a. The costs and benefits of coloniality in the cliff swallow. Ph.D. diss., Princeton University.

———. 1985b. Vocalizations of barn and cliff swallows. *Southwestern Nat.* 30:325–33.

———. 1986. Cliff swallow colonies as information centers. *Science* 234:83–85.

———. 1988a. Enhanced foraging efficiency through information centers: A benefit of coloniality in cliff swallows. *Ecology* 69:602–13.

———. 1988b. Social foraging in cliff swallows: Local enhancement, risk sensitivity, competition and the avoidance of predators. *Anim. Behav.* 36:780–92.

Brown, C. R., and E. J. Bitterbaum. 1980. Implications of juvenile harassment in purple martins. *Wilson Bull.* 92:452–57.

Brown, C. R., and M. B. Brown. 1986. Ectoparasitism as a cost of coloniality in cliff swallows (*Hirundo pyrrhonota*). *Ecology* 67:1206–18.

———. 1987. Group-living in cliff swallows as an advantage in avoiding predators. *Behav. Ecol. Sociobiol.* 21:97–107.

———. 1988a. A new form of reproductive parasitism in cliff swallows. *Nature* 331:66–68.

———. 1988b. Genetic evidence of multiple parentage in broods of cliff swallows. *Behav. Ecol. Sociobiol.* 23:379–87.

———. 1988c. The costs and benefits of egg destruction by conspecifics in colonial cliff swallows. *Auk* 105:737–48.

———. 1989. Behavioural dynamics of intraspecific brood parasitism in colonial cliff swallows. *Anim. Behav.* 37:777–96.

———. 1990. The great egg scramble. *Natural History*, no. 2:34–41.

———. 1991. Selection of high-quality host nests by parasitic cliff swallows. *Anim. Behav.* 41:457–65.

———. 1992. Ectoparasitism as a cause of natal dispersal in cliff swallows. *Ecology* 73:1718–23.

———. 1995. Cliff swallow. In *The birds of North America*, ed. A. Poole and F. Gill, no. 149. Philadelphia: Academy of Natural Sciences.

Brown, C. R., M. B. Brown, and A. R. Ives. 1992. Nest placement relative to food and its influence on the evolution of avian coloniality. *Amer. Nat.* 139:205–17.

Brown, C. R., M. B. Brown, and B. Rannala. 1995. Ectoparasites reduce long-term survivorship of their avian host. *Proc. R. Soc. London,* ser. B, 262:313–19.

Brown, C. R., M. B. Brown, and M. L. Shaffer. 1991. Food-sharing signals among socially foraging cliff swallows. *Anim. Behav.* 42:551–64.

Brown, C. R., and J. L. Hoogland. 1986. Risk in mobbing for solitary and colonial swallows. *Anim. Behav.* 34:1319–23.

Brown, C. R., and B. Rannala. 1995. Colony choice in birds: Models based on temporally invariant site quality. *Behav. Ecol. Sociobiol.* 36:221–28.

Brown, C. R., B. J. Stutchbury, and P. D. Walsh. 1990. Choice of colony size in birds. *Trends Ecol. Evol.* 5:398–403.

Brown, D. 1988. Components of lifetime reproductive success. In *Reproductive success: Studies of individual variation in contrasting breeding systems,* ed. T. H. Clutton-Brock, 439–53. Chicago: University of Chicago Press.

Brown, J. L. 1964. The evolution of diversity in avian territorial systems. *Wilson Bull.* 76:160–69.

———. 1982. Optimal group size in territorial animals. *J. Theor. Biol.* 95:793–810.

———. 1987. *Helping and communal breeding in birds.* Princeton: Princeton University Press.

Brown, K. M., M. Woulfe, and R. D. Morris. 1995. Patterns of adoption in ring-billed gulls: Who is really winning the inter-generational conflict? *Anim. Behav.* 49:321–31.

Brownie, C., D. R. Anderson, K. P. Burnham, and D. S. Robson. 1985. *Statistical inference from band recovery data—a handbook,* 2d ed. Res. Publ. 156. Washington, D.C.: U.S. Department of the Interior.

Browning, M. R. 1992. Geographic variation in *Hirundo pyrrhonota* (cliff swallow) from northern North America. *Western Birds* 23:21–29.

Bruner, L., R. H. Wolcott, and M. H. Swenk. 1904. *A preliminary review of the birds of Nebraska.* Omaha: Klopp and Bartlett.

Bruton, L. 1975. *The swallows of San Juan Capistrano.* Los Angeles: San Juan.

Bryant, D. M. 1975. Breeding biology of house martins *Delichon urbica* in relation to aerial insect abundance. *Ibis* 117:180–216.

———. 1978a. Environmental influences on growth and survival of nestling house martins *Delichon urbica. Ibis* 120:271–83.

———. 1978b. Establishment of weight hierarchies in the broods of house martins *Delichon urbica*. *Ibis* 120:16–26.

———. 1988a. Energy expenditure and body mass changes as measures of reproductive costs in birds. *Funct. Ecol.* 2:23–34.

———. 1988b. Lifetime reproductive success in house martins. In *Reproductive success: Studies of individual variation in contrasting breeding systems,* ed. T. H. Clutton-Brock, 173–88. Chicago: University of Chicago Press.

———. 1989. House martin. In *Lifetime reproduction in birds,* ed. I. Newton, 89–106. London: Academic Press.

———. 1991. Constraints on energy expenditure by birds. *Acta XX Congr. Internatl. Ornithol.,* 1989–2001.

Bryant, D. M., and K. R. Westerterp. 1982. Evidence for individual differences in foraging efficiency amongst breeding birds: A study of house martins *Delichon urbica* using the doubly labelled water technique. *Ibis* 124:187–92.

Buckley, F. G., and P. A. Buckley. 1980. Habitat selection and marine birds. In *Behavior of marine animals,* vol. 4, *Marine birds: Current perspectives in research,* ed. J. Burger, B. L. Olla, and H. E. Winn, 69–112. New York: Plenum.

Buckley, P. A., and F. G. Buckley. 1977. Hexagonal packing of royal tern nests. *Auk* 94:36–43.

———. 1980a. What constitutes a waterbird colony? Reflections from the northeastern U.S. *Proc. 1979 Conf. Colonial Waterbird Group,* 1–15.

———. 1980b. Population and colony-site trends of Long Island waterbirds for five years in the mid 1970s. *Trans. Linn. Soc. N.Y.* 9:23–56.

Bullard, R. T., Jr. 1963. Banding notes on the Nickajack cliff swallows (*Petrochelidon pyrrhonota*). *Eastern Bird-Banding Assoc. News* 26:191–203.

Bulmer, M. G. 1984. Risk avoidance and nesting strategies. *J. Theor. Biol.* 106: 529–35.

Bunzel, M., and J. Druke. 1989. Kingfisher. In *Lifetime reproduction in birds,* ed. I. Newton, 107–16. London: Academic Press.

Burger, J. 1974a. Breeding adaptations of Franklin's gull (*Larus pipixican*) to a marsh habitat. *Anim. Behav.* 22:521–67.

———. 1974b. Breeding biology and ecology of the brown-hooded gull in Argentina. *Auk* 91:601–13.

———. 1981. A model for the evolution of mixed-species colonies of ciconiiformes. *Quart. Rev. Biol.* 56:143–67.

———. 1988a. Social attraction in nesting least terns: Effects of numbers, spacing, and pair bonds. *Condor* 90:575–82.

———. 1988b. Effects of age on foraging in birds. *Acta XIX Congr. Internatl. Ornithol.,* 1127–40.

Burger, J., and M. Gochfeld. 1990. *The black skimmer: Social dynamics of a colonial species.* New York: Columbia University Press.

———. 1991. *The common tern: Its breeding biology and social behavior.* New York: Columbia University Press.

———. 1993. When is a heronry crowded: A case study of Huckleberry Island, New York, U.S.A. *J. Coastal Res.* 9:221–28.

Burke, T. 1989. DNA fingerprinting and other methods for the study of mating success. *Trends Ecol. Evol.* 4:139–44.

Burke, T., and M. W. Bruford. 1987. DNA fingerprinting in birds. *Nature* 327: 149–52.

Burke, T., O. Hanotte, M. W. Bruford, and E. Cairns. 1991. Multilocus and single locus minisatellite analysis in population biological studies. In *DNA fingerprinting: Approaches and applications,* ed. T. Burke, G. Dolf, A. J. Jeffreys, and R. Wolff, 154–68. Basel: Birkhauser.

Burton, P. J. K., and M. H. Thurston. 1959. Observations on arctic terns in Spitsbergen. *British Birds* 52:149–61.

Burtt, E. H., Jr. 1977. Some factors in the timing of parent-chick recognition in swallows. *Anim. Behav.* 25:231–39.

Buskirk, R. E. 1975. Coloniality, activity patterns and feeding in a tropical orb-weaving spider. *Ecology* 56:1314–28.

Buskirk, W. H. 1976. Social systems in a tropical forest avifauna. *Amer. Nat.* 110: 293–310.

Buss, I. O. 1942. A managed cliff swallow colony in southern Wisconsin. *Wilson Bull.* 54:153–61.

Buss, L. W. 1981. Group living, competition, and the evolution of cooperation in a sessile invertebrate. *Science* 213:1012–14.

Bustamante, J., and F. Hiraldo. 1990. Adoptions of fledglings by black and red kites. *Anim. Behav.* 39:804–6.

Butler, R. W. 1982. Wing fluttering by mud-gathering cliff swallows: Avoidance of "rape" attempts? *Auk* 99:758–61.

Cairns, D. K. 1989. The regulation of seabird colony size: A hinterland model. *Amer. Nat.* 134:141–46.

Caraco, T. 1981. Risk-sensitivity and foraging groups. *Ecology* 62:527–31.

Caraco, T., and L.-A. Giraldeau. 1991. Social foraging: Producing and scrounging in a stochastic environment. *J. Theor. Biol.* 153:559–83.

Caraco, T., and H. R. Pulliam. 1984. Sociality and survivorship in animals exposed to predation. In *A new ecology: Novel approaches to interactive systems,* ed. P. W. Price, C. N. Slobodchikoff, and W. S. Gaud, 279–309. New York: Wiley.

Caraco, T., G. W. Uetz, R. G. Gillespie, and L.-A. Giraldeau. 1995. Resource consumption variance within and among individuals: On coloniality in spiders. *Ecology* 76:196–205.

Carleton, J. H. 1943. *The prairie logbooks: Dragoon campaigns to the Pawnee villages in 1844, and to the Rocky Mountains in 1845.* Chicago: Caxton Club.

Carter, L. R., and L. B. Spear. 1986. Costs of adoption in western gulls. *Condor* 88:253–56.

Chapman, B. R., and J. E. George. 1991. The effects of ectoparasites on cliff swallow growth and survival. In *Bird-parasite interactions: Ecology, evolution and behaviour,* ed. J. E. Loye and M. Zuk, 69–92. Oxford: Oxford University Press.

Chapman, J. A. 1954. Studies on summit-frequenting insects in western Montana. *Ecology* 35:41–49.

Chardine, J. W. 1986. Interference of copulation in a colony of marked black-legged kittiwakes. *Can. J. Zool.* 64:1416–21.

Chek, A. A., and R. J. Robertson. 1994. Weak mate guarding in tree swallows: Ecological constraint or female control? *Ethology* 98:1–13.

Chepko-Sade, B. D., and Z. T. Halpin, eds. 1987. *Mammalian dispersal patterns: The effects of social structure on population genetics.* Chicago: University of Chicago Press.

Chiera, J. W., R. M. Newson, and M. P. Cunningham. 1985. Cumulative effects of host resistance on *Rhipicephalus appendiculatus* Neumann (Acarina: Ixodidae) in the laboratory. *Parasitology* 90:401–8.

Christenson, T. E. 1984. Behaviour of colonial and solitary spiders of the theridiid species *Anelosimus eximius. Anim. Behav.* 32:725–34.

Clark, C. W., and M. Mangel. 1984. Foraging and flocking strategies: Information in an uncertain environment. *Amer. Nat.* 123:626–41.

———. 1986. The evolutionary advantages of group foraging. *Theor. Pop. Biol.* 30:45–75.

Clark, F., and D. A. C. McNeil. 1981. The variation in population densities of fleas in house martin nests in Leicestershire. *Ecol. Entomol.* 6:379–86.

———. 1991. Temporal variation in the population densities of fleas in house martin nests (*Delichon u. urbica* (Linnaeus)) in Leicestershire, U.K. *Entomol. Gazette* 42:281–88.

Clark, G. A., Jr. 1979. Body weights of birds: A review. *Condor* 81:193–202.

Clark, G. W., and B. Swinehart. 1966. Blood parasitism in cliff swallows from the Sacramento Valley. *J. Protozool.* 13:395–97.

Clarke, M. F., and G. F. Fitz-Gerald. 1994. Spatial organisation of the cooperatively breeding bell miner *Manorina melanophrys. Emu* 94:96–105.

Clench, M. H., and R. C. Leberman. 1978. Weights of 151 species of Pennsylvania birds analyzed by month, age, and sex. *Bull. Carnegie Mus. Nat. Hist.* 5:1–85.

Clobert, J., V. Bauchau, A. A. Dhondt, and C. Vansteenwegen. 1987. Survival of female starlings in relation to brood size. *Acta Oecologica* 8:427–33.

Clobert, J., J. D. Lebreton, and D. Allaine. 1987. A general approach to survival rate estimation by recaptures or resightings of marked birds. *Ardea* 75:133–42.

Clobert, J., J. D. Lebreton, M. Clobert-Gillet, and H. Coquillart. 1985. The estimation of survival in bird populations by recaptures or sightings of marked individuals. In *Statistics in ornithology,* ed. B. J. T. Morgan and P. M. North, 197–213. New York: Springer-Verlag.

Clobert, J., C. M. Perrins, R. H. McCleery, and A. G. Gosler. 1988. Survival rate in the great tit *Parus major* in relation to sex, age, and immigration status. *J. Anim. Ecol.* 57:287–306.

Clode, D. 1993. Colonially breeding seabirds: Predators or prey? *Trends Ecol. Evol.* 8:336–38.

———. 1994. Reply from D. Clode. *Trends Ecol. Evol.* 9:26.

Clutton-Brock, T. H. 1988a. Reproductive success. In *Reproductive success: Studies of individual variation in contrasting breeding systems,* ed. T. H. Clutton-Brock, 472–85. Chicago: University of Chicago Press.

———, ed. 1988b. *Reproductive success: Studies of individual variation in contrasting breeding systems.* Chicago: University of Chicago Press.

———. 1991. *The evolution of parental care*. Princeton: Princeton University Press.

Clutton-Brock, T. H., F. E. Guinness, and S. D. Albon. 1982. *Red deer: Behavior and ecology of two sexes*. Chicago: University of Chicago Press.

Colagross, A. M. L., and A. Cockburn. 1993. Vigilance and grouping in the eastern grey kangaroo, *Macropus giganteus*. *Aust. J. Zool.* 41:325–34.

Comstock, J. H. 1940. *An introduction to entomology*. Ithaca, N.Y.: Comstock.

Connor, R. C. 1986. Pseudo-reciprocity: Investing in mutualism. *Anim. Behav.* 34:1562–66.

Conrad, K. F., and R. J. Robertson. 1993. Clutch size in eastern phoebes (*Sayornis phoebe*). I. The cost of nest building. *Can. J. Zool.* 71:1003–7.

Cooke, F., C. S. Findlay, R. F. Rockwell, and K. F. Abraham. 1983. Life history studies of the lesser snow goose (*Anser caerulescens caerulescens*). II. Colony structure. *Behav. Ecol. Sociobiol.* 12:153–59.

Cooke, F., and R. F. Rockwell. 1988. Reproductive success in a lesser snow goose population. In *Reproductive success: Studies of individual variation in contrasting breeding systems*, ed. T. H. Clutton-Brock, 237–50. Chicago: University of Chicago Press.

Cooper, J. G. 1870. *Ornithology*. Vol. 1. *Land birds*. N.p.: Geological Survey of California.

Cooper, R. J., and R. C. Whitmore. 1990. Arthropod sampling methods in ornithology. *Stud. Avian Biol.* 13:29–37.

Coresh, J., and N. Goldman. 1988. The effect of variability in the fertility schedule on numbers of kin. *Math. Popul. Studies* 1:137–56.

Cote, I. M., and M. R. Gross. 1993. Reduced disease in offspring: A benefit of coloniality in sunfish. *Behav. Ecol. Sociobiol.* 33:269–74.

Coues, E. 1878. *Birds of the Colorado Valley*. Part 1. Washington, D.C.: U.S. Government Printing Office.

Coulson, J. C. 1968. Differences in the quality of birds nesting in the centre and on the edges of a colony. *Nature* 217:478–79.

———. 1971. Competition for breeding sites causing segregation and reduced young production in colonial animals. In *Dynamics of populations: Proceedings of the Advanced Study Institute on "Dynamics of numbers in populations," Oosterbeek, the Netherlands, 7–18 September 1970*, ed. P. J. den Boer and G. R. Gradwall, 257–68. Wageningen: Pudoc.

———. 1988. Lifetime reproductive success in the black-legged kittiwake (*Rissa tridactyla*). *Acta XIX Congr. Internatl. Ornithol.*, 2140–47.

Coulson, J. C., and G. N. De Mevergnies. 1992. Where do young kittiwakes *Rissa tridactyla* breed, philopatry or dispersal? *Ardea* 80:187–97.

Coulson, J. C., and F. Dixon. 1979. Colonial breeding in sea-birds. In *Biology and systematics of colonial organisms*, ed. G. Larwood and B. R. Rosen, 445–58. London: Academic Press.

Coulson, J. C., and J. Horobin. 1976. The influence of age on the breeding biology and survival of the arctic tern *Sterna paradisaea*. *J. Zool.* (London) 178:247–60.

Coulson, J. C., and J. M. Porter. 1985. Reproductive success of the kittiwake *Rissa tridactyla*: The roles of clutch size, chick growth rates and parental quality. *Ibis* 127:450–66.

Coulson, J. C., and E. White. 1958. The effect of age on the breeding biology of the kittiwake *Rissa tridactyla*. *Ibis* 100:40–51.
Covich, A. P. 1976. Analyzing shapes of foraging areas: Some ecological and economic theories. *Ann. Rev. Ecol. Syst.* 7:235–57.
Crofton, H. D. 1971. A quantitative approach to parasitism. *Parasitology* 62: 179–93.
Crook, J. H. 1965. The adaptive significance of avian social organizations. *Symp. Zool. Soc. London* 14:181–218.
Crook, J. R., and W. M. Shields. 1985. Sexually selected infanticide by adult male barn swallows. *Anim. Behav.* 33:754–61.
Cuthbert, F. J. 1988. Reproductive success and colony-site tenacity in Caspian terns. *Auk* 105:339–44.
Cyprich, D., M. Krumpal, and D. Hornychova. 1988. Annual cycle of *Ceratophyllus hirundinis* (Curtis, 1826) (Siphonaptera, Insecta) in the nests of *Delichon urbica* Linnaeus, 1758 in south-west Slovakia. *Biologia (Bratislava)* 43: 141–52.
Dale, S., T. Amundsen, J. T. Lifjeld, and T. Slagsvold. 1990. Mate sampling behaviour of female pied flycatchers: Evidence for active mate choice. *Behav. Ecol. Sociobiol.* 27:87–91.
Dale, S., and T. Slagsvold. 1994. Polygyny and deception in the pied flycatcher: Can females determine male mating status? *Anim. Behav.* 48:1207–17.
Danchin, E. 1990. L'hypothèse du "centre d'information": Enfin des resultats tangibles chez les oiseaux. *Alauda* 58:81–84.
———. 1992a. The incidence of the tick parasite *Ixodes uriae* in kittiwake *Rissa tridactyla* colonies in relation to the age of the colony, and a mechanism of infecting new colonies. *Ibis* 134:134–41.
———. 1992b. Food shortage as a factor in the 1988 kittiwake *Rissa tridactyla* breeding failure in Shetland. *Ardea* 80:93–98.
Danchin, E., B. Cadiou, J.-Y. Monnat, and R. R. Estrella. 1991. Recruitment in long-lived birds: Conceptual framework and behavioural mechanisms. *Acta XX Congr. Internatl. Ornithol.*, 1641–56.
Darling, F. F. 1938. *Bird flocks and the breeding cycle.* Cambridge: Cambridge University Press.
da Silva, J., and J. M. Terhune. 1988. Harbour seal grouping as an anti-predator strategy. *Anim. Behav.* 36:1309–16.
Davidar, P., and E. S. Morton. 1993. Living with parasites: Prevalence of a blood parasite and its effect on survivorship in the purple martin. *Auk* 110:109–16.
Davies, A. K., and G. K. Baggott. 1989. Egg-laying, incubation and intraspecific nest parasitism by the Mandarin duck *Aix galericulata*. *Bird Study* 36: 115–22.
Davies, C. R., J. M. Ayres, C. Dye, and L. M. Deane. 1991. Malaria infection rate of Amazonian primates increases with body weight and group size. *Funct. Ecol.* 5:655–62.
Davies, N. B. 1978. Territorial defense in the speckled wood butterfly (*Pararge aegeria*): The resident always wins. *Anim. Behav.* 26:138–47.
———. 1988. Dumping eggs on conspecifics. *Nature* 331:19.
Davis, J. W. F., and E. K. Dunn. 1976. Intraspecific predation and colonial breeding in lesser black-backed gulls *Larus fuscus*. *Ibis* 118:65–77.

Dawson, W. L. 1923. *The birds of California.* Vol. 2. Los Angeles: South Moulton.

De Steven, D. 1978. The influence of age on the breeding biology of the tree swallow *Iridoprocne bicolor*. *Ibis* 120:516–23.

Dexheimer, M., and W. E. Southern. 1974. Breeding success relative to nest location and density in ring-billed gull colonies. *Wilson Bull.* 86:288–90.

Dhondt, A. A. 1989. Blue tit. In *Lifetime reproduction in birds*, ed. I. Newton, 15–33. London: Academic Press.

Dobson, A. P. 1988. The population biology of parasite-induced changes in host behavior. *Quart. Rev. Biol.* 63:139–65.

Dominey, W. J. 1981. Anti-predator function of bluegill sunfish nesting colonies. *Nature* 290:586–88.

Donazar, J. A., and O. Ceballos. 1990. Acquisition of food by fledgling Egyptian vultures *Neophron percnopterus* by nest-switching and acceptance by foster adults. *Ibis* 132:603–17.

Downes, J. A. 1969. The swarming and mating flight of Diptera. *Ann. Rev. Entomol.* 14:271–98.

———. 1970. The feeding and mating behaviour of the specialized Empidinae (Diptera): Observations on four species of *Rhamphomyia* in the High Arctic and a general discussion. *Canadian Entomol.* 102:769–91.

Drake, V. A., and R. A. Farrow. 1988. The influence of atmospheric structure and motions on insect migration. *Ann. Rev. Entomol.* 33:183–210.

Draulans, D. 1988. The importance of heronries for mate attraction. *Ardea* 76:187–92.

Drent, R. H., and S. Daan. 1980. The prudent parent: Energetic adjustments in avian breeding. *Ardea* 68:225–52.

Drost, R. 1938. Geschlechtsbestimmung lebender Vogel nach der Form der Kloakengegend. *Vogelzug* 9:102–5.

DuBowy, P. J., and S. W. Moore. 1985. Weather-related mortality in swallows in the Sacramento Valley of California. *Western Birds* 16:49–50.

Ducey, J. E. 1988. *Nebraska birds: Breeding status and distribution.* Omaha: Simmons-Boardman.

Du Feu, C. R. 1992. How tits avoid flea infestation at nest sites. *Ring. Migr.* 13:120–21.

Duffy, D. C. 1983. The ecology of tick parasitism on densely nesting Peruvian seabirds. *Ecology* 64:110–19.

Duncan, P., and N. Vigne. 1979. The effect of group size in horses on the rate of attacks by blood-sucking flies. *Anim. Behav.* 27:623–25.

Dunn, P. O., R. J. Robertson, D. Michaud-Freeman, and P. T. Boag. 1994. Extrapair paternity in tree swallows: Why do females mate with more than one male? *Behav. Ecol. Sociobiol.* 35:273–81.

Dunn, P. O., L. A. Whittingham, J. T. Lifjeld, R. J. Robertson, and P. T. Boag. 1994. Effects of breeding density, synchrony, and experience on extrapair paternity in tree swallows. *Behav. Ecol.* 5:123–29.

Eadie, J. M. 1991. Constraint and opportunity in the evolution of brood parasitism in waterfowl. *Acta XX Congr. Internatl. Ornithol.*, 1031–40.

Eadie, J. M., F. P. Kehoe, and T. D. Nudds. 1988. Pre-hatch and post-hatch brood amalgamation in North American Anatidae: A review of hypotheses. *Can. J. Zool.* 66:1701–21.

Earle, R. A. 1985. Predators, parasites and symbionts of the South African cliff swallow *Hirundo spilodera* (Aves: Hirundinidae). *Navors. Nas. Mus. Bloemfontein* 5:1–18.

———. 1986. The breeding biology of the South African cliff swallow. *Ostrich* 57:138–56.

Earle, R. A., and L. G. Underhill. 1991. The effects of brood size on growth of South African cliff swallow *Hirundo spilodera* chicks. *Ostrich* 62:13–22.

Eberhard, W. G. 1978. Mating swarms of a South American *Acropygia* (Hymenoptera: Formicidae). *Entomol. News* 89:14–16.

Edelmann, K. G. 1990. A summary of the swarming habits of the insect prey of the cliff swallow (*Hirundo pyrrhonota*). Unpubl. 475b research thesis, Department of Biology, Yale University.

Eichholz, M. W., and W. D. Koenig. 1992. Gopher snake attraction to birds' nests. *Southwestern Nat.* 37:293–98.

Ekman, J. 1989. Group size in dominance-structured populations. *Ornis Scandinavica* 20:86–88.

Elgar, M. A. 1989. Predator vigilance and group size in mammals and birds: A critical review of the empirical evidence. *Biol. Rev.* 64:13–33.

Elkins, N. 1988. Dead hirundines in nests. *British Birds* 81:329.

Elliot, R. D. 1985. The exclusion of avian predators from aggregations of nesting lapwings (*Vanellus vanellus*). *Anim. Behav.* 33:308–14.

Elliott, B. 1983. House wren breeds in cliff swallow nests. *Western Birds* 14:206.

Emlen, J. T., Jr. 1941. Cliff swallow colonies of the central Sacramento Valley in 1941. *Condor* 43:248.

———. 1952. Social behavior in nesting cliff swallows. *Condor* 54:177–99.

———. 1954. Territory, nest building, and pair formation in the cliff swallow. *Auk* 71:16–35.

———. 1986. Responses of breeding cliff swallows to nidicolous parasite infestations. *Condor* 88:110–11.

Emlen, S. T. 1982a. The evolution of helping. I. An ecological constraints model. *Amer. Nat.* 119:29–39.

———. 1982b. The evolution of helping. II. The role of behavioral conflict. *Amer. Nat.* 119:40–53.

———. 1994. Benefits, constraints and the evolution of the family. *Trends Ecol. Evol.* 9:282–85.

Emlen, S. T., and N. J. Demong. 1975. Adaptive significance of synchronized breeding in a colonial bird: A new hypothesis. *Science* 188:1029–31.

Emlen, S. T., and P. H. Wrege. 1986. Forced copulations and intra-specific parasitism: Two costs of social living in the white-fronted bee-eater. *Ethology* 71:2–29.

Emms, S. K., and N. A. M. Verbeek. 1989. Significance of the pattern of nest distribution in the pigeon guillemot (*Cepphus columba*). *Auk* 106:193–202.

Erwin, R. M. 1978. Coloniality in terns: The role of social feeding. *Condor* 80:211–15.

———. 1983. Feeding habitats of nesting wading birds: Spatial use and social influences. *Auk* 100:960–70.
Estes, R. D. 1976. The significance of breeding synchrony in the wildebeest. *E. African Wildl. J.* 14:135–52.
Farner, D. S. 1945. The return of robins to their birthplaces. *Bird-Banding* 16: 81–99.
Farr, J. A. 1977. Social behavior of the golden silk spider, *Nephila clavipes* (Linnaeus) (Araneae, Araneidae). *J. Arachnol.* 4:137–44.
Fasola, M., and F. Barbieri. 1978. Factors affecting the distribution of heronries in northern Italy. *Ibis* 120:537–40.
Feare, C. J. 1976. The breeding of the sooty tern *Sterna fuscata* in the Seychelles and the effects of experimental removal of its eggs. *J. Zool.* (London) 179: 317–60.
Ferguson, J. W. H. 1987. Vigilance behaviour in white-browed sparrow-weavers *Plocepasser mahali*. *Ethology* 76:223–35.
Ferrer, M. 1993. Natural adoption of fledglings by Spanish imperial eagles *Aquila adalberti*. *J. Ornithol.* 134:335–37.
Finney, G., and F. Cooke. 1978. Reproductive habits of the snow goose: The influence of female age. *Condor* 80:147–58.
Fisher, J., and H. G. Vevers. 1944. The breeding distribution, history, and population of the North Atlantic gannet (*Sula bassana*). *J. Anim. Ecol.* 13:49–62.
Fitch, H. S. 1982. Resources of a snake community in prairie-woodland habitat of northeastern Kansas. In *Herpetological communities*, ed. N. J. Scott Jr., 83–97. Wildlife Research Report 13. Washington, D.C.: U.S. Fish and Wildlife Service.
FitzGibbon, C. D. 1994. The costs and benefits of predator inspection behaviour in Thomson's gazelles. *Behav. Ecol. Sociobiol.* 34:139–48.
Fitzpatrick, J. W., and G. E. Woolfenden. 1988. Components of lifetime reproductive success in the Florida scrub jay. In *Reproductive success: Studies of individual variation in contrasting breeding systems*, ed. T. H. Clutton-Brock, 305–20. Chicago: University of Chicago Press.
Fjeldså, J., and N. Krabbe. 1990. *Birds of the High Andes*. Svendborg, Denmark: Apollo Books.
Flasskamp, A. 1994. The adaptive significance of avian mobbing. V. An experimental test of the "move on" hypothesis. *Ethology* 96:322–33.
Fogden, M. P. L., and P. M. Fogden. 1979. The role of fat and protein reserves in the annual cycle of the grey-backed camaroptera in Uganda (Aves: Sylvidae). *J. Zool.* (London) 189:233–58.
Forbes, L. S. 1989. Coloniality in herons: Lack's predation hypothesis reconsidered. *Colonial Waterbirds* 12:24–29.
Foster, W. A. 1968. Total brood mortality in late-nesting cliff swallows. *Condor* 70:275.
Foster, W. A., and W. Olkowski. 1968. The natural invasion of artificial cliff swallow nests by *Oeciacus vicarius* (Hemiptera: Cimicidae) and *Ceratophyllus petrochelidoni* (Siphonaptera: Ceratophyllidae). *J. Med. Entomol.* 5:488–91.
Foster, W. A., and J. E. Treherne. 1981. Evidence for the dilution effect in the selfish herd from fish predation on a marine insect. *Nature* 293:466–67.

Fowler, J. A., and S. Cohen. 1983. A method for the quantitative collection of ectoparasites from birds. *Ring. Migr.* 4:185–89.

Francis, C. M., M. H. Richards, F. Cooke, and R. F. Rockwell. 1992. Long-term changes in survival rates of lesser snow geese. *Ecology* 73:1346–62.

Freed, L. A. 1981. Loss of mass in breeding wrens: Stress or adaptation? *Ecology* 62:1179–86.

Freeland, W. J. 1979. Primate social groups as biological islands. *Ecology* 60:719–28.

Freeman, J. A. 1945. Studies in the distribution of insects by aerial currents. *J. Anim. Ecol.* 14:128–54.

Freeman, S., and W. M. Jackson. 1990. Univariate metrics are not adequate to measure avian body size. *Auk* 107:69–74.

Fretwell, S. D., and H. L. Lucas Jr. 1970. On territorial behavior and other factors influencing habitat distribution in birds. I. Theoretical development. *Acta Biotheoretica* 19:1–36.

Frumkin, R. 1994. Intraspecific brood-parasitism and dispersal in fledgling sparrowhawks *Accipiter nisus*. *Ibis* 136:426–33.

Furness, R. W., and T. R. Birkhead. 1984. Seabird colony distributions suggest competition for food supplies during the breeding season. *Nature* 311:655–56.

Galef, B. G., Jr. 1991. Information centres of Norway rats: Sites for information exchange and information parasitism. *Anim. Behav.* 41:295–301.

Ganier, A. F. 1962. Snakes as climbers. *Migrant* 33:53.

Gaston, A. J., G. Chapdelaine, and D. G. Noble. 1983. The growth of thick-billed murre chicks at colonies in Hudson Strait: Inter- and intra-colony variation. *Can. J. Zool.* 61:2465–75.

Gauthier, M., and D. W. Thomas. 1993. Nest site selection and cost of nest building by cliff swallows (*Hirundo pyrrhonota*). *Can. J. Zool.* 71:1120–23.

George, J. E. 1987. Field observations on the life cycle of *Ixodes baergi* and some seasonal and daily activity cycles of *Oeciacus vicarius* (Hemiptera: Cimicidae), *Argas cooleyi* (Acari: Argasidae), and *Ixodes baergi* (Acari: Ixodidae). *J. Med. Entomol.* 24:683–88.

Gibbons, D. W. 1986. Brood parasitism and cooperative nesting in the moorhen, *Gallinula chloropus*. *Behav. Ecol. Sociobiol.* 19:221–32.

———. 1989. Seasonal reproductive success of the moorhen *Gallinula chloropus*: The importance of male weight. *Ibis* 131:57–68.

Gibbs, J. P. 1991. Spatial relationships between nesting colonies and foraging areas of great blue herons. *Auk* 108:764–70.

Gibbs, J. P., S. Woodward, M. L. Hunter, and A. E. Hutchinson. 1987. Determinants of great blue heron colony distribution in coastal Maine. *Auk* 104:38–47.

Gibson, N. H. E. 1945. On the mating swarms of certain Chironomidae (Diptera). *Trans. R. Entomol. Soc. London* 95:263–94.

Gillespie, J. H. 1974. Natural selection for within-generation variance in offspring number. *Genetics* 76:601–6.

———. 1977. Natural selection for variances in offspring numbers: A new evolutionary principle. *Amer. Nat.* 111:1010–14.

Gillespie, R. G. 1987. The role of prey availability in aggregative behaviour of the orb weaving spider *Tetragnatha elongata*. *Anim. Behav.* 35:675–81.

Giraldeau, L.-A. 1988. The stable group and the determinants of foraging group size. In *The ecology of social behavior*, ed. C. N. Slobodchikoff, 33–53. San Diego: Academic Press.

Giraldeau, L.-A., and T. Caraco. 1993. Genetic relatedness and group size in an aggregation economy. *Evol. Ecol.* 7:429–38.

Giraldeau, L.-A., and D. Gillis. 1985. Optimal group size can be stable: A reply to Sibly. *Anim. Behav.* 33:666–67.

Girard, O., and P. Yesou. 1991. Developpement spatial d'une colonie d'avocettes (*Recurvirostra avosetta*). *Gibier Faune Sauvage* 8:31–32.

Gladstone, D. E. 1979. Promiscuity in monogamous colonial birds. *Amer. Nat.* 114:545–57.

Glick, P. A. 1939. *The distribution of insects, spiders, and mites in the air.* Technical Bulletin 673. Washington, D.C.: U.S. Department of Agriculture.

Gochfeld, M. 1980. Mechanisms and adaptive value of reproductive synchrony in colonial seabirds. In *Behavior of marine animals*, vol. 4, *Marine birds: Current perspectives in research*, ed. J. Burger, B. L. Olla, and H. E. Winn, 207–70. New York: Plenum.

Godin, J.-G. J. 1986. Antipredator function of shoaling in teleost fishes: A selective review. *Naturaliste Can. (Rev. Ecol. Syst.)* 113:241–50.

Goodman, S. M. 1982. A test of nest cup volume and reproductive success in the barn swallow. *Jack-Pine Warbler* 60:107–12.

Gori, D. F. 1988. Colony-facilitated foraging in yellow-headed blackbirds: Experimental evidence for information transfer. *Ornis Scandinavica* 19:224–30.

Gotmark, F., and M. Andersson. 1984. Colonial breeding reduces nest predation in the common gull (*Larus canus*). *Anim. Behav.* 32:485–92.

Gowaty, P. A., and W. C. Bridges. 1991. Nestbox availability affects extra-pair fertilizations and conspecific nest parasitism in eastern bluebirds, *Sialia sialis*. *Anim. Behav.* 41:661–75.

Grafen, A. 1988. On the uses of data on lifetime reproductive success. In *Reproductive success: Studies of individual variation in contrasting breeding systems*, ed. T. H. Clutton-Brock, 454–71. Chicago: University of Chicago Press.

Grant, G. S., and T. L. Quay. 1977. Breeding biology of cliff swallows in Virginia. *Wilson Bull.* 89:286–90.

Grant, P. R. 1990. Reproductive fitness of birds. *Trends Ecol. Evol.* 5:379–80.

Graves, J. A., and A. Whiten. 1980. Adoption of strange chicks by herring gulls, *Larus argentatus* L. *Z. Tierpsychol.* 54:267–78.

Greene, E. 1987. Individuals in an osprey colony discriminate between high and low quality information. *Nature* 329:239–41.

Greenwood, P. J. 1980. Mating systems, philopatry and dispersal in birds and mammals. *Anim. Behav.* 28:1140–62.

Greenwood, P. J., and P. H. Harvey. 1982. The natal and breeding dispersal of birds. *Ann. Rev. Ecol. Syst.* 13:1–21.

Grinnell, J., J. S. Dixon, and J. M. Linsdale. 1930. *Vertebrate natural history of a section of northern California through the Lassen Peak region*. University of California Publications in Zoology 35. Berkeley: University of California.

Grinnell, J., and A. H. Miller. 1944. *The distribution of the birds of California.* Pacific Coast Avifauna 27. Berkeley, Calif.: Cooper Ornithological Society.

Gross, M. R., and A. M. MacMillan. 1981. Predation and the evolution of colonial nesting in bluegill sunfish (*Lepomis macrochirus*). *Behav. Ecol. Sociobiol.* 8:163–74.

Gustafsson, L. 1989. Collared flycatcher. In *Lifetime reproduction in birds,* ed. I. Newton, 75–88. London: Academic Press.

Gutzwiller, K. J., and S. H. Anderson. 1986. Use of abandoned cliff swallow nests by breeding house wrens. *Prairie Nat.* 18:53–54.

Haas, V. 1985. Colonial and single breeding in fieldfares, *Turdus pilaris* L.: A comparison of nesting success in early and late broods. *Behav. Ecol. Sociobiol.* 16:119–24.

Hadrys, H., M. Balick, and B. Schierwater. 1992. Applications of random amplified polymorphic DNA (RAPD) in molecular ecology. *Mol. Ecol.* 1:55–63.

Hagan, J. M., III, and J. R. Walters. 1990. Foraging behavior, reproductive success, and colonial nesting in ospreys. *Auk* 107:506–21.

Hails, C. J., and D. M. Bryant. 1979. Reproductive energetics of a free-living bird. *J. Anim. Ecol.* 48:471–82.

Halley, D. J., and M. P. Harris. 1993. Intercolony movement and behaviour of immature guillemots *Uria aalge. Ibis* 135:264–70.

Halpin, Z. T. 1983. Naturally occurring encounters between black-tailed prairie dogs (*Cynomys ludovicianus*) and snakes. *Amer. Midl. Nat.* 109:50–54.

Hamilton, G. D., and R. F. Martin. 1985. Investigator perturbation and reproduction of the cliff swallow. *Auk* 102:167–70.

Hamilton, W. D. 1964. The genetical evolution of social behavior. *J. Theor. Biol.* 7:1–52.

———. 1971. Geometry for the selfish herd. *J. Theor. Biol.* 31:295–311.

———. 1975. Innate social aptitudes of man: An approach from evolutionary genetics. In *Biosocial anthropology,* ed. R. Fox, 133–55. New York: Wiley.

Hamilton, W. D., and M. Zuk. 1982. Heritable true fitness and bright birds: A role for parasites? *Science* 218:384–87.

Hamilton, W. J., III, and G. H. Orians. 1965. Evolution of brood parasitism in altricial birds. *Condor* 67:361–82.

Hamilton, W. J., III, and K. E. F. Watt. 1970. Refuging. *Ann. Rev. Ecol. Syst.* 1:263–86.

Haramis, G. M., J. D. Nichols, K. H. Pollock, and J. E. Hines. 1986. The relationship between body mass and survival of wintering canvasbacks. *Auk* 103:506–14.

Hardy, K. R., and H. Ottersten. 1969. Radar investigations of convective patterns in the clear atmosphere. *J. Atmospheric Sci.* 26:666–72.

Harris, M. P., and S. Wanless. 1990. Breeding success of British kittiwakes *Rissa tridactyla* in 1986–88: Evidence for changing conditions in the northern North Sea. *J. Appl. Ecol.* 27:172–87.

Hart, B. L. 1990. Behavioral adaptations to pathogens and parasites: Five strategies. *Neurosci. Biobehav. Rev.* 14:273–94.

Harvey, P. H., P. J. Greenwood, C. M. Perrins, and A. R. Martin. 1979. Breeding success of great tits *Parus major* in relation to age of male and female parent. *Ibis* 121.216–19.

Hassell, M. P. 1968. The behavioural response of a tachinid fly (*Cyzenis albicans* (Fall.)) to its host, the winter moth (*Operophtera brumata* (L.)). *J. Anim. Ecol.* 37:627–39.

———. 1971. Parasite behaviour as a factor contributing to the stability of insect host-parasite interactions. In *Dynamics of populations: Proceedings of the Advanced Study Institute on "Dynamics of Numbers in Populations," Oosterbeek, the Netherlands, 7–18 September 1970*, ed. P. J. den Boer and G. R. Gradwall, 366–79. Wageningen: Pudoc.

Hasson, O. 1991. Pursuit-deterrent signals: Communication between prey and predator. *Trends Ecol. Evol.* 6:325–29.

Hatchwell, B. J. 1988. Intraspecific variation in extra-pair copulation and mate defence in common guillemots *Uria aalge*. *Behaviour* 107:157–85.

Hayes, R. O., D. B. Francy, J. S. Lazuick, G. C. Smith, and E. P. J. Gibbs. 1977. Role of the cliff swallow bug (*Oeciacus vicarius*) in the natural cycle of a western equine encephalitis-related alphavirus. *J. Med. Entomol.* 14:257–62.

Hebblethwaite, M. L., and W. M. Shields. 1990. Social influences on barn swallow foraging in the Adirondacks: A test of competing hypotheses. *Anim. Behav.* 39:97–104.

Heeb, P., and H. Richner. 1994. Seabird colonies and the appeal of the information center hypothesis. *Trends Ecol. Evol.* 9:25.

Helle, T., and J. Aspi. 1983. Does herd formation reduce insect harassment among reindeer? A field experiment with animal traps. *Acta Zool. Fennica* 175:129–31.

Henny, C. J., L. J. Blus, and C. J. Stafford. 1982. DDE not implicated in cliff swallow, *Petrochelidon pyrrhonota*, mortality during severe spring weather in Oregon. *Can. Field-Nat.* 96:210–11.

Henry, C. 1982. Etude du régime alimentaire des passereaux par la méthode des colliers. *Alauda* 50:92–107.

Hieber, C. S., and G. W. Uetz. 1990. Colony size and parasitoid load in two species of colonial *Metepeira* spiders from Mexico (Araneae: Araneidae). *Oecologia* 82:145–50.

Higashi, M., and N. Yamamura. 1993. What determines animal group size? Insider-outsider conflict and its resolution. *Amer. Nat.* 142:553–63.

Hill, G. E., R. Montgomerie, C. Roeder, and P. Boag. 1994. Sexual selection and cuckoldry in a monogamous songbird: Implications for sexual selection theory. *Behav. Ecol. Sociobiol.* 35:193–99.

Hinde, R. A. 1961. Behaviour. In *Biology and comparative physiology of birds*, ed. A. J. Marshall, 373–412. New York: Academic Press.

Hiruki, L. M., W. G. Gilmartin, B. L. Becker, and I. Stirling. 1993. Wounding in Hawaiian monk seals (*Monachus schauinslandi*). *Can. J. Zool.* 71:458–68.

Hiruki, L. M., I. Stirling, W. G. Gilmartin, T. C. Johanos, and B. L. Becker. 1993. Significance of wounding to female reproductive success in Hawaiian monk seals (*Monachus schauinslandi*) at Laysan Island. *Can. J. Zool.* 71:469–74.

Hochberg, M. E. 1991. Viruses as costs to gregarious feeding behaviour in the Lepidoptera. *Oikos* 61:291–96.

Hoffenberg, A. S., H. W. Power, L. C. Romagnano, M. P. Lombardo, and T. R.

McGuire. 1988. The frequency of cuckoldry in the European starling (*Sturnus vulgaris*). *Wilson Bull.* 100:60–69.

Hoglund, J., R. Montgomerie, and F. Widemo. 1993. Costs and consequences of variation in the size of ruff leks. *Behav. Ecol. Sociobiol.* 32:31–39.

Hoi, H., B. Schleicher, and F. Valera. 1994. Female mate choice and nest desertion in penduline tits, *Remiz pendulinus:* The importance of nest quality. *Anim. Behav.* 48:743–46.

Holloway, M. 1993. The variable breeding success of the little tern *Sterna albifrons* in south-east India and protective measures needed for its conservation. *Biol. Conservation* 65:1–8.

Hoogland, J. L. 1979a. Aggression, ectoparasitism, and other possible costs of prairie dog (Sciuridae, *Cynomys* spp.) coloniality. *Behaviour* 69:1–35.

———. 1979b. The effect of colony size on individual alertness of prairie dogs (Sciuridae: *Cynomys* spp.). *Anim. Behav.* 27:394–407.

———. 1981. The evolution of coloniality in white-tailed and black-tailed prairie dogs (Sciuridae: *Cynomys leucurus* and *C. ludovicianus*). *Ecology* 62: 252–72.

———. 1995. *The black-tailed prairie dog: Social life of a burrowing mammal.* Chicago: University of Chicago Press.

Hoogland, J. L., and P. W. Sherman. 1976. Advantages and disadvantages of bank swallow (*Riparia riparia*) coloniality. *Ecol. Monogr.* 46:33–58.

Hopla, C. E. 1965. *Alaskan hematophagous insects, their feeding habits and potential as vectors of pathogenic organisms. I. The Siphonaptera of Alaska.* Technical Report 64-12. N.p.: Arctic Aeromedical Laboratory.

Hopla, C. E., and J. E. Loye. 1983. The ectoparasites and microorganisms associated with cliff swallows in west-central Oklahoma. I. Ticks and fleas. *Bull. Soc. Vector Ecology* 8:111–21.

Horn, H. S. 1968. The adaptive significance of colonial nesting in Brewer's blackbird (*Euphages cyanocephalus*). *Ecology* 49:682–94.

Hotker, H. 1989. Meadow pipit. In *Lifetime reproduction in birds*, ed. I. Newton, 119–33. London: Academic Press.

Hotta, M. 1994. Infanticide in little swifts taking over costly nests. *Anim. Behav.* 47:491–93.

Houston, A. I., and J. M. McNamara. 1988. The ideal free distribution when competitive abilities differ: An approach based on statistical mechanics. *Anim. Behav.* 36:166–74.

Houston, D. C., P. J. Jones, and R. M. Sibly. 1983. The effect of female body condition on egg laying in lesser black-backed gulls *Larus fuscus*. *J. Zool.* (London) 200:509–20.

Hovi, M., and O. Ratti. 1994. Mate sampling and assessment procedures in female pied flycatchers (*Ficedula hypoleuca*). *Ethology* 96:127–37.

Howell, F. G., and B. R. Chapman. 1976. Acarines associated with cliff swallow communities in northwest Texas. *Southwestern Nat.* 21:275–80.

Hudson, W. H. 1951. *Letters on the ornithology of Buenos Ayres.* Ithaca, N.Y.: Cornell University Press.

Humphries, D. A. 1969. Behavioural aspects of the ecology of the sand martin flea *Ceratophyllus styx jordani* Smit (Siphonaptera). *Parasitology* 59:311–34.

Hunt, G. L., Jr., Z. A. Eppley, and D. C. Schneider. 1986. Reproductive performance of seabirds: The importance of population and colony size. *Auk* 103: 306–17.

Hunt, G. L., Jr., and D. C. Schneider. 1987. Scale-dependent processes in the physical and biological environment of marine birds. In *Seabirds: Feeding ecology and role in marine ecosystems,* ed. J. P. Croxall, 7–41. Cambridge: Cambridge University Press.

Hurlbert, S. H. 1984. Pseudoreplication and the design of ecological field experiments. *Ecol. Monogr.* 54:187–211.

Hussell, D. J. T., and T. E. Quinney. 1987. Food abundance and clutch size of tree swallows *Tachycineta bicolor. Ibis* 129:243–58.

Imms, A. D. 1951. *A general textbook of entomology.* London: Methuen.

Ims, R. A. 1990. On the adaptive value of reproductive synchrony as a predator-swamping strategy. *Amer. Nat.* 136:485–98.

Inglis, I. R., and J. Lazarus. 1981. Vigilance and flock size in brent geese: The edge effect. *Z. Tierpsychol.* 57:193–200.

Inman, A. J., and J. Krebs. 1987. Predation and group living. *Trends Ecol. Evol.* 2:31–32.

Iverson, S. S. 1988. Site tenacity in culvert-nesting barn swallows in Oklahoma. *J. Field Ornithol.* 59:337–44.

Jackson, W. M. 1992. Estimating conspecific nest parasitism in the northern masked weaver based on within-female variability in egg appearance. *Auk* 109:435–43.

———. 1993. Causes of conspecific nest parasitism in the northern masked weaver. *Behav. Ecol. Sociobiol.* 32:119–26.

Jaeger, M. M., R. L. Bruggers, and W. A. Erickson. 1989. Formation, sizes, and groupings of quelea nesting colonies. In Quelea quelea, *Africa's bird pest,* ed. R. L. Bruggers and C. C. H. Elliott, 181–97. Oxford: Oxford University Press.

James, F. C. 1970. Geographic size variation in birds and its relationship to climate. *Ecology* 51:365–90.

James, F. C., and C. E. McCulloch. 1990. Multivariate analysis in ecology and systematics: Panacea or Pandora's box? *Ann. Rev. Ecol. Syst.* 21:129–66.

Janetos, A. C. 1980. Strategies of female mate choice: A theoretical analysis. *Behav. Ecol. Sociobiol.* 7:107–12.

Janetos, A. C., and B. J. Cole. 1981. Imperfectly optimal animals. *Behav. Ecol. Sociobiol.* 9:203–9.

Janovy, J., Jr. 1978. *Keith County journal.* New York: St. Martin's Press.

———. 1981. *Back in Keith County.* New York: St. Martin's Press.

———. 1994. *Dunwoody Pond: Reflections on the High Plains wetlands and the cultivation of naturalists.* New York: St. Martin's Press.

Janson, C. H. 1988. Food competition in brown capuchin monkeys (*Cebus apella*): Quantitative effects of group size and tree productivity. *Behaviour* 105:53–76.

Janson, C. H., and C. P. Van Schaik. 1988. Recognizing the many faces of primate food competition: Methods. *Behaviour* 105:165–86.

Jeffreys, A. J., V. Wilson, and S. L. Thein. 1985. Individual-specific "fingerprint" regions in human DNA. *Nature* 316:76–79.

Jehl, J. R., Jr. 1994. Absence of nest density effects in a growing colony of California gulls. *J. Avian Biol.* 25:224–30.
Jenni, L. 1993. Structure of a brambling *Fringilla montifringilla* roost according to sex, age and body-mass. *Ibis* 135:85–90.
Jennings, T., and S. M. Evans. 1980. Influence of position in the flock and flock size on vigilance in the starling, *Sturnus vulgaris. Anim. Behav.* 28:634–35.
Johnsgard, P. A. 1984. *The Platte: Channels in time.* Lincoln: University of Nebraska Press.
Johnson, C. G. 1969. *Migration and dispersal of insects by flight.* London: Methuen.
Johnson, L. S., and D. J. Albrecht. 1993. Effects of haematophagous ectoparasites on nestling house wrens, *Troglodytes aedon:* Who pays the cost of parasitism? *Oikos* 66:255–62.
Johnson, L. S., M. D. Eastman, and L. H. Kermott. 1991. Effect of ectoparasitism by larvae of the blow fly *Protocalliphora parorum* (Diptera: Calliphoridae) on nestling house wrens, *Troglodytes aedon. Can. J. Zool.* 69:1441–46.
Johnson, M. L., and M. S. Gaines. 1990. Evolution of dispersal: Theoretical models and empirical tests using birds and mammals. *Ann. Rev. Ecol. Syst.* 21:449–80.
Johnston, R. F. 1961. Population movements of birds. *Condor* 63:386–89.
Jolly, G. M. 1965. Explicit estimates from capture-recapture data with both death and immigration-stochastic model. *Biometrika* 52:225–47.
Jones, G. 1987a. Parental foraging ecology and feeding behaviour during nestling rearing in the swallow. *Ardea* 75:169–74.
———. 1987b. Body condition changes of sand martins (*Riparia riparia*) during breeding, and a comparison with fledgling condition. *J. Zool.* (London) 213:263–81.
———. 1987c. Parent-offspring resource allocation in swallows during nestling rearing: An experimental study. *Ardea* 75:145–68.
———. 1987d. Colonization patterns in sand martins *Riparia riparia. Bird Study* 34:20–25.
———. 1988. Concurrent demands of parent and offspring swallows *Hirundo rustica* in a variable feeding environment. *Ornis Scandinavica* 19:145–52.
Jones, I. L. 1994. Mass changes of least auklets *Aethia pusilla* during the breeding season: Evidence for programmed loss of mass. *J. Anim. Ecol.* 63:71–78.
Jones, M. A. H. 1933. A large nesting colony of the eastern cliff swallow in Kearney County. *Nebraska Bird Review* 1:134–35.
Jones, P. J., and P. Ward. 1976. The level of reserve protein as the proximate factor controlling the timing of breeding and clutch-size in the red-billed quelea *Quelea quelea. Ibis* 118:547–74.
Kanyamibwa, S., A. Schierer, R. Pradel, and J. D. Lebreton. 1990. Changes in adult annual survival rates in a western European population of the white stork *Ciconia ciconia. Ibis* 132:27–35.
Karlsson, J., and S. G. Nilsson. 1977. The influence of nest-box area on clutch size in some hole-nesting passerines. *Ibis* 119:207–11.
Kaul, R. 1989. Plants. In *An atlas of the Sand Hills*, ed. A. Bleed and C. Flowerday, 127–42. Lincoln: Conservation and Survey Division, University of Nebraska.

Kaul, R. B., Challaiah, and K. H. Keeler. 1983. Effects of grazing and juniper-canopy closure on the prairie flora in Nebraska high-plains canyons. *Proc. Seventh North Amer. Prairie Conf.*, 95–105.

Kennedy, C. R. 1975. *Ecological animal parasitology.* New York: Wiley.

———. 1984. Host-parasite interrelationships: Strategies of coexistence and coevolution. In *Producers and scroungers: Strategies of exploitation and parasitism,* ed. C. J. Barnard, 34–60. London: Croom Helm.

Kenward, R. E., V. Marcstrom, and M. Karlbom. 1993. Post-nestling behaviour in goshawks, *Accipiter gentilis:* II. Sex differences in sociality and nest-switching. *Anim. Behav.* 46:371–78.

Keymer, A. E., and A. F. Read. 1991. Behavioural ecology: The impact of parasitism. In *Parasite-host associations: Coexistence or conflict?* ed. C. A. Toft, A. Aeschlimann, and L. Bolis, 37–61. Oxford: Oxford University Press.

Kharitonov, S. P., and D. Siegel-Causey. 1988. Colony formation in seabirds. In *Current ornithology,* vol. 5, ed. R. F. Johnston, 223–72. New York: Plenum.

Kilgore, D. L., Jr., and K. L. Knudsen. 1977. Analysis of materials in cliff and barn swallow nests: Relationship between mud selection and nest architecture. *Wilson Bull.* 89:562–71.

Kilpi, M. 1989. The effect of varying pair numbers on reproduction and use of space in a small herring gull *Larus argentatus* colony. *Ornis Scandinavica* 20:204–10.

Kimball, F. H. 1889. Mortality among eave swallows. *Auk* 6:338–39.

Klomp, H. 1970. The determination of clutch-size in birds: A review. *Ardea* 58:1–124.

Kluijver, H. N. 1951. The population ecology of the great tit. *Ardea* 39:1–135.

Knopf, F. L. 1979. Spatial and temporal aspects of colonial nesting of white pelicans. *Condor* 81:353–63.

Koenig, W. D. 1988. Reciprocal altruism in birds: A critical review. *Ethol. Sociobiol.* 9:73–84.

Koenig, W. D., P. A. Gowaty, and J. L. Dickinson. 1992. Boxes, barns, and bridges: Confounding factors or exceptional opportunities in ecological studies? *Oikos* 63:305–8.

Koenig, W. D., and R. L. Mumme. 1987. *Population ecology of the cooperatively breeding acorn woodpecker.* Princeton: Princeton University Press.

Kohls, G. M., and R. E. Ryckman. 1962. New distributional records of ticks associated with cliff swallows, *Petrochelidon* spp., in the United States. *J. Parasitol.* 48:507–8.

Kopachena, J. G. 1991. Food dispersion, predation, and the relative advantage of colonial nesting. *Colonial Waterbirds* 14:7–12.

Krapu, G. L. 1986. Patterns and causes of change in a cliff swallow colony during a 17-year period. *Prairie Nat.* 18:109–14.

Krebs, J. R. 1974. Colonial nesting and social feeding as strategies for exploiting food in the great blue heron (*Ardea herodias*). *Behaviour* 51:99–134.

———. 1978. Colonial nesting in birds, with special reference to the Ciconiiformes. In *Wading birds,* ed. A. Sprunt IV, J. C. Ogden, and S. Winckler, 299–314. New York: National Audubon Society.

Krebs, J. R., and C. J. Barnard. 1980. Comments on the function of flocking in birds. *Acta XVII Congr. Internatl. Ornithol.*, 795–99.

Krementz, D. G., J. D. Nichols, and J. E. Hines. 1989. Postfledging survival of European starlings. *Ecology* 70:646–55.

Krementz, D. G., J. R. Sauer, and J. D. Nichols. 1989. Model-based estimates of annual survival rate are preferable to observed maximum lifespan statistics for use in comparative life-history studies. *Oikos* 56:203–8.

Kruuk, H. 1964. Predators and anti-predator behaviour of the black-headed gull (*Larus ridibundus* L.). *Behav. Suppl.* 11:1–129.

Kunz, T. H. 1976. Observations on the winter ecology of the bat fly *Trichobius corynorhini* Cockerell (Diptera: Streblidae). *J. Med. Entomol.* 12:631–36.

Kuris, A. M., A. R. Blaustein, and J. J. Alio. 1980. Hosts as islands. *Amer. Nat.* 116:570–86.

Lack, D. 1967. Interrelationships in breeding adaptations as shown by marine birds. *Acta XIV Congr. Internatl. Ornithol.,* 3–42.

———. 1968. *Ecological adaptations for breeding in birds.* London: Methuen.

Lake, P. E. 1981. Male genital organs. In *Form and function in birds,* vol. 2, ed. A. S. King and J. McLelland, 1–61. London: Academic Press.

Lamey, T. C., and D. W. Mock. 1991. Nonaggressive brood reduction in birds. *Acta XX Congr. Internatl. Ornithol.,* 1741–51.

Languy, M., and C. Vansteenwegen. 1989. Influence of parental age on the growth of nestling swallows (*Hirundo rustica*). *Ardea* 77:227–32.

Lank, D. B., P. Mineau, R. F. Rockwell, and F. Cooke. 1989. Intraspecific nest parasitism and extra-pair copulation in lesser snow geese. *Anim. Behav.* 37:74–89.

Lank, D. B., and C. M. Smith. 1992. Females prefer larger leks: Field experiments with ruffs (*Philomachus pugnax*). *Behav. Ecol. Sociobiol.* 30:323–29.

Larimore, R. W. 1987. Synchrony of cliff swallow nesting and development of the tick, *Ixodes baergi. Southwestern Nat.* 32:121–26.

Layton, M. H., and W. H. Buckhannan. 1926. *Soil survey of Keith County, Nebraska.* Washington, D.C.: U.S. Department of Agriculture, Bureau of Chemistry and Soils.

Le Boeuf, B. J., and S. Mesnick. 1990. Sexual behavior of male northern elephant seals. I. Lethal injuries to adult females. *Behaviour* 116:143–62.

Lebreton, J. D., K. P. Burnham, J. Clobert, and D. R. Anderson. 1992. Modeling survival and testing biological hypotheses using marked animals: A unified approach with case studies. *Ecol. Monogr.* 62:67–118.

Lebreton, J. D., R. Pradel, and J. Clobert. 1993. The statistical analysis of survival in animal populations. *Trends Ecol. Evol.* 8:91–95.

Lehmann, T. 1993. Ectoparasites: Direct impact on host fitness. *Parasitol. Today* 9:8–13.

Leighton, M., and D. R. Leighton. 1982. The relationship of size of feeding aggregate to size of food patch: Howler monkeys (*Alouatta palliata*) feeding in *Trichilia cipo* fruit trees on Barro Colorado Island. *Biotropica* 14:81–90.

Lemmetyinen, R. 1971. Nest defense behaviour of common and arctic terns and its effects on the success achieved by predators. *Ornis Fennica* 48:13–24.

Lessells, C. M., M. I. Avery, and J. R. Krebs. 1994. Nonrandom dispersal of kin: Why do European bee-eater (*Merops apiaster*) brothers nest close together? *Behav. Ecol.* 5:105–13.

Lewis, T. 1965. The effects of an artificial windbreak on the aerial distribution of flying insects. *Ann. Appl. Biol.* 55:503–12.

Lewis, T., and L. R. Taylor. 1964. Diurnal periodicity of flight by insects. *Trans. R. Entomol. Soc. London* 116:393–469.

Lifjeld, J. T., and B. Marstein. 1994. Paternity assurance behaviour in the house martin *Delichon urbica. J. Avian Biol.* 25:231–38.

Lima, S. L., and L. M. Dill. 1990. Behavioral decisions made under the risk of predation: A review and prospectus. *Can. J. Zool.* 68:619–40.

Lincoln, F. C. 1934. The operation of homing instinct. *Bird-Banding* 5:149–55.

Lipetz, V. E., and M. Bekoff. 1982. Group size and vigilance in pronghorns. *Z. Tierpsychol.* 58:203–16.

Littrell, E. E. 1992. Swallow mortality during the "March miracle" in California. *Calif. Fish Game* 78:128–30.

Loesche, P., P. K. Stoddard, B. J. Higgins, and M. D. Beecher. 1991. Signature versus perceptual adaptations for individual vocal recognition in swallows. *Behaviour* 118:15–26.

Lohrl, V. H. 1971. Die Auswirkungen einer Witterungskatastrophe auf den Brutbestand der Mehlschwalbe (*Delichon urbica*) in verschiedenen Orten in Sudwestdeutschland. *Die Vogelwelt* 92:58–66.

Lombardo, M. P. 1986. Attendants at tree swallow nests. I. Are attendants helpers at the nest? *Condor* 88:297–303.

———. 1987. Attendants at tree swallow nests. II. The exploratory-dispersal hypothesis. *Condor* 89:138–49.

Lombardo, M. P., H. W. Power, P. C. Stouffer, L. C. Romagnano, and A. S. Hoffenberg. 1989. Egg removal and intraspecific brood parasitism in the European starling (*Sturnus vulgaris*). *Behav. Ecol. Sociobiol.* 24:217–23.

Loye, J. E. 1985. The life history and ecology of the cliff swallow bug, *Oeciacus vicarius* (Hemiptera: Cimicidae). *Cahiers Office de la Recherche Scientifique et Technique Outre-Mer, Série Entomologie Médicale et Parasitologie* 23:133–39.

Loye, J. E., and S. P. Carroll. 1991. Nest ectoparasite abundance and cliff swallow colony site selection, nestling development, and departure time. In *Bird-parasite interactions: Ecology, evolution and behaviour*, ed. J. E. Loye and M. Zuk, 222–41. Oxford: Oxford University Press.

Loye, J. E., and C. E. Hopla. 1983. Ectoparasites and microorganisms associated with the cliff swallow in west-central Oklahoma. II. Life history patterns. *Bull. Soc. Vector Ecology* 8:79–84.

Loye, J. E., and M. Zuk, eds. 1991. *Bird-parasite interactions: Ecology, evolution and behaviour.* Oxford: Oxford University Press.

Lubin, Y. D. 1974. Adaptive advantages and the evolution of colony formation in *Cyrtophora* (Araneae: Araneidae). *Zool. J. Linn. Soc.* 54:321–39.

Lueshen, W. 1962. Cliff swallow banding in Nebraska. *Eastern Bird-Banding Assoc. News* 25:107–9.

Lynch, M. 1988. Estimation of relatedness by DNA fingerprinting. *Mol. Biol. Evol.* 5:584–99.

———. 1990. The similarity index and DNA fingerprinting. *Mol. Biol. Evol.* 7:478–84.

Lyon, B. E. 1991. Brood parasitism in American coots: Avoiding the constraints of parental care. *Acta XX Congr. Internatl. Ornithol.*, 1023–30.

———. 1993. Tactics of parasitic American coots: Host choice and the pattern of egg dispersion among host nests. *Behav. Ecol. Sociobiol.* 33:87–100.

MacRoberts, M. H. 1973. Extramarital courting in lesser black-backed and herring gulls. *Z. Tierpsychol.* 32:62–74.

Magrath, R. D. 1991. Nestling weight and juvenile survival in the blackbird, *Turdus merula. J. Anim. Ecol.* 60:335–51.

Magurran, A. E. 1990. The adaptive significance of schooling as an anti-predator defence in fish. *Ann. Zool. Fennici* 27:51–66.

Malmer, N., and P. H. Enckell. 1994. Ecological research at the beginning of the next century. *Oikos* 71:171–76.

Marshall, A. G. 1981. *The ecology of ectoparasitic insects*. London: Academic Press.

Martin, T. E. 1987. Food as a limit on breeding birds: A life-history perspective. *Ann. Rev. Ecol. Syst.* 18:453–87.

Martinez, M. M. 1983. Nidificacion de *Hirundo rustica erythrogaster* (Boddaert) en la Argentina (Aves, Hirundinidae). *Neotropica* 29:83–86.

May, R. M. 1983. Parasitic infections as regulators of animal populations. *Amer. Sci.* 71:36–45.

May, R. M., and R. M. Anderson. 1979. Population biology of infectious diseases. Part 2. *Nature* 280:455–61.

Mayhew, W. W. 1958. The biology of the cliff swallow in California. *Condor* 60:7–37.

McCleery, R. H., and C. M. Perrins. 1988. Lifetime reproductive success of the great tit, *Parus major*. In *Reproductive success: Studies of individual variation in contrasting breeding systems*, ed. T. H. Clutton-Brock, 136–53. Chicago: University of Chicago Press.

McCracken, G. F. 1984. Communal nursing in Mexican free-tailed bat maternity colonies. *Science* 223:1090–91.

McRae, S. B. 1995. Temporal variation in responses to intraspecific brood parasitism in the moorhen. *Anim. Behav.* 49:1073–88.

Mead, C. J. 1979a. Mortality and causes of death in British sand martins. *Bird Study* 26:107–12.

———. 1979b. Colony fidelity and interchange in the sand martin. *Bird Study* 26:99–106.

Mead, C. J., and J. D. Harrison. 1979. Sand martin movements within Britain and Ireland. *Bird Study* 26:73–86.

Medvin, M. B., P. K. Stoddard, and M. D. Beecher. 1992. Signals for parent-offspring recognition: Strong sib-sib call similarity in cliff swallows but not barn swallows. *Ethology* 90:17–28.

———. 1993. Signals for parent-offspring recognition: A comparative analysis of the begging calls of cliff swallows and barn swallows. *Anim. Behav.* 45:841–50.

Mesnick, S. L., and B. J. Le Boeuf. 1991. Sexual behavior of male northern elephant seals. II. Female response to potentially injurious encounters. *Behaviour* 117:262–80.

Milinski, M. 1977. Do all members of a swarm suffer the same predation? *Z. Tierpsychol.* 45:373–88.
———. 1988. Games fish play: Making decisions as a social forager. *Trends Ecol. Evol.* 3:325–30.
Mills, J. A. 1994. Extra-pair copulations in the red-billed gull: Females with high-quality, attentive males resist. *Behaviour* 128:41–64.
Minot, H. D. 1877. *The land-birds and game-birds of New England.* Boston: Estes and Lauriat.
Mock, D. W. 1994. Brood reduction: Narrow sense, broad sense. *J. Avian Biol.* 25:3–7.
Mock, D. W., T. C. Lamey, and D. B. A. Thompson. 1988. Falsifiability and the information centre hypothesis. *Ornis Scandinavica* 19:231–48.
Moen, S. M. 1991. Morphologic and genetic variation among breeding colonies of the Atlantic puffin (*Fratercula arctica*). *Auk* 108:755–63.
Mohler, L. L. 1946. Albino cliff swallow and English sparrow. *Nebraska Bird Review* 14:47.
———. 1952. Cliff swallows at Kingsley Dam, Keith County. *Nebraska Bird Review* 20:58.
Moller, A. P. 1982. Clutch size in relation to nest size in the swallow *Hirundo rustica*. *Ibis* 124:339–43.
———. 1985. Mixed reproductive strategy and mate guarding in a semi-colonial passerine, the swallow *Hirundo rustica*. *Behav. Ecol. Sociobiol.* 17:401–8.
———. 1987a. Advantages and disadvantages of coloniality in the swallow, *Hirundo rustica*. *Anim. Behav.* 35:819–32.
———. 1987b. Intraspecific nest parasitism and anti-parasite behaviour in swallows, *Hirundo rustica*. *Anim. Behav.* 35:247–54.
———. 1989. Parasites, predators and nest boxes: Facts and artefacts in nest box studies of birds? *Oikos* 56:421–23.
———. 1990. Effects of parasitism by a haematophagous mite on reproduction in the barn swallow. *Ecology* 71:2345–57.
———. 1992a. Nest boxes and the scientific rigour of experimental studies. *Oikos* 63:309–11.
———. 1992b. Frequency of female copulations with multiple males and sexual selection. *Amer. Nat.* 139:1089–1101.
———. 1994. Parasites as an environmental component of reproduction in birds as exemplified by the swallow *Hirundo rustica*. *Ardea* 82:161–72.
Moller, A. P., and T. R. Birkhead. 1989. Copulation behaviour in mammals: Evidence that sperm competition is widespread. *Biol. J. Linn. Soc.* 38:119–31.
———. 1993a. Cuckoldry and sociality: A comparative study of birds. *Amer. Nat.* 142:118–40.
———. 1993b. Certainty of paternity covaries with parental care in birds. *Behav. Ecol. Sociobiol.* 33:261–68.
Moore, J., D. Simberloff, and M. Freehling. 1988. Relationships between bobwhite quail social-group size and intestinal helminth parasitism. *Amer. Nat.* 131:22–32.
Moore, W. S. 1977. An evaluation of narrow hybrid zones in vertebrates. *Quart. Rev. Biol.* 52:263–77.

Moorhouse, D. E., and M. H. Colbo. 1973. On the swarming of *Austrosimulium pestilens* MacKerras and MacKerras (Diptera: Simuliidae). *J. Aust. Entomol. Soc.* 12:127–30.

Mooring, M. S., and B. L. Hart. 1992. Animal grouping for protection from parasites: Selfish herd and encounter-dilution effects. *Behaviour* 123:173–93.

Moreau, R. E. 1947. Relations between number in brood, feeding-rate and nestling period in nine species of birds in Tanganyika Territory. *J. Anim. Ecol.* 16:205–9.

Moreno, J. 1989. Strategies of mass change in breeding birds. *Biol. J. Linn. Soc.* 37:297–310.

Morris, R. D., and R. A. Hunter. 1976. Factors influencing desertion of colony sites by common terns (*Sterna hirundo*). *Can. Field-Nat.* 90:137–43.

Morrison, M. L. 1981. Population trends of the loggerhead shrike in the United States. *Amer. Birds* 35:754–57.

Morse, D. H., and R. S. Fritz. 1987. The consequences of foraging for reproductive success. In *Foraging behavior*, ed. A. C. Kamil, J. R. Krebs, and H. R. Pulliam, 443–55. New York: Plenum.

Morton, E. S. 1987. Variation in mate guarding intensity by male purple martins. *Behaviour* 101:211–24.

Morton, E. S., and K. C. Derrickson. 1990. The biological significance of age-specific return schedules in breeding purple martins. *Condor* 92:1040–50.

Morton, E. S., L. Forman, and M. Braun. 1990. Extrapair fertilizations and the evolution of colonial breeding in purple martins. *Auk* 107:275–83.

Morton, E. S., and R. M. Patterson. 1983. Kin association, spacing, and composition of a post-breeding roost of purple martins. *J. Field Ornithol.* 54:36–41.

Moss, R., A. Watson, R. A. Parr, I. B. Trenholm, and M. Marquiss. 1993. Growth rate, condition and survival of red grouse *Lagopus lagopus scoticus* chicks. *Ornis Scandinavica* 24:303–10.

Moss, W. W., and J. H. Camin. 1970. Nest parasitism, productivity, and clutch size in purple martins. *Science* 168:1000–1003.

Muldal, A., H. L. Gibbs, and R. J. Robertson. 1985. Preferred nest spacing of an obligate cavity-nesting bird, the tree swallow. *Condor* 87:356–63.

Mumme, R. L., W. D. Koenig, and F. A. Pitelka. 1983. Reproductive competition in the communal acorn woodpecker: Sisters destroy each other's eggs. *Nature* 306:583–84.

Mumme, R. L., W. D. Koenig, R. M. Zink, and J. M. Marten. 1985. An analysis of genetic variation and parentage in a California population of acorn woodpeckers. *Auk* 102:305–12.

Munn, C. A. 1986. Birds that "cry wolf." *Nature* 319:143–45.

Munro, J., and J. Bedard. 1977. Gull predation and crèching behaviour in the common eider. *J. Anim. Ecol.* 46:799–810.

Murphy, E. C., and E. Haukioja. 1986. Clutch size in nidicolous birds. In *Current ornithology*, vol. 4, ed. R. F. Johnston, 141–80. New York: Plenum.

Murray, B. G., Jr. 1985. Population growth rate as a measure of individual fitness. *Oikos* 44:509–11.

———. 1990. Population dynamics, genetic change, and the measurement of fitness. *Oikos* 59:189–99.

———. 1992. The evolutionary significance of lifetime reproductive success. *Auk* 109:167–72.
Murray, M. G. 1981. Structure of association in impala, *Aepyceros melampus*. *Behav. Ecol. Sociobiol.* 9:23–33.
———. 1985. Estimation of kinship parameters: The island model with separate sexes. *Behav. Ecol. Sociobiol.* 16:151–59.
Murton, R. K., and N. J. Westwood. 1977. *Avian breeding cycles*. Oxford: Oxford University Press.
Myers, L. E. 1928. The American swallow bug, *Oeciacus vicarius* Horvath (Hemiptera, Cimicidae). *Parasitology* 20:159–72.
Myres, M. T. 1957. Clutch size and laying dates in cliff swallow colonies. *Condor* 59:311–16.
Nelson, B. 1978. *The gannet*. London: Poyser.
Nelson, B. C. 1972. Fleas from the archaeological site at Lovelock Cave, Nevada. *J. Med. Entomol.* 9:211–14.
Nelson, J. B. 1988. Age and breeding in seabirds. *Acta XIX Congr. Internatl. Ornithol.*, 1081–97.
Nettleship, D. N. 1972. Breeding success of the common puffin (*Fratercula arctica* L.) on different habitats at Great Island, Newfoundland. *Ecol. Monogr.* 42:239–68.
Newman, J. A., and T. Caraco. 1989. Co-operative and non-co-operative bases of food-calling. *J. Theor. Biol.* 141:197–209.
Newton, I., ed. 1989a. *Lifetime reproduction in birds*. London: Academic Press.
———. 1989b. Introduction. In *Lifetime reproduction in birds*, ed. I. Newton, 1–11. London: Academic Press.
———. 1989c. Synthesis. In *Lifetime reproduction in birds*, ed. I. Newton, 441–69. London: Academic Press.
Newton, I., and C. R. G. Campbell. 1975. Breeding of ducks at Loch Leven, Kinross. *Wildfowl* 26:83–103.
Nielsen, E. T., and H. Greve. 1950. Studies on the swarming habits of mosquitos and other Nematocera. *Bull. Entomol. Res.* 41:227–58.
Nilsson, S. G. 1975. Clutch size and breeding success of birds in nest boxes and natural cavities. *Vår Fagelvarld* 34:207–11.
Nisbet, I. C. T. 1989. Long-term ecological studies of seabirds. *Colonial Waterbirds* 12:143–47.
Nisbet, I. C. T., J. M. Winchell, and A. E. Heise. 1984. Influence of age on the breeding biology of common terns. *Colonial Waterbirds* 7:117–26.
Nolan, V. 1978. *The ecology and behavior of the prairie warbler* Dendroica discolor. Lawrence, Kans.: American Ornithologists' Union.
Norberg, R. A. 1981. Temporary weight decrease in breeding birds may result in more fledged young. *Amer. Nat.* 118:838–50.
North, P. M. 1988. A brief review of the (lack of) statistics of bird dispersal. *Acta Ornithol.* 24:63–74.
Nowak, M. A., and R. M. May. 1992. Evolutionary games and spatial chaos. *Nature* 359:826–29.
Nur, N. 1984a. The consequences of brood size for breeding blue tits. II. Nestling weight, offspring survival and optimal brood size. *J. Anim. Ecol.* 53:497–517.

———. 1984b. The consequences of brood size for breeding blue tits. I. Adult survival, weight change and the cost of reproduction. *J. Anim. Ecol.* 53: 479–96.

———. 1984c. Fitness, population growth rate and natural selection. *Oikos* 42: 413–14.

———. 1987a. Parents, nestlings and feeding frequency: A model of optimal parental investment and implications for avian reproductive strategies. In *Foraging behavior,* ed. A. C. Kamil, J. R. Krebs, and H. R. Pulliam, 457–75. New York: Plenum.

———. 1987b. Population growth rate and the measurement of fitness: A critical reflection. *Oikos* 48:338–41.

Nur, N., and J. Clobert. 1988. Measuring Darwinian fitness in birds: A field guide. *Acta XIX Congr. Internatl. Ornithol.,* 2121–30.

O'Connor, R. J. 1984. *The growth and development of birds.* Chichester, England: Wiley.

Okubo, A. 1980. *Diffusion and ecological problems: Mathematical models.* Berlin: Springer-Verlag.

Oliver, D. R. 1971. Life history of the Chironomidae. *Ann. Rev. Entomol.* 16: 211–30.

Oliver, G. V., Jr. 1970. Black ratsnake predation upon nesting barn and cliff swallows. *Bull. Okla. Ornithol. Soc.* 3:17–20.

Ollason, J. C., and G. M. Dunnet. 1988. Variation in breeding success in fulmars. In *Reproductive success: Studies of individual variation in contrasting breeding systems,* ed. T. H. Clutton-Brock, 263–78. Chicago: University of Chicago Press.

Olson, M. 1956. Reports of nests, nestlings and fledglings. *Nebraska Bird Review* 24:14.

Orians, G. H. 1961. Social stimulation within blackbird colonies. *Condor* 63: 330–37.

———. 1971. Ecological aspects of behavior. In *Avian biology,* vol. 1, ed. D. S. Farner and J. R. King, 513–46. New York: Academic Press.

———. 1980. *Some adaptations of marsh-nesting blackbirds.* Princeton: Princeton University Press.

Orians, G. H., and H. S. Horn. 1969. Overlap in foods and foraging of four species of blackbirds in the potholes of central Washington. *Ecology* 50:930–38.

Owen, M., and W. A. Cook. 1977. Variations in body weight, wing length and condition of mallard *Anas platyrhynchos platyrhynchos* and their relationship to environmental changes. *J. Zool.* (London) 183:377–95.

Packer, C., D. Scheel, and A. E. Pusey. 1990. Why lions form groups: Food is not enough. *Amer. Nat.* 136:1–19.

Pamilo, P., and R. H. Crozier. 1982. Measuring genetic relatedness in natural populations: Methodology. *Theor. Pop. Biol.* 21:171–93.

Parker, G. A., and W. J. Sutherland. 1986. Ideal free distributions when individuals differ in competitive ability: Phenotype-limited ideal free models. *Anim. Behav.* 34:1222–42.

Parker, P. G., T. A. Waite, and M. D. Decker. 1995. Kinship and association in communally roosting black vultures. *Anim. Behav.* 49:395–401.

Parrish, J. K. 1989. Re-examining the selfish herd: Are central fish safer? *Anim. Behav.* 38:1048–53.

Parsons, J. 1976. Nesting density and breeding success in the herring gull. *Ibis* 118:537–46.

Part, T., and L. Gustafsson. 1989. Breeding dispersal in the collared flycatcher (*Ficedula albicollis*): Possible causes and reproductive consequences. *J. Anim. Ecol.* 58:305–20.

Partridge, L. 1989. Lifetime reproductive success and life-history evolution. In *Lifetime reproduction in birds,* ed. I. Newton, 421–40. London: Academic Press.

Patterson, I. J. 1965. Timing and spacing of broods in the black-headed gull *Larus ridibundus. Ibis* 107:433–59.

Payne, R. B. 1977. The ecology of brood parasitism in birds. *Ann. Rev. Ecol. Syst.* 8:1–28.

———. 1984. *Sexual selection, lek and arena behavior, and sexual size dimorphism in birds.* Ornithol. Monogr. 33.

———. 1989. Indigo bunting. In *Lifetime reproduction in birds,* ed. I. Newton, 153–72. London: Academic Press.

———. 1990. Natal dispersal, area effects, and effective population size. *J. Field Ornithol.* 61:396–403.

Pena, S. D. J., R. Chakraborty, J. T. Epplen, and A. J. Jeffreys, eds. 1993. *DNA fingerprinting: State of the science.* Basel: Birkhauser.

Peng, R. K., S. L. Sutton, and C. R. Fletcher. 1992. Spatial and temporal distribution patterns of flying diptera. *J. Zool.* (London) 228:329–40.

Perrins, C. M. 1965. Population fluctuation and clutch size in the great tit *Parus major* L. *J. Anim. Ecol.* 34:601–47.

Perrins, C. M., and D. Moss. 1974. Survival of young great tits in relation to age of female parent. *Ibis* 116:220–24.

Peterson, B. V. 1959. Observations on mating, feeding, and oviposition of some Utah species of black flies (Diptera: Simuliidae). *Can. Entomol.* 91:147–55.

Petit, D. R., and K. L. Bildstein. 1987. Effect of group size and location within the group on the foraging behavior of white ibises. *Condor* 89:602–9.

Petrie, M. 1983. Female moorhens compete for small fat males. *Science* 220:413–15.

Phillips, J. B. 1990. Lek behaviour in birds: Do displaying males reduce nest predation? *Anim. Behav.* 39:555–65.

Pienkowski, M. W., and P. R. Evans. 1982a. Clutch parasitism and nesting interference between shelducks at Aberlady Bay. *Wildfowl* 33:159–63.

———. 1982b. Breeding behaviour, productivity and survival of colonial and non-colonial shelducks *Tadorna tadorna. Ornis Scandinavica* 13:101–16.

Pierotti, R. 1980. Spite and altruism in gulls. *Amer. Nat.* 115:290–300.

———. 1991. Infanticide versus adoption: An intergenerational conflict. *Amer. Nat.* 138:1140–58.

Pigage, H. K., and O. R. Larson. 1983. The detection of glycerol in overwintering purple martin fleas, *Ceratophyllus idius. Comp. Biochem. Physiol.* 75A:593–95.

Pleasants, J. M., and B. Y. Pleasants. 1979. The superterritory hypothesis: A critique, or why there are so few bullies. *Amer. Nat.* 114:609–14.

Podolsky, R. H., and S. W. Kress. 1989. Factors affecting colony formation in Leach's storm-petrel. *Auk* 106:332–36.

Poiani, A. 1991. Anti-predator behaviour in the bell miner *Manorina melanophrys*. *Emu* 91:164–71.

———. 1992. Ectoparasitism as a possible cost of social life: A comparative analysis using Australian passerines (Passeriformes). *Oecologia* 92:429–41.

Poole, A. 1982. Breeding ospreys feed fledglings that are not their own. *Auk* 99:781–84.

Post, W. 1982. Why do grey kingbirds roost communally? *Bird Behav.* 4:46–49.

———. 1994. Are female boat-tailed grackle colonies neutral assemblages? *Behav. Ecol. Sociobiol.* 35:401–7.

Potvin, C., and D. A. Roff. 1993. Distribution-free and robust statistical methods: Viable alternatives to parametric statistics? *Ecology* 74:1617–28.

Poulin, R. 1991a. Group-living and infestation by ectoparasites in passerines. *Condor* 93:418–23.

———. 1991b. Group-living and the richness of the parasite fauna in Canadian freshwater fishes. *Oecologia* 86:390–94.

Poulin, R., and G. J. FitzGerald. 1989. Shoaling as an anti-ectoparasite mechanism in juvenile sticklebacks (*Gasterosteus* spp.). *Behav. Ecol. Sociobiol.* 24:251–55.

Poulin, R., M. E. Rau, and M. A. Curtis. 1991. Infection of brook trout fry, *Salvelinus fontinalis*, by ectoparasitic copepods: The role of host behaviour and initial parasite load. *Anim. Behav.* 41:467–76.

Power, H. W., E. D. Kennedy, L. C. Romagnano, M. P. Lombardo, A. S. Hoffenberg, P. C. Stouffer, and T. R. McGuire. 1989. The parasitism insurance hypothesis: Why starlings leave space for parasitic eggs. *Condor* 91:753–65.

Poysa, H. 1992. Group foraging in patchy environments: The importance of coarse-level local enhancement. *Ornis Scandinavica* 23:159–66.

Pradel, R. J., J. Clobert, and J. D. Lebreton. 1990. Recent developments for the analysis of multiple capture-recapture data sets: An example concerning two blue tit populations. *Ring* 13:193–204.

Pradel, R. J., and J. D. Lebreton. 1993. *User's manual for program SURGE: Version 4.2*. Saint-Georges-D'Orques, France: Avenix.

Price, P. W. 1980. *Evolutionary biology of parasites*. Princeton: Princeton University Press.

Pruett-Jones, M. A., and S. G. Pruett-Jones. 1982. Spacing and distribution of bowers in Macgregor's bowerbird (*Amblyornis macgregoriae*). *Behav. Ecol. Sociobiol.* 11:25–32.

Pulliam, H. R. 1973. On the advantages of flocking. *J. Theor. Biol.* 38:419–22.

Pulliam, H. R., and T. Caraco. 1984. Living in groups: Is there an optimal group size? In *Behavioural ecology*, ed. J. R. Krebs and N. B. Davies, 122–47. Sunderlund, Mass.: Sinauer.

Pulliam, H. R., and G. C. Millikan. 1982. Social organization in the nonreproductive season. In *Avian biology*, vol. 6, ed. D. S. Farner, J. R. King, and K. C. Parkes, 169–97. New York: Academic Press.

Pullum, T. W. 1982. The eventual frequencies of kin in a stable population. *Demography* 19:549–65.

Queller, D. C., and K. F. Goodnight. 1989. Estimating relatedness using genetic markers. *Evolution* 43:258–75.

Quinn, T. W., J. S. Quinn, F. Cooke, and B. N. White. 1987. DNA marker analysis detects multiple maternity and paternity in single broods of the lesser snow goose. *Nature* 326:392–94.

Rabenold, P. P. 1987. Recruitment to food in black vultures: Evidence for following from communal roosts. *Anim. Behav.* 35:1775–85.

Ramo, C. 1993. Extra-pair copulations of grey herons at high densities. *Ardea* 81:115–20.

Randolph, S. E. 1979. Population regulation in ticks: The role of acquired resistance in natural and unnatural hosts. *Parasitology* 79:141–56.

Rannala, B. H. 1995. Demography and genetic structure in island populations. Ph.D. diss., Yale University.

Rannala, B. H., and C. R. Brown. 1994. Relatedness and conflict over optimal group size. *Trends Ecol. Evol.* 9:117–19.

Ranta, E. 1992. Gregariousness versus solitude: Another look at parasite faunal richness in Canadian freshwater fishes. *Oecologia* 89:150–52.

———. 1993. There is no optimal foraging group size. *Anim. Behav.* 46:1032–35.

Ranta, E., and K. Lindstrom. 1990. Assortative schooling in three-spined sticklebacks? *Ann. Zool. Fennici* 27:67–75.

Ranta, E., K. Lindstrom, and N. Peuhkuri. 1992. Size matters when three-spined sticklebacks go to school. *Anim. Behav.* 43:160–62.

Ranta, E., H. Rita, and K. Lindstrom. 1993. Competition versus cooperation: Success of individuals foraging alone and in groups. *Amer. Nat.* 142:42–58.

Ranta, E., H. Rita, and N. Peuhkuri. 1995. Patch exploitation, group foraging, and unequal competitors. *Behav. Ecol.* 6:1–5.

Rayor, L. S., and G. W. Uetz. 1990. Trade-offs in foraging success and predation risk with spatial position in colonial spiders. *Behav. Ecol. Sociobiol.* 27:77–85.

Real, L., and T. Caraco. 1986. Risk and foraging in stochastic environments. *Ann. Rev. Ecol. Syst.* 17:371–90.

Redondo, T., F. S. Tortosa, and L. A. de Reyna. 1995. Nest switching and alloparental care in colonial white storks. *Anim. Behav.* 49:1097–1110.

Reed, J. M., and A. P. Dobson. 1993. Behavioural constraints and conservation biology: Conspecific attraction and recruitment. *Trends Ecol. Evol.* 8:253–56.

Reiter, J., K. J. Panken, and B. J. LeBoeuf. 1981. Female competition and reproductive success in northern elephant seals. *Anim. Behav.* 29:670–87.

Rendell, W. B. 1993. Intraspecific killing observed in tree swallows, *Tachycineta bicolor. Can. Field-Nat.* 107:227–28.

Rendell, W. B., and R. J. Robertson. 1993. Cavity size, clutch-size and the breeding ecology of tree swallows *Tachycineta bicolor. Ibis* 135:305–10.

Reville, B. J. 1988. Effects of spacing and synchrony on breeding success in the great frigatebird (*Fregata minor*). *Auk* 105:252–59.

———. 1991. Nest spacing and breeding success in the lesser frigatebird (*Fregata ariel*). *Condor* 93:555–62.
Rheinwald, G. 1975. The pattern of settling distances in a population of house martins *Delichon urbica*. *Ardea* 63:136–45.
Richner, H., and P. Heeb. 1995. Is the information center hypothesis a flop? *Adv. Study Behav.* 24:1–45.
Richner, H., and C. Marclay. 1991. Evolution of avian roosting behaviour: A test of the information centre hypothesis and of a critical assumption. *Anim. Behav.* 41:433–38.
Ricklefs, R. E. 1965. Brood reduction in the curve-billed thrasher. *Condor* 67:505–10.
———. 1974. Energetics of reproduction in birds. In *Avian energetics*, ed. R. A. Paynter Jr., 152–292. Cambridge, Mass.: Nuttall Ornithological Club.
Ricklefs, R. E., and D. J. T. Hussell. 1984. Changes in adult mass associated with the nesting cycle in the European starling. *Ornis Scandinavica* 15:155–61.
Ridgely, R. S., and G. Tudor. 1989. *The birds of South America*. Austin: University of Texas Press.
Riedman, M. L. 1982. The evolution of alloparental care and adoption in mammals and birds. *Quart. Rev. Biol.* 57:405–35.
Riedman, M. L., and B. J. Le Boeuf. 1982. Mother-pup separation and adoption in northern elephant seals. *Behav. Ecol. Sociobiol.* 11:203–15.
Riggs, C. D. 1947. Purple martins feeding on emerging may-flies. *Wilson Bull.* 59:113–14.
Riley, H. T., D. M. Bryant, R. E. Carter, and D. T. Parkin. 1995. Extra-pair fertilizations and paternity defence in house martins, *Delichon urbica*. *Anim. Behav.* 49:495–509.
Rising, J. D. 1983. The Great Plains hybrid zones. In *Current ornithology*, vol. 1, ed. R. F. Johnston, 131–57. New York: Plenum.
Rising, J. D., and K. M. Somers. 1989. The measurement of overall body size in birds. *Auk* 106:666–74.
Roberts, B. D., and S. A. Hatch. 1993. Behavioral ecology of black-legged kittiwakes during chick rearing in a failing colony. *Condor* 95:330–42.
Robertson, G. 1986. Population size and breeding success of the gentoo penguin, *Pygoscelis papua*, at Macquarie Island. *Aust. Wildl. Res.* 13:583–87.
Robertson, J. McB. 1926. Some notes on the cliff swallow. *Condor* 28:244–45.
Robertson, R. J. 1973. Optimal niche space of the red-winged blackbird: Spatial and temporal patterns of nesting activity and success. *Ecology* 54:1085–93.
Robertson, R. J., and W. B. Rendell. 1990. A comparison of the breeding ecology of a secondary cavity nesting bird, the tree swallow (*Tachycineta bicolor*), in nest boxes and natural cavities. *Can. J. Zool.* 68:1046–52.
Robidoux, Y. P., and A. Cyr. 1989. Selection granulométrique pour la construction du nid chez l'hirondelle à front blanc, *Hirundo pyrrhonota*. *Can. Field-Nat.* 103:577–83.
Robinson, J. G. 1988. Group size in wedge-capped capuchin monkeys *Cebus olivaceus* and the reproductive success of males and females. *Behav. Ecol. Sociobiol.* 23:187–97.

Robinson, S. K. 1985. Coloniality in the yellow-rumped cacique as a defense against nest predators. *Auk* 102:506–19.

———. 1986. Competitive and mutualistic interactions among females in a neotropical oriole. *Anim. Behav.* 34:113–22.

Rodgers, J. A., Jr. 1987. On the antipredator advantages of coloniality: A word of caution. *Wilson Bull.* 99:269–71.

Roeloffs, R., and S. E. Riechert. 1988. Dispersal and population-genetic structure of the cooperative spider, *Agelena consociata*, in West African rainforest. *Evolution* 42:173–83.

Rogers, C. A., R. J. Robertson, and B. J. Stutchbury. 1991. Patterns and effects of parasitism by *Protocalliphora sialia* on tree swallow nestlings. In *Bird-parasite interactions: Ecology, evolution and behaviour*, ed. J. E. Loye and M. Zuk, 123–39. Oxford: Oxford University Press.

Rohwer, F. C., and S. Freeman. 1989. The distribution of conspecific nest parasitism in birds. *Can. J. Zool.* 67:239–53.

Rohwer, S. 1986. Selection for adoption versus infanticide by replacement "mates" in birds. In *Current ornithology*, vol. 3, ed. R. F. Johnston, 353–95. New York: Plenum.

Romagnano, L., T. R. McGuire, and H. W. Power. 1989. Pitfalls and improved techniques in avian parentage studies. *Auk* 106:129–36.

Rood, J. P. 1983. Banded mongoose rescues pack member from eagle. *Anim. Behav.* 31:1261–62.

Rosche, R. C., and P. A. Johnsgard. 1984. Birds of Lake McConaughy and the North Platte River valley, Oshkosh to Keystone. *Nebraska Bird Review* 52:26–35.

Rothstein, S. I., and R. Pierotti. 1988. Distinctions among reciprocal altruism, kin selection, and cooperation and a model for the initial evolution of beneficent behavior. *Ethol. Sociobiol.* 9:189–209.

Rowe, L., D. Ludwig, and D. Schluter. 1994. Time, condition, and the seasonal decline of avian clutch size. *Amer. Nat.* 143:698–722.

Royama, T. 1966. Factors governing feeding rate, food requirement and brood size of nestling great tits *Parus major*. *Ibis* 108:313–47.

———. 1971. Evolutionary significance of predators' response to local differences in prey density: A theoretical study. In *Dynamics of populations: Proceedings of the Advanced Study Institute on "Dynamics of Numbers in Populations," Oosterbeek, the Netherlands, 7–18 September 1970*, ed. P. J. den Boer and G. R. Gradwall, 344–57. Wageningen: Pudoc.

Rubenstein, D. I. 1982. Risk, uncertainty and evolutionary strategies. In *Current problems in sociobiology*, ed. King's College Sociobiology Group, 91–111. Cambridge: Cambridge University Press.

———. 1984. Resource acquisition and alternative mating strategies in water striders. *Amer. Zool.* 24:345–53.

Rubenstein, D. I., R. J. Barnett, R. S. Ridgely, and P. H. Klopfer. 1977. Adaptive advantages of mixed-species feeding flocks among seed-eating finches in Costa Rica. *Ibis* 119:10–21.

Rubenstein, D. I., and M. E. Hohmann. 1989. Parasites and social behavior of island feral horses. *Oikos* 55:312–20.

Rutberg, A. T. 1987. Horse fly harassment and the social behavior of feral ponies. *Ethology* 75:145–54.

Ryder, J. P. 1970. A possible factor in the evolution of clutch size in Ross' goose. *Wilson Bull.* 82:5–13.

———. 1980. The influence of age on the breeding biology of colonial nesting seabirds. In *Behavior of marine animals*, vol. 4, *Marine birds: Current perspectives in research*, ed. J. Burger, B. L. Olla, and H. E. Winn, 153–68. New York: Plenum.

Rypstra, A. L. 1979. Foraging flocks of spiders: A study of aggregate behavior in *Cyrtophora citricola* Forskål (Araneae; Araneidae) in West Africa. *Behav. Ecol. Sociobiol.* 5:291–300.

———. 1983. The importance of food and space in limiting web-spider densities; a test using field enclosures. *Oecologia* 59:312–16.

———. 1985. Aggregations of *Nephila clavipes* (L.) (Araneae, Araneidae) in relation to prey availability. *J. Arachnol.* 13:71–78.

———. 1989. Foraging success of solitary and aggregated spiders: Insights into flock formation. *Anim. Behav.* 37:274–81.

Rypstra, A. L., and R. S. Tirey. 1991. Prey size, prey perishability and group foraging in a social spider. *Oecologia* 86:25–30.

Saether, B.-E. 1990. Age-specific variation in reproductive performance of birds. In *Current ornithology*, vol. 7, ed. D. M. Power, 251–83. New York: Plenum.

Saitou, T. 1991. Comparison of breeding success between residents and immigrants in the great tit. *Acta XX Congr. Internatl. Ornithol.*, 1196–1203.

Samuel, D. E. 1969. House sparrow occupancy of cliff swallow nests. *Wilson Bull.* 81:103–4.

———. 1971a. Vocal repertoires of sympatric barn and cliff swallows. *Auk* 88:839–55.

———. 1971b. The breeding biology of barn and cliff swallows in West Virginia. *Wilson Bull.* 83:284–301.

Sasvari, L., and Z. Hegyi. 1994. Colonial and solitary nesting choice as alternative breeding tactics in tree sparrow *Passer montanus*. *J. Anim. Ecol.* 63:265–74.

Schaefer, G. W. 1976. Radar observations of insect flight. *Symp. R. Entomol. Soc. London* 7:157–97.

Schaffner, F. C. 1991. Nest-site selection and nesting success of white-tailed tropicbirds (*Phaethon lepturus*) at Cayo Luís Peña, Puerto Rico. *Auk* 108:911–22.

Schelhaas, D. P., and O. R. Larson. 1989. Cold hardiness and winter survival in the bird flea, *Ceratophyllus idius*. *J. Insect Physiol.* 35:149–53.

Schleicher, B., F. Valera, and H. Hoi. 1993. The conflict between nest guarding and mate guarding in penduline tits (*Remiz pendulinus*). *Ethology* 95:157–65.

Schmutz, J. K., R. J. Robertson, and F. Cooke. 1983. Colonial nesting of the Hudson Bay eider duck. *Can. J. Zool.* 61:2424–33.

Schneider, S. S. 1989. Spatial foraging patterns of the African honey bee, *Apis mellifera scutellata*. *J. Insect Behav.* 2:505–21.

Schoener, T. W. 1971. Theory of feeding strategies. *Ann. Rev. Ecol. Syst.* 2:369–404.

Scolaro, J. A. 1990. Effects of nest density on breeding success in a colony of Magellanic penguins (*Spheniscus magellanicus*). *Colonial Waterbirds* 13:41–49.
Scott, M. E. 1987. Temporal changes in aggregation: A laboratory study. *Parasitology* 94:583–95.
Scott, T. W., G. S. Bowen, and T. P. Monath. 1984. A field study of the effects of Fort Morgan virus, an arbovirus transmitted by swallow bugs, on the reproductive success of cliff swallows and symbiotic house sparrows in Morgan County, Colorado, 1976. *Amer. J. Trop. Med. Hygiene* 33:981–91.
Seber, G. A. F. 1982. *The estimation of animal abundance and related parameters.* 2d ed. New York: Macmillian.
Seeley, T. D. 1985. *Honeybee ecology.* Princeton: Princeton University Press.
Seger, J. 1977. A numerical method for estimating coefficients of relationship in a langur troop. In *The langurs of Abu: Female and male strategies of reproduction,* by S. B. Hrdy, 317–26. Cambridge: Harvard University Press.
Seibt, U., and W. Wickler. 1988. Why do "family spiders," *Stegodyphus* (Eresidae), live in colonies? *J. Arachnol.* 16:193–98.
Semel, B., and P. Sherman. 1986. Dynamics of nest parasitism in wood ducks. *Auk* 103:813–16.
Senar, J. C., J. L. Copete, J. Domenech, and G. Von Walter. 1994. Prevalence of louse-flies Diptera, Hippoboscidae parasiting [*sic*] a cardueline finch and its effect on body condition. *Ardea* 82:157–60.
Shapiro, D. Y., and R. H. Boulon Jr. 1987. Evenly dispersed social groups and intergroup competition for juveniles in a coral-reef fish. *Behav. Ecol. Sociobiol.* 21:343–50.
Sharpe, R. B., and C. W. Wyatt. 1885–94. *A monograph of the Hirundinidae or family of swallows.* London: Taylor and Francis.
Sheldon, F. H., and D. W. Winkler. 1993. Intergeneric phylogenetic relationships of swallows estimated by DNA-DNA hybridization. *Auk* 110:798–824.
Shields, W. M. 1982. *Philopatry, inbreeding, and the evolution of sex.* Albany: State University of New York Press.
———. 1984a. Barn swallow mobbing: Self-defence, collateral kin defence, group defence, or parental care? *Anim. Behav.* 32:132–48.
———. 1984b. Factors affecting nest and site fidelity in Adirondack barn swallows (*Hirundo rustica*). *Auk* 101:780–89.
Shields, W. M., and J. R. Crook. 1987. Barn swallow coloniality: A net cost for group breeding in the Adirondacks? *Ecology* 68:1373–86.
Shields, W. M., J. R. Crook, M. L. Hebblethwaite, and S. S. Wiles-Ehmann. 1988. Ideal free coloniality in the swallows. In *The ecology of social behavior,* ed. C. N. Slobodchikoff, 189–228. San Diego: Academic Press.
Shine, R. 1986. Ecology of a low-energy specialist: Food habits and reproductive biology of the Arafura filesnake (Acrochordidae). *Copeia* 1986:424–37.
Sibly, R. M. 1983. Optimal group size is unstable. *Anim. Behav.* 31:947–48.
Siegel, S. 1956. *Nonparametric statistics for the behavioral sciences.* New York: McGraw-Hill.
Siegel-Causey, D., and G. L. Hunt Jr. 1986. Breeding-site selection and colony formation in double-crested and pelagic cormorants. *Auk* 103:230–34.
Siegel-Causey, D., and S. P. Kharitonov. 1990. The evolution of coloniality. In *Current ornithology,* vol. 7, ed. D. M. Power, 285–330. New York: Plenum.

———. 1996 History, ecology, and the evolution of coloniality in waterbirds. In *Coloniality in waterbirds*, ed. F. Cezilly, H. Hafner, and D. N. Nettleship. Oxford: Oxford University Press, in press.

Siegfried, W. R. 1972. Breeding success and reproductive output of the cattle egret. *Ostrich* 43:43–55.

Sikes, P. J., and K. A. Arnold. 1984. Movement and mortality estimates of cliff swallows in Texas. *Wilson Bull.* 96:419–25.

Sillen-Tullberg, B. 1990. Do predators avoid groups of aposematic prey? An experimental test. *Anim. Behav.* 40:856–60.

Silver, M. 1993. Second-year management of a cliff swallow colony in Massachusetts. *Bird Observer* 21:150–55.

Simpson, K., J. N. M. Smith, and J. P. Kelsall. 1987. Correlates and consequences of coloniality in great blue herons. *Can. J. Zool.* 65:572–77.

Sjoberg, G. 1994. Early breeding leads to intra-seasonal clutch size decline in Canada geese. *J. Avian Biol.* 25:112–18.

Skead, D. M. 1979. Feeding associations of *Hirundo spilodera* with other animals. *Bokmakierie* 31:63.

Skead, D. M., and C. J. Skead. 1970. Hirundinid mortality during adverse weather, November 1968. *Ostrich* 41:247–51.

Skutch, A. F. 1976. *Parent birds and their young.* Austin: University of Texas Press.

Slobodchikoff, C. N., and W. C. Schulz. 1988. Cooperation, aggression, and the evolution of social behavior. In *The ecology of social behavior*, ed. C. N. Slobodchikoff, 13–32. San Diego: Academic Press.

Smith, C. C. 1968. The adaptive nature of social organization in the genus of three squirrels (*Tamiasciurus*). *Ecol. Monogr.* 38:31–63.

Smith, D. R. R. 1982. Reproductive success of solitary and communal *Philoponella oweni* (Araneae: Uloboridae). *Behav. Ecol. Sociobiol.* 11:149–54.

———. 1983. Ecological costs and benefits of communal behavior in a presocial spider. *Behav. Ecol. Sociobiol.* 13:107–14.

———. 1985. Habitat use by colonies of *Philoponella republicana* (Araneae, Uloboridae). *J. Arachnol.* 13:363–73.

Smith, G. C., and R. B. Eads. 1978. Field observations on the cliff swallow, *Petrochelidon pyrrhonota* (Vieillot), and the swallow bug, *Oeciacus vicarius* Horvath. *J. Wash. Acad. Sci.* 68:23–26.

Smith, H. M. 1943. Size of breeding populations in relation to egg-laying and reproductive success in the eastern red-wing (*Agelaius p. phoeniceus*). *Ecology* 24:183–207.

Smith, J. N. M. 1988. Determinants of lifetime reproductive success in the song sparrow. In *Reproductive success: Studies of individual variation in contrasting breeding systems*, ed. T. H. Clutton-Brock, 154–72. Chicago: University of Chicago Press.

Smith, M. J., and H. B. Graves. 1978. Some factors influencing mobbing behavior in barn swallows (*Hirundo rustica*). *Behav. Biol.* 23:355–72.

Smyth, A. P., B. K. Orr, and R. C. Fleischer. 1993. Electrophoretic variants of egg white transferrin indicate a low rate of intraspecific brood parasitism in colonial cliff swallows in the Sierra Nevada, California. *Behav. Ecol. Sociobiol.* 32:79–84.

Snapp, B. D. 1976. Colonial breeding in the barn swallow (*Hirundo rustica*) and its adaptive significance. *Condor* 78:471–80.

Snyder, N. F. R., S. R. Beissinger, and R. E. Chandler. 1989. Reproduction and demography of the Florida Everglade (snail) kite. *Condor* 91:300–316.

Sokal, R. R., and F. J. Rohlf. 1969. *Biometry: The principles and practice of statistics in biological research*. San Francisco: Freeman.

Sordahl, T. A. 1990. The risks of avian mobbing and distraction behavior: An anecdotal review. *Wilson Bull.* 102:349–52.

Sorenson, M. D. 1991. The functional significance of parasitic egg laying and typical nesting in redhead ducks: An analysis of individual behaviour. *Anim. Behav.* 42:771–96.

———. 1992. Is parasitic egg laying the avian equivalent of "don't put all your eggs in one basket?" Abstract, 4th International Behavioral Ecology Congress, Princeton University.

Southwood, T. R. E. 1957. Observations on swarming in Braconidae and Coniopterygidae. *Proc. R. Entomol. Soc. London* 32:80–82.

Speich, S. M., H. L. Jones, and E. M. Benedict. 1986. Review of the natural nesting sites of the barn swallow in North America. *Amer. Midl. Nat.* 115:248–54.

Spiller, D. A. 1992. Relationship between prey consumption and colony size in an orb spider. *Oecologia* 90:457–66.

Spiller, D. A., and T. W. Schoener. 1989. Effect of a major predator on grouping of an orb-weaving spider. *J. Anim. Ecol.* 58:509–23.

Stacey, P. B. 1986. Group size and foraging efficiency in yellow baboons. *Behav. Ecol. Sociobiol.* 18:175–87.

Stacey, P. B., and W. D. Koenig, eds. 1990a. *Cooperative breeding in birds*. Cambridge: Cambridge University Press.

———. 1990b. Introduction. In *Cooperative breeding in birds*, ed. P. B. Stacey and W. D. Koenig, ix–xviii. Cambridge: Cambridge University Press.

Stamps, J. A. 1988. Conspecific attraction and aggregation in territorial species. *Amer. Nat.* 131:329–47.

Steeger, C., and R. C. Ydenberg. 1993. Clutch size and initiation date of ospreys: Natural patterns and the effect of a natural delay. *Can. J. Zool.* 71:2141–46.

Stephens, D. W., and J. R. Krebs. 1986. *Foraging theory*. Princeton: Princeton University Press.

Sternberg, H. 1989. Pied flycatcher. In *Lifetime reproduction in birds*, ed. I. Newton, 55–74. London: Academic Press.

Stewart, P. A. 1972. Mortality of purple martins from adverse weather. *Condor* 74:480.

Stewart, R. M. 1972. Nestling mortality in swallows due to inclement weather. *Calif. Birds* 3:69–70.

Still, E., P. Monaghan, and E. Bignal. 1987. Social structuring at a communal roost of choughs *Pyrrhocorax pyrrhocorax*. *Ibis* 129:398–403.

Stoddard, P. K. 1983. Violation of ideal nest placement: Cliff swallows entombed by their own excrement. *Wilson Bull.* 95:674–75.

———. 1988. The "bugs" call of the cliff swallow: A rare food signal in a colonially nesting bird species. *Condor* 90:714–15.

Stoddard, P. K., and M. D. Beecher. 1983. Parental recognition of offspring in the cliff swallow. *Auk* 100:795–99.

Stoner, D. 1941. Homing instinct in the bank swallow. *Bird-Banding* 12:104–9.

———. 1942. Behavior of young bank swallows after first leaving the nest. *Bird-Banding* 13:107–10.

———. 1945. Temperature and growth studies of the northern cliff swallow. *Auk* 62:207–16.

Storer, T. I. 1927. Three notable nesting colonies of the cliff swallow in California. *Condor* 29:104–8.

Stutchbury, B. J. 1991. Coloniality and breeding biology of purple martins (*Progne subis hesperia*) in saguaro cacti. *Condor* 93:666–75.

Stutchbury, B. J., and R. J. Robertson. 1988. Within-season and age-related patterns of reproductive performance in female tree swallows (*Tachycineta bicolor*). *Can. J. Zool.* 66:827–34.

Sullivan, K. A. 1989. Predation and starvation: Age-specific mortality in juvenile juncos (*Junco phaenotus*). *J. Anim. Ecol.* 58:275–86.

Sullivan, R. T. 1981. Insect swarming and mating. *Florida Entomol.* 64:44–65.

Summers, R. W., G. E. Westlake, and C. J. Feare. 1987. Differences in the ages, sexes and physical condition of starlings *Sturnus vulgaris* at the centre and periphery of roosts. *Ibis* 129:96–102.

Sutherland, D. M., and S. B. Rolfsmeier. 1989. An annotated list of the vascular plants of Keith County, Nebraska. *Trans. Neb. Acad. Sci.* 17:83–101.

Sutton, G. M. 1967. *Oklahoma birds*. Norman: University of Oklahoma Press.

———. 1986. *Birds worth watching*. Norman: University of Oklahoma Press.

Svensson, B. G., and E. Petersson. 1994. Mate choice tactics and swarm size: A model and a test in a dance fly. *Behav. Ecol. Sociobiol.* 35:161–68.

Swinehart, J. B., and R. F. Diffendal Jr. 1989. Geology of the pre-dune strata. In *An atlas of the Sand Hills*, ed. A. Bleed and C. Flowerday, 29–42. Lincoln: Conservation and Survey Division, University of Nebraska.

Sydeman, W. J., J. F. Penniman, T. M. Penniman, P. Pyle, and D. G. Ainley. 1991. Breeding performance in the western gull: Effects of parental age, timing of breeding and year in relation to food availability. *J. Anim. Ecol.* 60:135–49.

Szep, T. 1991. A Tisza magyarországi szakaszán fészkelő partifecske (*Riparia riparia* L., 1758) állomány eloszlása és egyedszáma. *Aquila* 98:111–24.

———. 1995. Relationship between west African rainfall and the survival of central European sand martins *Riparia riparia*. *Ibis* 137:162–68.

Szep, T., and Z. Barta. 1992. The threat to bank swallows from the hobby at a large colony. *Condor* 94:1022–25.

Tate, J., Jr. 1986. The Blue List for 1986. *Amer. Birds* 40:227–35.

Taylor, L. R., and R. A. J. Taylor. 1977. Aggregation, migration and population mechanics. *Nature* 265:415–21.

Taylor, R. A. J. 1981. The behavioural basis of redistribution. I. The Δ-model concept. *J. Anim. Ecol.* 50:573–86.

Taylor, R. H. 1962. The Adelie penguin *Pygoscelis adeliae* at Cape Royds. *Ibis* 104:176–204.

Tenaza, R. R. 1971. Behavior and nesting success relative to nest location in Adelie penguins (*Pygoscelis adeliae*). *Condor* 73:81–92.

Terborgh, J. 1983. *Five New World primates: A study in comparative ecology*. Princeton: Princeton University Press.
Terborgh, J., and C. H. Janson. 1986. The socioecology of primate groups. *Ann. Rev. Ecol. Syst.* 17:111–35.
Thomas, C. S., and J. C. Coulson. 1988. Reproductive success of kittiwake gulls, *Rissa tridactyla*. In *Reproductive success: Studies of individual variation in contrasting breeding systems*, ed. T. H. Clutton-Brock, 251–62. Chicago: University of Chicago Press.
Thompson, B. C., and C. L. Turner. 1980. Bull snake predation at a cliff swallow nest. *Murrelet* 61:35–36.
Thompson, C. F., and J. E. C. Flux. 1988. Body mass and lipid content at nest-leaving of European starlings in New Zealand. *Ornis Scandinavica* 19:1–6.
Thompson, C. F., J. E. C. Flux, and V. T. Tetzlaff. 1993. The heaviest nestlings are not necessarily the fattest nestlings. *J. Field Ornithol.* 64:426–32.
Thompson, D. W. 1910. *The works of Aristotle*. Vol. 4. *Historia animalium*. Oxford: Oxford University Press.
Thorpe, W. H. 1956. *Learning and instinct in animals*. London: Methuen.
Tinbergen, J. M., and M. C. Boerlijst. 1990. Nestling weight and survival in individual great tits (*Parus major*). *J. Anim. Ecol.* 59:1113–27.
Tinkle, D. W. 1979. Long-term field studies. *Bioscience* 29:717.
Traub, R., M. Rothschild, and J. F. Haddow. 1983. *The Rothschild collection of fleas. The Ceratophyllidae: Key to the genera and host relationships*. London: Academic Press.
Trivers, R. L. 1971. The evolution of reciprocal altruism. *Quart. Rev. Biol.* 46:35–57.
———. 1972. Parental investment and sexual selection. In *Sexual selection and the descent of man, 1871–1971*, ed. B. Campbell, 136–79. Chicago: Aldine.
Turner, A. K. 1982. Optimal foraging by the swallow (*Hirundo rustica* L.): Prey size selection. *Anim. Behav.* 30:862–72.
Turner, A. K., and C. Rose. 1989. *Swallows and martins: An identification guide and handbook*. Boston: Houghton Mifflin.
Turner, G. F., and T. J. Pitcher. 1986. Attack abatement: A model for group protection by combined avoidance and dilution. *Amer. Nat.* 128:228–40.
Tyler, W. A., III. 1995. The adaptive significance of colonial nesting in a coral-reef fish. *Anim. Behav.* 49:949–66.
Uetz, G. W., and C. S. Hieber. 1994. Group size and predation risk in colonial web-building spiders: Analysis of attack abatement mechanisms. *Behav. Ecol.* 5:326–33.
Uetz, G. W., T. C. Kane, and G. E. Stratton. 1982. Variation in the social grouping tendency of a communal web-building spider. *Science* 217:547–49.
Underwood, R. 1982. Vigilance behaviour in grazing African antelopes. *Behaviour* 79:81–107.
Usinger, R. L. 1966. *Monograph of Cimicidae (Hemiptera—Heteroptera)*. College Park, Md.: Thomas Say Foundation.
Van Balen, J. H. 1967. The significance of variations in body weight and wing length in the great tit, *Parus major*. *Ardea* 55:1–59.

Van Dyke, E. C. 1919. A few observations on the tendency of insects to collect on ridges and mountain snowfields. *Entomol. News* 30:241–44.
Van Noordwijk, A. J., and J. H. Van Balen. 1988. The great tit, *Parus major*. In *Reproductive success: Studies of individual variation in contrasting breeding systems*, ed. T. H. Clutton-Brock, 119–35. Chicago: University of Chicago Press.
Van Schaik, C. P. 1983. Why are diurnal primates living in groups? *Behaviour* 87: 120–44.
Van Schaik, C. P., and M. Horstermann. 1994. Predation risk and the number of adult males in a primate group: A comparative test. *Behav. Ecol. Sociobiol.* 35:261–72.
Van Schaik, C. P., M. A. Van Noordwijk, R. J. De Boer, and I. Den Tonkelaar. 1983. The effect of group size on time budgets and social behaviour in wild long-tailed macaques (*Macaca fascicularis*). *Behav. Ecol. Sociobiol.* 13: 173–81.
Van Schaik, C. P., M. A. Van Noordwijk, B. Warsono, and E. Sutriono. 1983. Party size and early detection of predators in Sumatran forest primates. *Primates* 24:211–21.
Van Vessem, J., and D. Draulans. 1986. The adaptive significance of colonial breeding in the grey heron *Ardea cinera*: Inter- and intra-colony variability in breeding success. *Ornis Scandinavica* 17:356–62.
Veen, J. 1977. Functional and causal aspects of nest distribution in colonies of the sandwich tern (*Sterna s. sandvicensis* Lath.). *Behav. Suppl.* 20:1–193.
———. 1980. Waarom broeden vogels in kolonies? *Limosa* 53:37–48.
Vehrencamp, S. L. 1977. Relative fecundity and parental effort in communally nesting anis, *Crotophaga sulcirostris*. *Science* 197:403–5.
Vehrencamp, S. L., R. R. Koford, and B. S. Bowen. 1988. The effect of breeding-unit size on fitness components in groove-billed anis. In *Reproductive success: Studies of individual variation in contrasting breeding systems*, ed. T. H. Clutton-Brock, 291–304. Chicago: University of Chicago Press.
Venier, L. A., and R. J. Robertson. 1991. Copulation behaviour of the tree swallow, *Tachycineta bicolor*: Paternity assurance in the presence of sperm competition. *Anim. Behav.* 42:939–48.
Vickery, W. L., L.-A. Giraldeau, J. J. Templeton, D. L. Kramer, and C. A. Chapman. 1991. Producers, scroungers, and group foraging. *Amer. Nat.* 137: 847–63.
Vine, I. 1971. Risk of visual detection and pursuit by a predator and the selective advantage of flocking behaviour. *J. Theor. Biol.* 30:405–22.
Visscher, P. K., and T. D. Seeley. 1982. Foraging strategy of honeybee colonies in a temperate deciduous forest. *Ecology* 63:1790–1801.
von Frisch, K. 1967. *The dance language and orientation of bees*. Cambridge: Harvard University Press.
Waddington, K. D., P. K. Visscher, T. J. Herbert, and M. R. Richter. 1994. Comparisons of forager distributions from matched honey bee colonies in suburban environments. *Behav. Ecol. Sociobiol.* 35:423–29.
Wagner, R. H. 1992a. Extra-pair copulations in a lek: The secondary mating system of monogamous razorbills. *Behav. Ecol. Sociobiol.* 31:63–71.

———. 1992b. The pursuit of extra-pair copulations by monogamous female razorbills: How do females benefit? *Behav. Ecol. Sociobiol.* 29:455–64.

———. 1993. The pursuit of extra-pair copulations by female birds: A new hypothesis of colony formation. *J. Theor. Biol.* 163:333–46.

Wakelin, D. 1978. Genetic control of susceptibility and resistance to parasitic infection. *Adv. Parasitol.* 16:219–308.

Wakelin, D., and J. M. Blackwell, eds. 1988. *Genetics of resistance to bacterial and parasitic infection.* London: Taylor and Francis.

Wallington, C. E. 1961. *Meteorology for glider pilots.* London: Murray.

Waloff, N. 1973. Dispersal by flight of leafhoppers (Auchenorrhyncha: Homoptera). *J. Appl. Ecol.* 10:705–30.

Waltz, E. C. 1981. The information-center hypothesis and colonial nesting behavior. Ph.D. diss., State University of New York, Syracuse.

———. 1982. Resource characteristics and the evolution of information centers. *Amer. Nat.* 119:73–90.

———. 1983. On tolerating followers in information-centers, with comments on testing the adaptive significance of coloniality. *Colonial Waterbirds* 6:31–36.

———. 1987. A test of the information-centre hypothesis in two colonies of common terns, *Sterna hirundo. Anim. Behav.* 35:48–59.

Ward, P., and A. Zahavi. 1973. The importance of certain assemblages of birds as "information centers" for food-finding. *Ibis* 115:517–34.

Waser, P. M., and R. H. Wiley. 1980. Mechanisms and evolution of spacing in animals. In *Handbook of behavioural neurobiology,* vol. 3, ed. P. Marler and J. G. Vandenbergh, 159–223. New York: Plenum.

Watts, D. P. 1985. Relations between group size and composition and feeding competition in mountain gorilla groups. *Anim. Behav.* 33:72–85.

Weatherhead, P. J. 1983. Two principal strategies in avian communal roosts. *Amer. Nat.* 121:237–43.

Weatherhead, P. J., and M. R. L. Forbes. 1994. Natal philopatry in passerine birds: Genetic or ecological influences? *Behav. Ecol.* 5:426–33.

Weigmann, C., and J. Lamprecht. 1991. Intraspecific nest parasitism in bar-headed geese, *Anser indicus. Anim. Behav.* 41:677–88.

Weimerskirch, H. 1990. The influence of age and experience on breeding performance of the antarctic fulmar, *Fulmarus glacialoides. J. Anim. Ecol.* 59:867–75.

Weimerskirch, H., O. Chastel, and L. Ackermann. 1995. Adjustment of parental effort to manipulated foraging ability in a pelagic seabird, the thin-billed prion *Pachyptila belcheri. Behav. Ecol. Sociobiol.* 36:11–16.

Wenzel, R. L., and V. J. Tipton. 1966. Some relationships between mammal hosts and their ectoparasites. In *Ectoparasites of Panama,* ed. R. L. Wenzel and V. J. Tipton, 677–723. Chicago: Field Museum of Natural History.

Westneat, D. F. 1992. Nesting synchrony by female red-winged blackbirds: Effects on predation and breeding success. *Ecology* 73:2284–94.

Westneat, D. F., P. C. Frederick, and R. H. Wiley. 1987. The use of genetic markers to estimate the frequency of successful alternative reproductive tactics. *Behav. Ecol. Sociobiol.* 21:35–45.

Westneat, D. F., P. W. Sherman, and M. L. Morton. 1990. The ecology and evo-

lution of extra-pair copulations in birds. In *Current ornithology,* vol. 7, ed. D. M. Power, 331–69. New York: Plenum.
Wheeler, T. A., and W. Threlfall. 1986. Observations on the ectoparasites of some Newfoundland passerines (Aves: Passeriformes). *Can. J. Zool.* 64:630–36.
Whittingham, L. A., P. O. Dunn, and R. J. Robertson. 1993. Confidence of paternity and male parental care: An experimental study in tree swallows. *Anim. Behav.* 46:139–47.
Whittingham, L. A., P. D. Taylor, and R. J. Robertson. 1992. Confidence of paternity and male parental care. *Amer. Nat.* 139:1115–25.
Wickler, W., and U. Seibt. 1993. Pedogenetic sociogenesis via the "sibling-route" and some consequences for *Stegodyphus* spiders. *Ethology* 95:1–18.
Wiklund, C. G. 1982. Fieldfare (*Turdus pilaris*) breeding success in relation to colony size, nest position and association with merlins (*Falco columbarius*). *Behav. Ecol. Sociobiol.* 11:165–72.
Wiklund, C. G., and M. Andersson. 1980. Nest predation selects for colonial breeding among fieldfares *Turdus pilaris. Ibis* 122:363–66.
———. 1994. Natural selection of colony size in a passerine bird. *J. Anim. Ecol.* 63:765–74.
Wiley, R. H., and M. S. Wiley. 1980. Spacing and timing in the nesting ecology of a tropical blackbird: Comparison of populations in different environments. *Ecol. Monogr.* 50:153–78.
Wilhite, D. A., and K. G. Hubbard. 1989. Climate. In *An atlas of the Sand Hills,* ed. A. Bleed and C. Flowerday, 17–28. Lincoln: Conservation and Survey Division, University of Nebraska.
Wilkinson, G. S. 1985. The social organization of the common vampire bat. I. Pattern and cause of association. *Behav. Ecol. Sociobiol.* 17:111–21.
———. 1992. Information transfer at evening bat colonies. *Anim. Behav.* 44:501–18.
Wilkinson, G. S., and G. M. English-Loeb. 1982. Predation and coloniality in cliff swallows (*Petrochelidon pyrrhonota*). *Auk* 99:459–67.
Willadsen, P. 1980. Immunity to ticks. *Adv. Parasitol.* 18:293–313.
Wilson, E. O. 1971. *The insect societies.* Cambridge: Harvard University Press.
Winkler, D. W., and F. H. Sheldon. 1993. Evolution of nest construction in swallows (Hirundinidae): A molecular phylogenetic perspective. *Proc. Natl. Acad. Sci. USA* 90:5705–7.
Withers, P. C. 1977. Energetic aspects of reproduction by the cliff swallow. *Auk* 94:718–25.
Wittenberger, J. F. 1981. *Animal social behavior.* Boston: Duxbury.
Wittenberger, J. F., and M. B. Dollinger. 1984. The effect of acentric colony location on the energetics of avian coloniality. *Amer. Nat.* 124:189–204.
Wittenberger, J. F., and G. L. Hunt Jr. 1985. The adaptive significance of coloniality in birds. In *Avian biology,* vol. 8, ed. D. S. Farner and J. R. King, 1–78. San Diego: Academic Press.
Witter, M. S., and I. C. Cuthill. 1993. The ecological costs of avian fat storage. *Phil. Trans. R. Soc. Lond.,* ser. B, 340:73–92.
Wolff, J. O. 1994. Reproductive success of solitary and communally nesting white-footed mice and deer mice. *Behav. Ecol.* 5:206–9.

Wolfson, A. 1952. The cloacal protuberance: A means for determining breeding condition in live male passerines. *Bird-Banding* 23:159–65.

Woodland, D. J., Z. Jaafar, and M. L. Knight. 1980. The "pursuit deterrent" function of alarm signals. *Amer. Nat.* 115:748–53.

Woolfenden, G. E., and J. W. Fitzpatrick. 1984. *The Florida scrub jay.* Princeton: Princeton University Press.

Wrege, P. H., and S. T. Emlen. 1987. Biochemical determination of parental uncertainty in white-fronted bee-eaters. *Behav. Ecol. Sociobiol.* 20:153–60.

Wunderle, J. M., Jr. 1991. Age-specific foraging proficiency in birds. In *Current ornithology*, vol. 8, ed. D. M. Power, 273–324. New York: Plenum.

Wynne-Edwards, V. C. 1962. *Animal dispersion in relation to social behaviour.* Edinburgh: Oliver and Boyd.

Yom-Tov, Y. 1980. Intraspecific nest parasitism in birds. *Biol. Rev.* 55:93–108.

Zack, R. S. 1990. Swallow bug (Heteroptera: Cimicidae) in Washington with an unusual overwintering site. *Pan-Pacific Entomol.* 66:251–52.

Index

abandonment. *See* colony desertion; nest abandonment
acaricide, 32. *See also* fumigation
Acarina, 72
Accipiter striatus, 483. *See also* hawk, sharp-shinned
Acer negundo, 483. *See also* boxelder
Acer saccharinum, 483. *See also* maple, silver
Acrididae, table 9.1, 249
active nest, 34–35, 38, 113, fig. 4.12, 155, 218, 356, table 11.9. *See also* nest
   definition of, 13
   identifying, 18
   and population size, 57, fig. 3.7
   and swallow bugs, fig. 4.11, 108, 111, 112
adult
   age of, and ectoparasite load, 91–92
   banding of, 20–22, fig. 2.3, 38, 397
   body mass of, 14, 22, 37, 290–303, figs. 10.9–12, table 10.3, 305, 316, 319–20, 327, 380, 384–86, fig. 12.12, 398, 415–18, fig. 13.3 (*see also* body mass)
   determining sex of, 22–24, 39
   ectoparasites of, 102–6, fig. 4.10, 116–17
      on feet, 102–3, fig. 4.11, 108–9, 117
      sampling of, 26, 28–30, fig. 2.8, 39, 73, 102
   emigration of, 368
   number of marked, 363–64, table 12.1
   recapture of, 363–64, table 12.1, 365, 397
   survivorship of, 105–6, 362, 370–73, figs. 12.1–3, 375–80, figs. 12.6–8, 384–86, 395–98, 457–58
Africa, 10
age
   classes, 356, 369
   and clutch size, 323, 326, 328, 355–56, 359
   and colony settlement patterns, 379–80, 409–12, fig. 13.1, 471
   and colony size choice, 400, 426, 431–32, table 13.6, 447, 465
   and costs and benefits of coloniality, 5–6
   determining, for nestlings, 19
   and differences in arrival time, 62–65, fig. 3.9, 67
   and dispersal, 424–25, table 13.3, 447, 474
   established by banding birds, 20, 22
   of female, and clutch size, 332–33, 355–56, 359
   of mates, 357
   of nest, 35, 39, 79–84, 86–89, figs. 4.3–4, 112, 114–17, fig. 4.12, 168, 327–28, 329, 331, fig. 11.5, 343, 350–53, 359, 391, 395–97, 398, 425, table 13.3, 473–74
   and brood parasitism, 168
   and nest ectoparasitism, 91–92
   of nestlings
      and food delivery, 281–83, fig. 10.2, table 10.2
      and swallow bug load, 111
      at data collection, 93

529

530 • *Index*

age (*continued*)
  of nest owners, 393
  of parents, and nestling survivorship, 393
  and plumage dimorphism, 24
  and sex-specific survivorship, 370, fig. 12.2, 379, 397
  and sorting of birds among colonies, 409–12, fig. 13.1
  and structure of colony, 376, 379–80, 409–12, fig. 13.1
  and survival or recapture probabilities, 365–66
  and survivorship, 368–70, fig. 12.1
  and wing chord, 24
age classes, 369
Ageneotettix deorum, 24
age-specific differences
  brood parasitism, 356, 358, table 11.9, 359
  clutch size, 355–56, table 11.9, 359
  egg laying date, 356–58, table 11.9, 359
  egg laying (breeding) synchrony, 356, 358, table 11.9, 359
  egg tossing (destruction), 356, 358, table 11.9
  egg transfer, 356, 358, table 11.9
  flea parasitism, 358
  nearest neighbor distance, 356, 358, table 11.9
  nest building, 356, 358, table 11.9, 359
  nest diameter (size), 356, 358, table 11.9
  number of young surviving, 356–57
  parental experience, 355–56
  position of nest, 356–58, table 11.9, 359
  reproductive success, 355–59, table 11.9, 431–32
  survivorship, 368–70, fig. 12.1, 397
  use of new nest, 356, 358, table 11.9, 359
age structure
  of breeders in colony, 376, 379–80, 409–12, fig. 13.1
  and colony size, 379–80
aggregation. *See also* coloniality
  of conspecifics, 412
  of ectoparasites, 92, 472
  of foragers for predator avoidance, 244, 251
  and habitat limitation, 2–3, 5, 11, 119
  of juveniles in crèche, 175–79, figs. 6.12–13
  and reproductive success, 208
  and resource abundance, 2–3, 5, 11, 119, 450
  and shortage of nest sites, 187–88, 208, 450
  signal, 267
  at traditional sites, 206–8, 209
aggression
  and competition, 118
  and fighting at nests, 119–26, figs. 5.1–3
alarm calls, 16, 219, 241
  deceptive, 142–43
  and nest trespassing, 142–43, 144
  and relatedness, 474
  as signal to predators, 214, 225
alarm response, 137, 142, 144. *See also* predator alarm response
  to different predators, 213–16
  used in defining colony, 16
Alaska, 8, 10, 71
Alberta, 368
Alca torda, 484. *See also* razorbill
allozyme parentage studies, 164–65, 171, 185
Alphavirus, 97
alternative reproductive tactics. *See also* conspecific brood parasitism; extrapair copulation; misdirected parental care
  and coloniality, 4–6, 11–12, 150–51, 164–65, 167, 175
  evolution of, 166
  and group size, 150–51
altricial young, 281
altruism, 445
Alydidae, table 9.1, 249
Amaranthus, 42
Amblycera, 73
Amphitornus coloradus, 248
Andes, 8
Andropogon, 42
ani, groove-billed, 361, 484
annual colony site use. *See also* colony use patterns
  and colony size, 197–202, figs. 7.6–9, 209
  and ectoparasitism, 114, 195, 196–98, 201
  predictability of, 201–2, fig. 7.9, 209
  and substrate size, 192, 195–96, fig. 7.5
  variation in, 201–2, fig. 7.9
annual reproductive success, 6–7, 12, 150, 471. *See also* reproductive success
  age-specific variation in, 323

and climate, 472–73
and colony size, 323, 333, 456, 467, 480
as component of lifetime reproductive success, 360
and fitness, 317
as number of young surviving to fledge, 360, 460
patterns in, 321–23, 371
peaks at intermediate colony size, 375
yearly variation in, 323, 472–73
annual survivorship or survival, 360, 362, 386, 396, 451, table 14.1, 456, 460, 473, 474. *See also* survivorship
and body mass, 376, 379, 380, 385, fig. 12.12, 398
and colony size, 361, 378, fig. 12.8, 397–98
and costs and benefits of coloniality, 373
difficulty of studying, 361
and fitness, 317
and foraging efficiency, 373
variance in, 369, 371–73, 472
Antarctica, 10
Anthomiidae, table 9.1, 249
Aphididae, table 9.1, 249
Apidae, table 9.1, 249
Apis mellifera, 483. *See also* honeybee
arbovirus, 97
Ardea herodias, 483. *See also* heron, great blue
Argasidae, 72
Argentina, 8
Argidae, table 9.1, 249
Arizona, 10, 233
arrival times, spring
of females, fig. 3.9, 62–65, 67, 161, 407
of known-age individuals, 62–65, fig. 3.9, 67
of males, fig. 3.9, 62–65, 67, 161, 407
of yearlings, fig. 3.9, 62–65, 67
Arthur, Nebraska, 44, tables 3.1–2
artificial nesting sites, 9, 12, 16, 47, 49, fig. 3.3, 192, 201–2
assessing nest site limitation on, 204–5, 209, 450
and behavior of colony residents, 56–57, 67, 204–5
compared with natural sites, 56–57, 67, 204–5, 209
and ectoparasitism, 75
historical use of, 50
measuring substrate of, 192, 204

positioning of nests on, figs. 3.5–6, 56
predation on, 205, 216
predictability of use of, 201–2, fig. 7.9, 205
size of colonies on, 50–51, 205
spacing of nests on, 205
types of, 49–57, figs. 3.5–6, 67, 204–5
ash, green, 42, 483
Ash Hollow, Nebraska, 41, 46–47
Ash Hollow State Park, 49–50
Asia, 10
Asilidae, table 9.1, 249
assessment
of colony sites, 401, 404, 406, 407–9, 446, 456, 467, 474 (*see also* colony choice)
by kleptoparasitic juveniles, 185
by visiting active colonies, 208
of host nests for brood parasitism, 168, 174
of nest ectoparasite loads, 112–13, 133, 148, 455, 474
of nest sites, 135, 401
of parental quality of neighbors, 133, 174
Aster, 42
Astragalus, 42
asymmetries
in costs and benefits of group living, 469–74, 481
in reproductive success, 481
spatial, 469–71, 481
temporal, 471–73, 481
Aulocara elliotti, 248
Australia, 10

badger, 216, 484
Baetidae, table 9.1, 249
Bahamas, 8
banding, 362–67, table 12.1, 402, 420–25
of adults, 20–22, fig. 2.3, 38, 397
of juveniles, 22, 183, 374, 409, 433
of nestlings, 19–20, 27, 38, 180, 183, 356, 374, 397, 409, 433
bands, U.S. Fish and Wildlife Service, 19
bat, 71, 251
bat, evening, 244, 484
bat bug, 70
bathing, communal, 217
bedbug, 70. *See also* swallow bug
begging
calls, 176, 178–79
by juveniles, 176

532 • Index

behavior
  and artificial nest sites, 56–57, 67, 204–5
  of barn swallows, 202–3
  during colony selection, 400–409, table 13.1
  and ectoparasitism, 68–69, 103–5, 110, 112–17, fig. 4.12
  epideictic display, 445
  foraging, 245–47, 251, 268, 315
  intraspecific competition, 118–20
  postfledging, 175–85, figs. 6.12–16
  of radio-tagged birds, 401
  and relatedness of individuals, 474
  in response to predators, 213–17
  roosting, 442, 445
Berytidae, table 9.1, 249
Bibionidae, table 9.1, 249
bird flea. *See* flea
bison, 247, 484
Bison bison, 484. *See also* bison
blood, 93
blood collecting, 19
blood meal
  of amblyceran louse, 73
  of bird flea, 72, 82, 102
  of soft-bodied tick, 73
  of swallow bug, 70, 82, 102, 111
Blue Creek, 49
bluffs, 246
body mass, 275, 281, 315, 360, 373, fig. 12.9, 401, 465
  of adults, 14, 22, 37, 290–303, figs. 10.9–12, table 10.3, 305, 316, 319–20, 327, 380, 384–86, 398, 415–18, fig. 13.3
  and adult survival, 291, 297, 376, 379, 384–86, fig. 12.12, 398
  and colony reduction experiment, 305
  and colony settlement patterns, 293, 319, 415–18, fig. 13.3
  and condition for migration, 384
  cost of maintaining, 316–17
  and effect of
    brood size on nestlings, 94–95, figs. 4.6–9, table 4.7, 98–102, 286–88, figs. 10.6–7, 317, 319
    colony size on adult, 291, 293–97, figs. 10.10–11, table 10.3, 299, 301–3, 305, 319–29, 379, 380, 384, 415–18, fig. 13.3
    colony size on nestling, 288–90, figs. 10.7–8, 317, 319, 380
    ectoparasitism on nestling, 93–102, figs. 4.6–9, table 4.7, 116, 286–90, figs. 10.6–8, 319, 386
    year and resource availability on nestling, 286–88, fig. 10.7
  and fat reserves and foraging ability, 291
  of female
    and clutch size, 332–33
    and energy reserves, 333
    and food delivery rate, 281
    and foraging efficiency, 290–91, 294, 297–300, 316, 319, 386
  of juveniles, 22, 382–84, figs. 12.10–11, 398
  loss of, table 14.1
  as measure of energy reserves, 380, 416
  of nestlings, 20, 27, 93–96, figs. 4.6–9, 98–102, table 4.7, 116, 286–90, figs. 10.6–8, 305, 316–17, 319, 360, 380–83, figs. 12.9–10, 386, 391, 398
    as index of food availability, 286
    as index of parental foraging efficiency, 286
  recording and analyzing, 291–93
  reduction in adult, 297–303, fig. 10.12, 319–20, 327, 451, table 14.1
  and season, 292–98, figs. 10.9–11, table 10.3, 319, 327, 417
  sex differences in, 293–97, figs. 10.9–11, table 10.3, 319, 417
  and social foraging, 275
  sorting among colonies by, 415–18, fig. 13.3, 447, 465, 481
  survival-related advantages of, 317
  and time of day, 292
  variation in, 293–98, fig. 10.11, 385–86
body size, fig. 4.8
  and colony size choice, 24, 465
  and competitive ability, 418
  differences among colonies, 418–19, fig. 13.4, 465
  and flea infestation of nestlings, 95
  as measured by wing chord, 24, 418–19, fig. 13.4
  skeletal, 418
  sorting among colonies by, 409, 418–19, fig. 13.4, 447, 465, 481
body weight. *See* body mass
bolus, food. *See also* food, amount delivered

collecting, 248, 284
  insects contained in, 248–49, table 9.1
  mass of, 284, fig. 10.4
Bouteloua, 42
boxelder, 42, 483
Braconidae, table 9.1, 249
Brazil, 8
breeding dispersal. *See* dispersal
breeding life span, 457–58, 460. *See also* life span
breeding phenology, 38, 62–65, figs. 3.9–10, 67. *See also* date
breeding site, 118–29. *See also* nest site
breeding synchrony, 211, figs. 8.6–9, 239, 242, 327, 363, 412, 451, table 14.1. *See also* synchrony
bridges. *See also* artificial nesting sites
  colony sizes on, 51, 205
  ectoparasites on, 75
  as nesting sites, 50–57, figs. 3.5–6, 67, 204–5, 209, 450
  patterns of use of, 114, 201–2, fig. 7.9, 205
  predation on, 205, 216, 235
  spacing of nests on, 205
  substrate area of, 189, 192–93, 195, 204
Britain, 72, 81
Broadwater, Nebraska, 49–50
broken-stick model, 203
Bromus, 42
brood age, fig. 10.2. *See also* nestling, age of
brood amalgamation. *See* brood mixing
brood loss
  and brood size, 314–15, fig. 10.15, 320
  and colony size, 314–15, fig. 10.15, 320
  from nestling starvation, 313–15, fig. 10.15, 320
  partial, 314–15, fig. 10.15, 320
  scoring, 314
brood mixing, 151, 175–86, figs. 6.12–16
brood parasitism. *See* conspecific brood parasitism
brood patch, 23, 28, 39, 103
brood reduction, 314–15, fig. 10.15, 320. *See also* brood loss, from nestling starvation; brood loss, partial
brood size, 77, 381
  after brood parasitism, 172
  after egg tossing, 145
  definition of, 20
  and food deliveries, 281–83, figs. 10.2–3, table 10.2

and nestling body mass, 94–95, figs. 4.6–7, 99, table 4.7, 286, fig. 10.6, 319
and parental foraging, 94, 96, 281
and partial brood loss, 314–15, fig. 10.15, 320
Brueelia longa, 73. *See also* louse, chewing
Bubo virginianus, 484. *See also* owl, great horned
Bubulcus ibis, 483. *See also* egret, cattle
Buchloe, 42
Buenos Aires, 8
bug. *See also* swallow bug
buildings, 259, 450. *See also* artificial nesting sites
  colony sizes on, 51
  as nesting sites, 50–57, 67, 197, 450
  predation on, 216
bunting, 42
Buteo
  jamaicensis, 483. *See also* hawk, red-tailed
  swainsoni, 483. *See also* hawk, Swainson's

Cacicus cela, 484. *See also* cacique, yellow-rumped
cacique, yellow-rumped, 188, 418, 484
Calamovilfa, 42
California, 8, 175, 183, 236, 304
Calliphoridae, table 9.1, 249
Canada, 10
canyons, 246, 301
capture effort, 291
  for mark-recapture, 362–64
  and number of known-age birds at colony, 409
  and synchrony, 363
  variation among colony sites, 362–64, 461
  and weather, 363
capture history, 362, 364
capuchin, brown, 307, 484
Carabidae, table 9.1, 249
Carleton, J. Henry, 46–49
cat, house, 216, 484
Cathartes aura, 483. *See also* vulture, turkey
cavity size, 331
Cebus apella, 484. *See also* capuchin, brown
cedar, eastern red, 41–42, 483

Cedar Point Biological Station, 16, 40, 48, fig. 3.3, 177, 215, 246, 401
Celtis occidentalis, 483. *See* hackberry
center nests, 35, 39, 84, 119, 345, 357, 473, 477. *See also* nest position in colony
 clutch size of, 329–31, fig. 11.5, 347
 competition for, 329
 date occupied, 329
 ectoparasites in, 88–89, fig. 4.4, 349, 470–71
 fighting at, 123–26, fig. 5.3, 147
 and first-year survivorship, 393, 396, 398
 predation risk at, 211, 233–34, 345, 470–71
 reproductive success in, 347–50, fig. 11.12, 359, 471, 481
 trespassing at, 138
center of foraging area, 307, 310, 320
center of foraging flock, 269, 271–73
Central America, 8, 10
central place foraging, 319
Cerambycidae, table 9.1, 249
Ceratophyllidae, 71
Ceratophyllus, 71–72, 81, 414
 celsus celsus, 71, 116, 483. *See also* flea
 hirundinis, 72
 petrochelidoni, 71
 styx, 72
Cercopidae, table 9.1, 249
Chamaemyiidae, table 9.1, 249
chasing. *See* fighting
Chelydra serpentina, 483. *See also* turtle, snapping
Chen caerulescens, 483. *See also* goose, snow
chicken bug, 70
Chironomidae, table 9.1, 249, 252
chokecherry, 42, 483
Chorispora, 42
Chrysomelidae, table 9.1, 249
Chrysopidae, table 9.1, 249
chur call, 266–67. *See also* food call; vocalizations
Cicadellidae, table 9.1, 249, 251
Cimicidae, 26, 70. *See also* swallow bug
cliff. *See* natural or cliff colony sites
cliff swallow. See *Hirundo;* swallow, cliff
climate, 11, 40, 42–46, 60, 67, 275, 334, 472, 473. *See also* rainfall; temperature (air)
cloacal protuberance, 23, 39, 407

clutch initiation date, 36, 39, 82, 227–29, 326, 327, 424–25, table 13.3
clutch loss, partial, 344
clutch size, 20, 352
 and age of birds, 323, 326, 328, 332–33, 355–56, 359
 and brood parasitism, 168, 172, 324
 of center nests, 329, 331, fig. 11.5, 359
 and colony size, 323–28, fig. 11.3, table 11.1, 358
 and competition for nest sites, 327, 328
 and date, 326–28, fig. 11.4, table 11.1, 340, 358
 definition of, 324
 and dispersal, 424–25, table 13.3
 and distance from colony center, 329–30, table 11.2, 359
 and ectoparasitism, 324, 333
 of edge nests, 329, 331, fig. 11.5
 and egg destruction, 143, 145–47, 172
 and fecundity, 323
 and female body mass, 332–33
 and individual condition, 323
 and individual phenotypic quality, 323, 326
 and nearest neighbor distance, 329–30, table 11.2, 359
 and nest age, 327, 329, fig. 11.5, 359
 and nestling survival, table 4.6, 98–99
 and nest position, 327–31, table 11.2, fig. 11.5, 359
 and nest size, 327, 329–32, fig. 11.6, 359
 and resource availability, 325–26
 and seasonal reduction in reproductive success, 340, fig. 11.4, fig. 11.10
 variation in, 323–33, fig. 11.1
 and year, 324–26, figs. 11.2–3, 358
Coccinellidae, table 9.1, 249
coefficient of variation
 for predictability of site use, 201–2, fig. 7.9
 for variance in ectoparasite load, 90–91, fig. 4.5
 for variance in reproductive success, 353–55, fig. 11.13
Coenagrionidae, table 9.1, 249
cohorts of birds, 364, 367, 369, 397, 456
Coleoptera, table 9.1, 249
coleopterans, 251
coloniality. *See also various aspects of coloniality*
 asymmetries in expected costs and benefits of, 469–74

in bank swallow, 454
in barn swallow, 454
benefits of, 2–6, 11–12, 90, 130, 141, 145, 147–48, 150–51, 164–67, 171–75, 185–88, 191, 210–12, 239–43, 245, 260, 263, 266, 272, 275, 315–17, 321, 323, 358, 360–62, 373, 379, 380, 386, 398, 399, 449–56, table 14.1, 464, fig. 14.2, 467–70, 472, 478, 479, 481
and competition for breeding sites, 118–20
consequences of, 452–54, 469, 479, 481
cost-benefit approach to studying, 449–54, table 14.1
costs of, 2–6, 11–12, 69, 90, 93, 102, 105–6, 112, 116, 118–20, 125–26, 132, 141, 145, 147–48, 150–51, 164–67, 171–75, 185–87, 208, 212, 236, 239, 245, 306, 311, 315, 317, 321, 323, 358, 360–62, 373, 380, 399, 449–56, table 14.1, 464, fig. 14.2, 467–70, 472, 480, 481
definition of, 3–4
dynamics of, 452
evolution of, 1, 3, 5, 7, 12, 167, 316, 319, 321, 339, 399, 449–82
and extrapair copulation, 167, 454–56
and geometrical model, 245, 317–19
fitness consequences of, 7, 317, 456–63, fig. 14.1
historical causes of, 449–56, table 14.1
and lifetime reproductive success, 456–63, fig. 14.1
and local resource availability, 468
and net increase in fitness, 453
phenotypic variation in costs and benefits of, 478
phylogenetic study of, 453, 455
prevalence of among birds, 4
in purple martin, 454
and retort nesting, 453
selective pressures for, 187–89, 208
yearly variation in costs and benefits of, 472
colony
on artificial sites, 9, 12, 16, 47, 49–57, figs. 3.5–6, 204–5
criteria for including in study, 16–17, 26, 49
definition of, 15–16, 38
and geometrical model, 318

as information center, 243–44, 272
initiation date, 15, 65–66, fig. 3.10, 67, 127
on natural sites, 49–50, 56–57, fig. 3.2, fig. 3.4, 124, 204–5
number in study area, 466
and population size, 57–61, fig. 3.7
proximity to mud sources, 129–30, fig. 5.6, 147–48
reasons for, 1, 449, 478
reduction experiment, 304–6
selection behavior, 400–409, fig. 13.1
use patterns, 15, 32, 49, 71, 76, 108, 113–14, 117, 189, 192, 195–202, table 7.1, figs. 7.5–9, 204, 206, 208–9, 435
variation in size of, 1, 449, 478
colony abandonment. *See* colony desertion
colony choice, 11–12, 24, 64–66, 301, 353, 362, 399–448
and age, 400, 409–12, fig. 13.1, 426, 447
assessing sites for, 401, 404, 406, 407, 446, 447
as behavior, 400–409, table 13.1, 446–48
and body mass, 293, 319, 415–18, fig. 13.3, 447, 465
and body size, 24, 409, 418–19, fig. 13.4, 447, 465
and colony size, 405, 414, 425–32, tables 13.4–6, figs. 13.7–8, 435–37, figs. 13.10–11, 445–48, 461–66, fig. 14.2
and colony size variation, 400
constraints on, 399–401, 446, 467–69, 479
difficulties in studying, 402
distance traveled for, 403–5, 408–9, table 13.1, 447–48
and ectoparasitism, 90, 103–4, 110, 112–13, 117, 400, 409, 412–15, fig. 13.2, 417–18, 433–35, 447, 465
heritability of, 460
histories of, 419–37, tables 13.2–6, figs. 13.5–11, 461–62
hypotheses for, 400, 445–47, 463–66, fig. 14.2
and ideal free distribution, 399–400, 446–47, 467, 468
individual patterns in, 4, 7, 12, 24, 399–401

colony choice (*continued*)
  information used to make, 400, 406, 407, 408–9, 424, 445–47, 463
  and mate choice by females, 407
  and nest choice, 407, 447
  number of colonies visited for, 404–5, 447
  and optimal colony size, 399–401, 439, 446, 447–48, 463–66, table 14.2
  and phenotypic differences, 400, 415, 432, 439, 445–48, 463–69, table 14.2, 479
  and postbreeding assessment of sites, 408–9, 447
  and predictability of colony size, 400, 441–45, fig. 13.13, 446, 448
  and predicting position of nest, 400
  and radio telemetry, 401–9, table 13.1, 447
  and relatedness of individuals, 474–78
  and reproductive success, 353, 399–400, 424, 446, 447
  sex differences in, 405–7, 433
  and sleeping patterns, 402, 404–7
  social constraints on, 479
  sorting of birds, 409–19, figs. 13.1–4, 447, 465
  spatial constraints on, 479
  and switching between colonies, 400, 420–25, tables 13.2–3, figs. 13.5–6, 437–41, fig. 13.12, 445, 448, 460
  theoretical framework for, 399–400, 447
  time available for, 408, 479
  by yearlings, 432–37, figs. 13.9–11, 447–48
  between years, 419, 425–32, tables 13.4–6, figs. 13.7–11, 432–37, 445, 447–48, 461–62
colony density, 34, 69, 79, 86. *See also* density; nest density
  and alternative reproductive tactics, 150, 166
  and competition for nest sites, 119
  and entombment inside nests, 130–32, fig. 5.7, 148
  and flea movement, 110
  and kleptoparasitism, 180
  and nest building time, 128–29, 148
  and nest spacing, 189–91, figs. 7.1–2
  and shortage of breeding habitat, 187
  and swallow bug infestation, 73–77, fig. 4.2, 84, 116
colony desertion, 14, 107–8, 113, 117
colony geometry, 50, 56, fig. 3.4, fig. 3.6, 124, 131, 211, 329, 330, table 11.2, 357, 471
colony initiation date, 15, 65–67, fig. 3.10, 79
colony mapping, 34–35, 39, 190, 202
colony reduction experiment, 14, 32, 304–6, 320
colony selection. *See* colony choice
colony settlement pattern. *See also* colony choice
  and age of birds, 326, 376, 379–80, 409–12, fig. 13.1, 447
  and location of nest in colony, 477–78, fig. 14.3
  and neighbors, 477–78, fig. 14.3
  and relatedness, 474–75
  and reproductive success, 323
  by yearlings, 432–37, figs. 13.9–11, 447–48
colony site
  assessing, 185, 358, 401–9, 456
  determining active status of, 15, 38, 49, 57, 112, 120
  fidelity to, 420–25, tables 13.2–3, figs. 13.5–6, 433, 439, fig. 13.9
  food availability around, 205–6, 209, 304–6, 446–47, 468
  information transfer at, 245
  initiation dates at, 65–67, fig. 3.10
  as measure of dispersal, 420–25, tables 13.2–3, figs. 13.5–6, 433, 439, fig. 13.9, fig. 13.12
  naming, 37
  pattern of visiting, 15, 38, 49
  population trends at, 57–61, fig. 3.7
  selection of, 400–409, table 13.1 (*see also* colony choice)
  structures supporting, 8–9, 15–16, 47–57, figs. 3.4–6, 67
  substrate size of, 192–96
  use of, 15, 32, 49, 71, 76, 108, 113–14, 117, 189, 192, 195–202, table 7.1, figs. 7.5–9, 206, 208–9, 435
  variation in age structure of, 409–12, fig. 13.1
colony size
  in analyses, 14–15
  on artificial sites, 50–51, 205
  average, 338
  best, 399–400, 469, 479

choice of, 414, 425–32, tables 13.4–6,
  figs. 13.7–8, 435–37, figs. 13.10–11,
  441, 445, 447–48, 460–61
classes, 37, 198, 375, 426–27, table
  13.4, 441, 461, 466
constraints choice of, 399–400, 467–69
cyclical pattern in, 198–200
daily changes in, 11, 14, 442–45,
  fig. 13.13
definition of, 13, 38
determining, 13–15, table 2.1, 49, 442
environmental correlates of, 7
heritability of, 460
histories, 427–32, tables 13.4–6, figs.
  13.7–8, 435–37, figs. 13.10–11,
  441, 461–62
maximum nest packing, 191
on natural or cliff sites, 50, 205
optimal, 7, 299–300, 317, 375, 399–
  401, 408, 418, 432, 439, 446–48,
  456, 464–66
percentage of birds occupying each, 11,
  466
predictability of, 441–45, fig. 13.13,
  448, 467
predicting, 192, 209, 400, 448, 456
reduction of, 300–306, table 10.4, 320
reflecting population size, 57–61, fig. 3.7
seasonal changes in, 14
variation in, 1–7, 9, 12, 66–67, fig. 3.11,
  198–202, fig. 7.9, 317, 399–400,
  425–26, 449, 463–69, fig. 14.2, 479,
  481
colony stage, 22, 77, 79, 102–5, 112, 120–
  21. *See also* nest stage
colony switching, 400, 420–25, tables
  13.2–3, figs. 13.5–6, 437–41,
  fig. 13.12, 445, 448, 460, 467
colony synchrony. *See* breeding synchrony;
  synchrony
colony use patterns, 15, 32, 49, 71, 76, 108,
  113–14, 117, 189, 192, 195–202, 204,
  table 7.1, figs. 7.5–9, 206, 208–9, 435
Colorado, 40, 97–98
color marking, 22, 24–26, fig. 2.4, 39, 362,
  469
Colydiidae, table 9.1, 249
communal breeding. *See* cooperative
  breeding
communal roosts, 471
competition, 92, 118–48, 327, 328, 329,
  345, 418, 465, 466

and aggression, 118–19
behavioral manifestations of, 118–20
and body size, 418
as cost of coloniality, 5, 11, 118, 120,
  125–26
and fighting, 119–32, figs. 5.1–3, 147
for food, 5, 120, 306–7, 313, 315, 465
for limited resources, 2–3, 5, 11, 118–19
for mates, 5, 120, 465
for nest material, 5, 119, 133
for nest sites, 5, 11, 90, 118–48, 327–
  29, 345, 465
and reproductive interference, 118–20
and trespassing, 132–47, figs. 5.8–12,
  148
conspecific aggregation, 412
conspecific brood parasitism, 20, 26, 57,
  113, 119, 133–36, 167–75, 358, 395,
  table 14.1, 454, 469, 471, 474–75, 479
and age, 5–6
assessing host nest, 168, 174
as benefit of coloniality, 5–6, 11, 150–
  51, 167, 171–75, 185, 451, table
  14.1
benefits of, 171–75, 185–86
and clutch size, 168, 172, 324
and colony size, 5, 6, 141, 150–51, 167,
  169–71, figs. 6.9–10, 174–75, 185–
  86, 236, 462
as cost of coloniality, 5–6, 11, 150–51,
  167, 171–75, 185, 451, table 14.1
costs of, 171–75, 185–86
criteria for inferring from nest checking,
  169
and ectoparasitism, 113, 168
egg acceptance, 168
egg appearance, 149
egg destruction or egg tossing, 146–48,
  167–70, 172
by egg laying, 135, 149, 167–68, 356,
  table 11.9, 425, table 13.3
by egg transfer, 135, 167–69, 356, table
  11.9, 425, table 13.3
and egg white protein electrophoresis,
  175
versus extrapair copulation, 164–65
frequency of, 171–72, 174–75, 186
identity of females in, 167, 169, 174,
  186, 480
location of nests parasitized, 167, 170–
  71, fig. 6.10, 186
male involvement in, 146–48, 167

conspecific brood parasitism (*continued*)
  methods of detecting, 18, 164–65, 169
  as misdirected parental care, 149–51, 164–65, 167, 172, 185
  and multiple parasitism, 172
  natural history of, 167–68
  by nestling transfer, 168
  opportunities for, 5–6, 11
  phenotypes of individuals in, 174, 480
  as quasi-parasitism, 167–68
  recruitment of birds for, 236
  and reproductive success, 173–75, fig. 6.11, 186, 462
  as reproductive tactic, 5–6, 12, 150, 164–65, 167, 174–75
  resulting from nest trespassing, 133–36, 146–48, 167
  and risk spreading, 172–74
  spatial and temporal aspects of, 170, 186
  and sphere of choice, 170–71, fig. 6.10
  and synchrony, 168, 170
  timing of, 168
  variation in, 174–75
Conthasidae, table 9.1, 249
convection currents and thermals, 249, 251, 259–60, 273
cooperative breeding, 2, 4, 361, 465
Copromorphidae, table 9.1, 249
copulation. *See also* extrapair copulation
  by members of pair, 161–62
  resulting in quasi-parasitism, 167
  used in determining sex, 22–23
Cordillacris occipitalis, 248
cost-benefit approach, 2–5, 449–54, table 14.1, 470, 479, 481
costs of coloniality. *See* coloniality
costs of foraging, 306–11, figs. 10.13–14, 319
cottonwood, eastern, 42, 483
crèche or crèching, 175–79, figs. 6.12–13, 185–86, 236, 451, table 14.1, 480, 483
Crotophaga sulcirostris, 484. *See also* ani, groove-billed
Cuba, 8
cuckoldry, 149–51, 157, 161–62, 164–67, 171, 185, 454, 463. *See also* extrapair copulation; misdirected parental care
Culicidae, table 9.1, 249
culverts, 442, 450, 477. *See also* artificial nesting sites
  barn and cliff swallows nesting in, 202–4
  as colony sites, 50–57, figs. 3.5–6, 67, 152, 155, 209, 450

colony sizes of, 51, 205
ectoparasites in, 75, 107–8, 112–14
mice as nest site competitors in, 217–18
mud gathering at, 152
patterns of use of, 114, 201
predation at, 205, 216, 235
predictability of size and use of, 201–2, fig. 7.9
spacing of nests on, 205
substrate area of, 189, 192–93
Curculionidae, table 9.1, 249
Cynomys ludovicianus, 484. *See also* prairie dog
Cyrtophora citricola, 2
Czechoslovakia, 72

Dalea, 42
dam, 9
Darwinian fitness, 456, 469, 474. *See also* fitness
date, 20, 62–66, 77–79, 81, 96, 111, 326–27, 339–45
  and assigning sex, 23
  and clutch size, 326–27, fig. 11.4, 340
  of colony initiation, 15, 65–66, fig. 3.10, 67, 127
  and colony synchrony, 35–36, 39
  and determining colony size, 15
  and food availability, 339
  and food call, 267
  of hatching, 19
  and individuals' arrival, 62–65, 67
  and laying in incomplete nests, 343–45, fig. 11.11
  and nest ectoparasite loads, 77–79, table 4.1, 81–82, 90, 111, 339–40
  and reproductive success, 96, 339–45, figs. 11.10–11, tables 11.3–4
  and survivorship, 391–92, fig. 12.15, 398
death, 125, 132, 147, 148, 300–304, table 10.4, 320, 362. *See also* mortality, causes of
deciduous woodlands, 42
deformities, of nestlings, 20
Delichon, 10, 70
Delichon urbica, 484. *See also* martin, house
Delphacidae, table 9.1, 249
demography, 6–7, 11–12, 67, 105–6, 361, 368–73, figs. 12.1–3, 379, 399, 476, 480. *See also* survivorship
density, 34, 56, 69, 73–77, 79, 84, 86,

fig. 4.2, 110–12, 116, 118–19, 128–32, fig. 5.7, 148, 150, 166, 187, 189–90, fig. 7.1, 322. *See also* colony density; nest clustering; nest density
density dependency, 92
departure frequencies, 276
despotic distribution, 469, 481
Dibrom, 32. *See also* fumigation
Dictyopharidae, table 9.1, 249
diet, 42, 46, table 9.1, 247–51. *See also* food
Diptera, table 9.1, 249
disease, 2–3, 116
dispersal
　age-specific, 474
　breeding, 380, 419, 433
　causes and correlates of, 424–25, table 13.3, 433–35
　and colony size, 379–80, 425–26, 429, 447
　differential, 380
　distances, 362, 390, 420–25, tables 13.2–3, figs. 13.5–6, 433–34, fig. 13.9, 447–48
　of ectoparasites, 68–69, 79, 89, 92
　and ectoparasitism, 388–90, 433–35, 459, 475, 480
　and egg laying date, 424–25, table 13.3
　of fleas, 71–72, 82, 84, 86, 106, 109–10, 116
　of juveniles, 361, 368, 432–37
　and long-distance movement, 362, 380, 389–90, 433, 447–48
　measured by site distance, 421
　natal, 114, 388–90, 400, 419, 432–37, figs. 13.9–11, 447–48, 474–75, 480
　and problems with emigration, 368
　and reproductive success, 424–25, table 13.3
　within season, 437–41, fig. 13.12, 447–48
　sex differences in, 421–23, figs. 13.5–6, 433
　short-range, 388–90
　of siblings, 474–75
　and site fidelity, 420–25, tables 13.2–3, figs. 13.5–6, 433, 447–48, 474–75
　and survival probability, 458–60
　of swallow bugs, 70–71, 82, 84, 86, 103, 106–12, fig. 4.11, 116–17
　of yearlings (*see* natal dispersal)
distance from center of colony, 350, 396
　and clutch size, 329–30, table 11.2
　and ectoparasitism, 84–89, table 4.3, 470
　as measure of nest position, 345, 393, 470–71
　and nearest neighbor distance, 470
　and reproductive success, 345–47, tables 11.5–6, 359, 471
distance traveled to forage, 306–11, fig. 10.13, 315, 318, 320, 321, 451, table 14.1
distribution of cliff swallows, 8, 12, 40, 47–49, figs. 3.1–3
disturbance, 17, 20–21, 25, 30, 32, 57, 107–8
DNA fingerprinting, 165–66, 172, 174, 185, 463
Dolichopodidae, table 9.1, 249
drop netting, 20, fig. 2.3. *See also* mist netting
Drosophilidae, table 9.1, 249
Dytiscidae, table 9.1, 249

early season, 22, 102, 113, 120–21, 245–46, 267, 291, 333
economic defensibility, 2
ectoparasite sampling jars, 22, fig. 2.8, 28–30, 39, 73, 102–5, 413
ectoparasitism (ectoparasites), 26–34, 39, 57, 66, 68–117, 185, 190, 207, 286, 288–90, 293, 297, 300, 314, 316, 317, 328, 331, 338, 342, 345, 350–51, 359, 373, 385–95, 408, 441, 459, 466, 480
　of adults, 102–5, fig. 4.10, 106–10, 293, 386
　and adult survivorship, 105–6, 369, 372, fig. 12.4, 376–78, figs. 12.6–7, 386
　on artificial versus natural sites, 57
　assessment of, 112–13, 117, 133, 135, 148, 168, 455, 473–74
　behavioral response to, 112–17, fig. 4.12
　and brood size, 94–95, figs. 4.6–7, 99
　and clutch size, 324
　and colony choice, 103–5, 110, 113–14, 117, 400–408, 412–15, fig. 13.2, 465
　and colony density, 73–77, fig. 4.2, 84, 86, 110–12, 116
　and colony or nest desertion, 14, 107–8, 113, 117
　and colony site usage, 196, 201–2, 205
　and colony size, 68–69, 73–77, fig. 4.1, 79–93, fig. 4.3, tables 4.2–7, fig. 4.5, 96–117, figs. 4.9–10, 338, 412–13, 465

ectoparasitism (ectoparasites) (*continued*)
  as cost of coloniality, 2–3, 5, 11, 69, 93, 101–2, fig. 4.8, 105–6, 112, 116, 451, table 14.1, 464
  and date, 77–79, table 4.1, 81–82, 103–5, 110–11, 334, 472
  and dispersal, 114, 388–90, 433–35, 459, 475
  dispersal of, 68–69, 71–73, 79, 82, 84, 86, 89, 92, 103, 106–12, fig. 4.11, 116
  distribution of among hosts, 90–93, 110, 117
  feeding habits of, 70–73, 93, 113, 116–17
  and first-year survivorship, 369, fig. 12.1, 386–90, figs. 12.13–14, 397–98
  and host behavior, 68–69, 103–5, 112–17, fig. 4.12
  and host population density, 68–69
  and host susceptibility, 90–91, 110, 455, 464
  immunological response to, 455
  of juveniles, 22, 102
  measurement of
    on adults, 22, 26, fig. 2.8, 28–30, 39, 73, 102–5, fig. 4.11, 106–12
    on juveniles, 22, 102
    on nestlings, 19–20, 26–28, figs. 2.6–7, 30, 39, 73–74, fig. 4.1, tables 4.1–5, 77–97, fig. 4.3–7, 110–12, 386
    in nests, 26–27, fig. 2.5, fig. 2.7, 30–31, fig. 2.9, 33, 39, 74–77, fig. 4.2, 107
  movement of, 92–93, 106–12, fig. 4.11, 116–17, 190
  natural history of, 69–73, 79, 82, 116
  and nearest neighbor distance, 84–89, table 4.4, 190
  and nest age, 79–84, figs. 4.3–4, 88–89, 116–17, 473
  and nest building, 115–17, fig. 4.12, 130
  and nestling body mass, 93–96, figs. 4.6–9, table 4.7, 98–102, 116, 286–90, figs. 10.6–8, 380–82, fig. 12.9
  and nestling growth, 93–95, figs. 4.6–7, 102–11
  and nestling health, 93, 102, fig. 4.8
  and nestling survivorship, 93, 96–97, tables 4.5–6, 98–102, fig. 4.9, 116, 369, fig. 12.1, 374–75, fig. 12.5, 380–82, fig. 12.9, 386–90, figs. 12.13–14, 398
  and nest position in colony, 79, 84–89, tables 4.3–4, fig. 4.4, 116, 470–71
  and nest reuse, 115–16, fig. 4.12, 117, 195
  and nest size, 79, 82–84, table 4.2, 116
  overdispersion of, 69, 92
  and parasite life cycle, 69, 79
  physiological effect on host, 93
  and reproductive success, 230, fig. 8.9, 334–36, fig. 11.7, 340, 353, 374–75, 455, 472
  sampling, 26–34, figs. 2.5–11, 39, 75–76, 93, 102, 107
  and sex of host, 103–5, fig. 4.10, 116
  and shortage of nesting habitat, 195, 196, 202, 205
  sorting among colonies by, 409, 412–15, fig. 13.2, 417, 418, 447, 465, 481
  statistical analysis of, 20, 77–78, 90
  and substrate type, 75, 114
  transmission of, 5, 68–69, 73, 79, 103, 106–12, fig. 4.11, 116–17, 408, 479
  travel on birds' feet, 102–3, 109–10, fig. 4.11, 116–17
  variance in, 90–93, fig. 4.5, 102, 110, 116
edge nests, 35, 39, 84, 119, 357, 473, 477. *See also* nest position in colony
  age of birds in, 358, 359
  clutch size of, 329–31, fig. 11.5, 347
  competition for, 329
  date occupied, 329
  ectoparasites in, 88–89, fig. 4.4, 349, 470–71
  fighting at, 123–26, fig. 5.3, 147
  predation risk at, 211, 233–34, 239, 242, 345, 349, 396, 470–71
  reproductive success in, 347–50, fig. 11.12, 471, 481–82
  and survivorship, 393, 395–96, 398
  trespassing at, 138
edge of flock, 222–23, fig. 8.4, 269, 271–73
Edmonton, Alberta, 368
egg, 216–17, 303
egg appearance, 149
egg cannibalism, 146
egg destruction, 5, 20, 119, 352, 356, table 11.9, 425, table 13.3
  and alarm calls, 142, 144

and brood parasitism, 146–48, 167–68, 172
and clutch size, 143, 172
and colony size, 142, 144–48, fig. 5.12
and completeness of nest, 343–44
as cost of coloniality, 451, table 14.1
costs and benefits of, 145–48
effect on perpetrators, 145–48
and egg transfer, 168, 172
and extrapair copulation, 146–48
hypotheses for, 146–48, 480
individuals involved in, 142
inferring, 143–44, 146, 169
location of nests in, 142, 148
method of, 142
natural history of, 142–44
and nest trespassing, 133, 135–36, 141–48, fig. 5.12
by other species, 212, 217–18
sex of individuals in, 142, 146–48
timing of, 143, 148
egg laying, 18, 26, 113, 126, 130, 141, 143, 155, 162, 170, fig. 8.9, 273, 291, 294, 305, 323–24, 333, fig. 11.10, tables 11.3–4, 351, 355, 359, 393
and brood parasitism, 135, 149, 167–68, 171, 172
and brood patch, 23
and colony size, 227–29, figs. 8.6–8
and completeness of nest, 18, 342–45, fig. 11.11, 472
date of, 20, 35, 39, 326, 356, 358, table 11.9
and dispersal, 424–25, table 13.3
pattern of, 227–29, figs. 8.6–9
prolonging by destroying eggs, 146–47
stage (early), 22, 291
synchrony, 36, 39, 227–29, figs. 8.6–9 (*see also* laying synchrony)
and trespassing, 133–37
egg loss. *See* egg destruction
egg tossing, 20, 133, 135–36, 141–44, 146–48, 168, 169, 343–44, 356, table 11.9, 358, 425, table 13.3. *See also* egg destruction
egg transfer, 135, 146, 148, 167–69, 356, 358, table 11.9. *See also* conspecific brood parasitsm
egg white protein electrophoresis, 175
egret, cattle, 119, 483
eider, common, 236, 483
El Salvador, 8

Elaeagnus commutata, 483. *See also* olive, Russian
Elaphe obsoleta, 483. *See also* snake, rat
electrophoresis, 164, 175
elm, American, 42, 483
emigration, 368, 421, 475
Empididae, table 9.1, 249
encephalitis, western equine, 97
Enicocephalidae, table 9.1, 249
entombment inside nest, 130–32, fig. 5.7, 148, 451–52, table 14.1
Ephemeridae, table 9.1, 249
Ephemeroptera, table 9.1, 249
epideictic display, 445
erythrocytes, 93
Euphorbia, 42
Europe, 10
evolution of coloniality, 1, 3, 5, 7, 12, 167, 316, 319, 321, 339, 399, 449–82. *See also* coloniality
experiment, 480
colony reduction, 14, 32, 304–6, 320
food call playback, 266–67
fumigation, 98–102, 112–13, tables 4.6–9
model snake predator, 218–20, figs. 8.1–2, 241
predator mobbing, 226
wing fluttering, 162–64, fig. 6.1
extrapair copulation, 149–67, 183, 357, 449, 479, 480
age of birds in, 5–6
as alternative reproductive tactic, 5–6, 12, 150, 164–67
in barn swallows, 454, 455
behavior of males in, 154–57
as benefit of coloniality, 5–6, 11, 150–51, 164–67, 185, 451, table 14.1
benefits of, 164–67, 454, 480
versus brood parasitism, 164–66, 171–72
and coloniality, 5–6, 12, 150–51, 164–67, 449, 454–56
and colony size, 150, 152, 157–61, figs. 6.5–7, 164–67, 185, 454, 462–63
as cost of coloniality, 5–6, 11, 150–51, 164–67, 185, 451, table 14.1
costs of, 164–67
and cuckoldry, 149–51, 157, 161–62, 164–67, 171–72, 185, 455, 462–63
defenses against, 161–64, fig. 6.8, 185
and egg destruction, 146–48

extrapair copulation (*continued*)
  evolution of, 166–67, 186, 453
  females' expectation of, 157, 161, fig. 6.7, 166–67, 455
  females' pursuit of, 449, 454–55
  fertilization from, 149–50, 164–67, 455, 462–63
  frequency of, 154–55, fig. 6.4
  at grass collecting sites, 151, 153, 155–57, 162
  and intrapair copulation, 162
  males' expectation of, 157, 161, fig. 6.7, 165–67
  and mate guarding, 161–62, 185
  at mud gathering sites, 151–67, figs. 6.1–8, 185
  natural history of, 151–53, figs. 6.1–2
  and nest guarding, 162
  opportunities for, 5–6, 11, 150, 151–53, 157–61
  and parental assistance, 166
  perpetrators of, 153–57, figs. 6.3–4
  in purple martins, 6, 454
  and quasi-parasitism, 167–68
  recruitment of males to colonies for, 159–61
  residency of birds in, 154–57, figs. 6.3–4, 166
  role of females in, 149, 157, 161, fig. 6.7, 166–67, 454–55
  seasonal pattern of, 166
  and sex ratio, 62, 159–61, fig. 6.7, 186
  studied with DNA methods, 164–66, 171–72, 455
  and trespassing, 133–37, 151–52
  and wing fluttering, 162–64, fig. 6.8, 185
extrapair fertilization. *See* cuckoldry; extrapair copulation

Falco
  mexicanus, 484 (*see also* falcon, prairie)
  sparverius, 483 (*see also* kestrel, American)
falcon, prairie, 216, 484
feather growth, 93, 102, fig. 4.8, 111
feathers, 218. *See also* plumage
fecundity, 323
feeding. *See* foraging
Felis catus, 484. *See also* cat, house
fertile period of females, 137, 146–48
fertilization, 149–50, 164–67
fidelity. *See* site fidelity
fieldfare, 361, 484

fighting, 118–32, 164, 172, 221, 396, 451, table 14.1
  as benefit of coloniality, 451, table 14.1
  and colony size, 119, 121–23, 125–29, fig. 5.2
  and completeness of nest, 120–21, fig. 5.1, 123–28, fig. 5.3, 147
  as cost of coloniality, 125–26, 147
  and extrapair copulation, 153, 163–64
  and mud availability, 129–30
  natural history of, 120–21
  and nest building, 126–32
  and nest density, 119
  and nest position, 123–26, fig. 5.3, 131, 147
  and resource competition, 118–19
  seasonal pattern of, 120–21, fig. 5.1
  versus trespassing, 133
first-year survivorship, 361, 369, fig. 12.1, 371–76, figs. 12.4–6, 380–82, fig. 12.9, 386–98, figs. 12.13–16, 458–60, 480
fish, 217
fitness, 360, 469
  and benefits of coloniality, 451
  and colonial nesting, 333
  and colony size, 460–63, 467–68
  and costs and benefits of coloniality, 451–53, 456
  curve varies with group size, 467–68
  Darwinian, 456, 469, 474
  decreased by costs of coloniality, 451
  in different-sized colonies, 317
  inclusive, 2, 474
  individual, 456
  at intermediate colony sizes, 467
  and lifetime reproductive success, 322–23, 456, 458, 460, 479
  Malthusian parameter as measure of, 456
  and nest age, 473
  and nest spacing, 471–72
  and phenotype, 464
flea, 28, 30–31, 33–34, fig. 2.9, 39, 69, 71–89, figs. 4.1–5, tables 4.1–5, 90–97, fig. 4.7, 98, 102–6, fig. 4.10, 109–17, 300, 317, 358, 386, 388–90, fig. 12.14, 398, 413–15, fig. 13.2, 424–25, table 13.3, 451, table 14.1, 483. *See also* ectoparasitism
fledging, 19, 22, 26, 32, 74, 93, 102, 114, 334, 360, 369, 375, 382, 383, 451, table 14.1, 458, 460
  behavior after, 175–86, figs. 6.12–16

and crèching, 175–79, figs. 6.12–13, 186
and kleptoparasitism, 179–86, figs. 6.14–16
and misdirected parental care, 175–86, figs. 6.12–16
returning to colonies after, 175
success (*see* reproductive success)
survivorship after, 96, 373, 386–95, figs. 12.13–16, 398
synchrony of, 175
fledgling. *See* juvenile
flicker, 42
flight, 299
flocking, 8, 210, 212, 221–24, figs. 8.3–5, 239, 241, 269, 271–72. *See also* predator avoidance; predator detection; vigilance
following foragers, 244, 260–64, fig. 9.4, tables 9.2–3, 273, 276–80, table 10.1, fig. 10.1, 451, table 14.1, 452. *See also* foraging; information center; information transfer
food, 245, 247–51, table 9.1, 325–26, 334, 342, 345, 373, 391, 399, 450, 463. *See also* foraging
  amount delivered, 283–86, figs. 10.4–5, 319, 451, table 14.1
    and brood size, 284
    and colony size, 284–85, figs. 10.4–5, 319
    and differences in distribution, 285–86
  annual differences in, 275, 316, 472
  availability of, 205–6, 209, 304–6, 446–47, 468, 480
  of barn swallow, 250
  bolus, 248, 250, 284
  and breeding synchrony, 233
  calls for discovery of, 266–68, 273, 301, 312, 480, 482
  carrying of, fig. 9.4, 260–61
  competition for, 5, 120, 306–7, 313, 314, 465
  and convection currents, 249, 259–60
  delivery rates, 275, 280–85, figs. 10.2–3, table 10.2, fig. 10.5, 300, 309, 315, 319
  depletion of, 265–66, 306, 311–15
  distance traveled to, 275, 306–11, fig. 10.13, 315
  local differences in, 275, 285–86, 304–6, 446, 450, 468, 480
  mass emergence of, 259, 273
  measuring availability of, 206, 252–53, 260, 304
  patchiness of, 244, 247, 249
  rates of capture of, 245, 268–72, table 9.4, fig. 9.5, 311–13, table 10.5
  sampling of, 248, 250–53, fig. 9.1, 260, 273, 284, 304, 468
  shown by foraging group, 251–53, fig. 9.1
  spatiotemporal variability of, 245, 247–48, 251–60, figs. 9.1–3, 273, 285–86, 307, 311, 318–19, 452
  stability of, 257
  swarming of, 249–51, table 9.1, 273
  tracking of, 267–68, 273
  and weather, 246, 267, 269, 272–73, 300–304, table 10.3, 314, 373
food call, 246, 266–68, 273, 301, 312, 452, 476, 480, 482. *See also* information transfer
food delivery
  amount of, 283–86, figs. 10.4–5, 319
  as benefit of coloniality, 451, table 14.1
  and brood size, 281–83, figs. 10.2–3, table 10.2, fig. 10.5
  and colony size, 281–86, table 10.2, figs. 10.2–5, 309, 319
  counting, 280
  and distance traveled, 309–10
  frequency of, 309
  and nestling age, 281–83, fig. 10.2, table 10.2
  rate of, 275, 280–86, figs. 10.2–3, table 10.2, fig. 10.5, 300, 315
foraging, 45, 46, 57, 136, 140, 143, 190, 191, 205–6, 212, 221, 243–320, 321, 347, 362, 401, 404, 408, 450, 465, 466, 476, 477. *See also* food; information center; information transfer; social foraging
  annual differences in, 296–97
  associates as source of information, 304
  during bad weather, 267, 273, 300–304, table 10.4
  behavior, 245–47, 254, 268, 315
  as benefit of coloniality, 2–3, 5, 187, 191, 212, 243–45, 260, 266, 275, 315–17, 319, 386, 450–52, table 14.1, 481
  benefits of, 315–20, 452
  at center of group, 268, 271–73, 319, 320
  by circling above colony, 278–79

544 • Index

foraging (continued)
  and colony size, 244–45, 255, 258, 264–65, 273, 275–77, table 10.1, 285, 304–6, 310, 312, 315, 316, 317, 319
  by convergence on insect source, 246, 267
  as cost of coloniality, 245, 306–15, figs. 10.13–15, table 10.5, 319, 451, table 14.1
  costs of, 300, 306–15, figs. 10.13–15, table 10.5, 319
  cues used by unsuccessful birds, 261–63, fig. 9.4, 273
  in defining colony, 16
  departure frequencies, 276–80, table 10.1, fig. 10.1
  disguising success, 265–66, 273
  distance traveled, 275, 306–11, fig. 10.13, 315, 318, 320
  early in season, 245–46, 267
  on edge of group, 269, 271–73
  efficiency (see foraging efficiency)
  and fat reserves, 291
  and fitness, 316
  following insect concentrations, 247
  following neighbors, 262–63, table 9.3, 273
  following successful bird, 244, 260–64, tables 9.2–3, 273, 276–77, 452
  food calls, 260, 266–68, 273, 301, 312, 452, 480
  group, 246, 247, 268–74, table 9.4, fig. 9.5, 307, 312–13, table 10.5, 452, 481
  habitat, 275, 310, 318, 468
  during incubation and nestling feeding, 246–47, 260, 265, 273, 352
  indexes of, 316, 319
  information available for, 276, 277, table 10.1, 319
  information transfer, 243–45, 247, 251, 257, 260–66, fig. 9.4, tables 9.2–3, 268, 270, 272–75, 285, 291, 296, 300, 304–6, 310, 315–17, 319, 333, 452, 453, 463, 471, 477, 480
  joining streams of departing birds, 264–65, 273, 276, 278
  kleptoparasitism, 185
  late in season, 245–46
  local enhancement, 244, 246, 268, 312
  locations around colony, 206, 247, 254–58, fig. 9.2, 275
  measured by food delivered, 283–84
  methods of
    by barn swallows, 250
    by cliff swallows, 245–47, 268
  monitoring neighbors, 262–63, table 9.3, 273
  natural history of, 245–47
  around natural or cliff colony sites, 206, 256–58, figs. 9.2–3, 275, 308–11, figs. 10.13–14
  and nestling growth and body mass, 94, 96, 286, 289–90, 319
  network, 245–46, 266
  of parents near crèche, 175
  and position in group, 268–69, 271–72, 274
  and predator avoidance, 268–74, table 9.4, fig. 9.5
  and presence of insects, 251–53, fig. 9.1
  prey capture rates, 245, 268–74, table 9.4, fig. 9.5, 311–13, table 10.5
  with purposeful flight, 260, 265
  and reproductive success, 306, 316
  and size of group, 268, 312–3, table 10.5, 320
  solitary, 268–71, table 9.4, 273–74
  success of, 263–64, 268, 312–13, table 10.5, 320, 386
  tracking insect movement, 266–67, 273
  variability in, 245, 247–48, 251–60, figs. 9.1–3, 273, 275, 285–86, 307, 318, 452
  variance in success of, 268, 273–74
  waiting at nest, 276–80, table 10.1, fig. 10.1, 315
  waiting intervals between trips, 276–80, table 10.1, fig. 10.1
  over water, 246, 273, 301
  and weather, 246–47, 267, 269, 273, 300–304, table 10.4, 314, 320
foraging area, 244–45, 255, 305–11, 319, fig. 10.14
foraging efficiency, 2–3, 11, 243–44, 267–68, 275, 283–86, 289–91, 294, 297–300, 304–6, 315–17, 333, 338, 345, 359, 373, 379, 381, 453, 471
  and adult body mass, 290–91, 294, 297–300, fig. 10.12, 305, 319–20
  annual differences in, 296–98, fig. 10.11
  and annual survival, 373
  as benefit of coloniality, 244, 453
  during cold weather, 300–304
  and colony reduction experiment, 304–6, 320

and colony size, 245, 258, 275, 283–85, fig. 10.5, 289, 296–300, fig. 10.12, 304–6, 319–20
and food call, 267–68, 301
as impetus for coloniality, 453
and information transfer, 275, 285, 304–6, 319–20, 452
measured by food delivered, 283–84, 319
nestling body mass as index of, 286, 289, 319
foraging groups, 246, 273–74, 307, 312–13, table 10.5
Formicidae, table 9.1, 249
Fort Leavenworth, Kansas, 47
Fort Morgan, Colorado, 97
Fort Morgan virus, 97–98
Fratercula arctica, 484. *See also* puffin, Atlantic
Fraximus pennsylvanica, 483. *See also* ash, green
Fulgoridae, table 9.1, 249
fumigation (of birds), 22, fig. 2.8, 28–30, 39, 73, 102–5, 413. *See also* ectoparasite sampling jars
fumigation (of nests), 92, 93, 96, 230–34, 240, fig. 8.12, 317, 368, 369. *See also* ectoparasitism
and clutch size, 324, 331
and colony choice, 112–13
and egg laying dates, 326
experiment, 98–102, tables 4.6–7, figs. 4.8–9, 112–13
and first-year survivorship, 368–75, fig. 12.1, figs. 12.4–5, 380–84, fig. 12.9, 386–95, figs. 12.13–16
for predation analysis, fig. 8.9, 230–34, fig. 8.12
methods for, 31–35, figs. 2.10–11, 39, 98

Garden County, Nebraska, 48–49, fig. 3.3, 73
Gasterosteus aculeatus, 483. *See also* stickleback, three-spined
genetic relatedness. *See* relatedness
geometrical model, 245, 317–19, 320
Germany, 304
goose, snow, 368, 477, 483
grackle, common, 212, 214–15, 224, 235–39, fig. 8.11, 241, 484
grass collecting, 151–53, 155–57, 162, 214–15. *See also* nest materials

grasshopper, 248–50
grass stealing, 148
Great Plains, 8, 43
group foraging
and capture success, 268, 273–74, 312–13, table 10.5, 320
as cause of coloniality, 452, 481
by circling above colony, 278–79
and information transfer, 268, 270
operational definition of, 269
and predator avoidance, 268, 270, 273–74
versus solitary foraging, 268–71, table 9.4, 273–74
and spatial position, 268–74, table 9.4, fig. 9.5
waiting intervals for, 276–80, table 10.1, fig. 10.1
group living, 118, 141, 151, 187–89, 211, 243, 373, 375, 399, 449. *See also* coloniality
group size. *See also* colony size
and alertness, 221–24, figs. 8.3–5, 239, 241
and annual survival, 361
and average life span, 361
of crèches, 175–79, fig. 6.13
effects of, 450–52, 463
and extrapair copulation, 157–61, figs. 6.5–6, 185
and fitness curve, 467–68
and foraging observations, 269, 273–74
and foraging success, 312–13, table 10.5, 320
and grass collecting, 153
and intraspecific competition, 118
and mobbing, 210, 225, 241
and mud gathering, 128, 140, 152, 157–61
nonoptimal, 439, 444, 467
optimal, 322, 399, 432, 439, 467–68
and parent-offspring recognition, 151
and predator avoidance, 210, 241
and predator detection, 218, 239, 241
and relatedness, 474
relation to colony size, 157–58
and vigilance, 210, 220–24, figs. 8.3–5, 239, 241
and wing fluttering, 162–64, fig. 6.8
growth of nestlings, 93–96, figs. 4.6–8, 102, 286
Gryllidae, table 9.1, 249
guillemot, common, 119, 484

gull, herring, 236, 484
gulls, 146

habitat, winter, 8
habitat limitation, 2–3, 5, 11, 119, 187–89. *See also* nest site, limitation of
hackberry, 42, 483
Halictidae, table 9.1, 249
Haliplidae, table 9.1, 249
haliplid water beetle, 251
hatching date, 19–20, 36, 39, 286, 391–92, fig. 12.15, 398
hatching synchrony, 36, 39, 229–30, fig. 8.9. *See also* synchrony
hawk
   red-tailed, 216, 483
   sharp-shinned, 212–13, 226, 241, 445, 483
   Swainson's, 216, 238, 483
helping. *See* cooperative breeding
hematophagous ectoparasites, 70–73, 116. *See also* ectoparasitism
Hemiptera, 70, table 9.1, 249. *See also* swallow bug
hemipterans, 251
hematocrit, 93
Hemerobiidae, table 9.1, 249
hemoglobin, 93
heritability, of colony choice, 460
heron, great blue, 205, 468, 483
Hesperotettix viridis, 248
High Plains Climate Center's Automated Weather Data Network, 43
Hirundo
   ariel, 10, 484 (*see also* martin, fairy)
   evolution of coloniality in, 453, 481
   fluvicola, 10, 484 (*see also* swallow, Indian cliff)
   fulva, 10, 484 (*see also* swallow, cave)
   hybridization among, 10
   as mud nest builders, 453, 481
   nest design of, 453
   preussi, 10, 484 (*see also* swallow, Preuss's cliff)
   pyrrhonota, 10, 484 (*see also* swallow, cliff)
   rufigula, 10, 484 (*see also* swallow, Angolan cliff)
   rustica, 10, 453, 484 (*see also* swallow, barn)
   spilodera, 10, 484 (*see also* swallow, South African cliff)

Histeridae, table 9.1, 249
historical causes of coloniality, 449–56, table 14.1. *See also* coloniality; evolution of coloniality
historical occurrence of cliff swallows, 46–50
histories of breeding colony choices, 419–32, tables 13.2–6, figs. 13.5–8
hole-nesting birds, 331
Holohil Systems, 401
home range, 361
Homoptera, table 9.1, 249, 251
honeybee, 257, 483
Horn's geometrical model. *See* geometrical model
hosts for ectoparasites, 90, 110. *See also* ectoparasitism
human-altered habitats, 56–57, 217
humidity, 43
Hungary, 192
hybrid zone, 42
Hydrophilidae, table 9.1, 249
Hymenoptera, table 9.1, 249

Ichneumonidae, table 9.1, 249
ideal free distribution, 7, 251, 399–400, 446–47, 450, 467, 468, 481
Illinoian glaciation, 40
immigration
   and estimating group relatedness, 475
   of ectoparasites into colony, 106–10
immunological response to parasites, 455
incubation, 26, 133, 155, 245, 334, 343, 352
   and egg destruction, 143
   and egg transfer, 167–68
   foraging during, 246–47, 260, 265, 352
   and nest checking, 19
   stage of, 22
   variable length of, 19
indeterminate laying, 147, 168, 172
indirect selection, 4, 476, 481
individual alertness, 210, 220–24, figs. 8.3–5, 239, 268
individual fitness, 456. *See also* fitness
infanticide, 144. *See also* nestling, death of
inference space, 37–38
information center, 11–12, 57, 243–45, 263–64, 266, 272. *See also* foraging; social foraging
information for colony choice, 400, 406–9, 424, 447

information parasitism, 260
information sharing. *See* information transfer
information transfer, 244, 247, 251, 270, 296, 333, 345, 346, 359, 375, 382, 386, 463, 471, 477
  by active signaling of information, 243, 260, 266–68, 273
  in barn swallows, 453
  as benefit of coloniality, 245, 260, 272, 275, 452
  benefits of, 315–17, 319–20, 480
  as cause of social foraging, 272, 275
  by circling above colony, 278–79
  during cold weather, 300–304
  at colony site, 260–66, fig. 9.4, tables 9.2–3, 268, 272–74
  and colony size, 264–65, 273, 278, 285, 300, 304–6, 310, 315, 317, 319–20
  cost of, 265
  cues used in, 262–63, fig. 9.4, 273
  and departure frequencies, 276
  and disguising success, 265–66, 273
  while feeding nestlings, 246–47, 265, 273, 277
  by following neighbors, 262–63, table 9.3, 273
  by following successful foragers, 260–64, table 9.2, table 9.4, 273, 276
  through food calls, 266–68, 273, 452, 480
  and foraging efficiency, 243, 275, 304–6, 319–20
  by information parasitism, 260
  by joining departing birds, 264–65, 273, 278
  measured by adult body mass, 290–91, 305
  methods of, 260
  and nest clustering, 452
  passive, 243, 260
  and purposeful flight, 260, 265
  and spatiotemporal variability in food, 251, 273
  and tracking insect swarm, 266–68, 270–71, 273
  and waiting intervals, 277–80, table 10.1, fig. 10.1
injuries, 20, 125, 144
insects. *See* food
interference among conspecifics, 5, 11, 118–20, 130–33

intermediate colony size
  costs and benefits at, 451
  and fitness, 467
  and food delivery rates, 300
  and foraging, 300, 317
  optimal, 299–300, 338, 467
  reproductive success at, 321–22, 338–39, figs. 11.8–9, 359, 458
  survivorship at, 374–75, fig. 12.5
internest distance, 203, fig. 7.10, 209
Interstate 80, 51
intraspecific brood parasitism. *See* conspecific brood parasitism
intraspecific competition. *See* competition; fighting
irrigation structures, 50–57, fig. 3.5, 215. *See also* artificial nesting sites
Ischnocera, 73

Juniperus virginiana, 483. *See also* cedar, eastern red
juvenile, 57, 222, 370, 373, 408, 458
  banding of, 22, 38, 409, 433
  begging calls of, 176, 178–79
  body mass of, 22, 382–84, figs. 12.10–11, 398
  colony assessment by, 408
  crèching of, 175–79, figs. 6.12–13, 185–86
  dependence on parents, 175–76
  dispersal, 361, 368, 432–37 (*see also* dispersal; natal dispersal)
  and ectoparasites, 102–3
  facial plumage of, 176
  kleptoparasitism by, 175, 179–86, figs. 6.14–16
  movement between colonies, 182–83, fig. 6.16, 186
  number banded, 363, table 12.1
  postfledging activities, 175–86, figs. 6.12–16
  predation on, 214–15
  recapture of, 363–64, table 12.1, 365
  recognition of by parents, 176–77, 184–86
  return to natal nest, 180
  survivorship, 360, 362, 382–84, fig. 12.11, 398, 458 (*see also* first-year survivorship)
juvenile body mass, 382–84, figs. 12.10–11, 398

548 • Index

Kearney County, Nebraska, 50
Keith County, Nebraska, 40, 48–50, figs. 3.3–4, 52–53, 67, 98
Kentucky, 50
kestrel, American, 212–14, 226, 238, 241, 483
Keystone, Nebraska, 41, fig. 3.1, 44, 48, fig. 3.3, 50
kin association, 475–76, 480
Kingsley Dam, 42, 50, 56, 404
kin selection, 4, 476, 481
kinship. *See* relatedness
kittiwake, black-legged, 396, 484
kleptoparasitism, 57, 151, 175, 179–86, figs. 6.14–16, 391
  and assessment of colony sites, 185
  and colony size, 181–84, fig. 6.15, 186
  as cost of coloniality, 185, 451, table 14.1
  distance moved from natal colony for, 182–83
  and postfledging survivorship, 185
  for predator avoidance, 185
  recruitment to colonies for, 182–84, fig. 6.16, 186
  and spatial distribution, 180, fig. 6.14, 477
  and synchrony, 181–82
  tolerance of by nest owners, 184–85
known-age birds, 20, 22, 363–64

Lake McConaughy, 41–42, fig. 3.1, 48, figs. 3.3–4, 50–53, 206, 257, 308
Lanius ludovicianus, 484. *See also* shrike, loggerhead
Larus argentatus, 484. *See also* gull, herring
late season, 102
  adult body mass during, 291, 293–98, fig. 10.11, 384–86, fig. 12.12
  definition of, 22
  foraging during, 245–46
  survivorship measured from, 384–86, fig. 12.12, 398
late-starting colonies, 336–39, figs. 11.8–9, 402, 472
Lauxaniidae, table 9.1, 249
laying synchrony, 36, 39, 227–30, figs. 8.6–9, 356, 358, table 11.9, 359. *See also* egg laying; synchrony
leks, 152, 454, 465
Lepidoptera, table 9.1, 249, 250

leucocytes, 93
Lewellen, Nebraska, 41, fig. 3.1, 46, 48–49, fig. 3.3, 50
Libellulidae, table 9.1, 249
life history trait or parameter, 175, 362
life span, 362, 379
  breeding, 457–58, 460
  and colony size, 361
  estimating, 457
  and fitness, 360, 457
  and lifetime reproductive success, 360, 397, 457–58
lifetime reproductive success, 6–7, 316, 322, 339, 361, 362, 379, 386, 399, 456–63, 467. *See also* reproductive success
  and brood parasitism, 462
  and colony size, 323, 362, 458–63, fig. 14.1, 464, 481
  and colony size choice, 460–63
  components of, 360, 397
  definition of, 456
  estimating, 456–58
  and evolution of coloniality, 456–63, fig. 14.1
  and extrapair fertilization, 462
  and fitness, 456
  implications of, 460–63
  as index of fitness, 456, 458
  influenced by survivorship, 323
  at intermediate colony sizes, 458, 462
  and life span, 457–58
  potential bias in, 458–460
  and uncertain parentage, 462–63
limited resources, 2–3, 5, 11, 118
Lincoln County, Nebraska, 47, 49
Lithospermum, 42
local enhancement, 244, 246, 268, 312
local food resources, 326, 424, 465, 480
  as cause of coloniality, 468
  and colony reduction experiment, 304–6, 320
  at colony sites, 468
  and colony size, 304–6, 399, 446, 450
  depletion of, 306, 312, 313, 314–15, 320
  foraging efficiency related to, 304–6
  ideal free distribution of, 468, 481
local resource availability, 206, 320, 468, 481
logarithmic scale, 38
longevity. *See* life span; survivorship
louse, chewing, 28, 69, 73, 300, 317, 385

Lower Sonoran Zone, 8
Lygaeidae, table 9.1, 249

Macaca fascicularis, 484. *See also* macaque, long-tailed
macaque, long-tailed, 307, 484
Machaerilaemus malleus, 73. *See also* louse, chewing
magpie, black-billed, 212–13, 226, 241, 484
Mallophaga, 73
Malthusian parameter of population growth, 456
maple, silver, 42, 483
mapping of colonies, 34–35, 39
marking. *See* banding; color marking
mark-recapture, 7, 16, 49, 362–64, 367, 368, 394, 397, 419, 427, 439, 477. *See also* methods; mist netting; survivorship
martin
 fairy, 10, 484
 house, 70, 81, 86, 137, 304, 418, 484
 purple, 5–6, 32, 146, 150, 180, 357, 370, 408, 454, 484
mass. *See* body mass
mass emergences of insects, 259, 273
mass reduction. *See* body mass
mate choice, 68, 407
mate guarding, 136, 161–62, 185
mates, 92, 465
 ages of, 357
 competition for, 5, 120, 465
maximum likelihood methods, 365
Maxwell, Nebraska, 49
Melanoplus
 augustipennis, 248
 confuscus, 248
 foedus, 248
 sanguinipes, 248, 250
Meloidae, table 9.1, 249
Menoponidae, 73
Metepeira incrassata, 236
methods, 10–12, 13–39, 43–44, 49, 56–62, 77, 93, 98–99, 102, 120–21, 127, 129, 362–68
 of analyzing rainfall and temperature data, 43–44
 of banding, 19–22, fig. 2.3, 38–39
 of calculating nest building time, 127
 of charting nest completeness, 17–18, fig. 2.2
 of checking nest, 13, 17–19, fig. 2.1

 of color marking, 22, 24–26, fig. 2.4, 39
 of counting food delivery rates, 280
 criteria for including colony in study, 16–17, 25, 49
 definition of colony, 15–16, 38
 of determining
  adult body mass, 291–93,
  adult body mass reduction, 297–99
  age of nestlings, 19
  arrival times, 62–64
  body mass, 20, 22, 291–93
  body size, 24
  brood size, 20
  center nests, 34–35, 39
  colony initiation date, 15, 65
  colony size and use, 13–17, 49, table 2.1, 442
  colony stage, 22
  colony status, 15, 49
  departure frequencies, 276
  ectoparasitism in nest, 27–28, fig. 2.7, 30–33, fig. 2.9, 39
  edge nests, 34–35, 39
  egg laying date, 18, 36, 39
  food, 248, 252–53, 260, 304
  foraging locations, 254–55, 257
  hatching date, 19
  hatching synchrony, 35–36, 39
  internest distance, 202–3
  laying synchrony, 35–36, 39
  lifetime reproductive success, 456–58
  nearest neighbor distance, 34–35
  nest age, 35, 39
  nestling growth, 93
  nestling survivorship, 96
  nest ownership, 21, 26, 39
  nest size, 20, 35
  nest status, 18
  nest synchrony, 20, 39
  sex ratio, 61–62
  synchrony, 35–36, 39
  waiting intervals, 277–80
 of drop netting, 20
 of estimating survivorship, 362–68
 of field sampling for mark-recapture, 362–64
 of fumigating nest, 31–34, figs. 2.10–11, 39, 98–99
 of identifying individual nests, 19
 of inferring egg destruction, 143
 of locating colonies, 15, 49
 of mapping colonies, 34–35, 39

550 • Index

methods (*continued*)
  of measuring
    colony density, 34
    colony substrate, 34, 192–93
    distance to mud source, 129
    nest diameter, 35
    nest size, 20, 35
    population size trends, 57–61
    travel distance and foraging areas, 307–8
  of mist netting, 20–21, fig. 2.3
  of naming colony sites, 37
  of observing
    fighting, 120–21
    following of foragers, 261–62
    nests, 26
  pattern of visiting study sites, 15, 49
  of plugging nest, 21–22
  of radio telemetry, 401–2
  of sampling ectoparasites, 20, 22, 26–34, figs. 2.5–11, 39, 77, 102, 107
  of scoring foraging group size and position, 268–69
  of scoring prey captures, 268–69
  of sexing birds, 22–24, 39
  of statistical analysis, 20, 36–38, 39, 90, 363–66
  SURGE, 365, 394, 397
  of weighing, 19–20, 93
  years included in data analysis, 11–12
Mexico, 8, 10
Michigan, 301, 331
midseason, 22, 102, 291
midge, 252. *See also* Chironomidae
migration, 8, 241, 294, 373, 379, 380, 384, 388, 391, 416, 475
Miridae, table 9.1, 249
mirror, dental, 17, fig. 2.1. *See also* nest checking
misdirected parental care, 149–86
  and annual reproductive success, 150
  as benefit of coloniality, 5, 11, 150–51, 164–67, 171–75, 185–86, 452
  brood mixing as, 151
  brood parasitism as, 149–51, 164–65, 167–75, figs. 6.9–11, 185–86
  brood parasitism versus extrapair copulation, 164–65, 171–72
  as cost of coloniality, 5, 11, 150–51, 164–67, 171–75, 185–86
  cuckoldry as, 149–51, 164–67, 171, 185
  egg transfer as, 167–68
  extrapair copulation as, 149–67, figs. 6.1–8, 171, 185–86
  feeding unrelated offspring as, 175, 178–79
  and group size, 150–51, 179
  kleptoparasitism as, 175, 179–86, figs. 6.14–16
  mixing of mobile offspring as, 175–86, figs. 6.12–16
  multiple parentage of brood as, 164–65, 171–72, 185–86
  nestling transfer as, 168
  postfledging juvenile behavior leading to, 175–86
  quasi-parasitism as, 167–68
mist netting, 20–21, fig. 2.3, 25, 49, 182, 362, 402
mixed prairie, 42
mixed reproductive strategy, 166. *See also* alternative reproductive tactics
mixed-species flocks, 142
mixing of mobile offspring, 175–86, figs. 6.12–16
mobbing. *See also* predator avoidance
  as benefit of coloniality, 188, 210, 212, 241, 451, table 14.1
  experiment to study, 226
  and group size, 210, 241
  and relatedness of birds, 474
  risk of, 225–26, 241, 451
molecular methods for parentage studies, 149, 165–66, 174, 185, 479
monogamy, 149
Morrill County, Nebraska, 49
mortality, causes of, 21, 42, 46, 62, 68, 93, 100–102, 116, 125, 130–32, 231, 300–304, table 10.4, 313–15, fig. 10.15, 320, 334, 360, 361, 366, 371. *See also* survivorship
mouse, deer, 217–18, 484
movement, long distance, 362
mud, 397, 445
  and determining nest status, 18
  distance traveled to collect, 11, 129–30, fig. 5.6, 139, 147–48, 152
  efficiency of gathering, 224, 473
  and entombment inside nest, 130–32, fig. 5.7, 148
  gathering (*see* mud gathering)
  location of, and colony size, 129–30, fig. 5.6, 147–48
  quality of, 129

mud gathering, 9, 126, 128, 136, 147–48, 152, figs. 6.1–2, 221, 239, 241, 268, 343, 397, 408, 445, 473. *See also* extrapair copulation; nest building
and colony and group size, 157–58
extrapair copulation during, 151–61, figs. 6.1–3, figs. 6.5–6, 185
group size for, 128, 140, 152, 224, 239, 241
mate guarding during, 161–62, 185
method of, 152, figs. 6.1–2
nest guarding during, 161–62
predation by grackles during, 214–15, 224
tolerance of human observers during, 154
vigilance during, 221, 223–24, fig. 8.5, 239, 241
wing fluttering during, 162–64, fig. 6.1, fig. 6.8, 152, 159, 185
mud nest builders, 453, 481
multivariate statistics, 36, 394
Muscidae, table 9.1, 249
Mustela frenata, 484. *See also* weasel, long-tailed
Mycetophilidae, table 9.1, 249
Myrmeleontidae, table 9.1, 249

Nabidae, table 9.1, 249
naled, 32, 39. *See also* fumigation
natal colony size, 361, 448
choice of, 435–37, figs. 13.10–11
fidelity to, 435–37, figs. 13.10–11
and first breeding colony size, 435–37, figs. 13.10–11
histories of, 435–37, figs. 13.10–11
and survivorship, 373–80, figs. 12.5–6, 397–98
natal dispersal, 114, 388–90, 400, 419, 432–37, figs. 13.9–11, 447–48, 458–60, 474–75, 480. *See also* dispersal; philopatry
natal dispersal distance, 433–34, fig. 13.9, 447–48
natal nest. *See also* nest
age of, 391, 395, 398
fumigation of, 368–75, fig. 12.1, figs. 12.4–5, 380–82, fig. 12.9, 386–95, figs. 12.13–16
position of, 391, 393–95, fig. 12.16, 398
return to by juveniles, 180–82
natal philopatry, 362, 364, 432–37, figs. 13.9–11, 474–75. *See also* dispersal; natal dispersal
natural or cliff colony sites, 67, 124, 197
compared with artificial sites, 56–57, 67, 204–5, 209
fighting at, 124
foraging area of, 255–58, figs. 9.2–3, 308–11, figs. 10.13–14
geometry of, 50, fig. 3.4, 471
location of within study area, 49–50, figs. 3.2–4
nest placement on, 50, fig. 3.4
predation at, 205, 216
predictability of size and use of, 201–2, fig. 7.9, 204–5
size of, 50, 205
substrate of, 189
nearest neighbor distance
and clutch size, 329–30, table 11.2
and colony size, 190–91, fig. 7.2, 208
and dispersal, 424–25, table 13.3
definition of, 35
and ectoparasitism, 84–89, table 4.4, 190
as internest distance, 202–3, fig. 7.10, 209
minimum, 191
and nest spacing, 190–91, fig. 7.2, 192
and reproductive success, 345–49, tables 11.7–8, 359
and shortage of nesting sites, 188–91, figs. 7.1–2, 194–95, fig. 7.4, 208–9
and spatial position, 345, 470–71
and spatial structure of colony, 470–71
substrate size and, 192, 194–95, fig. 7.4, 209
Nebraska, 1, 8, 15, 40–41, figs. 3.1–2, 45–47, 50, figs. 3.4–5, 67, 69, 72, 116, 166, 175, 202, 226, 233, 236, 241, 267, 272, 301, 362, 379, 460, 463, 475, 480
Nebraska Sand Hills. *See* Sand Hills
nest
abandonment of, 113, 131–32, 147
age of, 35, 39, 79–84, figs. 4.3–4, 88–89, 116, 168, 327–29, 331, fig. 11.5, 344, 350–52, 359, 391, 395–98, 425, table 13.3, 444–45, 473–74 (*see also* nest age)
birds captured in, 21–22, 362, 477
and brood parasitism, 146, 148, 168
bugs at entrance of, 108–9, fig. 4.11, 112
building of, 115–17, fig. 4.12, 126–32,

552 • Index

nest (continued)
  figs. 5.4–7, 136, 140, 147–48, 291, 294, 333, 352, 397–98, 441, 473–74
  center, 84, 88–89, fig. 4.4, 123–26, fig. 5.3, 147, 351–52, 393, 396, 398, 470–71, 473, 482
  clustering of, 188, 195, 202–4, fig. 7.10, 209, 212, 242, 345, 346, 352, 452, 470
  competition for, 118–20
  completeness of, 17–18, fig. 2.2, 35, 120–21, 123–26, 127, fig. 5.3, 147, 342–45, fig. 11.11, 472
  defense of, 120–21, 125–26, 130–32, 133–34, 453
  desertion of, 22, 113, 131–32, 135
  diameter of, 35, 82, 352, 425, table 13.3
  distance between, 202–3, fig. 7.10, 209
  distance from center of colony, 84–89, table 4.3, 329–30, 345–47, table 11.2, tables 11.5–6, 350, 359, 470–71
  ectoparasites in
    dispersal of, 69, 79, 84–89, tables 4.3–4, fig. 4.4, 111–12
    distribution of, 90–93, fig. 4.5, 102, 106–12, 114–16
    effect on nestlings, 93–102, figs. 4.6–9, tables 4.5–7, 116, 386–90, figs. 12.13–14
    sampling for, 26–27, fig. 2.5, fig. 2.7, 30–34, figs. 2.9–11, 39, 112–13, 117, 133, 148
    transmission of, 106–12, fig. 4.11, 116–17
  edge, 84, 88–89, fig. 4.4, 123–26, fig. 5.3, 147, 242, 351–52, 393, 396, 398, 470–71, 473, 481–82
  entombment in, 130–32, fig. 5.7, 148
  failure, 14, 36, 96, 106, 113, 119, 145, 148, 437, 441, 445
  fighting at, 120–30, figs. 5.1–3, 147
  fumigation of, 31–34, figs. 2.10–11, 35, 39, 92, 93, 96, 98–102, tables 4.6–7, figs. 4.8–9, 112–13, 386
  guarding of, 136–37, 140, 148, 162
  of Hirundo, 453
  identification and numbering of, 19
  methods of checking, 17–19, 38
  nearest neighbor distance of, 35, 84–89, table 4.4, 189–91, fig. 7.2, 192, 194–95, fig. 7.4, 202, 208–9, 359, 424–25, table 13.3, 470–71
  new, 35, 39, 79–84, figs. 4.3–4, 88–89, 115–17, 127, 207, 327, 329, 331, fig. 11.5, 350–52, 395, 397, 398, 425, table 13.3, 473–74 (see also nest age)
  observed in nineteenth century, 47
  old, 35, 39, 79–84, figs. 4.3–4, 88–89, 112, 115–17, fig. 4.12, 127, 130, 168, 192, 207, 327, 329, 331, fig. 11.5, 344, 350–52, 395, 397, 398, 444–45, 473–74 (see also nest age)
  owners of (see nest ownership)
  packing of, 234–36, fig. 8.10, 242
  parasites (see nest ectoparasitism)
  of Petrochelidon, 453
  position of, 20, 34–35, 39, 65, 79, 84–89, tables 4.3–4, fig. 4.4, 116, 119, 123–26, fig. 5.3, 233–34, 242, 357–58, 359, 395–96, 398, 424–25, table 13.3, 470, 481
  return to natal, 180
  reuse of, 9, 27, 32, 65, 80–82, 115–17, fig. 4.12, 130
  size of, 20, 35, 79, 82–84, table 4.2, 116, 324, 329, 331–32, fig. 11.6, 351, 352–53, 359, 424–25, table 13.3
  spacing of, 69, 79, 84–89, tables 4.3–4, fig. 4.4, 111–12, 188–89, 195, 202–4, fig. 7.10, 208–9, 471–72, 479
  stealing materials for, 119, 133, 140, 148
  success of, 173–74, fig. 6.11, 240, fig. 8.12, 242
  thermal advantages of, 351–53
  trespassing at, 132–41, figs. 5.8–11
  wall sharing, 128–30, fig. 5.5, 148, 234–36, fig. 8.10, 242, 331, 346
nest abandonment, 113, 131–32, 135, 147, 170
nest age, 39, 130, 425, table 13.3
  and adult survivorship, 395–98
  and brood parasitism, 168
  and clutch size, 327, 329, 331, fig. 11.5, 359
  and colony size, 473–74
  and ectoparasite loads, 79–84, figs. 4.3–4, 88–89, 112, 116–17, 351, 473
  and fitness, 473–74
  and predicting colony size, 444–45
  and reproductive success, 350–52, 359, 473–74
  and survivorship, 391, 395, 398
nest attachment, 192
nest boxes, 188

Index • 553

nest building, 11–12, 15, 17–18, 26, 46, 126–32, 136, 294, 329, 333, 347, 356, 358, table 11.9, 359, 395, 398, 453, 474. *See also* mud; mud gathering
  as benefit of coloniality, 451, table 14.1, 453
  and colony size, 126–30, 473
  and commitment to site, 441, 444
  and competition for materials, 5, 119
  cost of, 329, 331, 351, 352, 397
  and ectoparasitism, 82, 112, 114–16, fig. 4.12, 117
  and entombment, 130–32, fig. 5.7, 148
  and extrapair copulation, 151–53, figs. 6.1–2
  and fighting at nest, 126–30
  and mud availability, 129–30, fig. 5.6, 147–48
  and placement of nests, 50, 56, figs. 3.4–6
  predation during, 214–15
  quality of mud, 8–9, 129
  stage (early), 22, 291
  time for, 126–30, fig. 5.4, 140, 147–48, 473
  time of completion, 82
  and wall sharing, 128–30, fig. 5.5, 148, 331, 346, 473
nest checking, 15–19, 38, 127, 190, 337, 395
  for brood parasitism, 169
  with dental mirror, 17, fig. 2.1
  for egg destruction, 143, 169
  during incubation, 19
  for nestling survivorship, 334
  nests included in, 19
  and reproductive success, 17, 20
  termination of, 19
  timing of, 18
nest clustering, 188, 195, 202–4, fig. 7.10, 209, 345, 346, 347, 352, 393
  and information transfer, 452
  and nest predation, 211–12, 234–36, fig. 8.10, 242, 470
  as selfish herd effect, 211, 212
nest completeness, and egg laying, 342–45, fig. 11.11, 472
nest defense, 120–21, 124–26, 133–34, 453. *See also* nest guarding
nest density, 56, 69, 84, 86, 110–12, 116, 188, 238, 470–71, 477. *See also* colony density; density; nest clustering
  and colony size, 189–90, fig. 7.1, 208
  and competition for nest sites, 118–19
  and ectoparasitism, 73–77, 79, fig. 4.2
  and entombment inside nests, 130–32, fig. 5.7, 148
  and extrapair copulation, 150, 166
  and kleptoparasitism, 180
  measuring, 34
  and nest building time, 128–29, 147–48
  and nest predation, 238–39
  and reproductive success, 322
  and shortage of breeding habitat, 187
  and shortage of nesting sites, 189–91, figs. 7.1–2, 208
nest diameter. *See* nest size
nest distance from center of colony. *See* distance from center of colony
nest ectoparasitism, 26–28, fig. 2.5, fig. 2.7, figs. 2.9–11, 31–35, 39, 69–102, figs. 4.1–9, tables 4.1–7, 106–17, figs. 4.11–12, 133, 135, 148, 168, 195, 196, 201–2, 385–90, figs. 12.13–14, 398, 433–35, 470. *See also* ectoparasitism
nest failure, 14, 36, 96, 106, 110, 113, 119, 145, 148, 170, 173, 217, 230, 334, 386, 437, 441, 445
nest fumigation. *See* fumigation
nest guarding, 136–37, 140, 148, 162
nesting synchrony. *See* breeding synchrony; synchrony
nestling, fig. 4.8, 369, 374
  abandonment of, 113
  age of
    determining, 19
    and ectoparasitism, 111, 116, 117
    and food deliveries, 281–83, figs. 10.2–3, table 10.2
  banding of, 19–20, 27, 38, 356, 368, 409, 433
  body mass (*see* nestling body mass)
  body size, and ectoparasitism, 95
  brood reduction (*see* brood loss, partial)
  death of, 144, 212
    during cold weather, 303
    from competitors, 212, 217–18
    from entombment in nest, 132, 148
    from predators, 216–17
  dispersal of, 114, 400, 432–37, fig. 13.9
  and ectoparasitism, 19, 26–28, figs. 2.6–7, 30, 34, 39, 69–70, 73–74, fig. 4.1, tables 4.1–7, 77–93, fig. 4.3, figs. 4.5–9, 90–93, 96–102, 116 (*see also* ectoparasitism)

554 • Index

nestling (continued)
  facial plumage, 176
  feeding of, 14, 18, 22, 26, 45, 94, 96, 175–76, fig. 6.12, 245–47, 260–61, 265, 273, 276, 291–92, 294, 301, 319
  fledging of, 360
  growth of, 32, 69, 93–96, figs. 4.6–8, 248
  injuries to, 144
  and kleptoparasitism, 175, 179–86, figs. 6.14–16
  mortality of, 93, 96–97, table 4.5, 100–102, 116, 132, 148, 231, 303, 313–15, fig. 10.15, 320, 334, 360
  number banded, 363, table 12.1
  and parent-offspring recognition, 151, 176–79
  recapture of, 363–65, table 12.1
  ring collaring of, 248, 250
  starvation of, 311–16, fig. 10.15, 320, 451, table 14.1
  survival of, and reproductive success, 333–50, figs. 11.7–10, tables 11.3–8, fig. 11.12, 358, 386, 391
  survivorship of, 32, 96–97, tables 4.5–6, 98–102, fig. 4.9, 116, 174, 183, 368–75, fig. 12.1, figs. 12.4–5, 380–84, fig. 12.9, 386–95, figs. 12.13–16, 397–98 (see also survivorship)
  tossing of, 144 (see also nestling, death of)
  transfer of, 168
  weighing, 19, 27, 77, 93
nestling body mass, 20, 27, 93–96, figs. 4.6–9, 98–102, table 4.7, 116, 286–90, figs. 10.6–8, 305, 316–17, 319, 360, 380–83, figs. 12.9–10, 386, 391, 398. See also body mass
nest materials
  collecting, 151–53, figs. 6.1–2
  competition for, 119
  stealing, 119, 133–35, 140, 148
nest observations, 26, 120–21
nest ownership, 32, 39, 120–21, 124–26, 393
  assigning, 21, 26
  and extrapair copulation, 155–57
  and fighting at nest, 120–21, 124–26, 130–32
  and nest defense, 120–21, 124–26, 130–34
  and trespassing, 133–34

nest packing, 234–36, fig. 8.10
nest plugging, 21–22, 477
nest position in colony, 130–32, fig. 5.7, 351–52, 481–82. See also center nests; distance from center of colony; edge nests; nearest neighbor distance
  and age of nest owners, 393
  changes between years, 477
  and clutch size, 327–31, table 11.2, fig. 11.5, 359
  and competition for nest sites, 119
  and date occupied, 329
  and dispersal, 424–25, table 13.3
  and ectoparasite load, 79, 84–89, tables 4.3–4, fig. 4.4, 110–12, 116, 329–30, table 11.2, 470
  and fighting at nest, 123–26, fig. 5.3, 147
  method of measuring, 34–35, 39
  and predation risk, 211, 242, 345
  predicting final, 400
  and reproductive success, 345–52, tables 11.5–8, fig. 11.12, 359
  and selfish herd effects, 233–34
  and survivorship, 395–96, 398
  trespassing, 138
nest reuse, 115–16, fig. 4.12, 117, 130
nest site, 401, 465, 472, 477
  artificial, 49, 50–57, figs. 3.5–6, 204–5, 209
  competition for, 5, 11, 90, 118–20, 465
  competition from other species for, 212, 217–18
  fighting for, 120–26, figs. 5.1–3, 147
  limitation of, 187–88, 202–4, 207–9, 212, 318, 322
  location of in colony each year, 477–78, fig. 14.3
  natural, 40–42, 49–50, fig. 3.4, 56–57, 204–5, 209
  quality of, 118–19
  structure of, and ectoparasitism, 75
nest site competitors
  deer mouse, 217–18, 484
  house sparrow, 217, 484
  house wren, 217–18, 484
nest size
  and clutch size, 327, 329–32, fig. 11.6, 359
  and cost of nest building, 329, 352
  diameter, 20, 35, 352, 425, table 13.3
  and dispersal, 424–25, table 13.3

and ectoparasite load, 79, 82–84, table 4.2, 116, 352–53
and reproductive success, 351, 352–53, 359
thermal advantages of, 351, 352
nest spacing, 189–92
of barn swallows, 189, 202–4, fig. 7.10, 209
and colony size, 191, figs. 7.1–2, 208, 242
and fitness, 471–72
and internest distance, 202–3, fig. 7.10, 209
and nearest neighbor distance, 190–92, fig. 7.2, 194–95, fig. 7.4, 202, 208–9, 470
and nest density, 189
and selfish herd effect, 211, 212
and substrate area, 192, 194–95, fig. 7.4, 208–9
nest stage
definition of, 22
and ectoparasitism, 77, 102–5, 110–12
and extrapair copulation, 155
and fighting, 120–21, fig. 5.1, 123–25, fig. 5.3, 147
and trespassing, 133, 135
nest success, 173–74, fig. 6.11, 211, 240, fig. 8.12, 242
nest trespassing. *See* trespassing
netting. *See* mist netting
network foraging, 245–46, 266
Neuroptera, table 9.1, 249
New Mexico, 10
nonparametric statistical tests, 36, 39
North America, 8–10, 12, 213, 217, 248
North Platte, Nebraska, 41, fig. 3.1, 43, 44, 47, 49
North Platte River, 40–42, figs. 3.1–3, 47–51, 67, 204, 247, 401, 404, 406, 407. *See also* Platte River
Notonectidae, table 9.1, 249
Nycticeius humeralis, 484. *See also* bat, evening

observations
of artificial versus natural colonies, 56–57, 67
choosing nests for, 26, 120–21
of extrapair copulation, 152–57
of following by foragers, 261–62
of food deliveries, 280
of foraging locations, 254–55
of trespassing, 140
of waiting intervals, 277–78
Odonata, table 9.1, 249
Ogallala, Nebraska, 40–42, fig. 3.1, 48, fig. 3.3
Oeciacus, 70
hirundinis, 70
vicarius, 70, 116, 483 (*see also* swallow bug)
offspring. *See* reproductive success
Oklahoma, 76, 79, 236
olive, Russian, 42, 483
optimal colony size. *See also* optimal group size
and colony choice, 399–401, 408, 418, 432, 439, 446–48, 456
exclusion to maintain, 468–69, 481
and foraging efficiency, 299–300, fig. 10.12
for individual, 466
versus nonoptimal, 439, 444
and phenotypic differences, 400, 418, 432, 439, 445–48, 463–66, fig. 14.2
and reproductive success, 337–39, figs. 11.8–9, 399–400, 456, 465
and survivorship, 375
optimal group size, 322, 399, 439
achieving, 468
and colony size, 467–68
evidence for, 467
intermediate, 467
stability of, 467
theory of, 467–68
Opuntia, 42
Oregon Trail, 47
oriole, 42
Ornithodoros concanensis, 72. *See also* tick, soft-bodied
Orthoptera, table 9.1, 249
Oshkosh, Nebraska, 41, fig. 3.1, 48–49, fig. 3.3
osprey, 216, 483
overdispersion of parasites, 69, 92
owl
barn, 212, 215–16, 241, 484
great horned, 212, 215–16, 226, 241, 484

Pandion haliaetus, 483. *See also* osprey
Panicum, 42
Paraguay, 8

parasitism. *See* ectoparasitism
parentage studies, 19, 21, 92, 175, 362, 395
  allozyme-based, 164–65, 171–72, 185
  by egg white protein electrophoresis, 175
  DNA-based, 165–66, 172, 174, 185, 455, 463
parental
  ability, 373
  age, 393
  assistance, 166, 455
  care, 57, 149, 351, 380, 452, 455 (*see also* brood mixing; misdirected parental care)
  experience, 289, 355
  investment, 373
  quality, 133
parent-offspring association, 474
parent-offspring recognition, 151, 480
  in crèche, 176–79, figs. 6.12–13, 186
  by facial plumage of juveniles, 176
  and kleptoparasitism, 184–85
  by voice, 176, 178–79, 184–85
Passer domesticus, 484. *See also* sparrow, house
paternity, 166
Paxton, Nebraska, 41, fig. 3.1, 49, fig. 3.3, 51
pelagic species, 188
Penstemon, 42
Pentatomidae, table 9.1, 249
Peromyscus maniculatus, 484. *See also* mouse, deer
pesticide. *See* fumigation
Petrochelidon, 12, 453, 484. *See also Hirundo*
  characteristics of, 10
  distribution of, 10
  evolution of coloniality in, 453
  hybridization of with *Hirundo*, 10
  as mud nest builders, 453
  nest design of, 453
  species of, 10
Phaethon lepturus, 483. *See also* tropicbird, white-tailed
Phalacridae, table 9.1, 249
phenotype-based colony choice hypothesis, 463–69, fig. 14.2, 479, 481
phenotypic differences, 90–93, 174, 400, 415, 418, 432, 439, 445–48, 463
Philomachus pugnax, 484. *See also* ruff
philopatry, 362, 364, 420, 424, 432–33.
  See also dispersal
Philopteridae, 73
phoebe, eastern, 331–32, 484
phoebe, Say's, 144, 484
Phoetaliotes nebrascensis, 248
Phoridae, table 9.1, 249
Pica pica, 484. *See also* magpie, black-billed
pigeon bug, 70
Pipunculidae, table 9.1, 249
Pituophis catenifer, 483. *See also* snake, bull
Plantago, 42
Platte River, 40–42, 46, 50. *See also* North Platte River; South Platte River
Platte River Valley. *See* Platte River
Pliocene, 40
Plocepasser mahali, 484. *See also* sparrow-weaver, white-browed
plumage, 22, 24, 176
population change, 11
population density, 68–69, 118. *See also* colony density; density
population genetics, 475
population size, 367, 466
  changes of, 442–45, fig. 13.13
  of cliff swallows during study, 57–61, fig. 3.7, 67
  of ectoparasites, 110
  of swallow bug, 71, 107–8
Populus deltoides, 483. *See also* cottonwood, eastern
position. *See* nest position in colony
postbreeding colony assessment, 408–9, 441, 447
postfledging behavior, 175–85, figs. 6.12–16, 408–9
prairie dog, 4, 119, 210, 221, 243, 484
precipitation. *See* rainfall
predation, 88, 92, 96, 125–26, 128, 140, 142, 144, 191, 210–42, 373, 393, 396, 424, 452–54, 465. *See also* individual alertness; vigilance
  and age of victims, 215, 233, 234
  by American kestrel, 212–14, 219, 224, 226, 238, 241
  at artificial versus natural colony sites, 205
  attempts, 212–18, 236–41, figs. 8.10–11
  by avian (aerial) predators, 205, 211–16, 218, 236–39, fig. 8.11, 241–42
  by badger, 216

by barn owl, 212, 215–16, 241
behavioral response to, 213–17
by black-billed magpie, 212–13, 226, 241
by bull snake, 212, 216–17, 218, fig. 8.1, 225, 232, 233, figs. 8.10–11, 235–39, 241–42, 314, 470
as cause of coloniality, 454, 479
and colony size, 238–42, fig. 8.12
on common eider, 236
by common garter snake, 217
by common grackle, 212, 214–15, 224, 235–39, fig. 8.11, 241
as cost of coloniality, 236, 239, 451, table 14.1
and crèche, 185, 236
experiments to study, 218–20, figs. 8.1–2, 226, 241
by great horned owl, 212, 215–16, 226, 241
by herring gull, 236
by house cat, 216
by loggerhead shrike, 212–13, 241
by long-tailed weasel, 216, 226
by mammalian predators, 216
measuring, 229–30
methods of, 212–18
and mobbing, 210, 212, 225–26, 241
natural history of, 212–18
and nest clustering, 212, 234–36, fig. 8.10, 242, 470
versus nestling starvation, 314
and nest packing, 234–36, fig. 8.10, 242
and nest position, 119, 211, 233–34, 242, 345, 349, 470
by nest site competitors, 212, 217–18
by raccoon, 216
by rat snake, 216, 236
and relatedness of individuals in group, 474
by reptilian predators, 216–17, 218
risk of, 232, 239, 242
and selfish herd, 211, 212, 233–34, 239
by sharp-shinned hawk, 212–13, 226, 241, 445
on shelduck, 236
by snapping turtle, 217
social foraging to avoid, 268–72, table 9.4, fig. 9.5
on spiders, 236
strategies of, 212–18
by Swainson's hawk, 216, 238

on swallow bug, 111
synchronized flights to avoid, 445
by wasp, 236
on white-browed sparrow-weaver, 236
predator alarm response, 57, 142, 144, 213–16, 219, 241
to avian predators, 225
to bull snakes, 225
and defining colony, 16
predator attraction, 5, 126, 211, 212, 236–39, 241–42, fig. 8.11, 451, table 14.1
predator avoidance, 126, 128, 210–42, 318, 321, 375, 465, 470. See also individual alertness; predation; predator detection; vigilance
as benefit of coloniality, 2–3, 5, 11, 187–88, 191, 210–12, 239, 241–42, 450, 453, 479
and crèching, 179
documenting, 212
by group vigilance, 210, 212, 218–24, figs. 8.2–5, 239, 241
and kleptoparasitism, 185
by mobbing, 210, 212, 225–26, 239, 241
by reproductive synchrony and swamping, 210, 212, 226–33, figs. 8.6–9, 239, 242
by selfish herd, 211, 212, 233–34, 239
and social foraging, 251, 268–74, table 9.4, fig. 9.5
and synchronized flights, 445
predator detection, 210, 212. See also individual alertness; vigilance
as benefit of coloniality, 451, table 14.1
at center and edge of flock, 223
and colony size, 218–21, fig. 8.2, 239–40, 241
and distance from colony, 218–20, fig. 8.2, 239, 241
and group size, 218, 220–24, figs. 8.3–5, 239, 241
and individual alertness, 220–24, figs. 8.3–5, 239, 241
and model snake experiment, 218–20, figs. 8.1–2, 241
predator deterrence. See mobbing
predator dilution effect, 211
predator mobbing. See mobbing
predator recruitment to colonies. See predator attraction

predator swamping, 210, 212
   by breeding synchrony, 230–33, figs.
      8.6–7, fig. 8.9, 239, 242
   and colony size, 232, 242
preening, 9, 105, 111, 180, 221–23,
   fig. 8.4, 272
preroosting behavior, 442, 445
prey capture rate
   and foraging group size, 268–74, table
      9.4, fig. 9.5, 311–13, table 10.5
   and foraging success, 268
   and position in group, 268, 271–72, 274
   scoring, 268–69, 312
   variance in, 245, 268, 273–74
primary study site, 15–16, 48–49, figs.
   3.1–3, 60, 67, 362
primate, 306–7, 465
Procyon lotor, 484. *See also* raccoon
producer, 264
Progne subis, 484. *See also* martin, purple
Prunus virginiana, 483. *See also*
   chokecherry
Pselaphidae, table 9.1, 249
pseudoreplication, 37–38
Psychodidae, table 9.1, 249
Psyllidae, table 9.1, 249
Pteromalidae, table 9.1, 249
puffin, Atlantic, 418, 484
purposeful flight, 260, 265
Pyralidae, table 9.1, 249

quasi-parasitism, 167–68
quelea, 401, 484
Quelea quelea, 4. *See* quelea
Quiscalus quiscula, 484. *See also* grackle,
   common

raccoon, 216, 484
radio telemetry, 11, 108, 155–57, 401–9,
   413, 442, 447
   methods of, 401–2
   profiles of birds used in, 402–4, table
      13.1
   using barn swallows, 401
radio tracking. *See* radio telemetry
rainfall, 296, 363
   effect of on cliff swallows, 42–43, 46, 60,
      335
   measuring, 43
   and reproductive success, 334–36
   seasonal patterns in, 42–43, 46, table
      3.2, 67

razorbill, 152, 484
recapture, 397
   consecutive years of, 420–32, table 13.2,
      tables 13.4–6, figs. 13.5–8
   history of, 362, 364
   as nuisance parameter, 365
   probability of, 363–66, table 12.2
   time necessary for, 367
reciprocity
   and association of same individuals,
      477–78, fig. 14.3
   classical, 476
   food calling as, 476, 481
   and location of nest, 477–78, fig. 14.3
   versus pseudoreciprocity, 476–77
   and relatedness, 476–78, fig. 14.3
   and social foraging, 476
   and spatially based cooperation, 476
recruitment into breeding population, 360,
   373
regulation of colony size, 419, 469
relatedness, 2–4
   and behavior, 474
   calculating, 475–76
   and colony size, 475
   and indirect selection, 476, 482
   of individuals in colony, 474–76, 480,
      482
   and kin association, 475–76, 480
   from parent-offspring association, 474–
      75
   and patterns of colony choice, 474–78
   and reciprocity, 474–78, fig. 14.3
   from sibling-sibling association, 474–75
Remiz pendulinus, 484. *See also* tit,
   penduline
reproductive interference, 118–20, 133
reproductive options. *See* alternative reproductive tactics
reproductive strategy. *See* alternative reproductive tactics
reproductive success, 3, 11, 21, 38, 42, 65,
   66, 93, 126, 147, 207, 224, 243, 321–
   59, 371, 375, 395, 412, 439, 465, 467,
   479, 480. *See also* annual reproductive
   success; clutch size; lifetime reproductive success
   age-specific differences in, 355–59, table
      11.9, 431
   and aggregation at colony sites, 208
   and alternative reproductive tactics, 150,
      164–67, 175

and breeding synchrony, 229–30, fig. 8.9, 242, 356, 358, table 11.9
and climate, 334–36, 359, 371, 472–73
and coloniality, 5–6, 12, 120, 164–67
and colony choice, 399–400, 424–25, table 13.3, 446, 463
and colony geometry, 471
and colony site fidelity, 424–25, table 13.3
and colony size, 98–102, tables 4.5–6, fig. 4.9, 239–40, fig. 8.12, 242, 317, 321–23, 326, 333–39, figs. 11.8–9, 353–55, fig. 11.13, 358–59, 362, 375, 399–400, 449, 454, 456–63, 465, 470, 481
and competition and interference, 120
components of lifetime, 360
and conspecific brood parasitism, 173–75, fig. 6.11
and costs and benefits of coloniality, 321, 358
and date, 339–45, figs. 11.10–11, tables 11.3–4, 471–73, 481
definition of, 322
and ectoparasitism, 96–102, tables 4.5–6, fig. 4.9, 373–75, fig. 12.5, 386–90, figs. 12.13–14, 455, 472
and evolution of coloniality, 456–63, fig. 14.1
and foraging, 244, 306, 316
investigator-induced effect on, 17, 20
at late starting colonies, 336–39, figs. 11.8–9
and laying eggs in incomplete nests, 343–45, fig. 11.11
lifetime, 322–23, 360–62, 456–63, fig. 14.1, 481
maximum, 399–400, 463
and misdirected parental care, 150, 164–67
and multiple parentage, 164–75
and nest age, 327–29, 331, fig. 11.5, 350–52, 359, 473–74
and nest density, 322
as nestling survivorship to fledging, 334, 373, 391, 481
and nest size, 329–32, fig. 11.6, 351–53, 359
and nest spatial position, 327–31, table 11.2, fig. 11.2, fig. 11.5, 345–50, tables 11.5–8, 359, 469–71
as number of offspring reared in year, 322
as number of offspring surviving to breed, 373, 458, 481
and parental experience, 355
peaks at intermediate colony size, 321–22, 337–39, figs. 11.8–9, 456, 481
and predation, 229–30, fig. 8.9, 239–41, fig. 8.12
and quality of nest site, 118
and rainfall, 334–36, 359, 472–73
seasonal patterns in, 339–45, figs. 11.10–11, tables 11.3–4, 358, 472
and settlement patterns, 323
and spatial structure of colony, 469–71, 473, 481
and temperature, 334–36, 359, 371, 472–73
and thermal advantages of nest size, 351–53
and trespassing, 133
variance in, 323, 334–36, fig. 11.7, 353–55, fig. 11.13, 359, 469, 472–73
reproductive synchrony. *See* breeding synchrony; synchrony
reproductive tactics. *See* alternative reproductive tactics
research, future, 93, 138, 142–43, 157, 165, 174–75, 178, 208, 338, 378, 478–80
resource
  abundance, 466
  depletion, 265–66, 311–15, 320
  limitation, 118
retort nesting, 8, 453
Rhagionidae, table 9.1, 249
ring collaring, 248, 250, 273, 284
ringing. *See* banding
Rio Paraná, 8
Rio Uruguay, 8
Riparia riparia, 484. *See also* swallow, bank
risk, spreading of, 172–74
Rissa tridactyla, 484. *See also* kittiwake, black-legged
road cuts, 246, 259
Rocky Mountains, 40–41
roost and roosting, 8, 70, 445. *See also* preroosting behavior
  behavior at, 442
  communal, 471
  as information center, 243
  predation attempts at, 212, 215
ruff, 465, 484

Salix amygdaloides, 483. *See also* willow, peachleaf
sample sizes, 36–38, 78, 291, 368, 369
Sand Hills, 40, 41, 43, 44, 46, 49
São Paulo province, 8
Sarcophagidae, table 9.1, 249
Sayornis
  phoebe, 484 (*see also* phoebe, eastern)
  saya, 484 (*see also* phoebe, Say's)
Scaphidiidae, table 9.1, 249
Scarabaeidae, table 9.1, 249
scientific names, 38, 483–84
Scottsbluff, Nebraska, 41, 50
scrounger, 264
Scutelleridae, table 9.1, 249
search area, 306–11, fig. 10.14. *See also* foraging area
secondary study site, 15, 49, fig. 3.1, 60, 67, 362
selfish herd, 211, 212, 239
senescence, 369
sex, 39
  and adult body mass, 293–97, figs. 10.9–10, table 10.3, 319
  and arrival times, 62–65, fig. 3.9, 67, 407
  and colony choice, 404–5
  and colony switching, 439
  and destroying eggs, 142, 146–48
  determining, 22–24, 39
  and dispersal, 421–23, figs. 13.5–6, 427
  and extrapair copulation, 149, 157, 161, 164–65
  and fighting for nests, 120
  and flea loads of adults, 103–5, fig. 4.10, 116, 413
  and mortality during cold weather, 303
  and natal dispersal, 433
  and site fidelity, 421–23, figs. 13.5–6, 427
  and sleeping patterns, 405–6, 407
  and survivorship, 368, 370–71, fig. 12.2, 385, 397
  and trespassing, 133, 146–48
sex-dependent survivorship, 368, 370–71, fig. 12.2, 385, 397
sex ratio
  and colony choice, 407
  and extrapair copulation, 159–61, fig. 6.7, 166, 186
  seasonal changes in, 61, fig. 3.8
  of study population, 61–62, fig. 3.8, 67, 159, 166, 186, 433

shelduck, 236, 483
shortage of suitable nesting sites, 187–209
  and advantages of coloniality, 188–89, 450, 454
  and colony use patterns, 192, 194–202, figs. 7.5–9, table 7.1, 204, 209
  comparison with barn swallows, 202–4, fig. 7.10, 206–8
  difficulty in studying, 188, 192, 208
  and food availability, 205–6
  and internest distances, 202–3, fig. 7.10
  and nearest neighbor distance, 190–91, fig. 7.2, 194–95, fig. 7.4, 208
  and nest clustering, 188, 195, 202–4, fig. 7.10, 208–9
  and nest spacing, 188–92, figs. 7.1–2, 194, 202–4, fig. 7.10, 208–9
  as pressure for coloniality, 187–89, 450, 453, 478
  and relation between
    colony size and nest density, 189, fig. 7.1, 208
    colony size and total substrate size, 193–94, fig. 7.3, 208–9
    nearest neighbor distance and colony size, 190–91, fig. 7.2, 208
    substrate size and nearest neighbor distance, 194–95, fig. 7.4, 209
  types of, 187–88
shortgrass prairie, 42
shrike, loggerhead, 212–13, 241, 484
siblings, 474–75
Sierra Nevada, 175
Simuliidae, table 9.1, 249
Siphonaptera, 71
Siricidae, table 9.1, 249
site dispersal distances, 420–24, figs. 13.5–6, 437–41, fig. 13.12, 447–48. *See also* colony site; dispersal; natal dispersal
site fidelity, 420–41, figs. 13.5–6, tables 13.2–3, figs. 13.8–9, 447–48. *See also* colony site; dispersal; natal dispersal
Skin Bond cement, 401
sleeping
  on cliff face, 442
  and colony choice, 402, 404, 406, 407
  at colony site, 442
  by early settlers, 405–7, 442
  in nest, 21, 404, 442
  sex differences in, 405–6, 407
  in trees, 442
Smithsonian Institution, 47

snake, bull, 205, 212, 314, 470, 483
  alarm response to, 217
  attraction of to colonies, 236–39, fig. 8.11
  colony types attacked, 216
  compared with avian predators, 216
  mobbing of, 225
  predation by, 216–17, 229, 232, 233, fig. 8.10, 235–36, 241–42
snake, common garter, 217, 483
snake, rat, 216, 236, 483
social foraging, 11, 212, 243–320, 386. *See also* foraging; information center; information transfer
  benefits of, 275, 319
  as cause of coloniality, 275, 452–54, 479
  costs of, 245, 319
  and predator avoidance, 268–74, table 9.4, fig. 9.5
  as pseudoreciprocity or reciprocity, 476
  as response to food distribution and information transfer, 272, 285, 452
  throughout year, 452
social interaction, and defining colony, 16
social systems, types of, 2
sociality, evolution of, 3
solar radiation, 43
Solidago, 42
solitary foragers, 268–74, table 9.4. *See also* foraging; group foraging
Somateria mollissima, 483. *See also* eider, common
sorting of birds among colonies, 463
  by age, 409–12, fig. 13.1, 447, 465, 481
  by body mass, 415–18, fig. 13.3, 447, 465, 481
  by body size, 409, 418–19, fig. 13.4, 447, 465, 481
  by ectoparasite load, 409, 412–15, fig. 13.2, 417, 418, 447, 465, 481
South America, 8, 10, 12, 72, 241, 379, 475
South Pass, Wyoming, 47
South Platte River, 40–41, fig. 3.1, 48–49, fig. 3.3, 51, 67, 97, 193, 247, 406. *See also* Platte River
sparrow, house, 71, 173, 217, 484
sparrow-weaver, white-browed, 236, 484
spatial effects, 469–74, 481. *See also* nest position in colony
spatiotemporal variability of food locations, 245, 247–48, 251–60, 268, 307
  and association between insects and foraging birds, 252–53, fig. 9.1
  and behavior of foragers, 251–53, fig. 9.1
  between-day and within-day, fig. 9.3, 257–58
  as cause of coloniality, 452
  as cause of social foraging, 272
  and colony size, 255–58, figs. 9.2–3, 285–86, 310
  and convection currents and thermals, 249, 251, 259–60, 273
  determining, 254–55, 257
  and food delivery, 285–86
  in foraging arenas, fig. 9.2, 255–58
  generating and maintaining, 259–60, 273
  and geometrical model, 318–19
  and information transfer, 251, 285
  and mass emergences of insects, 259, 273
  and sampling insects, 248, 251–53, fig. 9.1
  and swarming of insects, 249–51, table 9.1, 273
  and wind, 259, 273
Spharagemon cellare, 248
sphere of choice, 170–71, fig. 6.10
spiders, 111, 129, 236, 304–5, 471
spite, 146
squeak call, 246, 266–67, 273. *See also* food call
stage. *See* colony stage; nest stage
Staphylinidae, table 9.1, 249
starvation
  during cold weather, 300–304, table 10.4, 314, 320, 373
  of nestlings, 313–15, fig. 10.15, 320
statistical analyses, 20, 36–39, 43–46, 78, 90–91
statistical models for estimating survivorship, 363–68, 394, 397. *See also* survivorship
stealing
  and colony size, 119
  kleptoparasitism, 179–86
  of nest material, 119, 133, 135–36, 140, 148
Stelgidopteryx serripennis, 484. *See also* swallow, northern rough-winged
Sterna hirundo, 484. *See also* tern
stickleback, three-spined, 465, 466, 483
Stipa, 42
Stratiomyidae, table 9.1, 249

562 • Index

stream foraging, 264–65, 273, 276, 278
study area. *See* study site
study population, 11, 40, 50, 57–67
   arrival time of, 57, 62–65, fig. 3.9, 67
   colony initiation times of, 57, 65–66, fig. 3.10, 67
   colony sizes of, 57, 66–67, fig. 3.11
   demography of, 367–73, figs. 12.1-3
   ectoparasites affecting, 69–73
   sex ratio of, 57, 61–62, fig. 3.8, 67, 186
   size of, 57–61, fig. 3.7, 67
study site, 11, 15–16, 40–57, 362, 364, 367, 370. *See also* colony site
   artificial and natural colonies within, 56–57, 67
   artificial nest sites within, 49, 50–57, fig. 3.3, figs. 3.5–6
   boundaries of, 15, 40–42, 48–49, fig. 3.1, fig. 3.3, 67
   climate of, 42–46, tables 3.1–2, 67
   colony site use within, 196–97, table 7.1
   colony sizes within, 66–67, fig. 3.11
   elevation of, 40
   flora of, 41–42
   humidity of, 43
   hybridization within, 42
   land use in, 42
   natural colony sites within, 49–50, fig. 3.2, fig. 3.4, 56–57, 67
   North Platte River within, 40–42, 48–51, figs. 3.1–3, 67, 240, 247, 401, 404, 406, 407
   occurrence of cliff swallows within, 46–50, 67
   physiographic characteristics of, 40–46, 67
   Platte River within, 40–42, 48–49, fig. 3.1, fig. 3.3
   primary, 15–16, 48–49, fig. 3.3, 60, 67, 362
   rainfall in, 42–46, table 3.2, 67
   secondary, 15, 49, 60, 67, 362
   South Platte River within, 40–41, 48–51, fig. 3.1, fig. 3.3, 67, 202, 247, 406
   Sutherland Canal within, 51, 56, 401
   temperatures of, 42–46, table 3.1, 67
   wind speed and direction in, 43
substrate, colony, 34, 56–57, 75, 110–11, 187–89
   artificial versus natural sites, 204–5, 209
   and colony site use, 195–96, fig. 7.5, 209
   and colony size, 193–94, fig. 7.3, 204–5, 209, 446

   identifying suitable, 192, 204–5, 208
   and nearest neighbor distance, 194–95, fig. 7.4, 209
   and predictability of colony size and use, 201–2, fig. 7.9, 209
   size of, 192–96, figs. 7.3–5, 209
Suckley, George, 46–49
sunbathing, 9, 221–23
SURGE, 365, 394, 397
survival. *See* survivorship
survivorship, 11, 32, 42, 211, 301, 360–98, 418, 420
   of adults
      and body mass, 291, 297, 376, 384–86, fig. 12.12, 395, 398
      and cold weather, 300–304, table 10.4
      and colony size, 375–80, figs. 12.6–8, 395, 398, 451, table 14.1
      and ectoparasitism, 105–6, 386
      and foraging, 244, 306, 316, 373, 386
      and nest age, 395–98, 473
      and nest position, 395–96, 398
   age dependence in estimating, 365–66
   age-specific, 368–70, fig. 12.1, 397
   annual, 317, 360, 369, 386, 396, 397–98, 456, 472
   as benefit of coloniality, 451, table 14.1
   and breeding life span, 457
   and capture history, 362
   cohorts used in estimating, 364, 367, 397, 456
   and coloniality, 4, 6–7, 12, 361
   and differential dispersal, 380
   difficulty in studying, 361
   versus dispersal, 368, 390
   and emigration, 368
   estimating, 362–68
   of first-year birds
      and body mass, 380–84, fig. 12.9, fig. 12.11, 386, 398
      and colony size, 373–75, fig. 12.5, 397, 451, table 14.1, 458–60
      and ectoparasitism, 369, fig. 12.1, 386–90, figs. 12.13–14, 397, 398, 480
      and hatching date, 391–92, fig. 12.15, 398
      and nest age, 391, 395, 398
      and nest position, 391, 393–95, fig. 12.16, 398
      and parental age, 393
   and lifetime reproductive success, 323, 397, 457–62

and mass at fledging, 382–84, fig. 12.11, 398
of nestlings
and cold weather, 303
and colony size, 93, 96–102, tables 4.5–7, fig. 4.9, 116
and ectoparasitism, 69, 93, 96–102, tables 4.5–7, fig. 4.9, 116
and synchrony and predator avoidance, 211
notation for, 366
number of birds marked each year, 363, table 12.1
of parasitic nestlings, 174
and predation, 373
and recapture probability, 365–66, table 12.2
and self-maintenance, 380
and senescence, 369
sex-specific, 368, 370–71, fig. 12.2, 385, 397
and statistical models, 363–66
SURGE, 365, 394, 397
and temperature, 371–73, 397
time dependence in estimating, 365–66
time needed for estimating, 367
variance in, 369
year-dependent, 368, 371–73, figs. 12.3–4
yearly differences in 377–78, 472
Sutherland, Nebraska, 51
Sutherland Canal, 51, 56, 401
swallow, Angolan cliff, 10, 484
swallow, bank, 3, 71, 119, 192, 226, 300–301, 408, 454, 484
swallow, barn, 453, 454, 484,
and aggregation hypothesis, 206–9
breeding of on winter range, 8
brood parasitism in, 454
coloniality in, 454
cost of building nest, 353
dominated by cliff swallows, 202–3
extrapair copulation in, 454, 455
food of, 250, 453
foraging methods of, 250
historical distribution of, 47
hybridization with cave swallow, 10
information transfer in, 453
internest distance of, 202–4, fig. 7.10, 209
nest guarding in, 137
nesting sites of, 9
nest site limitation of, 189, 202–4, 209
nest size and clutch size in, 331–32,
nest trespassing, 137
predator mobbing by, 226
radio telemetry of, 401
swallow, cave, 10, 484
swallow, cliff, 10, 484. *See also main entries for many topics below*
altitudinal range, 8
artificial nest sites, 50–57, figs. 3.5–6, 67
and barn swallows, 202–3
breeding phenology, 38, 43, 62–65, figs. 3.9–10, 67
breeding range, 8, 12
clutch size, 323–24, fig. 11.1
colony initiation dates, 65–67, fig. 3.10
colony size, 7, 9, 50–51, 60, 66–67, fig. 3.11, 338, 400
demography, 368–73, figs. 12.1–4
determining sex of, 22–24
distribution of, 8, 12, 47–57, fig. 3.3, 67
ectoparasites of, 26–34, 69–73, 116
egg laying by, 18
field marks, frontispiece, 7, fig. 4.8, fig. 6.12, fig. 9.4
food, 9, 42, 247–51, table 9.1
foraging behavior, 245–47, 268
historical distribution, 40, 46–50, 67
incubation period, 19
migration, 8
morphological variation, 9–10, 12
mortality, 21, 42, 46, 62, 68, 96, 100–102, 125, 300–304
mud gathering, 128–30, fig. 5.6, fig. 6.1
natural history, 7–10, 12, 40, 46–67
natural nest sites, 40–42, 49–50, 56–57, fig. 3.2, fig. 3.4, 67
nest building, 8–9, 11–12, 47, 126–30, figs. 5.4–6
nest guarding, 136–37
nest reuse, 9, 27, 65
number marked each year, 363, table 12.1
plumage, 22, 24
population size, 57–61, fig. 3.7, 67
predator mobbing, 226
predators of, 212–18, 241
range expansion, 9
related species, 10
sex ratio, 61–62, fig. 3.8, 67
spring arrival, 62–65, fig. 3.9, 67
subspecies, 10, 12
vocalizations, 9, 266
winter range, 8, 12, 70

swallow, Indian cliff, 10, 484
swallow, northern rough-winged, 47, 226, 484
swallow, Preuss's cliff, 10, 484
swallow, South African cliff, 10, 247, 484
swallow, tree, 162, 188, 408, 484
swallow bug, 26–30, figs. 2.5–11, 39, 69–71, figs. 4.1–9, 73–103, tables 4.1–7, 106–17, fig. 4.11, 126, 130, 231–32, 334–37, 340, 345, 347, 350–52, 369, fig. 12.1, fig. 12.4, 372, 373, 375, 382, 386–88, fig. 12.13, 391, 393–94, 397–98, 408, 413, 435, 451, 452, table 14.1, 464, 472, 473, 480, 483. *See also* ectoparasitism
swarming of insects, table 9.1, 249–51, 265–68, 273
synchronized preroosting flights, 442, 445
synchrony, 9, 20, 22, 226–33
  and antipredator advantages, 226–33, figs. 8.6–9, 239, 242
  and capture effort, 363
  and clutch initiation, 327
  and colony size, 90, 227–29, 232, figs. 8.6–8, 242, 412
  and conspecific brood parasitism, 168, 170
  as cost of coloniality, 451, table 14.1
  and ectoparasitism, 230–33, fig. 8.9
  and egg destruction, 143
  fledging, 175
  and food abundance, 233
  hatching, 35–36, 39, 229–30, fig. 8.9
  and kleptoparasitism, 181–82
  laying, 35–36, 39, 227–30, figs. 8.6–9
  measuring, 35–36, 39
  nesting or breeding, 210–12, 242
  and predator swamping, 210–12, 226–33, figs. 8.6–9, 239, 242
  reasons for, 232–33
  and reproductive success, 211, 229–30, fig. 8.9, 242, 356, 358–59, table 11.9
  and sphere of choice for brood parasitism, 170–71, fig. 6.10
  and survival, 211
Syrphidae, table 9.1, 249

Tabanidae, table 9.1, 249
Tachinidae, table 9.1, 249
Tachycineta bicolor, 484. *See also* swallow, tree
Tadorna tadorna, 483. *See also* shelduck

tallgrass prairie, 42
tagging. *See* banding
Taxidea taxus, 484. *See also* badger
temperature (air), 42–46, table 3.1, 296, 363
  and annual survivorship, 371–73, 397
  and cliff swallows, 42–43, 46, 60, 300–304, table 10.4, 334–36
  and insects, 42, 300, 309, 335
  inside nest, 351–53
  and reproductive success, 334–36
  seasonal patterns, 42–46, table 3.1, 67, 334–36
  and swallow bugs, 111, 334–36
temporal effects, 469, 471–74, 481
Tenebrionidae, table 9.1, 249
Tenethredinidae, table 9.1, 249
Tephritidae, table 9.1, 249
tern, 226, 306, 468, 484
territoriality, 2, 4
territory, 361
Tetrigidae, table 9.1, 249
Texas, 10
Thamnophis sirtalis, 483. *See also* snake, common garter
Thereridae, table 9.2, 249
thermals and convection currents, 249, 251, 259–60, 273
thermoregulation, of swallow bug, 111
tick, soft-bodied, 69, 72–73, 93
Tierra del Fuego, 8
time dependence in survival or recapture probabilities, 365–66
Tingidae, table 9.1, 249
Tipulidae, table 9.1, 249
tit, penduline, 351, 484
Togaviridae, 97
Trachyhachys kiowa, 248
traditional aggregation hypothesis, 206–9
Transition Zone, 8
transmission of ectoparasites, 103, 106–12, fig. 4.11, 116–17, 408
travel distance. *See* distance traveled to forage
Tree Tanglefoot, 33, 35, fig. 2.11, 252. *See also* fumigation
trespassing, 26, 56, 110, 119–20, 132–48, 190, 395, 453, 469
  as benefit of coloniality, 451, table 14.1
  and colony geometry, 471
  and colony size, 139–41, figs. 5.10–11, 148

and colonywide alarms, 142–43
and conspecific brood parasitism, 133, 135–36, 148, 167, 170
costs and benefits of, 134–38, 141, 148
and ectoparasitism, 133, 135, 148
effects on individuals, 134–37, 148
and egg destruction, 133, 135–36, 141–48, fig. 5.12, 167, 170
and egg transfer, 135, 148, 167
and extrapair copulation, 133, 134–36, 148, 151–52
versus fighting, 133
frequency of, 137–38, fig. 5.9, 148
frequency of success of, 134–35, 148
natural history of, 133–34
and nest guarding, 136–37, 148
and nesting stage, 133, 135, 148
and parental quality of neighbors, 133
reasons for, 133, 134–37, 148
sex of birds committing, 133, 148
and spatial position of nests, 133–34, fig. 5.8, 148
and stealing nest material or mud, 133, 135, 140, 148
timing of, 133
Troglodytes aedon, 484. *See also* wren, house
tropicbird, white-tailed, 119, 483
Trupaneidae, table 9.1, 249
Tucumán province, 8
Turdus pilaris, 484
turtle, snapping, 217, 483
Tyto alba, 484. *See also* owl, barn

Ulmus americana, 483. *See also* elm, American
United States, 8, 9, 72
univariate statistics, 36, 39
Uria aalge, 484. *See also* guillemot, common
U.S. Highway 26, 51
U.S. Highway 30, 51
Utah, 10
Utah Territory, 47

Valentine, Nebraska, 43
variance in
ectoparasite load, 90–93, fig. 4.5, 102, 110, 116
prey capture success, 268, 273–74
reproductive success, 353–55, fig. 11.13, 359 (*see also* age-specific differences)
variation in colony size, 66–67, fig. 3.11, 463–69, fig. 14.2
Vespidae, table 9.1, 249
vigilance. *See also* individual alertness; predation; predator avoidance; predator detection
as benefit of coloniality, 451, table 14.1
at colony, 218–20, fig. 8.2
and colony size, 210
while foraging, 268–69, 272
and foraging efficiency, 243, 269, 272
of group, 220–24, figs. 8.3–5, 239, 241
and group size, 159, 210, 239, 241
measuring, 221–22
and position in group, 272, 274
and predator avoidance, 210, 239, 241, 268, 269, 272
and predator detection, 220, 239, 241
viremia, 97–98
Virgin Islands, 8
virus, Fort Morgan, 97–98
vocalizations. *See also* food call
begging calls, 176, 178–79
chur calls, 266–67
food calls, 266–68, 273
and recognition of offspring, 178–79, 184–85
squeak calls, 246, 266–67, 273
vulture, turkey, 216, 483

waiting
at nest, 276–80
circling above colony site, 278–79
intervals of foragers, 277–80, table 10.1, fig. 10.1, 315
wall sharing, 128–32, fig. 5.5, fig. 5.7, 148, 234–36, fig. 8.10, 242, 331, 346, 359, 473
Washington State, 184
wasp, 236
water beetle, haliplid, 251
weaning, 380
weasel, long-tailed, 216, 226, 484
weather, 380
and capture effort, 363
and insect prey, 42, 246–47, 249, 269, 275, 300
and foraging, 267
and starvation, 300–304, table 10.4, 314, 320
weight. *See* body mass
western equine encephalitis, 96

Wildcat Hills, 41
willow, peachleaf, 42, 483
windbreaks, 259, 273
wind speed and direction, 43
wing chord, 24, 418–19, fig. 13.4
wing fluttering
   experiment with stuffed models, 162–64, fig. 6.1
   and extrapair copulation, 162–64, fig. 6.8
   and group size, 163, fig. 6.8
   during mud gathering, 152, 159, fig. 6.1
Wisconsin, 198
wounds. *See* injuries
wren, house, 217–18, 484

Xanthium, 42

yearling, 409–12, 472
   colony choice by, 432–37, figs. 13.9–11, 447–48
   dispersal by, 388–90, 432–37, figs. 13.9–11, 447–48
   spring arrival times of, 62–65, fig. 3.9, 67, 472
Yucca, 42